ELECTRICAL MACHINES–II

J.B. GUPTA

ELECTRICAL MACHINES-II

(For Engineering Students)

By

J.B. Gupta

B.Sc. (Elec. Engg.); C. Engg. (I);

F.I.E. (India); M.I.S.T.E.

S.K. KATARIA & SONS®

Publisher of Engineering Books

4885/109, Prakash Mahal, Dr. Subhash Bhargav Lane,

Opposite Delhi Medical Association, Daryaganj, New Delhi-110002 (INDIA)

Phone: +91-11-43551243 Mobile: +91-9871775858, 7678425209

e-mail: katariabook@gmail.com, katariabooks@yahoo.com

Website: www.skkatariaandsons.com

(e-books, subscriptions and bulk buying also available on our website)

Published by:

S.K. Kataria & Sons®

4885/109, Prakash Mahal, Dr. Subhash Bhargav Lane,
Opposite Delhi Medical Association, Daryaganj, New Delhi-110002 (INDIA)
Phone: +91-11-43551243; Mobile: +91-9871775858, 7678425209
e-mail: katariabook@gmail.com, katariabooks@yahoo.com
Website: www.skkatariaandsons.com

Edited By:

Rajeev Manglik
B.E. (Elec.) Pb. University

Rohit Manglik
B.E. (Elec.) University of Roorkee

First Edition: July, 2011; Reprint: 2026

Price: ₹ 595/-

Laser Typesetting by : Plus Computers, Delhi

Printed at : New A.S. Printing Press, Delhi

PREFACE

The primary aim of presenting this book is to meet the needs of Third Year (Fifth Semester) students of B.Tech. (Electrical Engineering and Electrical and Electronics Engineering) studying in Engineering Colleges affiliated to Maharishi Dayanand University (MDU) Rohtak, (Haryana). The book has been prepared keeping in view the latest revised syllabus (W.E.F. Session 2011-12). The subject matter has been presented in an easily comprehensible manner, logical sequence, using a simple language and to provide detailed exposition to fundamentals. A large number of workedout examples, theoretical questions, short answer type questions with answers and numerical problems included in the text are from latest questions papers of universities/Deemed Universities/Technical Universities including MDU or are of equivalent standards.

The book consists of 6 chapters divided into 4 sections according to the syllabus (W.E.F. Session 2011-12), prescribed by Maharishi Dayanand University, for the subject of Electrical Machines–II.

The solution to illustrative examples covering almost all topics/subtopics of different chapters makes the entire presentation easy to follow. Selected exercises and numerical problems with answers given at the end of each chapter help the students in evaluating their understanding of the subject. Short answer type questions with answers help the students in preparing for viva-voce/practical examinations.

Every care has been taken to eliminate misprints, omissions and errors but it is too much to expect that no inaccuracy/ misprint/ error has crept in and the author would be grateful to the readers for bringing to his notice any such errors/ misprints/ omissions they may come across while going through the book.

The author shall feel satisfied if the book meets the needs of the students for whom it is meant. Suggestions and criticism for the improvement of the book shall be welcomed.

The author will like to appreciate the sincere efforts and contribution made by his sons Rajeev Manglik and Rohit Manglik in presentation of the book in this form.

The author will like to thank all the Scientists, Engineers, Authors and Publishers whose works and text have been the source of enlightenment, inspiration and guidance in presenting this new edition.

Lastly, but not the least, the author is thankful to his publisher, Sanjeev Kumar Kataria for bringing out the book in such a short time and for pricing it moderately despite heavy cost of paper and printing.

July 1, 2011 **J.B. GUPTA**

PREFACE

To My Loving Grand Daughter

RUSHIL

OTHER USEFUL BOOKS

SECTION D :

SYNCHRONOUS MOTOR .... 273—329

SECTION **A**

POLYPHASE INDUCTION MACHINES

CONTENTS OF SECTION-A

POLYPHASE INDUCTION MACHINES : Constructional Features; Production of Rotating Magnetic Field; Induction Motor Action; Torque Production; Testing; Development of Equivalent Circuit; Performance Characteristics; Circle Diagram; Starting Methods; Methods of Speed Control—Stator Voltage Control, Stator Resistance Control, Frequency Control, Rotor Resistance Control, Slip Power Recovery Control, Double Cage and Deep Bar Motors; Grid Excited and Self Excited Induction Generators.

1

CHAPTER

Rotating Magnetic Fields

1.1. INTRODUCTION

A rotating magnetic field is that field which is constant in magnitude but whose axis of direction rotates in space as field system of a dc machine. With a stationary field system the magnetic field is stationary in space, but if the field system is rotated at a certain speed, its magnetic field will rotate with it at the same speed.

INSIDE THIS CHAPTER

The magnetic field produced by single phase alternating current is an *alternating magnetic field*, the field acting along a fixed axis, varying in magnitude periodically and changing its direction alternately positive and negative.

1.2. THE FERRARIS PRINCIPLE

The magnetic field produced by single phase alternating current is an alternating magnetic field which, by Ferraris principle, can be resolved into two rotating fields of half its amplitude and rotating in opposite directions at synchronous speed.

Consider two rotating magnetic fields OA_1 and OA_2, each having magnitude of H units and travelling in opposite directions with angular velocity ω, as shown in Fig. 1.1.

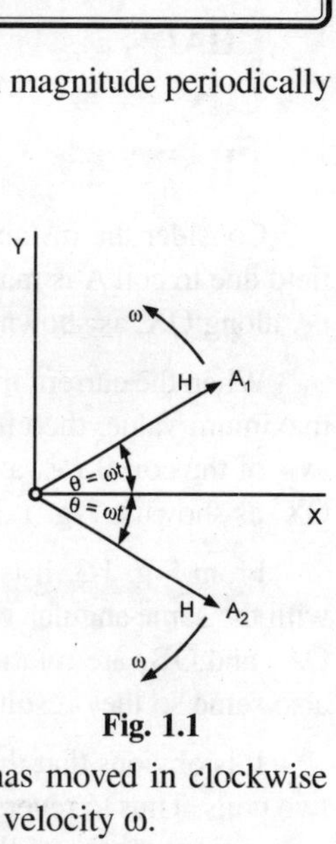

Fig. 1.1

Let both the fields start travelling from axis OX at time $t = 0$.

After time t seconds the angle through which fields OA_1 and OA_2 have rotated, $\theta = \omega t$ radians.

i.e. Field represented by vector OA_1 has moved in counter-clockwise direction through an angle $\theta = \omega t$ from axis OX and field represented by vector OA_2 has moved in clockwise direction through same angle θ, because both fields are travelling with same angular velocity ω.

Resolving the magnetic fields represented by vectors OA_1 and OA_2 along axis X and axis Y, we get

X-component, $H_x = OA_1 \cos\theta + OA_2 \cos\theta = H\cos\theta + H\cos\theta = 2H\cos\theta = 2H\cos\omega t$

Y-component, $H_y = OA_1 \sin\theta - OA_2 \sin\theta = H\sin\theta - H\sin\theta = 0$

Hence resultant magnetic field is 2H cos ωt along X-axis. It is, therefore, obvious that two rotating magnetic fields, travelling in opposite directions with the same angular velocity, result in an alternating field of twice their amplitude. *Conversely an alternating field can be replaced exactly by two rotating fields of half its amplitude travelling in opposite directions at synchronous speed.*

1.3. PRODUCTION OF TWO PHASE ROTATING MAGNETIC FIELD

Consider two similar coils placed 90° (electrical) apart and connected to two phase supply, as shown in Fig. 1.2. An alternating field is set up by each coil along its own axis *i.e.* coil A sets up an alternating field along X-axis and coil B sets up an alternating field along Y-axis. Let the fields set up by coils A and B, each be resolved into two components rotating in opposite directions with angular velocity $\omega = 2\pi f$, f being the supply frequency. Let the components of fields due to coil A be OA_1 and OA_2 and the components of field due to coil B be OB_1 and OB_2, as shown in Fig. 1.2.

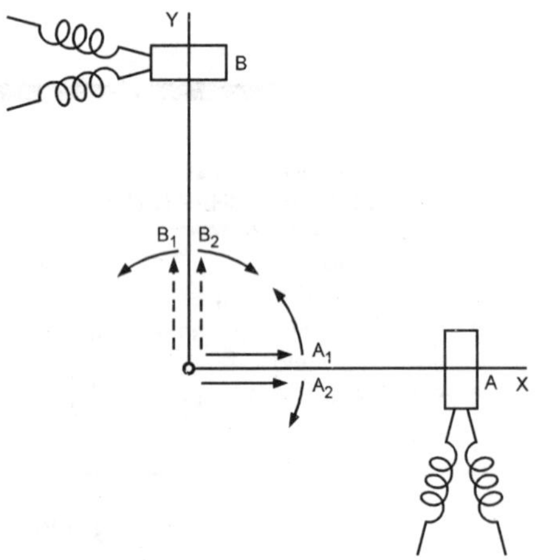

Fig. 1.2

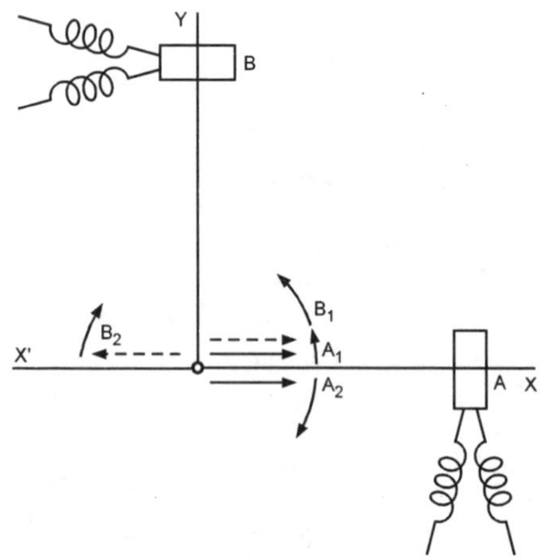

Fig. 1.3

Consider the instant when current in coil A is maximum. *i.e.* I_A is maximum. At this instant magnetic field due to coil A is maximum, therefore, both the components OA_1 and OA_2 are along the axis of the coil A *i.e.* along OX, as shown in Fig. 1.3.

When the current in coil A is maximum, the current in coil B is 90 electrical degrees behind the instant of maximum value, therefore, its components OB_1 and OB_2 will have to rotate through 90° to come along the axis of the coil B *i.e.*, axis Y. Hence at this instant component OB_1 is along OX and component OB_2 is along OX′ as shown n Fig. 1.3.

From Fig. 1.3. it is obvious that rotating components OA_2 and OB_2 are rotating in clockwise direction with the same angular velocity but opposing each other, therefore, their resultant is always zero. Components OA_1 and OB_1 are rotating in anti-clockwise direction with the same angular velocity. Since their direction is also same so they result in a pure rotating magnetic field.

It is obvious that the direction of resultant rotating field is same as the direction of phase sequence in the two coils. Thus to reverse the direction of the rotating field, phase sequence must be reversed by reversing the connections of one of the coils.

The manner in which the rotating magnetic field is established by a two phase stator winding can also be illustrated by considering the direction of current flow through the two phases at several successive instants.

Consider a simplified stator core with two equally spaced coils, as shown in Fig. 1.4. Normally the conductors are evenly spread round the core, but they are here shown in separate groups to enable the individual phases to be distinguished. The dots and crosses show the directions of the currents in the conductors when they are positive. Of course, they are never all positive at the same instant, nor are they ever all equal at any one instant. Phase relationship of the currents flowing through the coils is shown in Fig. 1.5. The different times are marked off at 45° intervals on the current waves in Fig. 1.5.

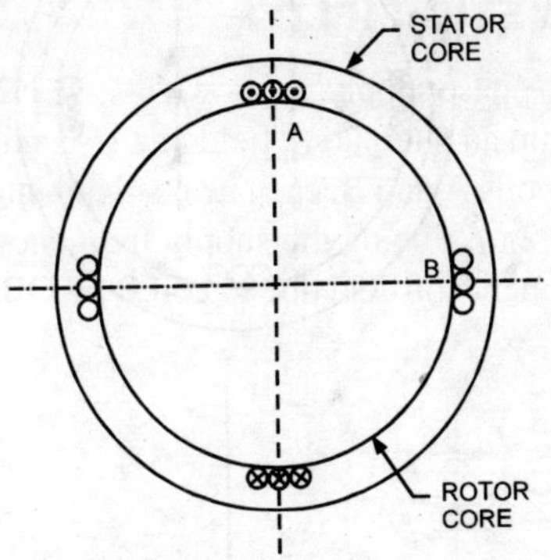

Fig. 1.4

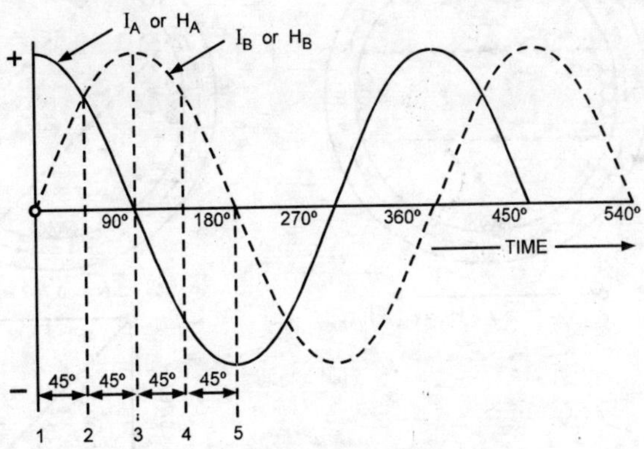

Fig. 1.5

At instant 1, the current in phase A, I_A is positive and at its maximum value and the current in phase B, I_B is zero. The directions of the currents in the conductors are indicated by crosses and dots. The field produced is towards the right, creating a north pole on the stator surface at the left and the south pole on the right, as shown in Fig. 1.6. (1).

At instant 2, the current in each phase is positive and is 0.707 of the maximum value and the resultant field due to these currents is shown in Fig. 1.6 (2). It will be seen that the axis of the resultant field has turned through 45° from that of Fig. 1.6 (1). At instant 3, the current in phase A has fallen to zero and the current in phase B is positive and at its maximum value; the resultant field is vertically downward as shown. Thus we see that the resultant field has rotated further through 45°. The remaining figs in Fig. 1.6 show the resultant field at the instants 4 and 5.

It is evident from Fig. 1.6 that the result field due to the currents flowing in the two phase windings is of the magnitude equal to the peak value of the field produced by individual phase windings and shifts around the stator surface, moving through a definite distance in each of the time interval considered.

From Fig. 1.6, it is also evident that in a 2-pole winding the field makes 1/2 revolution in 1/2 cycle of the current, and, therefore, one revolution in each cycle of current. In a 4-pole windng in which each phase has two separate groups of coils connected in series, it may be shown that the rotating field makes one revolution in two cycles of the current. In general, the field makes one revolution in P/2 cycles where P is the number of poles produced by the stator winding. If N is the speed of the resultant magnetic field then

$$\frac{N}{60} = \frac{f}{P/2}$$

$$\text{or } N = \frac{60 \times 2f}{P} = \frac{120f}{P}$$

The speed at which the rotating field revolves is called the *synchronous speed.*

By redrawing Figs. 1.6 (2), (3), (4) and (5) with the current in, say, phase B reversed, it can be shown that the direction of rotation of the resultant magnetic field is then counter-clockwise *i.e. the direction of the rotation of the resultant magnetic field can be reversed by changing over the connection to any one of the two phases.*

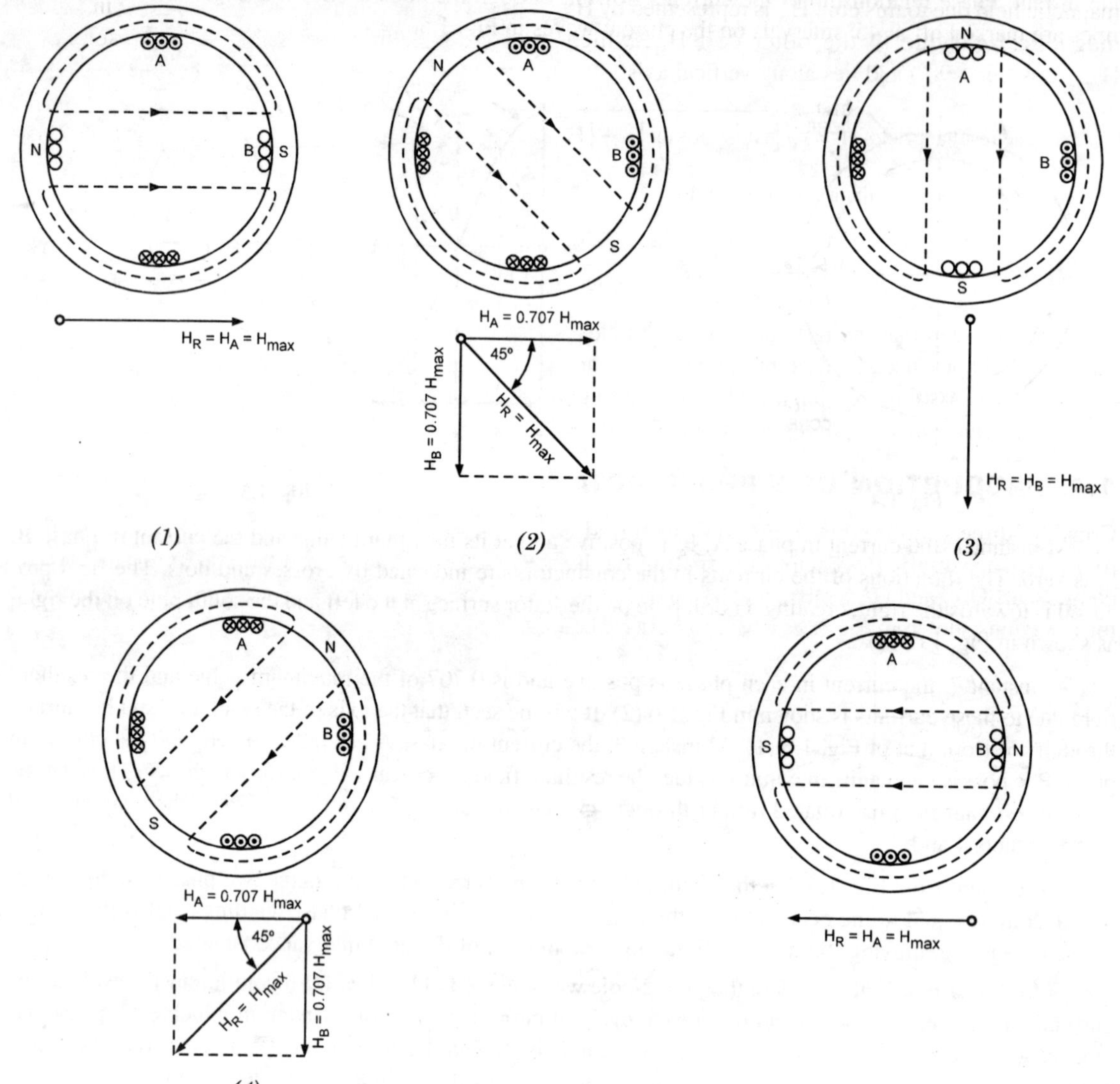

Fig. 1.6

Mathematical Proof. The above fact can be proved mathematically as below :

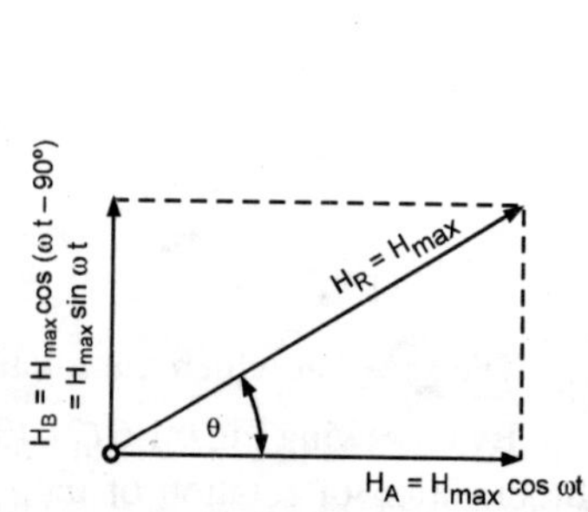

Fig. 1.7

The time diagrams for currents in coils A and B are shown in Fig. 1.5. Coils A and B establish alternating fields along their respective axis *i.e.* along X-axis and Y-axis respectively. Since phase difference between the currents I_A and I_B flowing through coils A and B respectively is 90°, hence phase difference between the alternating fields H_A and H_B produced is 90°.

Let each of the coil produce an alternating field of maximum strength H_{max} along its own axis. Considering an instant when the current in coil A has advanced through an angle ωt from its maximum value, the magnetic field due to this coil, H_A is represented by $H_{max} \cos \omega t$ and lies along X-axis. At the same instant the magnetic field due to the other coil H_B, which is lagging behind H_A by 90° is represented by $H_{max} \cos(\omega t - 90°)$ and lies along vertical axis.

$$\text{The resultant field, } H = \sqrt{\{H_{max} \cos \omega t\}^2 + \{H_{max} \cos(\omega t - 90)\}^2} = H_{max}\sqrt{\cos^2 \omega t + \sin^2 \omega t} = H_{max}$$

Inclination of the resultant field is given by

$$\therefore \theta = \text{Tan}^{-1} \frac{H_{max} \cos(\omega t - 90°)}{H_{max} \cos \omega t} = \text{Tan}^{-1} \frac{\sin \omega t}{\cos \omega t} = \omega t$$

It is thus seen that the resultant magnetic field due to two phases is equal to the maximum strength of magnetic field produced by each phase separately, and is constant in magnitude and independent of the angle of rotation ωt. Also the resultant field rotates through angle ωt corresponding to an advance in phase of the current ωt. *Hence the resultant field is of constant magnitude and rotates at synchronous speed.*

1.4. PRODUCTION OF 3-PHASE ROTATING MAGNETIC FIELD

Consider three similar coils A, B and C displaced in space by 120°, as shown in Fig. 1.8, and connected to 3-phase ac supply. Let each of alternating fields due to currents in coils A, B and C be resolved into two components OA_1 and OA_2, OB_1 and OB_2 and OC_1 and OC_2 respectively travelling with angular velocity ω in the directions shown in Fig. 1.8, where $\omega = 2\pi f$, f being the supply frequency.

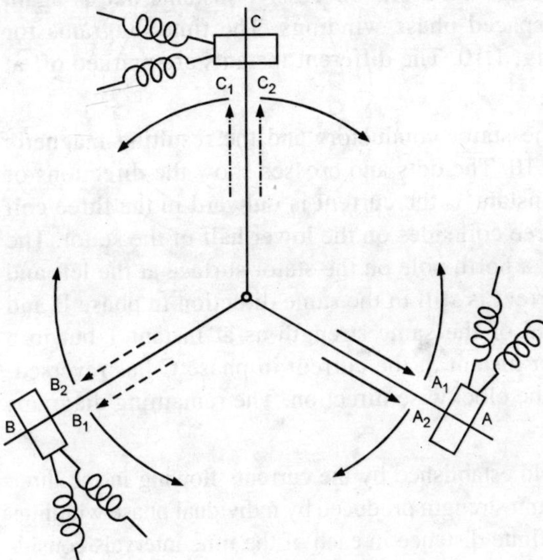

Fig. 1.8

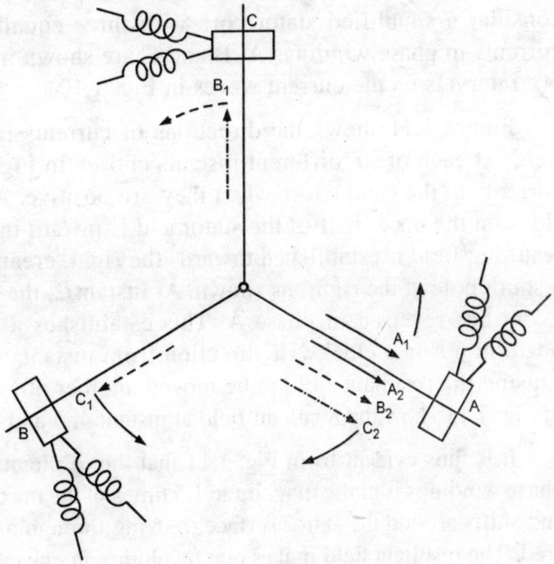

Fig. 1.9

Consider the instant when the current in coil A is maximum. At this instant both of the components OA_1 and OA_2 of the alternating field produced by the current in coil A are along the axis of the coil A *i.e.* along OA. The current in coil B is 120° behind the instant of its maximum value, therefore, each of its components OB_1 and OB_2 will have to rotate through 120° in order to reach the axis of the coil B *i.e.* along OB. Hence components OB_1 and OB_2 are along OC and OA respectively. The current in coil C is 240° being the instant of its maximum value, therefore, each of components OC_1 and OC_2 of the field produced by current in coil C will have to rotate through 240° in order to reach the axis of coil C *i.e.* along OC. Hence at this instant components OC_1 and OC_2 are along OB and OA respectively.

From Fig. 1.9. it is clear that components OA_1, OB_1 and OC_1 are rotating in counter-clockwise direction with same angular velocity ω but are 120° apart, therefore, their resultant is always zero. Components OA_2, OB_2 and OC_2 are rotating in clockwise direction with same angular velocity ω and also points in same direction, therefore, resultant gives a pure rotating field rotating at synchronous speed.

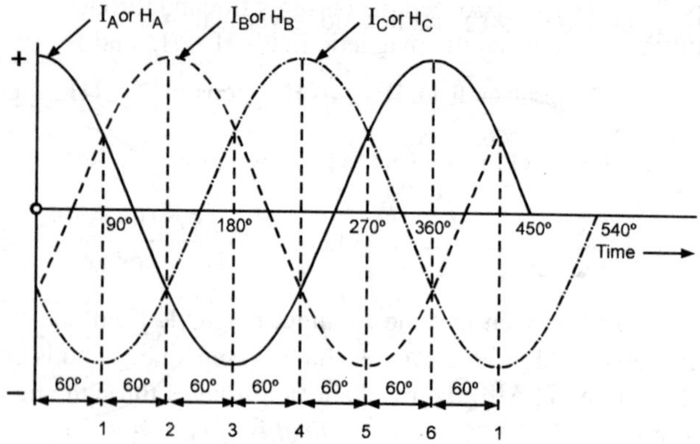

Fig. 1.10

The direction of rotation is the same as the direction of phase sequence *i.e.* first current in coil A attains its maximum value, then current in coil B attains its maximum value and then current in coil C attains its maximum value, as shown in Fig. 1.10. Hence to reverse the direction of rotation of the magnetic field, it is necessary to change the phase sequence which can be done by interchanging any two of the terminals.

For proving the establishment of rotating magnetic field by three phase currents by considering the direction of flow of currents through three phase windings at several successive instants, let us again consider a simplified stator core with three equally spaced phase windings. The time diagrams for currents in phase windings A, B and C are shown in Fig. 1.10. The different instants are marked off at 60° intervals on the current waves in Fig. 1.10.

Figure 1.11 shows the directions of currents in the stator conductors and the resulting magnetic fields at each of six different instants chosen in Fig. 1.10. The dots and crosses show the directions of currents in the conductors when they are positive. At instant 1, the current is outward in the three coil sides on the upper half of the stator and is inward in three coil sides on the lower half of the stator. The resulting field is established towards the right, creating a north pole on the stator surface at the left and a south pole at the right, as shown. At instant 2, the current is still in the same direction in phase B and C but has reversed in phase A. This establishes a field of the same strength as at instant 1 but in a position 60° in a clockwise direction from instant 1. At instant 3, the current in phase C has reversed, causing the resultant field to be moved another 60° in the clockwise direction. The remaining diagrams in Fig. 1.11 show the resultant field at instant 4, 5 and 6.

It is thus evident from Fig. 1.11 that the resultant field established by the currents flowing in the three phase windings is of the magnitude 1.5 times of the maximum strength produced by individual phase windings and shifts around the stator surface, moving through a definite distance in each of the time intervals considered. The resultant field makes one revolution in one cycle. If the stator is wound for P poles, the field makes one revolution in $\frac{P}{2}$ cycles or $\frac{2f}{P}$ revolutions in one second or $\frac{120f}{P}$ revolutions in one minute.

If any two of the three supply lines to the stator winding are reversed, thereby reversing the phase sequence of the stator currents, the analysis will show the rotation of the resultant magnetic field to be in the reverse direction, or counter-clockwise.

Mathematical Proof. The above fact can also be proved mathematically as below:

The time diagrams for currents in coils A, B and C are shown in Fig. 1.10. Coils A, B and C establish alternating (pulsating) fields along their respective axis *i.e.* along axis OA, OB and OC respectively. Since phase difference between the currents flowing through coils A, B and C respectively is 120º, therefore, phase difference between the magnetic fields H_A, H_B and H_C produced is 120º (electrical).

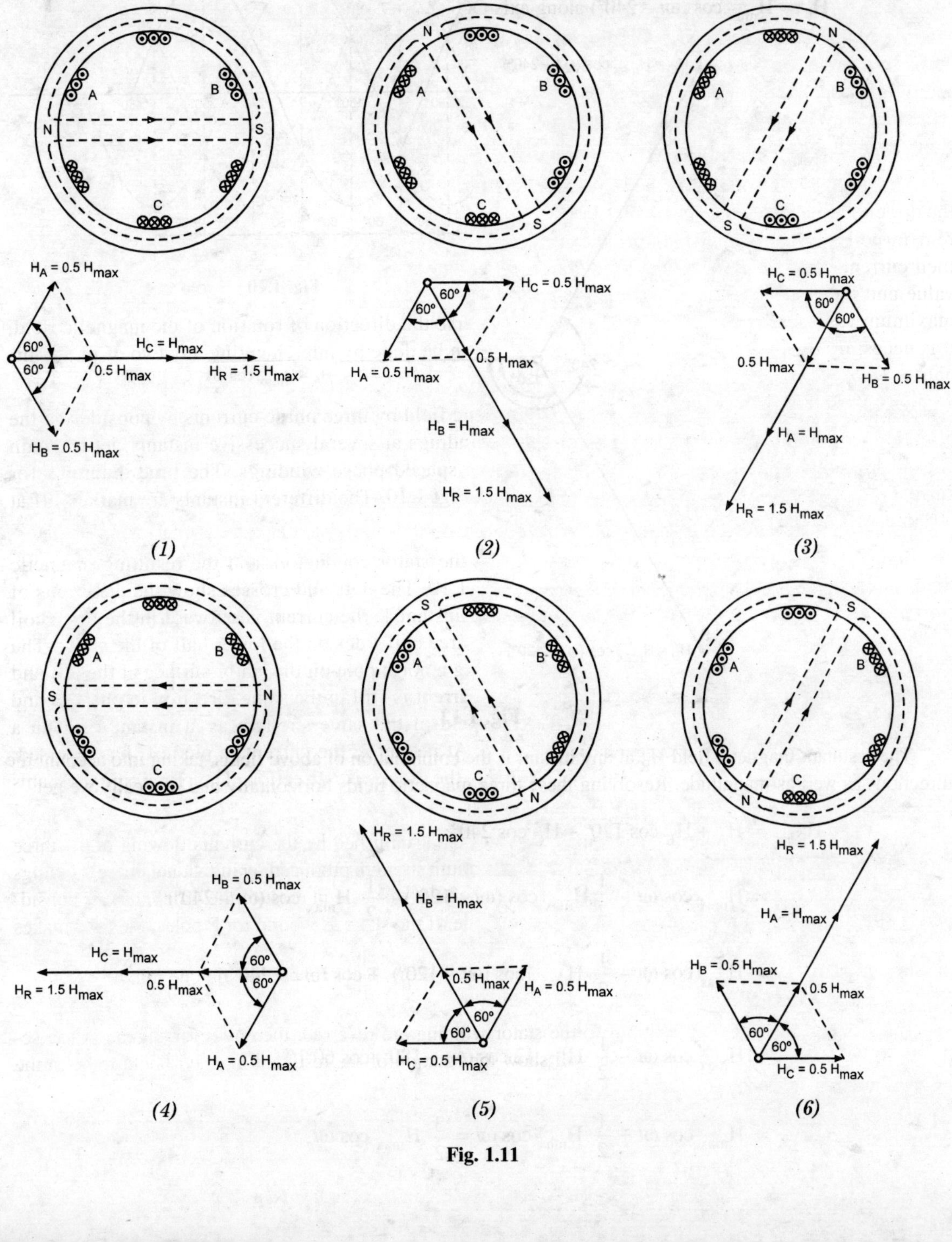

Fig. 1.11

Let each of the coil produce an alternating field of maximum strength H_{max} along its own axis then instantaneous values of magnetic fields produced by coils A, B and C with respect to time are given as below:

$$H_A = H_{max} \cos \omega t \text{ along axis OA}$$

$$H_B = H_{max} \cos (\omega t - 120°) \text{ along axis OB}$$

$$H_C = H_{max} \cos (\omega t - 240°) \text{ along axis OC}$$

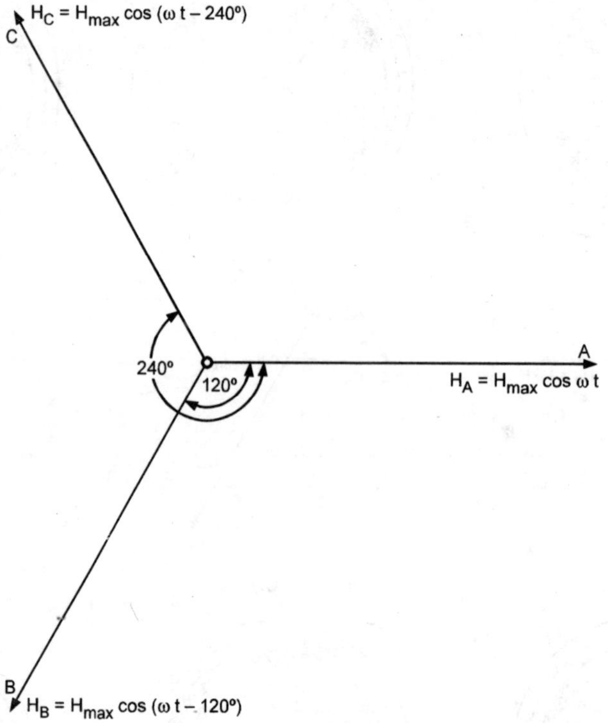

Fig. 1.12

The resultant magnetic field H_R at any instant is the combination of above fields, taking into account the directions as well as magnitude. Resolving these three magnetic fields horizontally and vertically we get

$$H_{Rh} = H_A + H_B \cos 120° + H_C \cos 240°$$

$$= H_{max} \cos \omega t - \frac{1}{2} H_{max} \cos (\omega t - 120°) - \frac{1}{2} H_{max} \cos (\omega t - 240°)$$

$$= H_{max} \cos \omega t - \frac{1}{2} H_{max} [\cos (\omega t - 120°) + \cos (\omega t - 240°)]$$

$$= H_{max} \cos \omega t - \frac{1}{2} H_{max} [2 \cos (\omega t - 180°) \cos 60°]$$

$$= H_{max} \cos \omega t + \frac{1}{2} H_{max} \cos \omega t = \frac{3}{2} H_{max} \cos \omega t$$

$$H_{Rv} = H_A \sin 0^\circ + H_B \sin(-120^\circ) + H_C \sin(-240^\circ)$$

$$= 0 - \frac{\sqrt{3}}{2} H_{max} \cos(\omega t - 120^\circ) + \frac{\sqrt{3}}{2} H_{max} \cos(\omega t - 240^\circ)$$

$$= \frac{\sqrt{3}}{2} H_{max} [\cos(\omega t - 240^\circ) - \cos(\omega t - 120^\circ)]$$

$$= \frac{\sqrt{3}}{2} H_{max} [2 \sin(\omega t - 180^\circ) \sin 60^\circ = \frac{3}{2} H_{max} \sin \omega t$$

$$\text{The resultant field, } H_R = \sqrt{(H_{Rh})^2 + (H_{Rv})^2} = \sqrt{\left(\frac{3}{2} H_{max} \cos \omega t\right)^2 + \left(\frac{3}{2} H_{max} \sin \omega t\right)} = \frac{3}{2} H_{max}$$

Inclination of the resultant field is given by

$$\theta = \text{Tan}^{-1} \frac{H_{Rv}}{H_{Rh}} = \text{Tan}^{-1} \frac{\frac{3}{2} H_{max} \sin \omega t}{\frac{3}{2} H_{max} \cos \omega t} = \omega t$$

It shows that the resultant magnetic field due to three phases is of constant magnitude (3/2 times the maximum strength of magnetic field produced by each phase separately) and travels in space with a uniform angular velocity ω radians per second. *Hence resultant field is of constant magnitude and is rotating at synchronous speed.*

The above principle is not only applicable to synchronous machines but to all machines with polyphase stator windings.

HIGHLIGHTS

1. The magnetic field produced by single phase alternating current is an *alternating magnetic field*, the field acting along a fixed axis, varying in magnitude periodically and changing its direction alternately positive and negative.
2. A rotating magnetic field is that field which is constant in magnitude but whose axis of direction rotates in space as field system of a dc machine.
3. When two identical coils placed at 90° in space are excited from 2-phase balanced supply, the resultant mmf will be rotating in nature with an amplitude equal to that cf each phase.
4. When three identical coils placed at 120° in space are excited from 3-phase balanced currents, the resultant mmf will be rotating in nature with an amplitude 1.5 times that of each phase.

EXERCISES

1. Prove that a single phase motor winding when excited by a single phase supply produces two equal and opposite revolving fields.
2. Show that when two identical coils displaced in space by 90° are connected across 2-phase ac supply, a uniformly rotating field of constant amplitude rotating at synchronous speed is produced.
3. Show that three phase current in a three phase winding always produces a rotating magnetic field.
4. Show that when 3-phase sinusoidally varying ac supply is fed to the stator of a 3-phase induction motor, a rotating magnetic field is produced that rotates in the air gap at the synchronous speed and whose magnitude is 3/2 times the amplitude of phase emf.

5. When a 3-ϕ supply is given to a 3-phase winding, a rotating field of constant magnitude will be proved. Justify the above statement.

6. Show that a rotating magnetic field can be produced by the use of 3-phase currents of equal magnitude.
[M.D. Univ. Electrical Machines-II, Dec. 2010]

SHORT ANSWER TYPE QUESTIONS WITH ANSWERS

Q. 1. What is a rotating magnetic field ?

Ans. A rotating magnetic field is that field which is constant in magnitude but whose axis of direction rotates in space like field system of a dc machine.

Q. 2. How can the direction of the stator revolving field be reversed ?

Ans. The direction of the stator revolving field can be reversed by reversing any two of the three supply lines to the stator winding.

Q. 3. When is a rotating magnetic field produced ? What is its importance in electrical machinery ?

Ans. When a 3-phase wound machine is energized by 3-phase supply, a rotating magnetic field is developed that tends the rotor of the machine to move in its direction of rotation. Energy transfer phenomenon occurs due to such rotating magnetic field.

Q. 4. Explain what is rotating magnetic field and how this is made use of in electrical machinery ?

Ans. A rotating magnetic field is that which is constant in magnitude but whose axis of direction rotates in space as field system of a dc machine. When a 3-phase winding is energized by 3-phase supply, a rotating magnetic field is developed that tends the rotor of the machine to move in the direction of rotation.

The operation of ac rotating machines (such as synchronous machines and induction machines) is based on the production of rotating magnetic field.

❑❑❑

required to prevent the rotor from rubbing against the stator, the air gap between the stator and rotor being very small. The radial ventilating ducts are provided along the length of the stator core, spaced every 5 or 7 cm. These are provided by the use of spacers placed between the laminations. The air gap between the stator and rotor is made as small as practicable (0.3 to 0.35 mm in small machines and 1.0 to 1.5 mm in high power machines) so as to make air-gap reluctance minimum. The slots are punched out on the inner periphery of the stator laminations, as shown in Fig. 2.1, to accommodate the stator winding, also sometimes known as *primary winding*. Some motors, especially of large ratings, have open type slots so that former wound coils can be used but usually the motors are provided with semi-closed type slots to make the air-gap reluctance minimum and to prevent excessive tooth loss. In large, high speed motors it is not essential to employ semi-closed slots to reduce the reluctance. The reason of it is that in large, high speed motors, pole pitch is large, therefore, the number of conductors per pole is large, so the desired flux density in the air gap can be attained without an excessive magnetizing current, even if open type slots are employed.

2.2.3. Stator or Primary or Field Winding

Motors up to 250 kW operate usually at voltages in the range 380-440 V. For 3.3 kV the minimum economical rating is 250-300 kW, and for 11 kV about 750 kW.

In a polyphase induction motor the stator winding is usually 3-phase winding which is usually supplied from a 3-phase supply mains. The three phases of the winding can be connected in either star or delta depending upon the methods of starting used. The squirrel cage motors are usually started by star-delta starters and, therefore, their stators are designed for delta connections and six terminals (two from each phase) are brought out to be connected to the starter. The wound rotor motors are started by inserting resistance in the rotor circuit and, therefore, the stator winding can be connected in either star or delta as desired. It is wound for a definite number of poles as per requirement of speed. Greater the number of poles, lesser the speed or vice versa for a supply of given frequency.

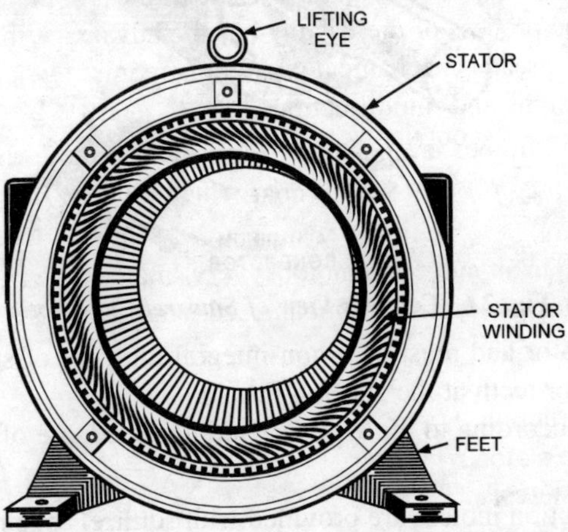

Fig. 2.2. ***Stator of a 3-Phase Induction Motor***

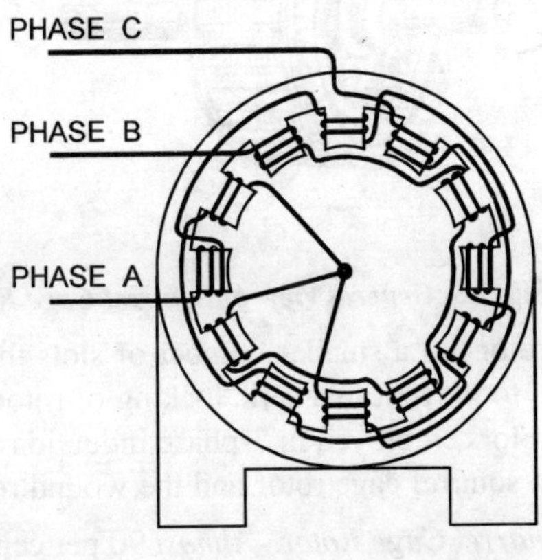

Fig. 2.3. ***3-Phase, 4-Pole, Star-Connected Stator Winding***

In general, the same windings can be used on the stator of induction motor as were suitable for synchronous generators or motors. The double layer winding is mostly used in medium size motors because of its greater ease of manufacture, assembly, and repair. Moreover, stator windings are almost always short-pitched because of reduced copper weight and winding resistance as well as reduced leakage reactance and harmonic torque disturbances that result. Small motors, operating at ordinary supply voltages, with a small number of slots having a large number of turns per phase may use single layer mush winding. For large motors, or high voltage motors the stator windings may be formed by single layer concentric coils. Double layer coils are placed in open slots so that former wound coils can be used while mush winding is housed in semi-closed slots.

Stator of a 3-phase induction motor and 3-phase, 4-pole, star-connected stator winding are illustrated in Figs. 2.2 and 2.3 respectively.

2.2.4. Rotor

The rotor comprises a cylindrical laminated iron core, much the same as that of a dc machine, with slots, around the core, carrying the rotor conductors. In general the same sheet steel laminations are employed for the rotor core as for the stator, but owing to the lower frequencies of the rotor flux, thicker laminations can be employed without excessive iron loss. The rotor stamping is shown in Fig. 2.4. Like stator, the rotor laminations are punched in one piece for small machines; in larger machines the laminations are segmented and dovetailed to a fabricated spider. If there are ventilating ducts on the stator core, an equal number of such ducts is provided on the rotor core also. Fan blades are usually employed on the ends of the rotor core to force circulating air through the motor for cooling.

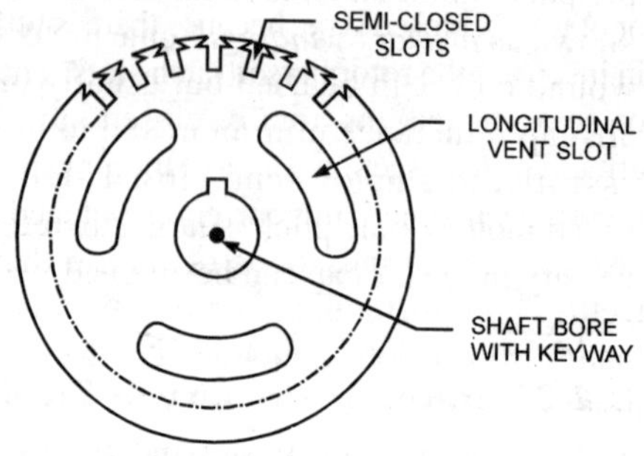

Fig. 2.4. *Rotor Stamping*

Fig. 2.5. *General View of Squirrel Cage Rotor*

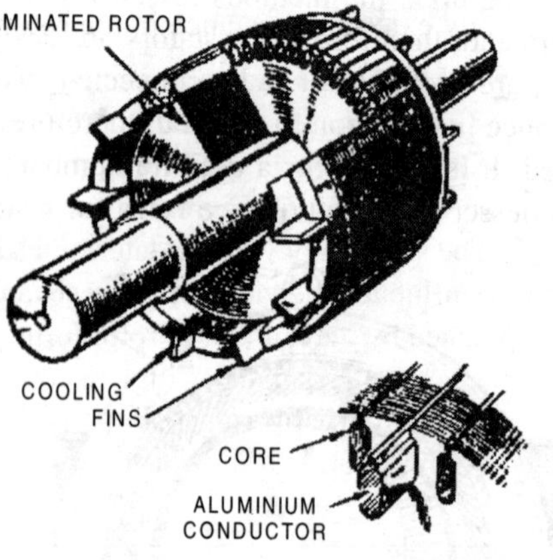

Fig. 2.6. *Cutaway View of Squirrel Cage Rotor*

The rotor has a smaller number of slots than the stator and must be a non-integral multiple of stator slots so as to prevent magnetic locking of rotor and stator teeth at the starting instant.

The rotors employed in 3-phase induction motors, according to the type of windings used, are of two types, viz; squirrel cage rotor and the wound rotor.

1. *Squirrel Cage Rotor. Almost* 90 per cent of induction motors are provided with squirrel cage rotor because of its very simple, robust and almost instructible construction. In cage construction, copper, brass, or aluminium bars are placed, as the rotor conductors, parallel or approximately parallel to the shaft (one bar in each slot) and close to the rotor surface. The conductors are not insulated from the core, since the rotor currents naturally follow the path of least resistance, *i.e.*, the rotor conductors. At both ends of the rotor, the rotor conductors are all short circuited by the continuous end rings of similar material to that of the rotor conductors. The rotor conductors and their end rings form a complete closed circuit in itself, resembling a squirrel cage, thus explaining the name. In motors with ratings up to 100 kW, the squirrel cage structure is formed by aluminium cast (under pressure) into the slots of the rotor (Fig. 2.6). In large motors the rotor bars, instead of being cast, are wedged into the rotor slots and are then welded securely to

the end rings. The slots on the rotor are always not parallel to the motor shaft but are usually skewed in order to obtain a uniform torque, reduce the magnetic locking of the stator and rotor and reduce the magnetic humming noise while running.

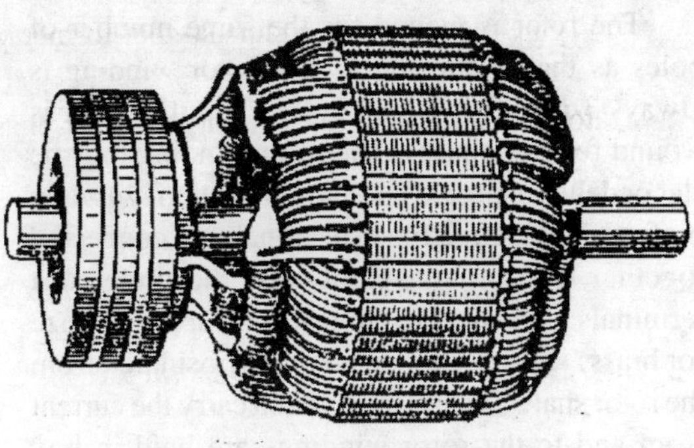

Fig. 2.7. *Rotor With Slip rings*

The slots on rotor are either of semi-closed type or of totally closed type, because there is little difficulty in inserting the rotor bars in such slots. The advantage of semi-closed and totally closed slots is that the effective cross-sectional area of the air gap is increased, therefore, magnetizing current is reduced. Such slots also reduce the pulsations of flux in the individual teeth, therefore, tooth loss is reduced. The disadvantage of such slots is that these give higher slot inductance than the open slots, which lowers the power factor and reduces the starting and breakdown torques of the motor.

The squirrel cage rotor windings are perfectly symmetrical and have the advantage of being adaptable to any number of pole pairs. The distribution of current due to electromagnetic induction in the rotor bars varies from bar to bar sinusoidally and depends upon the position and time, assuming sinusoidal distribution of radial flux density in space and also the applied voltage to be varying sinusoidally with time.

Since the rotor winding is permanently short circuited in cage construction, so there is no possibility of adding any external resistance in the rotor circuit.

2. *Wound Rotor. As* the name implies, such a rotor is wound with an insulated winding similar to that of the stator except that the number of slots is smaller and fewer turns per phase of a heavier conductor are used. Bar, strap, or wire is used for rotor windings, the last being used where many turns are desired. A large number of rotor turns increases the secondary voltage and reduces the current that flows through the slip rings. The secondary voltage determines the insulation that must be provided; furthermore, the voltage and current influence the value of the resistance to be employed across the slip rings. The motor operation is not influenced by the number of rotor turns, but the ratio of transformation is determined by consideration of secondary current, danger of high secondary emf at starting, and distance to secondary resistors. The standstill open-circuit slip-ring voltage is usually 100 to 400 V for small machines using hand operated gear and maximum up to 1 kV for large machines.

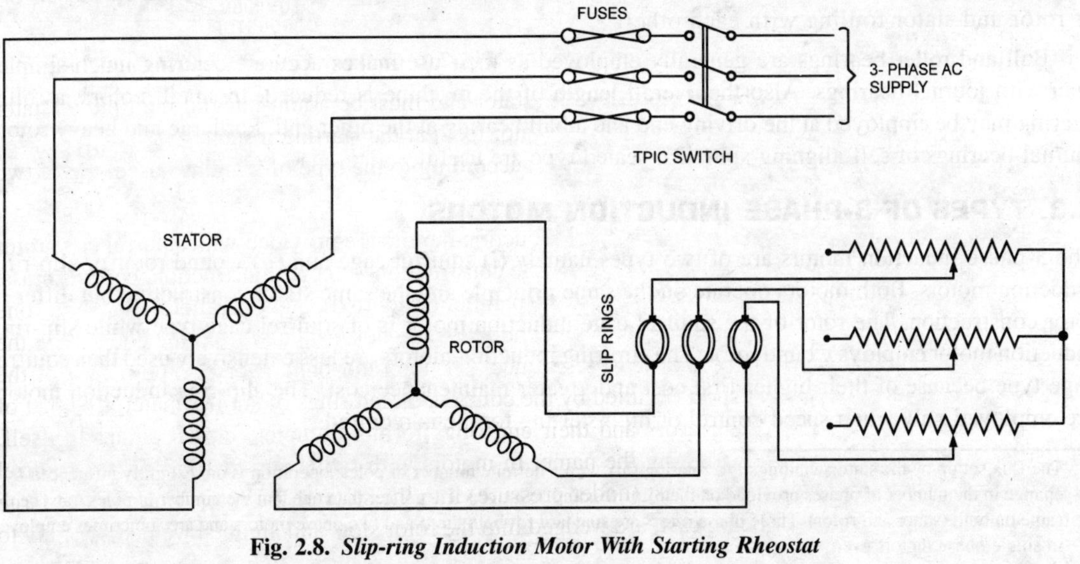

Fig. 2.8. *Slip-ring Induction Motor With Starting Rheostat*

The rotor is wound for the same number of poles as that of the stator. The rotor winding is always 3-phase* winding even when the stator is wound for two phases. The rotor winding may be star or delta connected but star-connection is usually preferred. The three finish terminals are connected together to form star-point and the three start terminals are connected to three phosphor-bronze (or brass) slip rings mounted on but insulated from the rotor shaft. The brushes, which carry the current from and to the rotor windings are held in box type holders mounted on insulated steel rods, securely bolted to the end shield. Each brush is fed forward by a lever held in tension by an adjustable spring. These brushes are further externally connected to a 3-phase star-connected rheostat for the purpose of starting and speed control (Fig. 2.8). At the time of starting, the entire resistance is included in the rotor circuit and this resistance is gradually cut out as the rotor picks up the speed. For the normal running condition the entire external resistance is cut out and the rotor windings are short circuited automatically through the slip rings by means of a metal collar which is pushed along the shaft and connects all the rings together. Some machines are provided with brush lifting and slip ring short-circuiting arrangement for running condition. However, brush lifting and short-circuiting gear is usually not used in modern machines. The rotor is skewed in this case also. Since the connection of the wound secondary to the external terminals is made through slip rings and brushes, so wound secondary motors often are called *slip-ring induction* motors. A sectional diagram of a slip-ring induction motor is shown in Fig. 2.9.

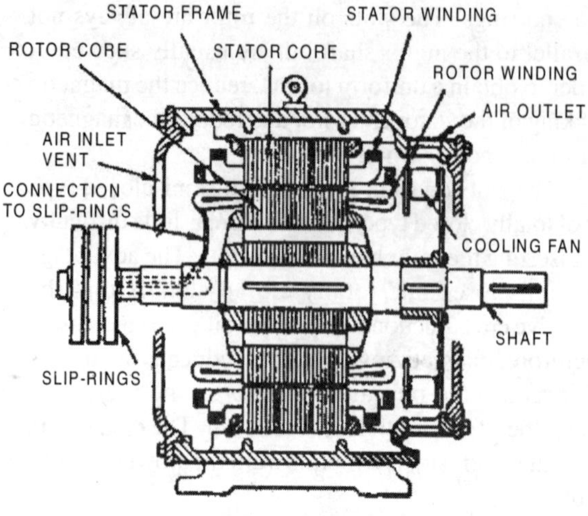

Fig. 2.9. *Section of a Slip-ring Induction Motor*

2.2.5. Shafts and Bearings

The rotor shaft is supported by bearings housed in the end shield. The air gap of an induction motor in kept necessarily small, therefore, the rotor shaft must be made short and stiff so that the rotor may not have any significant deflection. Even a minor deflection in the shaft would develop large irregularities in the air gap which would lead to production of an *unbalanced magnetic pull.* There is also a possibility of rotor and stator fouling with each other.

Ball and roller bearings are generally employed as their use makes accurate centring much simpler than with journal bearings. Also the overall length of the machine is reduced. In small motors, a roller bearing may be employed at the driving end and a ball bearing at the other end. For large and heavy rotors journal bearings of self aligning spherical seated type are useful.

2.3. TYPES OF 3-PHASE INDUCTION MOTORS

The 3-phase induction motors are of two types namely (*i*) squirrel cage and (*ii*) wound rotor or slip-ring induction motors. Both motors operate on the same principle and the same stator construction but differ in rotor construction. The rotor of the squirrel cage induction motor is of squirrel cage type while slip-ring induction motor employs wound rotor. The slip-ring induction motors are less extensively used than squirrel cage type because of their higher first cost and greater maintenance cost. The slip-ring induction motors are employed only when speed control or high starting torque is required.

* The flux set up by the stator windings is a rotating flux with a definite number of poles; operation is only slightly influenced by a change in the number of phases provided on the rotor winding so long as it is greater than one, but the number of poles must be the same on both (stator and rotor). Three phase rotors are standard for both three-and two phase motors and are sometimes employed in single phase motors even.

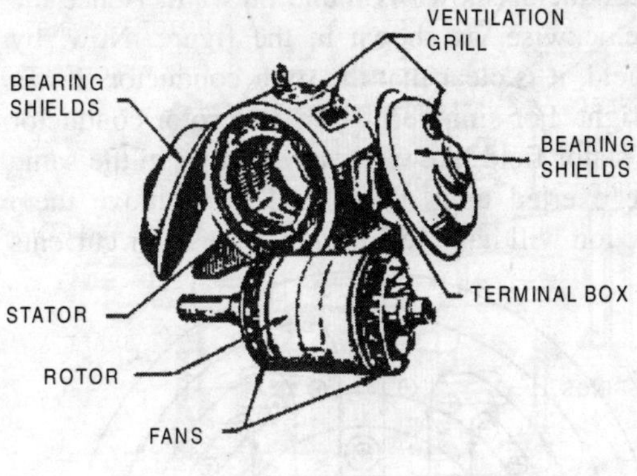

(a) *Induction Motor With Squirrel Cage Rotor (Disassembled)*

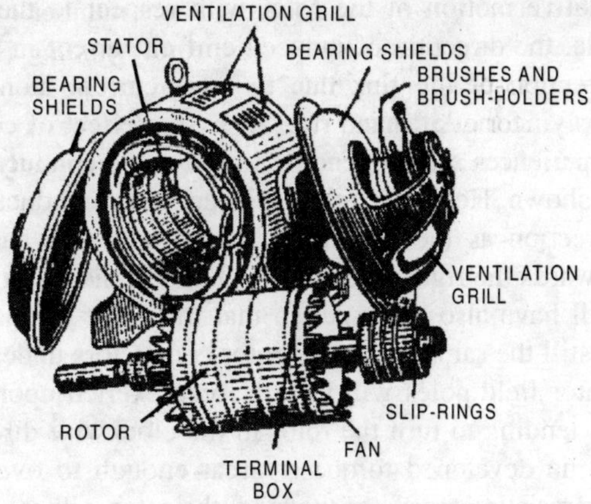

(b) *Induction Motor With Wound Rotor (Disassembled)*

Fig. 2.10

Induction motor with squirrel cage rotor and that with wound rotor, in disassembled form are shown in Fig. 2.10 (*a*) and 2.10 (*b*) respectively.

2.4. PRINCIPLE OF OPERATION

In a direct current motor, current is drawn from the supply and conducted into the armature conductors through the brushes and commutator. When the armature conductors carry current in the magnetic field established by the field, a force is exerted on the conductors which tends to move them at right angles to the field.

Though in an induction motor, there is no electrical connection to the rotor, but currents are induced in the rotor circuit, and therefore, the same condition exists as in the dc motors *i.e.*, the rotor conductors carry current in the stator magnetic field and thereby have a force exerted upon them tending to move them at right angles to the field.

When the stator or primary winding of a 3-phase induction motor is connected to a 3-phase ac supply, a rotating magnetic field is established which rotates at synchronous speed. The direction of revolution of this field will depend upon the phase sequence of the primary currents and, therefore, will depend upon the order of connection of the primary terminals to the supply. The direction of rotation of the field can be reversed by interchanging the connection to the supply of any two leads of a 3-phase induction motor. (Refer to Art. 1.4). The number of magnetic poles of the revolving field will be the same as the number of poles for which each phase of the primary or stator winding is wound. The speed at which the field produced by the primary currents will revolve is called the *synchronous speed* of the motor and is given by an expression, $N_s = \frac{120f}{P}$ where f is supply frequency and P is the number of poles on stator.

As the rotating magnetic field produced by the primary currents sweeps across the rotor conductors, an emf is induced in these conductor just as an emf is induced in the secondary winding of a transformer by the flux set up by the primary current. Since the rotor winding is either directly shorted or closed through some external resistance, the emf induced in the secondary by the revolving field causes a current to flow in the rotor conductors.

The setting up of the torque for causing the rotor to rotate is explained below:

A section of an induction motor stator and rotor, with the magnetic field assumed to be rotating in a clockwise direction and with the rotor stationary, as at starting, is shown in Fig. 2.11 (*a*). The relative motion of the rotor with respect to the stator field is anti-clockwise. By applying right hand rule, the direction of induced emf or current in the rotor conductor shown is found outward. Hence the direction of the flux due to rotor current alone is anti-clockwise, as shown in the figure. Now, by applying the left hand rule, or by the effect of combined field, it is clear that the rotor conductor shown experiences a force tending to move the conductor to the right. For simplicity, only one rotor conductor is shown. However, other adjacent rotor conductors in the stator field likewise carry current in the same direction as the conductor shown and also have a force exerted upon them tending to move them towards the right. One half cycle later, the stator field direction will have reversed, but the rotor currents will have also reversed, so that the force on the rotor is still the same. Likewise rotor conductors under other stator field poles will have a force exerted upon them all tending to turn the rotor in the clockwise direction. If the developed torque is great enough to overcome the resisting torque of the load, the rotor will accelerate in the clockwise direction or in the same direction as the rotation of the stator field.

The above facts have been illustrated diagrammatically in Fig. 2.11 (*b*).

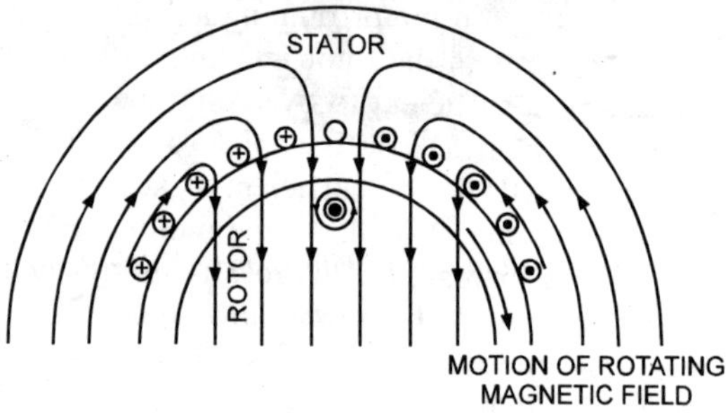

(*a*)

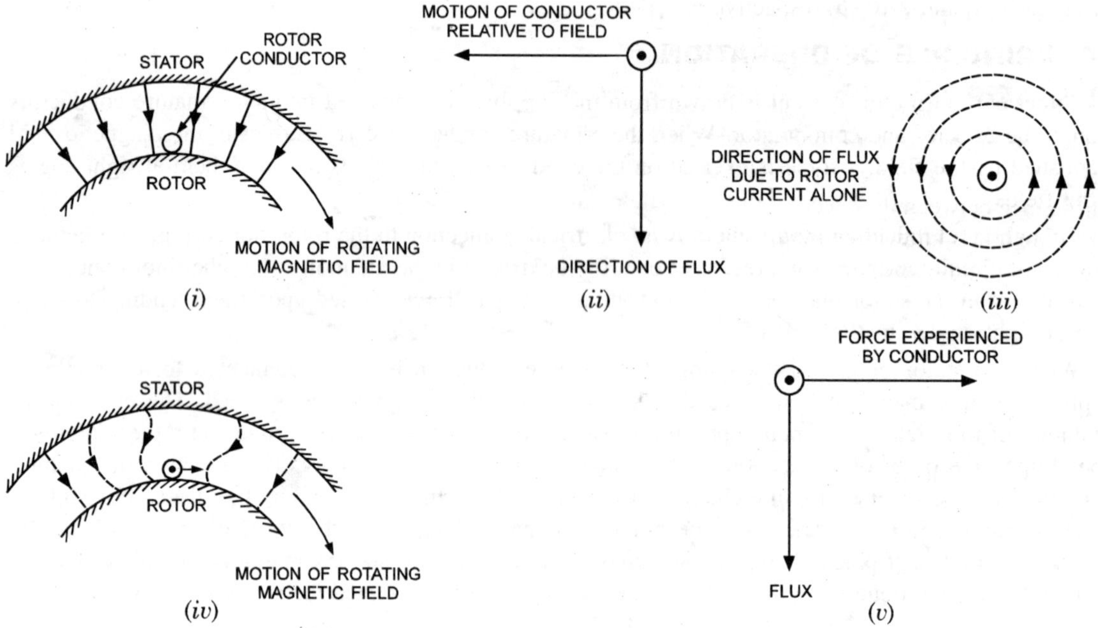

(*b*) *Production of Torque*

Fig. 2.11

This may also be explained in terms of Lenz's law. According to Lenz's law we know that the direction of induced emf would be in such a direction that it would try to oppose the very cause for which it is due. Because the cause producing the induced currents is the relative speed between the rotating magnetic field and stationary rotor conductors, therefore, they circulate in such a way that a torque is produced in the rotor tending to cause it to follow the rotating magnetic field and thus reducing the relative speed.

When the rotor is stationary and about to start, the frequency of induced emf in the rotor is equal to that of the supply fed to the stator, because the relative motion is at synchronous speed. As the motor picks up the speed, the relative motion between the rotor and the synchronously rotating magnetic field becomes less and the frequency of emf induced in the rotor decreases. The magnitude of rotor induced emf, induced rotor current and so that of torque developed depends upon the relative motion. In case the relative motion is zero, *i.e.*, the rotor runs at synchronous speed, there would be no induced emf and no current in the rotor conductors, no rotor field and hence no torque. Thus we see that an *induction motor cannot run at synchronous speed.*

An induction motor running at no load will have a speed very close to the synchronous speed and, therefore, emfs in the rotor winding will be very small. This small emf gives a small current producing a torque just sufficient to overcome the losses such as due to friction and maintain the rotor in motion. As the mechanical load is applied on the motor shaft, it must slow down because the torque developed at no load will not be sufficient to keep the rotor revolving at the no-load speed against the additional opposing torque of load. As the motor slows down, the relative motion between the magnetic field and the rotor is increased. This results in greater rotor currents and greater developed torque. Thus, as the load is increased, the motor slows down until the relative motion between the rotor and the rotating magnetic field is just sufficient to result in the development of the torque necessary for that particular load. The decrease in speed from no-load to full load is usually 4 to 5 per cent in case of small and medium size induction motors and in case of large size motors it varies from 2 to 2.25 per cent. In respect of speed-load characteristics, a 3-phase induction motor is quite similar to a dc shunt motor.

2.5. SLIP

In preceding article it has been determined that the speed of a polyphase induction motor must always be less than the synchronous speed and that, as the load is increased, the speed of the motor will decrease. The difference between the speed of the stator field, known as *synchronous speed* (N_s), and the actual speed of the rotor (N) is known as the *slip* and is denoted by *s*. Though the slip can be expressed in rpm or in radians per second, but usually it is expressed as a fraction or percentage of synchronous speed.

$$\text{Fraction slip, } s = \frac{\text{Synchronous speed} - \text{rotor speed}}{\text{Synchronous speed}} = \frac{N_s - N}{N_s} \quad ...(2.1)$$

$$\text{and percentage slip} = \frac{N_s - N}{N_s} \times 100 \quad ...(2.2)$$

At normal load the slip of an induction motor is usually between 2 and 5 per cent. At no load the slip is as small as 0.5 per cent and for many purposes the machine may then be considered to be running at synchronous speed. As the load is applied, the natural effect of the load torque is to cause the machine to slow down. As it does so, slip increases, and with it current and torque until the driving torque of the machine balances the retarding torque of the load. This determines the speed at which the machine runs on load.

The induction motor is thus a motor with substantially constant speed and plays the same role as a dc shunt motor.

Example 2.1. A 3-phase, 4-pole, 50 Hz, induction motor runs at 1,460 rpm. Determine its percentage slip.

Solution : Synchronous speed, $N_s = \frac{120 f}{P} = \frac{120 \times 50}{4} = 1{,}500$ rpm

Running speed of motor, N = 1,460 rpm

$$\text{Slip, } s = \frac{N_s - N}{N_s} \times 100 = \frac{1{,}500 - 1{,}460}{1{,}500} \times 100 = 2.667\% \text{ **Ans.**}$$

Example 2.2. Power to an induction motor is supplied by a 12-pole, 3-phase, 500 rpm alternator. The full-load speed of the motor is 1,450 rpm. Find the percentage slip and number of poles in the motor. [*M.D. Univ. May 2008*]

Solution : Frequency of generated emf, $f = \frac{P_a \times N_a}{120} = \frac{12 \times 500}{120} = 50$ Hz

The number of poles even in number and to give a synchronous speed nearest higher than actual speed of motor (*i.e.* 1,450 rpm) that the motor must have is 4. **Ans.**

$$\because \quad P_m = \frac{120 f}{N_m} = \frac{120 \times 50}{1{,}450} \simeq 4$$

$$\text{Synchronous speed of motor, } N_s = \frac{120 f}{P} = \frac{120 \times 50}{4} = 1{,}500 \text{ rpm}$$

$$\text{Slip, } s = \frac{N_s - N}{N_s} = \frac{1{,}500 - 1{,}450}{1{,}500} = 0.0333 \text{ or } 3.33\ \% \text{ **Ans.**}$$

Example 2.3. A motor-generator set used for providing variable frequency ac supply consists of a 3-phase, 10-pole synchronous motor and a 24-pole, 3-phase synchronous generator. The motor-generator set is fed from a 25 Hz, 3-phase ac supply. A 6-pole, 3-phase induction motor is electrically connected to the terminals of the synchronous generator and runs at a slip of 5%. Determine (*i*) the frequency of the generated voltage of the synchronous generator, (*ii*) the speed at which the induction motor is running.

Solution : Speed of motor-generator set, $N_s = \frac{120 \times \text{supply frequency}, f_1}{\text{Number of poles on synchronous motor}} = \frac{120 \times 25}{10} = 300$ rpm

(*i*) Frequency of generated voltage, $f_2 = \frac{\text{Speed of motor-generator set} \times \text{number of poles on synchronous generator}}{120}$

$$= \frac{300 \times 24}{120} = 60 \text{ Hz} \quad \textbf{Ans.}$$

(*ii*) Speed of induction motor, $N_m = N_s (1 - s) = \frac{120 f_2}{P_m} (1 - s) = \frac{120 \times 60}{6} (1 - 0.05) = 1{,}140$ rpm **Ans.**

2.6. FREQUENCY OF ROTOR CURRENT (OR EMF)

It has been shown in Art. 2.4 that the rotor runs in the direction of rotating magnetic field. When the rotor is stationary, rotor conductors are being cut by the rotating flux at synchronous speed N_s $\left(N_s \text{ being equal to } \frac{120 f}{P}\right)$, therefore, the frequency of rotor current (or emf) is the same as that of supply frequency. But when the rotor starts revolving, the rate at which the rotor conductors are being cut by the rotating flux depends upon the relative speed between the rotor and the stator revolving magnetic field, called the *slip speed.* The frequency of rotor current (or emf) induced by the relative motion between rotor conductors and the stator revolving magnetic field is given by the equation:

$$\text{Rotor emf frequency, } f' = \frac{\text{Relative speed in rpm}}{120/P} = \frac{N_s - N}{120/P} \quad ...(2.3)$$

$$\text{Since slip, } s = \frac{N_s - N}{N_s}$$

$$\text{or } N_s - N = s\,N_s = s \times \frac{120 f}{P}$$

Substituting $N_s - N = s\,N_s = s \times \frac{120f}{P}$ in Eq. (2.3) we have

$$\text{Rotor emf frequency, } f' = s \times \frac{120f}{P} \times \frac{P}{120} = sf \qquad \text{...(2.4)}$$

Thus the frequency of rotor current (or emf) in an induction motor is given by the product of slip s and the supply frequency f. That is why, it is also called the *slip frequency*.

Example 2.4. A 3-ϕ, 6-pole, 50 Hz, induction motor has a slip of 1% at no load and 3% at full load. Determine synchronous speed, no-load speed, full-load speed, frequency of rotor current at standstill and frequency of rotor current at full load.

Solution :

Supply frequency, $f = 50$ Hz

Synchronous speed, $N_s = \frac{120f}{6} = \frac{120 \times 50}{6} = 1{,}000$ rpm

No-load speed, $N_o = N_s(1 - s_o) = 1{,}000\,(1 - 0.01) = 990$ rpm **Ans.**

Full-load speed, $N_f = N_s(1 - s_f) = 1{,}000\,(1 - 0.03) = 970$ rpm **Ans.**

At standstill slip, $s = 1.0$

Frequency of rotor current at standstill, $f' = sf = 1.0 \times 50 = 50$ Hz **Ans.**

Frequency of rotor current at full load, $f' = sf = 0.03 \times 50 = 1.5$ Hz **Ans.**

Example 2.5. A 4-pole, 50 Hz induction motor runs with a slip of 0.01 pu on full load. Calculate the frequency of rotor current (*a*) at standstill (*b*) on full load. *[M.D. Univ. Electrical Machines-II, May 2009]*

Solution :

Supply frequency, $f = 50$ Hz

At full load, slip $s_2 = 0.01$

At standstill, slip $s_1 = 1.0$

(*a*) Frequency of rotor current at standstill $f'_1 = s_1 f = 1.0 \times 50 = 50$ Hz **Ans.**

(*b*) Frequency of rotor current on full load $f'_2 = s_2 f = 0.01 \times 50 = 0.5$ Hz **Ans.**

2.7. SPEED OF ROTOR FIELD (OR MMF)

The rotating field set up by stator currents rotates at synchronous speed N_s revolutions per minute, relative to the stator surface. Similarly the currents having a frequency of f' (f' being equal to sf) when flowing through the rotor winding set up the rotor magnetic field which rotates at a speed of $s\,N_s$ rpm, relative to the rotor surface, in the direction of rotation of the rotor. However, the rotor itself is running at a speed of N rpm with respect to stator surface.

So the speed of rotor magnetic field with respect to stator surface is equal to the sum of actual speed of rotor N and rotor field speed sN_s, with respect to rotor surface. Thus the speed of rotor field (or mmf) with respect to stator surface is given as

$$N + s\,N_s = N_s(1 - s) + s\,N_s = N_s \text{ rpm}$$

Thus the rotor magnetic field also rotates, in space, at the same speed and in the same direction as that of the stator field. This concludes that both the rotor and stator fields rotate synchronously, which means that they are stationary w.r.t. each other at all possible rotor speeds. These two synchronously rotating magnetic fields, in fact, superimpose on each other and give rise to the actually existing rotating field, which corresponds to the magnetizing current of the stator winding.

Since the two fields set up by stator (primary) and rotor (secondary) currents are stationary relative to each other, so the polyphase induction motor can be considered as being equivalent to a transformer having an air gap separating the iron portions of the magnetic circuit carrying the primary and secondary windings.

Owing to this air gap, the magnetizing current and the magnetic leakage for an induction motor are large compared with the corresponding values for a transformer of the same kVA rating. Also, the friction and windage losses contribute towards making the efficiency of the induction motor less than that of the corresponding transformer. On the other hand, the stator mmf has to balance the rotor mmf and also provide the magnetizing and no-load loss components of the stator current, as in a transformer . Hence, an increase in slip due to increase in load is accompanied by an increase in rotor currents and, therefore, by a corresponding increase in the stator currents. Because of distributed windings in induction motors the ratio of stator and rotor currents is not equal to the ratio of turns per phase in the rotor and stator winding.

Example 2.6. A 3-phase induction motor runs at a speed of 1,485 rpm at no load and at 1,350 rpm at full load when supplied from a 50 Hz, 3-phase line:

(*i*) How many poles does the motor have?

(*ii*) What is the slip at no load and at full load?

(*iii*) What is the frequency of rotor voltages at no load and at full load?

(*iv*) What is the speed at both no load and full load of, the rotor field with respect to rotor conductors, the rotor field with respect to the stator and the rotor field with respect to the stator field.

Solution : (*i*) The number of poles, even in number, to give a synchronous speed nearest higher than no-load speed of motor (*i.e.* 1,485 rpm) that the motor must have is 4. **Ans.**

$$\because \quad P = \frac{120f}{N} = \frac{120 \times 50}{1{,}485} \simeq 4$$

$$\text{Synchronous speed, } N_s = \frac{120f}{P} = \frac{120 \times 50}{4} = 1{,}500 \text{ rpm}$$

$$(ii) \text{ No-load slip, } s_0 = \frac{N_S - N_0}{N_S} = \frac{1{,}500 - 1{,}485}{1{,}500} = 0.01 \text{ or } 1\% \quad \textbf{Ans.}$$

$$\text{Full-load slip, } s_f = \frac{N_S - N_f}{N_S} = \frac{1{,}500 - 1{,}350}{1{,}500} = 0.1 \text{ or } 10\% \quad \textbf{Ans.}$$

(*iii*) Frequency of rotor voltages at no load, $f_0' = s_0 f = 0.01 \times 50 = 0.5$ Hz **Ans.**

Frequency of rotor voltages at full load, $f_f' = s_f f = 0.1 \times 50 = 5$ Hz **Ans.**

(*iv*) The speed of rotor field with respect to rotor conductors

$$\text{at no load} = \frac{120 \times f_0'}{P} = \frac{120 \times 0.5}{4} = 15 \text{ rpm} \quad \textbf{Ans.}$$

$$\text{at full load} = \frac{120 \times f_f'}{P} = \frac{120 \times 5}{4} = 150 \text{ rpm} \quad \textbf{Ans.}$$

Speed of rotor field with respect to stator, $N_S = 1{,}500$ rpm **Ans.**

Both of the rotor field and stator field are revolving at the same speed (synchronous speed of 1,500 rpm), therefore, speed of rotor field w.r.t. stator field is zero. **Ans.**

Example 2.7. A 3-phase, 6-pole, 50 Hz induction motor has a slip of 1% at no-load and 3% at full-load. Find (*i*) synchronous speed, (*ii*) no-load speed (*iii*) full-load speed (*iv*) frequency of rotor current at standstill (*v*) Frequency of rotor current at full-load.

[M.D. Univ. May 2005]

Solution :

Number of poles, $P = 6$

Supply frequency, $f = 50$ Hz

No-load slip, $s_0 = 1\%$ or 0.01

Full-load slip, $s_f = 3\%$ or 0.03

$$(i) \text{ Synchronous speed of motor, } N_s = \frac{120f}{P} = \frac{120 \times 50}{6} = 1{,}000 \text{ rpm} \quad \textbf{Ans.}$$

(*ii*) No-load speed, $N_0 = N_s(1 - s_0) = 1{,}000\,(1 - 0.01) = 990$ rpm **Ans.**

(*iii*) Full-load speed, $N_f = N_s(1 - s_f) = 1{,}000\,(1 - 0.03) = 970$ rpm **Ans.**

(*iv*) Frequency of rotor current at standstill = Supply frequency $f = 50$ Hz **Ans.**

(*v*) Frequency of rotor current at full-load, $f = s_f f = 0.03 \times 50 = 1.50$ Hz **Ans.**

2.8. ROTOR EMF

When the rotor is at standstill, the motor is equivalent to a 3-phase transformer with secondary short-circuited. So induced emf per phase E_2 in the rotor, when it is at standstill *i.e.* at the instant of starting is given by

$$E_2 = E_1 \times \frac{N_2}{N_1} \qquad ...(2.5)$$

where E_1 is applied voltage per phase to primary *i.e* stator winding, N_2 and N_1 are the numbers of turns per phase on rotor and stator respectively.

When the rotor starts running, the relative speed of the rotor with respect to stator flux drops in direct proportion with the relative speed or slip *s* and induced emf in the rotor is given by $s\,E_2$.

Hence for slip *s*, the induced emf in the rotor is *s* times the induced emf in the rotor at standstill.

2.9. ROTOR CURRENT AND POWER FACTOR

The circuit diagram of an induction motor rotor is given in Fig. 2.12.

Let the resistance and inductance per phase of rotor be R_2 and L_2 respectively and induced emf per phase in rotor at standstill be E_2.

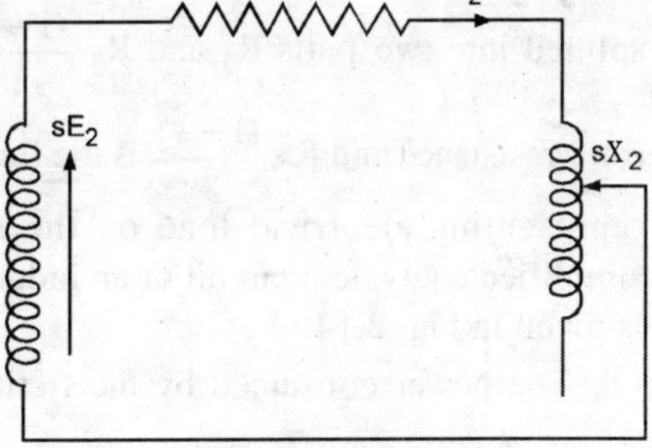

Fig. 2.12

At Standstill

Induced emf per phase in rotor = E_2

Rotor winding resistance / phase = R_2

Rotor winding reactance / phase, $X_2 = 2\pi f L_2$

where *f* is supply frequency

Rotor winding impedance / phase $Z_2 = \sqrt{R_2^2 + X_2^2}$ ...(2.6)

$$\text{Rotor current, } I_2 = \frac{E_2}{Z_2} = \frac{E_2}{\sqrt{R_2^2 + X_2^2}} \qquad ...(2.7)$$

$$\text{Power factor of rotor current, } \cos\phi_2 = \frac{R_2}{\sqrt{R_2^2 + X_2^2}} \qquad ...(2.8)$$

At Slip *s*

Induced emf per phase in rotor winding = sE_2

Rotor winding resistance / phase = R_2

Rotor winding reactance / phase = $2\pi f' L_2 = 2\pi s f L_2 = s(2\pi f L_2) = sX_2$

$$\text{Rotor winding impedance / phase} = \sqrt{R_2^2 + (sX_2)^2} \qquad ...(2.9)$$

Rotor current per phase, $I_2 = \frac{sE_2}{\sqrt{R_2^2 + (sX_2)^2}}$

$$= \frac{sE_2}{\sqrt{R_2^2 + s^2X_2^2}} \quad ...(2.10)$$

Power factor of rotor current, $\cos \phi_2 = \frac{R_2}{\sqrt{R_2^2 + s^2X_2^2}} \quad ...(2.11)$

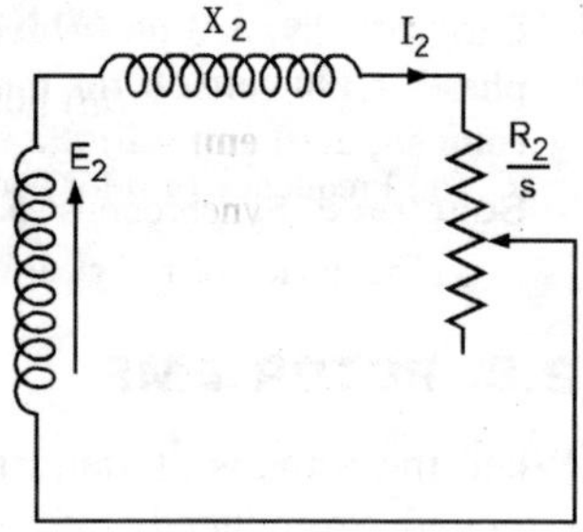

Fig. 2.13. ***Equivalent Circuit of Rotor***

The Eq. (2.10) for rotor current I_2 may be rewritten as

$$I_2 = \frac{E_2}{\sqrt{\left(\frac{R_2}{s}\right)^2 + X_2^2}} \quad ...(2.12)$$

The above equation gives convenient form of rotor equivalent circuit, as shown in Fig. 2.13.

The resistance $\frac{R_2}{s}$ is a function of slip and can be splitted into two parts R_2 and $R_2 \frac{(1-s)}{s}$ where R_2 is rotor resistance and $R_2 \frac{(1-s)}{s}$ is the fictitious resistance representing electrical load on the rotor. The final simplified equivalent circuit of an induction motor rotor is given in Fig. 2.14.

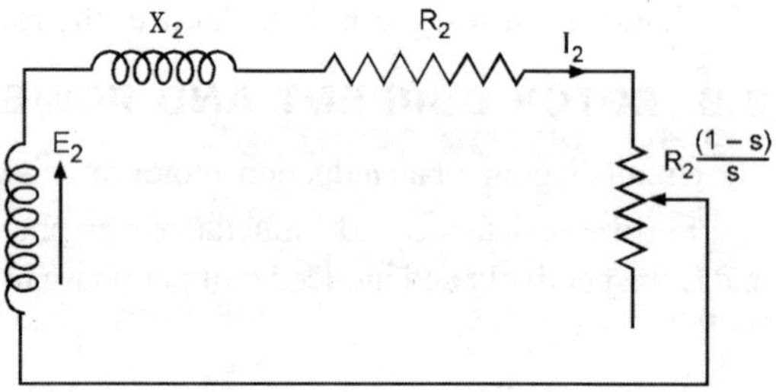

Fig. 2.14

The power consumed by the fictitious resistance representing the rotor electrical load is given as $I_2^2 R_2 \frac{(1-s)}{s}$ and is converted into mechanical power.

Example 2.8. A 3-phase, 440 V, 50 hp, 50 Hz induction motor runs at 1,450 rpm when it delivers rated output power. Determine :

(*i*) Number of poles in the machine.

(*ii*) Speed of rotating air-gap field.

(*iii*) Rotor induced voltage if stator to rotor turns ratio is 1 : 0.80. Assume the winding factors are the same.

(*iv*) Frequency of rotor current.

Solution : (*i*) The number of poles even in number and to give synchronous speed nearest higher than actual speed of motor (*i.e.* 1,450 rpm) that the motor must have is 4. **Ans.**

(*ii*) Speed of rotating air-gap field. $N_s = \frac{120 \times f}{P} = \frac{120 \times 50}{4} = 1{,}500$ rpm **Ans.**

(*iii*) Induced emf in rotor at standstill $= K E_1 = K V_1 = 0.8 \times 440 = 352$ V

Slip, $s = \frac{N_s - N}{N_s} = \frac{1{,}500 - 1{,}450}{1{,}500} = 0.03333$

Rotor induced voltage $= sKV_1 = 0.03333 \times 352 = 11.733$ V **Ans.**

(*iv*) Frequency of rotor current, $f' = sf = 0.03333 \times 50 = 1.6666$ Hz **Ans.**

Example 2.9. A 3-phase delta connected 440 volts, 50 Hz, 4-pole induction motor has a rotor standstill emf per phase of 130 volts. If the motor is running at 1,440 rpm, calculate for this speed: (*i*) the slip, (*ii*) the frequency of rotor induced emf, (*iii*) the value of the rotor induced emf per phase, and (*iv*) stator to rotor turn ratio.

Solution : Synchronous speed of induction motor,

$$N_s = \frac{120 \times \text{supply frequency}}{\text{Number of poles}} = \frac{120 \times 50}{4} = 1{,}500 \text{ rpm}$$

$$\text{Motor speed, } N = 1{,}440 \text{ rpm}$$

$$(i) \text{ Slip, } s = \frac{N_s - N}{N_s} = \frac{1{,}500 - 1{,}440}{1{,}500} = 0.04 \text{ or } 4\% \textbf{ Ans.}$$

(*ii*) Frequency of rotor induced emf = Slip × supply frequency = $sf = 0.04 \times 50 = 2$ Hz **Ans.**

(*iii*) Rotor induced emf per phase at the speed of 1,440 rpm or at slip of 0.04

= Slip × induced emf per phase in rotor winding at standstill

= $s \times E_2 = 0.04 \times 130 = 5.2$ V **Ans.**

$$(iv) \text{ Stator-to-rotor turn-ratio} = \frac{\text{Induced emf per phase in stator}}{\text{Induced emf per phase in rotor at standstill}}$$

$$= \frac{E_{P_1}}{E_{P_2}} = \frac{440}{130} = \frac{44}{13} \textbf{ Ans.}$$

2.10. ROTOR TORQUE

The torque of an induction motor, being due to interaction of the rotor and stator fields, depends on the strength of those fields and the phase relation between them. Thus the torque developed by the rotor of an induction motor is directly proportional to (*i*) rotor current, I_2 (*ii*) stator flux per pole, ϕ and (*iii*) power factor of the rotor circuit, $\cos\phi_2$

$$i.e.\ T \propto \phi I_2 \cos\phi_2$$

Since rotor emf per phase at standstill, $E \propto \phi$

$\therefore$ Torque developed, $T \propto E_2 I_2 \cos\phi_2$

or $T = K E_2 I_2 \cos\phi_2$ where K is any constant ...(2.13)

2.10.1. Torque Under Running Conditions

Substituting $I_2 = \dfrac{sE_2}{\sqrt{R_2^2 + s^2X_2^2}}$ from Eq. (2.10) and $\cos\phi_2 = \dfrac{R_2}{\sqrt{R_2^2 + s^2X_2^2}}$ from Eq. (2.11) in Eq. (2.13) we have torque developed by an induction motor rotor,

$$T = KE_2 \times \frac{sE_2}{\sqrt{R_2^2 + s^2X_2^2}} \times \frac{R_2}{\sqrt{R_2^2 + s^2X_2^2}} = \frac{KsR_2E_2^2}{R_2^2 + s^2X_2^2} \quad ...(2.14)$$

It may be seen from Eq. (2.14) that the running torque is proportional to the square of supply voltage because rotor induced emf E_2 is proportional to supply voltage.

2.10.2. Starting Torque

At start the rotor is stationary and so the slip $s = 1.0$ and the starting torque may be obtained by substituting $s = 1.0$ in Eq. (2.14).

$$\text{Starting torque, } T_{st} = \frac{KR_2E_2^2}{R_2^2 + X_2^2} \quad ...(2.15)$$

Generally, the stator supply voltage V is constant, so the rotating flux ϕ set up by the stator currents and rotor induced emf E_2 are also constant.

$$\therefore T_{st} = \frac{K_1 R_2}{R_2^2 + X_2^2} = \frac{K_1 R_2}{Z_2^2} \quad ...(2.16)$$ where K_1 is another constant.

2.10.3. Condition For Maximum Running Torque

Torque under running condition given by Eq. (2.14) is reproduced here

$$T = \frac{K R_2 s E_2^2}{R_2^2 + s^2 X_2^2}$$

Torque T for fixed supply voltage will be maximum when $\frac{sR_2}{R_2^2 + s^2 X_2^2}$ or $\frac{R_2}{\frac{R_2^2}{s} + sX_2^2}$ or $\frac{R_2}{\left(\frac{R_2}{\sqrt{s}} - X_2\sqrt{s}\right)^2 + 2R_2X_2}$ is maximum

$$\text{viz. } \frac{R_2}{\sqrt{s}} - X_2\sqrt{s} = 0 \text{ or } s = \frac{R_2}{X_2}$$

Substituting $s = \frac{R_2}{X_2}$ in Eq. (2.14) we have

$$T_{max} = \frac{KR_2^2 E_2^2/X_2}{R_2^2 + R_2^2} = \frac{KE_2^2}{2X_2} \quad ...(2.17)$$

From Eq. (2.17) it is obvious that

(*i*) Maximum torque is independent of rotor circuit resistance.

(*ii*) The slip at which maximum torque occurs is governed by the rotor resistance. As seen from above, the torque becomes equal when rotor reactance equals the rotor resistance. So by varying the rotor circuit resistance, which is possible with wound rotors, maximum torque can be made to occur at any desired slip or motor speed.

(*iii*) Maximum torque varies inversely as standstill reactance of the rotor, hence to have maximum torque, standstill reactance *i.e.* inductance of the rotor should be kept as small as possible. This is achieved by placing the rotor conductors very close to the surface of the rotor and reducing the air gap between stator and rotor to the smallest possible value.

(*iv*) Maximum torque varies directly as the square of supply voltage.

2.10.4. Condition For Maximum Starting Torque

Starting torque T_{st} will be maximum when $\frac{R_2}{X_2} = s = 1$ or $R_2 = X_2$

Since the rotor resistance is not more than 1 or 2 per cent of leakage reactance otherwise efficiency will be low, so to increase the starting torque, extra resistance is required to be inserted in the rotor circuit at start and cut out gradually as the motor picks up speed. This is possible in case of wound rotor motors. This is the reason that wound rotor motors are used where heavy loads are to be accelerated at start such as cranes, lifts, elevators etc.

2.11. TORQUE-SLIP AND TORQUE-SPEED CURVES

Torque developed by an induction motor rotor is given by the equation $T = \frac{KsR_2E_2^2}{R_2^2 + s^2X_2^2}$.

From the above equation it is revealed that

(*i*) When speed is synchronous *i.e.* when slip *s* is zero, the torque T is zero so the torque-slip curve starts from origin O.

(*ii*) When the speed is very near to synchronous speed N_s *i.e.* when slip s is very low the value of term sX_2 is very small and is negligible in comparison with rotor resistance R_2, therefore, torque T is approximately proportional to slip s, if rotor resistance R_2 is constant *i.e.* at speeds near synchronous speed the torque-speed and torque-slip curves are approximately straight lines, as shown in Fig. 2.15 (*a*) and 2.15 (*b*) respectively.

(*iii*) As the slip increases , *i.e.* as the speed drops with the increase in load, torque increases, reaches its maximum value when slip $s = \frac{R_2}{X_2}$. The maximum torque is also known as *breakdown or pull-out torque.* The slip corrresponding to breakdown torque is called the breakdown slip s_b.

(*iv*) With the further increase in slip or drop in speed due to increase in load beyond the point of maximum torque the torque begins to decrease. The result is that the motor slows down and eventually stops. The motor operates for the value of slip between zero and that corresponding to breakdown or pull-out torque.

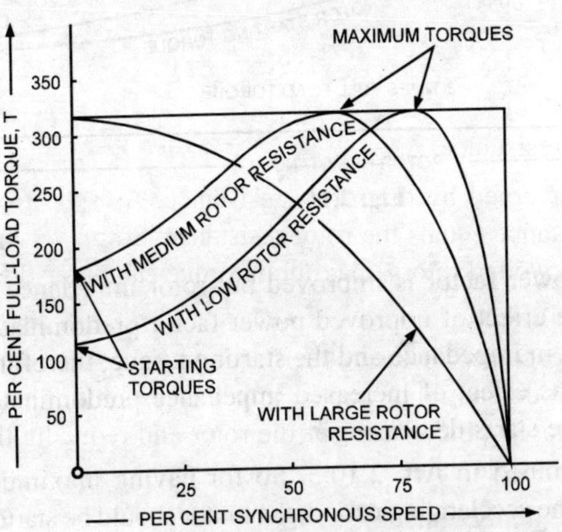

(*a*) *Torque-Speed Curves*

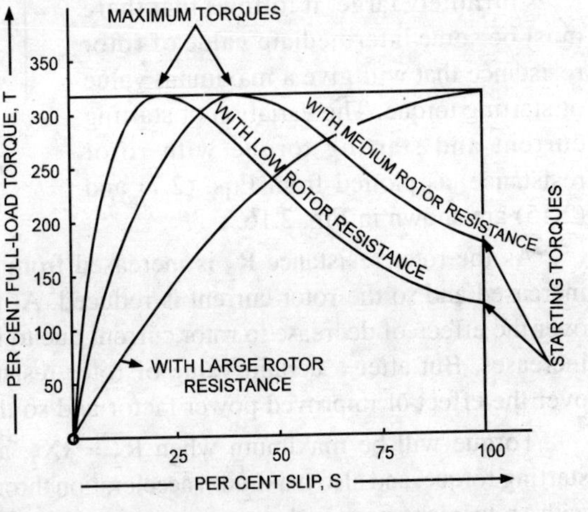

(*b*) *Torque-Slip Curves*

Fig. 2.15

With higher values of slip, R_2 becomes negligible as compared to $s\,X_2$ and the torque varies according to the relation $T \; \alpha \; \frac{s}{s^2X_2^2} \; \alpha \; \frac{1}{s}$ if standstill reactance X_2 is constant. It means that torque-speed curves are rectangular hyperbola with the speed or slip beyond that corresponding to breakdown or pull-out torque.

Torque-speed and torque-slip curves are shown in Figs. 2.15 (*a*) and 2.15 (*b*) respectively for different values of rotor resistance R_2.

2.11.1. Effect of Rotor Resistance Upon Torque-Slip or Torque-Speed Relationship

From the torque equation it is obvious that, for a constant supply voltage, when rotor resistance R_2 is very small compared with $s X_2$, the torque for a given slip is directly proportional to R_2; but when rotor resistance R_2 is large compared with $s X_2$, the torque for a given slip is inversely proportional to R_2.

From the Fig. 2.15 (*b*) it is seen that variation of rotor resistance does not change the magnitude of maximum torque T_{max}; merely changes the value of slip at which it occurs. Larger the rotor resistance the greater is the slip at which the maximum torque occurs.

For a given value of torque T, slip *s* is proportional to rotor resistance R_2, so addition of external resistance in the rotor circuit does not lower the torque curve but merely stretches it so that the same torque values occur at lower speeds (or higher slips) as shown in Fig.. 2.15 (*a*) and 2.15 (*b*).

2.11.2. Variation of Starting Torque With Rotor Resistance

If the rotor circuit resistance R_2 were zero, then the rotor current at standstill would lag 90° behind the rotor induced emf and the induction motor would, therefore, develop no torque at start. On the other hand, if the rotor resistance were infinitely large, there would be no rotor current, and, therefore, no torque. Since the starting torque is zero when $R_2 = 0$, and also when R_2 is infinitely large, it follows that there must be some intermediate value of rotor resistance that will give a maximum value of starting torque. The variation of starting current and starting torque with rotor resistance, as plotted from Eqs. (2.7) and (2.15) are shown in Fig. 2.16.

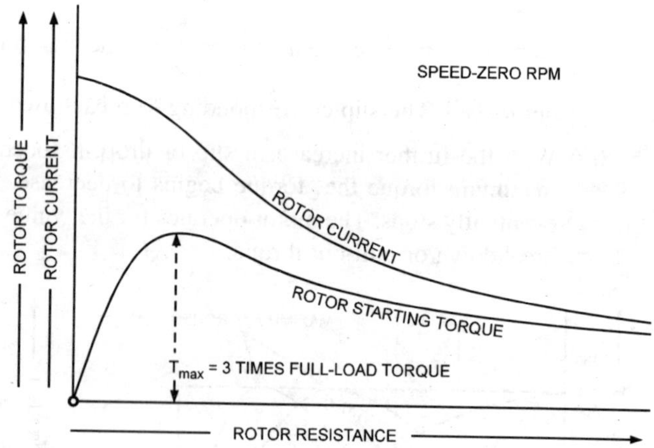

Fig. 2.16

As the rotor resistance R_2 is increased from zero, power factor is improved but rotor impedance is increased and so the rotor current is reduced. At first, the effect of improved power factor predominates over the effect of decrease in rotor current due to increase in impedance and the starting torque, therefore, increases. But after a certain value of rotor resistance, the effect of increased impedance predominates over the effect of improved power factor and so the torque starts decreasing.

Torque will be maximum when $R_2 = sX_2$, as determined in Art. 2.10.3, So for having maximum starting torque, and also maximum acceleration throughout the accelerating period, the motor should be started with a rotor resistance equal to rotor reactance at standstill, and rotor resistance R_2 should be reduced continuously, as the motor gains speed, always keeping the rotor resistance equal to the rotor reactance.

2.12. EFFECT OF CHANGE IN SUPPLY VOLTAGE ON STARTING TORQUE

From Eq. (2.15), Starting torque, $$T_{st} = \frac{KR_2E_2^2}{R_2^2 + X_2^2}$$

Since emf induced in rotor at standstill, $E_2 \propto \phi \propto V$, the supply voltage

So starting torque, $$T_{st} = \frac{K'R_2V^2}{R_2^2 + X_2^2}$$

where K′ is another constant or $T_{st} \propto V^2$, Rotor resistance R_2 and standstill reactance X_2 being fixed for a given machine.

i.e. starting torque is proportional to the square of the applied voltage. It means that the starting torque is very sensitive to any change in the applied voltage. For example, a change of 10 per cent in supply voltage will cause a change of 19 % in the rotor torque. This fact is very important in connection with starting methods of squirrel cage induction motors.

2.13. EFFECT OF CHANGE IN SUPPLY VOLTAGE ON TORQUE AND SLIP

When the motor is running with slip *s*, then torque acting on the rotor is given by

$$T = \frac{K s R_2 E_2^2}{R_2^2 + s^2 X_2^2} \qquad \text{...See Eq. (2.14)}$$

Since emf induced in the rotor at standstill, $E_2 \propto V$

$$\text{So torque developed } T = \frac{K' s\, R_2 V^2}{R_2^2 + s^2 X_2^2} \qquad \text{where } K' \text{ is another constant}$$

Since at full load, slip *s* is very low, therefore, $s^2 X_2^2$ can be neglected in comparison with R_2^2 and equation for torque may be written as $T = \dfrac{K' s\, R_2 V^2}{R_2^2} = \dfrac{K' s V^2}{R_2}$

$$\text{or } T \propto s V^2$$

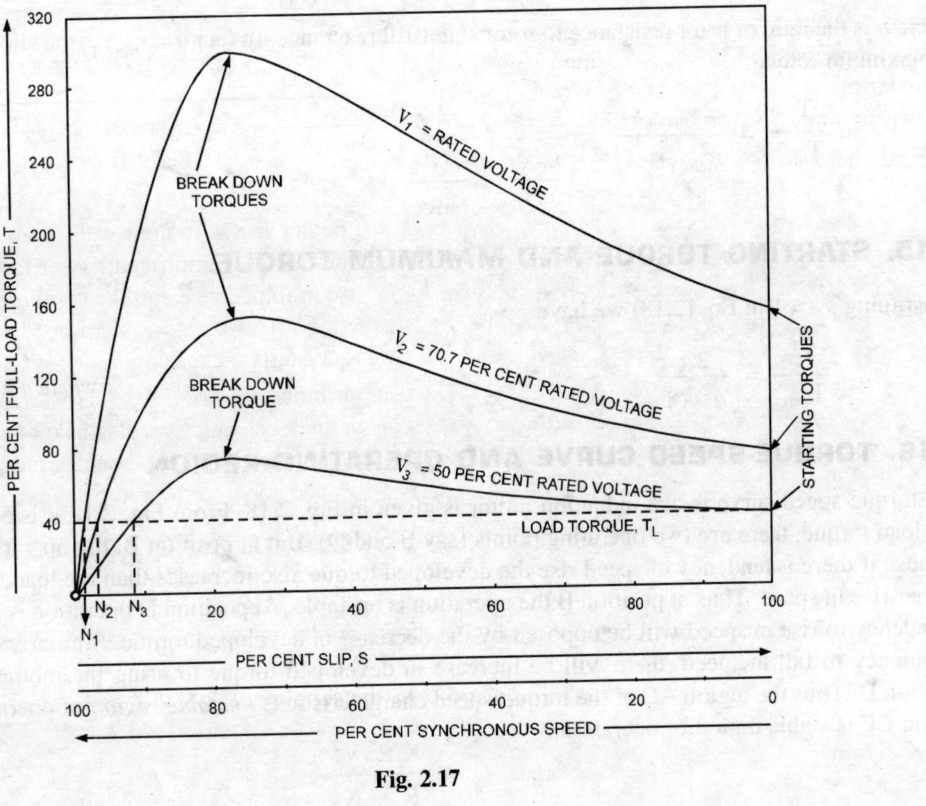

Fig. 2.17

If the supply voltage is changed, it changes the torque under running conditions also. With the decrease in supply voltage, torque under running condition decreases. As the motor now develops a reduced torque, the motor slows down (*i.e.* slip increases) so as to cause increase in torque to meet the load torque and the motor may come to a standstill (if the load, torque exceeds the pull-out torque of the motor corresponding to reduced supply voltage) unless the load is removed. Figure 2.17 shows the variation of torque with slip for three different values of line voltage.

2.14. FULL-LOAD TORQUE AND MAXIMUM TORQUE

Let the full-load slip of the motor be s_f.

Since full-load torque T_f and maximum torque T_{max} are given by equation

$$T_f = \frac{Ks_f R_2 E_2^2}{R_2^2 + s_f^2 X_2^2} \quad \text{...Refer to Eq. (2.14)}$$

$$\text{and } T_{max} = \frac{KE_2^2}{2X_2} \quad \text{...Refer to Eq. (2.17)}$$

$$\text{so } \frac{T_f}{T_{max}} = \frac{\dfrac{Ks_f R_2 E_2^2}{R_2^2 + s_f^2 X_2^2}}{\dfrac{KE_2^2}{2X_2}} = \frac{2s_f R_2 X_2}{R_2^2 + s_f^2 X_2^2} = \frac{2s_f \dfrac{R_2 X_2}{X_2^2}}{\dfrac{R_2^2 + s_f{}^2 X_2^2}{X_2^2}} = \frac{2s_f \dfrac{R_2}{X_2}}{\dfrac{R_2^2}{X_2^2} + s_f^2} = \frac{2as_f}{a^2 + s_f^2} \quad \text{...(2.18)}$$

where a is the ratio of rotor resistance to rotor standstill reactance. In fact $a = s_{max\,T}$, the slip corresponding to maximum torque.

$$\text{So } \frac{T_f}{T_{max}} = \frac{2s_{max\,T}\, s_f}{s_{max\,T}^2 + s_f^2} = \frac{2}{\dfrac{s_{max\,T}}{s_f} + \dfrac{s_f}{s_{max\,T}}} \quad \text{...(2.19)}$$

2.15. STARTING TORQUE AND MAXIMUM TORQUE

Substituting $s_f = 1$ in Eq. (2.18) we have

$$\frac{T_{st}}{T_{max}} = \frac{2a}{a^2 + 1} \quad \text{where } a = \frac{R_2}{X_2} \quad \text{... (2.20)}$$

2.16. TORQUE-SPEED CURVE AND OPERATING REGION

The torque-speed curve for an induction motor is given in Fig. 2.18. From Fig. 2.18 it is evident that for any load torque, there are two operating points (say B and D). But at position B, the operation is unstable because if there is tendency of speed rise the developed torque also increases than the load torque causing further rise in speed. Thus at position B the operation is unstable. At position D operation is stable because a tendency to rise in speed will be opposed by the decrease in developed torque. Similarly when there is a tendency to fall in speed, there will be increase in developed torque to bring the motor to operating position D. Thus the region AC of the torque-speed characteristic is *unstable region* of operation while the region CF is stable region of operation.

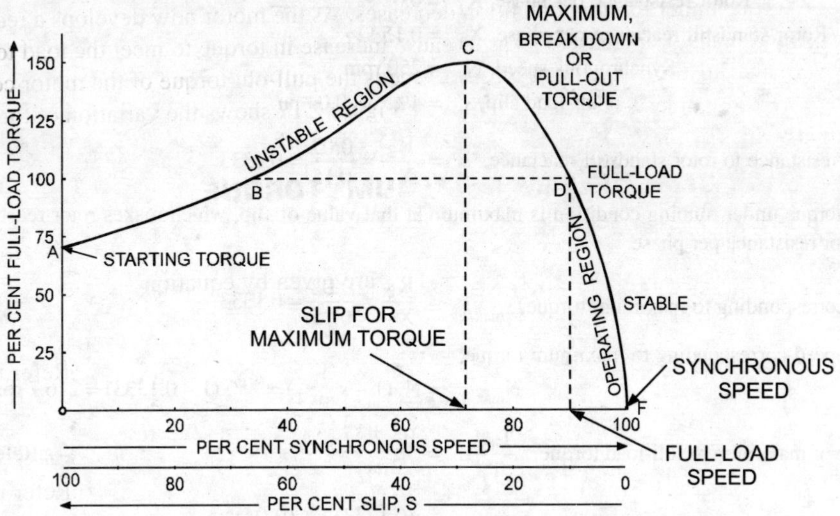

Fig. 2.18

At full load, the motor runs at a speed of N rpm. If the mechanical load applied to the motor is increased, motor speed drops till the torque developed by the motor equals the load torque. As long as the two torques are in balance, the motor will run at constant speed. However, if the load torque exceeds the breakdown torque of the motor, the motor will immediately stop.

The value of breakdown torque varies with the design of the motor but ranges from 200 to 300 per cent of full-load torque in standard squirrel cage motors.

The torque-speed characteristics of an induction motor in the operating region are quite similar to those of a dc shunt motor. Even at full load, the drop in speed from no-load does not exceed 5 per cent.

Example 2.10. A 4-pole, 50 Hz, 3-phase induction motor has a rotor resistance of 0.02 Ω per phase and standstill reactance of 0.5 Ω per phase. Determine the speed at which the maximum torque is developed.

[M.D. Univ. Electrical Machine-II, Dec. 2010]

Solution : Rotor resistance per phase, $R_2 = 0.02\ \Omega$

Rotor standstill reactance per phase, $X_2 = 0.5\ \Omega$

Since the torque under running is maximum at that value of slip, which makes rotor reactance per phase equal to the rotor resistance per phase

$$\therefore \text{ Slip corresponding to maximum torque, } s_{\max T} = \frac{R_2}{X_2} = \frac{0.02}{5} = 0.04$$

$$\text{Speed corresponding to maximum torque, } N_{\max T} = \frac{120 f}{P}(1 - s_{\max T})$$

$$= \frac{120 \times 50}{4}(1 - 0.04) = 1{,}440 \text{ rpm} \quad \textbf{Ans.}$$

Example 2.11. A 3-phase induction motor has a synchronous speed of 250 rpm and 4% slip at full load. The rotor has a resistance of 0.02 Ω per phase and a standstill leakage reactance of 0.15 Ω per phase. Calculate :

(*i*) Speed at which maximum torque is developed.

(*ii*) Ratio of maximum to full-load torque.

(*iii*) Ratio of maximum to starting torque. *[M.D. Univ. Electrical Machine-II, May 2009]*

Solution : Rotor resistaance per phase, $R_2 = 0.02\ \Omega$

Rotor standstill reactance per phase, $X_2 = 0.15\ \Omega$

Synchronous speed, $N_s = 250$ rpm

Full-load slip, $s_f = 4\%$ or 0.04 pu

Ratio of rotor resistance to rotor standstill reactance, $a = \dfrac{R_2}{X_2} = \dfrac{0.02}{0.15} = 0.1333$

Since the torque under running condition is maximum at that value of slip, which makes rotor reactance per phase equal to the rotor resistance per phase.

$\therefore$ Slip corresponding to maximum torque, $s_{\max T} = \dfrac{R_2}{X_2} = \dfrac{0.02}{0.15} = 0.1333$

(*i*) Speed corresponding to maximum torque,

$$N_{\max T} = N_s (1 - s_{\max T}) = 250\ (1 - 0.1333) = 216.7 \text{ rpm} \quad \textbf{Ans.}$$

(*ii*) Ratio of maximum to full-load torque, $\dfrac{T_{\max}}{T_f} = \dfrac{a^2 + s_f^2}{2as_f}$ ...Refer to Eq. (2.18)

$$= \frac{(0.1333)^2 + (0.04)^2}{2 \times 0.1333 \times 0.04} = 1.8167 \quad \textbf{Ans.}$$

(*iii*) Ratio of maximum to starting torque $\dfrac{T_{\max}}{T_{st}} = \dfrac{a^2 + 1}{2a}$ ...Refer to Eq. (2.20)

$$= \frac{(0.1333)^2 + 1}{2 \times 0.1333} = 3.817 \quad \textbf{Ans.}$$

Example 2.12. A 6-pole, 50 Hz, 3-phase slip-ring induction motor has resistance and reactance of 0.5 Ω and 5 Ω per phase respectively. Calculate (*i*) at what speed the torque is maximum? (*ii*) the ratio of maximum torque/starting torque. What must be the external resistance per phase so that the starting torque is half of the maximum torque.

Solution : Rotor resistance per phase, $R_2 = 0.5\ \Omega$

Rotor standstill reactance per phase, $X_2 = 5\ \Omega$

Since the torque under running condition is maximum at that value of slip, which makes rotor reactance per phase equal to the rotor resistance per phase.

$\therefore$ Slip corresponding to maximum torque, $s_{\max T} = \dfrac{R_2}{X_2} = \dfrac{0.5}{5} = 0.1$

Speed corresponding to maximum torque, $N_{\max T} = N_s (1 - s_{\max T})$

$$= \frac{120 f}{P} (1 - s_{\max T}) = \frac{120 \times 50}{6} (1 - 0.1) = 900 \text{ rpm} \quad \textbf{Ans.}$$

(*ii*) Ratio of rotor resistance to rotor standstill reactance,

$$a = \frac{R_2}{X_2} = \frac{0.5}{5} = 0.1$$

(*ii*) Ratio of $\dfrac{\text{maximum torque}}{\text{starting torque}} = \dfrac{a^2 + 1}{2a} = \dfrac{(0.1)^2 + 1}{2 \times 0.1} = 5.05$ **Ans.** ...See Eq. (2.20)

(*iii*) Ratio of starting torque T_{st} and maximum torque $T_{\max}$, according to Eq. (2.20) is given as

$$\frac{T_{st}}{T_{\max}} = \frac{2a}{a^2 + 1} = 0.5 \qquad \because \text{It is given that } T_{st} = \frac{1}{2} T_{\max}$$

or $a^2 - 4a + 1 = 0$

$$\text{or } a = \frac{4 \pm \sqrt{16-4}}{2} = 0.268 \text{ rejecting higher value.}$$

If r is the additional resistance introduced per phase in the rotor circuit to give the starting torque equal to half of the maximum torque then

$$\frac{R_2 + r}{X_2} = 0.268$$

$$\text{or } r = 0.268\, X_2 - R_2 = 0.268 \times 5 - 0.5 = 0.84\ \Omega \textbf{ Ans.}$$

Example 2.13. An 8-pole, 50 Hz, 3-phase induction motor develops a maximum torque of 150 N-m at 650 rpm. The rotor resistance is 0.6 Ω/phase. Find torque at 4 % slip. Neglect stator impedance.

Solution :

$$\text{Synchronous speed, } N_s = \frac{120 f}{P} = \frac{120 \times 50}{8} = 750 \text{ rpm}$$

$$\text{Speed corresponding to maximum torque, } N = 650 \text{ rpm}$$

$$\text{Slip corresponding to maximum torque, } s_{\max T} = \frac{750 - 650}{750} = 0.1333$$

$$\text{Ratio of rotor resistance to standstill reactance, } a = s_{\max T} = 0.1333$$

$$\text{Rotor standstill reactance, } X_2 = \frac{R_2}{s_{\max T}} = \frac{0.6}{0.1333} = 4.5\ \Omega$$

From Eq. (2.18)

$$\frac{T_f}{T_{\max}} = \frac{2 a s_f}{a^2 + s_f^2}$$

$$\text{or } T_f = \frac{2 a s_f}{a^2 + s_f^2} T_{\max} = \frac{2 \times 0.1333 \times 0.04}{(0.1333)^2 + (0.04)^2} \times 150 \times 150 = 82.6 \text{ Nm} \quad \textbf{Ans.}$$

Example 2.14. A 746 kW, 3-phase, 50 Hz, 16-pole induction motor has a rotor impedance of (0.02 + *j* 0.15) Ω at standstill. Full-load torque is obtained at 360 rpm. Calculate (*i*) the ratio of maximum to full-load torque (*ii*) the speed for maximum torque and (*iii*) the rotor resistance to be added to get maximum starting torque.

Solution :

$$\text{Synchronous speed, } N_s = \frac{120 f}{P} = \frac{120 \times 50}{16} = 375 \text{ rpm}$$

$$\text{Full-load speed, } N_f = 360 \text{ rpm}$$

$$\text{Full-load slip, } s_f = \frac{N_s - N_s}{N_s} = \frac{375 - 360}{375} = 0.04$$

$$\text{Slip corresponding to maximum torque, } s_{\max T} = \frac{R_2}{X_2} = \frac{0.02}{0.15} = 0.133$$

(*i*) From Eq. (2.19) the ratio of maximum to full-load torque,

$$\frac{T_{\max}}{T_f} = \frac{s_{\max T}/s_f + s_f/s_{\max T}}{2} = \frac{\frac{0.133}{0.04} + \frac{0.04}{0.133}}{2} = 1.817 \textbf{ Ans.}$$

$$\text{Speed corresponding to maximum torque, } N_{\max T} = N_s (1 - s_{\max T}) = 375 (1 - 0.133) = 325 \text{ rpm } \textbf{Ans.}$$

(*iii*) For obtaining maximum starting torque,

$$R_2 + r = X_2$$

$$\text{or } r = X_2 - R_2 = 0.15 - 0.02 = 0.13\ \Omega \textbf{ Ans.}$$

Example 2.15. A delta-connected, 6-pole, 50 Hz, 3-phase induction motor has a rotor resistance of 0.15 Ω per phase and exerts maximum torque at 880 rpm. Calculate the percentage maximum torque that would be exerted (*i*) at standstill and (*ii*) at 940 rpm.

Solution : Synchronous speed, $N_s = \frac{120f}{P} = \frac{120 \times 50}{6} = 1{,}000$ rpm

Speed corresponding to maximum torque, $N_{max\,T} = 880$ rpm

Slip corresponding to maximum torque, $s_{max\,T} = \frac{N_s - N}{N_s} = \frac{1{,}000 - 880}{1{,}000} = 0.12$

Ratio of rotor resistance to standstill reactance, $a = s_{max\,T} = 0.12$

(*i*) Standstill torque, $T_{st} = \frac{2a}{a^2+1} \times T_{max}$ ...Refer to Eq. (2.20)

$$= \frac{2 \times 0.12}{(0.12)^2 + 1} T_{max} = 0.2366\ T_{max}$$ or 23.66 per cent maximum torque **Ans.**

(*ii*) At speed of 940 rpm

$$\text{Slip, } s = \frac{1{,}000 - 940}{1{,}000} = 0.06$$

$$\text{Torque, } T = \frac{2\ T_{max}}{\frac{s_{max\,T}}{s} + \frac{s}{s_{max\,T}}}$$ ...Refer to Eq. (2.19)

$$= \frac{2\ T_{max}}{\frac{0.12}{0.06} + \frac{0.06}{0.12}} = 0.8\ T_{max}$$ or 80 per cent maximum torque **Ans.**

Example 2.16. A 3-phase induction motor has a starting torque of 100 % and maximum torque of 200 % of the full-load torque. Determine (*i*) slip at which maximum torque occurs (*ii*) full-load slip.

Solution : (*i*) Starting torque, T_{st} = 100 % of T_f or equal to T_f.

Maximum torque, T_{max} = 200 % of T_f or equal to 2 T_f

$$\text{so } \frac{T_{st}}{T_{max}} = \frac{2a}{a^2+1} = \frac{T_f}{2T_f} = \frac{1}{2}$$ ...Refer to Eq. (2.20)

$$\text{or } a^2 - 4a + 1 = 0$$

$$\text{or } a = \frac{4 \pm \sqrt{16-4}}{2} = 0.268$$ neglecting higher value

Hence slip corresponding to maximum torque,

$$s_{max\,T} = 0.268$$ **Ans.**

(*ii*) From Eq. (2.18) we have

$$\frac{T_f}{T_{max}} = \frac{2as_f}{a^2 + s_f^2}$$

$$\text{or } \frac{1}{2} = \frac{2 \times 0.268 \times s_f}{(0.268)^2 + s_f^2}$$

$$\text{or } s_f^2 - 1.072\ s_f + 0.071824$$

$$\text{or Full-load slip, } s_f = \frac{1.072 \pm \sqrt{(1.072)^2 - 4 \times 0.071824}}{2} = 0.072$$ **Ans.**

Example 2.17. The starting and maximum torques of a 3-phase induction motor are 1.5 times and 2.5 times its full-load torque. Determine the percentage change in rotor circuit resistance to obtain a full-load slip of 0.03. Neglect stator impedance.

Solution : Starting torque, $T_{st} = 1.5\ T_f$

Maximum torque, $T_{max} = 2.5\ T_f$

$$\frac{T_{st}}{T_{max}} = \frac{2a}{a^2+1} = \frac{1.5T_f}{2.5T_f} = \frac{1.5}{2.5} = 0.6 \quad \text{...Refer to Eq. (2.20)}$$

$$\text{or } 3a^2 - 10a + 3 = 0$$

$$\text{or } (3a-1)(a-3) = 0$$

$$\text{or } a = \frac{1}{3} \text{ rejecting higher value}$$

$$\text{or } \frac{R_2}{X_2} = a = \frac{1}{3}$$

$$\text{or rotor resistance, } R_2 = \frac{1}{3} X_2$$

From Eq. (2.18)

$$\frac{T_f}{T_{max}} = \frac{2as_f}{a^2+s_f^2}$$

$$\text{or } \frac{1}{2.5} = \frac{2a \times 0.03}{a^2+0.03^2} \quad \because T_{max} = 2.5\ T_f \text{ and } s_f = 0.03$$

$$\text{or } a^2 - 0.15a + 0.0009 = 0$$

$$\text{or } a = \frac{0.15 \pm \sqrt{0.15^2 - 4 \times 0.0009}}{2} = 0.1437, \text{ other value is not feasible.}$$

So percentage reduction in rotor resistance

$$= \frac{R_2 - R_2'}{R_2} \times 100 = \frac{\frac{X_2}{3} - 0.1437\ X_2}{\frac{X_2}{3}} \times 100 = 56.9\ \% \ \textbf{Ans.}$$

Example 2.18. A squirrel cage induction motor has a slip of 4 % at full load. Its starting current is 5 times the full-load current. The stator impedance and magnetising current may be neglected; the rotor resistance is assumed constant.

(*i*) Calculate the maximum torque and the slip at which it would occur

(*ii*) Calculate the starting torque. Express torque in pu of the full-load torque

Solution : From Eq. (2.8) rotor starting current,

$$I_{st} = \frac{E_2}{\sqrt{R_2^2 + X_2^2}}$$

and from Eq. (2.12) rotor current at full-load,

$$I_f = \frac{E_2}{\sqrt{\left(\frac{R_2}{s_f}\right)^2 + X_2^2}}$$

$$\text{so } \frac{I_{st}}{I_f} = \sqrt{\frac{(R_2/s_f)^2 + X_2^2}{R_2^2 + X_2^2}} = \sqrt{\frac{\left(\frac{R_2/X_2}{s_f}\right)^2 + 1}{\left(\frac{R_2}{X_2}\right)^2 + 1}} = \sqrt{\frac{\left(\frac{a}{s_f}\right)^2 + 1}{a^2 + 1}} \quad \text{where } a = \frac{R_2}{X_2}$$

$$\text{or } 5 = \sqrt{\frac{\left(\frac{a}{0.04}\right)^2 + 1}{a^2 + 1}} \quad \because I_{st} = 5\, I_f \text{ and } s_f = 0.04$$

$$\text{or } a = 0.2$$

(*i*) The slip at which maximum torque occur,

$$s_{\max T} = a = 0.2 \textbf{ Ans.}$$

From Eq. (2.19)

$$T_{\max} = \frac{1}{2}\left(\frac{s_{\max T}}{s_f} + \frac{s_f}{s_{\max T}}\right) \times T_f = \frac{1}{2}\left(\frac{0.2}{0.04} + \frac{0.04}{0.2}\right) \times T_f = 2.6 \text{ pu of full-laod torque } \textbf{Ans.}$$

(*ii*) From Eq. (2.20) starting torque

$$T_{st} = \frac{2a}{a^2 + 1} T_{\max} = \frac{2 \times 0.2}{0.2^2 + 1} T_{\max} = 0.3846\, T_{\max} = 0.3846 \times 2.6\, T_f$$

$$= T_f \text{ or } 1.0 \text{ pu of full-load torque } \textbf{Ans.}$$

2.17. POWER LOSSES AND EFFICIENCY

The power losses in a induction motor fall in the same general classifications as in any dynamo. There are electrical losses due to heat produced by the current in the stator and rotor windings (usually called the *copper losses*), magnetic losses in the iron due to hysteresis and eddy currents (usually called the *iron* or *core losses*), mechanical losses due to friction in the bearings and ventilation (usually called the *friction* and *windage losses*) and stray load losses. There are copper losses and core losses in the stator, and copper losses and mechanical losses in the rotor. Actually there is some core loss in the rotor but under normal operating conditions, the rotor frequency is so low that rotor core loss may be neglected.

The iron and mechanical losses are considered to be constant and are based on no-load conditions.

Friction losses actually vary slightly with the load, especially if the load is belted to the motor. These losses also vary with the speed. But since the speed varies slightly over the operating range, these losses can be considered to remain constant without involving noticeable error.

The core losses are due to both the main and leakage fluxes. The magnitude and distribution of such losses vary with the load. With the increase in load, the leakage fluxes increase due to increase in currents and the mutual flux decrease. Thus the loss due to mutual flux is reduced and that due to leakage fluxes is increased with the increase in load. Because of this partially neutralization effect; it is quite accurate for practical purposes to assume the core losses as constant at all loads and speeds within the operating range of the motor under constant voltage conditions*. Except for the rotor tooth losses, practically all the core losses occur in the stator, owing to the low frequency of the rotor flux.

The copper losses, of course, vary with the load. The stator and rotor copper losses vary with the square of their respective currents.

---

*For small changes in supply voltage, such losses are assumed to vary in direct proportion of supply voltage.

The stray load losses include rotor iron loss, additional fundamental frequency and high frequency losses in the iron, strand and circulating current losses in the stator winding and harmonic losses in rotor conductors under load. Such losses are approximately 0.5 per cent of power supplied to the motor.

The standard definition of efficiency (output /input) applies to the induction motors and it is given as

$$\eta_m = \frac{\text{Mechanical power output in watts (or kW)}}{\text{Mechanical power output in watts (or kW)} + \text{Fixed losses in watts (or kW)} + \text{Copper losses in watts (or kW)}} \times 100$$

$$\text{or } \eta_m = \frac{P_{mech}}{P_{mech} + P_{const} + P_{cu}} \times 100 \qquad \text{... (2.21)}$$

2.18. POWER FLOW DIAGRAM

Figure 2.19 gives a power flow diagram for a typical induction motor and clearly indicates how the electrical power supplied to the stator winding is converted into useful mechanical power available at the shaft.

Power input to the stator winding $= 3V_1 I_1 \cos \phi_1$ where V_1 is phase supply voltage, I_1 is stator current per phase and $\cos \phi_1$ is the stator power factor,

Some of the power supplied to the stator winding is lost as heat in the iron and in the copper coils. Stator copper loss is about 3.5% and stator iron loss is about 2.5% of power input to the motor. The remaining power (about 94% of power input to the motor) is transferred magnetically across the air gap to the rotor as to a transformer secondary. Part of this rotor power received (about 3.5% of power input to motor) is lost as heat in the rotor conductors and the rest is converted into mechanical power called the mechanical power developed in the rotor. This is about 90.5% of power input to the motor. Then there are friction and windage losses (about 2% of power input to the motor), as usual and the remaining power (about 88.5 per cent of power input to the motor) is available at the motor shaft, called the mechanical power output.

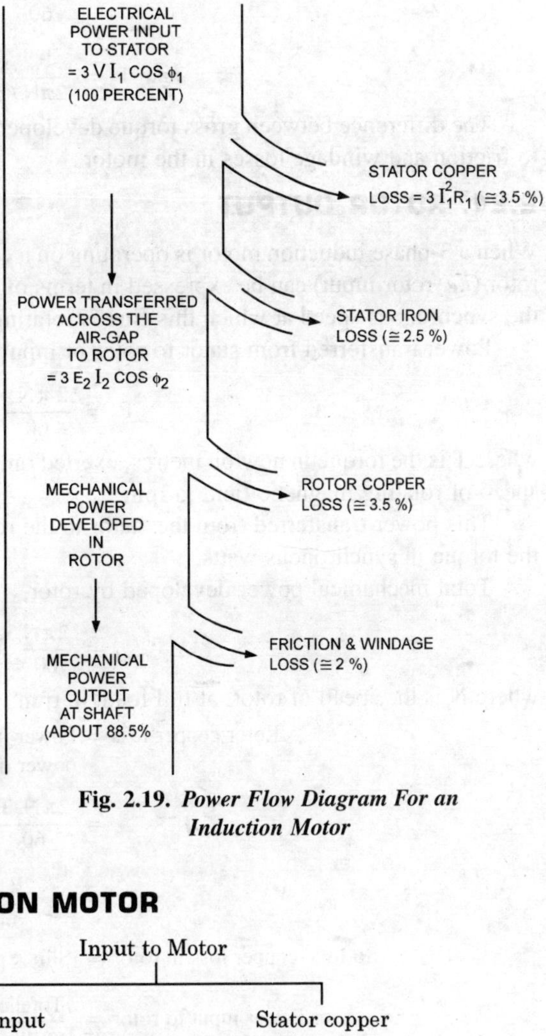

Fig. 2.19. ***Power Flow Diagram For an Induction Motor***

2.19. POWER STAGES IN AN INDUCTION MOTOR

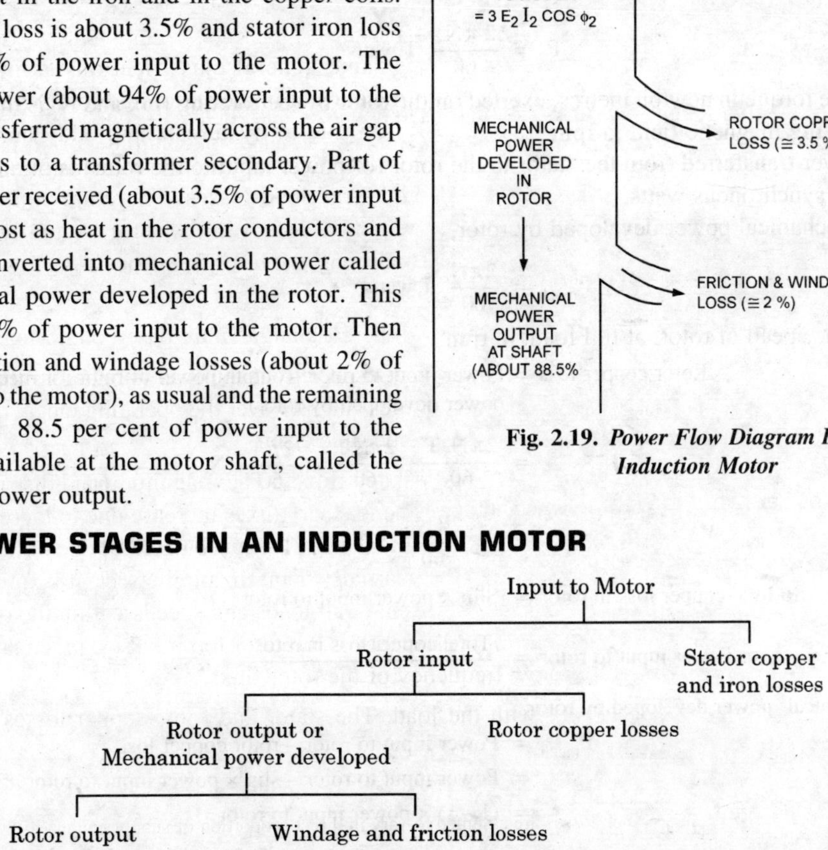

2.20. INDUCTION MOTOR TORQUE

If T_{sh} and T_g are the shaft torque and the gross torque developed in newton metres and N is the rotational speed in rpm, then

$$\text{Power output, } P_{out} = T_{sh} \times \frac{2\pi N}{60} \text{ watts} \quad ...(2.22)$$

and the mechanical power developed,

$$P_{mech} = T_g \times \frac{2\pi N}{60} \text{ watts} \quad ...(2.23)$$

$$\text{or } T_{sh} = \frac{P_{out}}{\frac{2\pi N}{60}} = 9.55 \frac{P_{out}}{N} \text{ Nm} \quad ...(2.24)$$

$$\text{and } T_g = \frac{P_{mech}}{2\pi N/60} = 9.55 \frac{P_{mech}}{N} \text{ Nm} \quad ...(2.25)$$

The difference between gross torque developed, T_g and shaft torque T_{sh} is equal to the torque lost due to friction and windage losses in the motor.

2.21. ROTOR OUTPUT

When a 3-phase induction motor is operating on a certain load, the power transferred from the stator to the rotor (*i.e.* rotor input) can be expressed in terms of the torque exerted on the rotor by the rotating flux and the synchronous speed at which this flux is rotating.

Power transferred from stator to rotor or input power to rotor or air–gap power,

$$P_2 = \frac{2\pi N_s}{60} \text{ T watts} \quad ...(2.26)$$

where T is the torque in newton metres, exerted on the rotor by the rotating flux and N_s is the synchronous speed of rotating magnetic field in rpm.

This power transferred from the stator to the rotor (or power input to the rotor) is often referred to as the torque in synchronous watts.

Total mechanical power developed by rotor,

$$P_{mech} = \frac{2\pi N}{60} \text{ T watts} \quad ...(2.27)$$

where N is the speed of rotor, at full load, in rpm.

Rotor copper loss = Power input to rotor – output power of rotor (or mechanical power developed by rotor)

$$= \frac{2\pi N_s T}{60} - \frac{2\pi N T}{60} = \frac{2\pi T}{60}(N_s - N) = \frac{2\pi T}{60} s N_s$$

$$= s \times \frac{2\pi N_s}{60} \times T = s \times \text{power input to rotor}$$

So total copper loss in rotor = Slip × power input to rotor ...(2.28)

$$\text{or Power input to rotor} = \frac{\text{Total copper loss in rotor}}{s} \quad ...(2.29)$$

or Mechanical power developed by rotor,

P_{mech} = Power input to rotor – rotor copper loss

= Power input to rotor – slip × power input to rotor

= $(1 - s)$ × power input to rotor ...(2.30)

From Eqs. (2.28) and (2.30) we have

$$\frac{\text{Total copper loss in rotor}}{\text{Gross rotor output, } P_{mech}} = \frac{s}{1-s} \qquad \text{...(2.31)}$$

Hence from above equations

Power input to rotor : Mechanical power developed by rotor (gross rotor output power) : rotor copper loss

$$i.e.\ P_2 : P_{mech} : P_{cu} :: 1 : (1-s) : s \qquad \text{...(2.32)}$$

2.22. SYNCHRONOUS WATT

It is a new unit of torque and is defined as the torque which develops a power of 1 watt at the synchronous speed of the motor. When we say that the motor is developing a torque of 100 synchronous watts, we mean that the rotor input is 100 watts and that the torque is such that the power developed would be 100 watts provided the rotor is running synchronously and developing the same torque.

$$\text{Rotor input} = T_{sy} \times 2\pi N_s$$

$$\text{or } T_{sy} = \frac{\text{Rotor input in watts}}{2\pi N_s} \qquad \text{...(2.33)}$$

2.23. FACTORS DETERMINING TORQUE

When a 3-phase induction motor is operating, on a certain load, with slip s, then

Induced emf per phase in rotor = sE_2 volts where E_2 is standstill rotor induced emf

Rotor impedance per phase, $\mathbf{Z}_2 = R_2 + sX_2$

where R_2 and X_2 are rotor resistance and standstill reactance per phase

$$\text{Rotor current per phase, } I_2 = \frac{sE_2}{Z_2} = \frac{sE_2}{\sqrt{R_2^2 + s^2 X_2^2}}$$

$$\text{Rotor circuit power factor, } \cos\phi_2 = \frac{R_2}{\sqrt{R_2^2 + s^2 X_2^2}}$$

Total rotor copper loss = Number of phases × rotor induced emf per phase × rotor current per phase × rotor circuit power factor

$$= 3 \times sE_2 \times \frac{sE_2}{\sqrt{R_2^2 + s^2 X_2^2}} \times \frac{R_2}{\sqrt{R_2^2 + s^2 X_2^2}} = \frac{3R_2 s^2 E_2^2}{R_2^2 + s^2 X_2^2} \qquad \text{...(2.34)}$$

Also from Eq. (2.28)

$$\text{Total rotor copper loss} = s \times \text{power input to rotor} = \frac{s \times T \times 2\pi N_s}{60}$$

$$\text{So } s \times T \frac{2\pi N_s}{60} = \frac{3R_2 s^2 E_2^2}{R_2^2 + s^2 X_2^2}$$

$$\text{or Torque developed by rotor, } T = \frac{3 \times 60}{2\pi N_s} \frac{R_2 sE_2^2}{R_2^2 + s^2 X_2^2} \qquad \text{...(2.35)}$$

$$\text{or } T \propto \frac{R_2 sE_2^2}{R_2^2 + s^2 X_2^2} \propto \frac{R_2 sV_1^2}{R_2^2 + s^2 X_2^2} \qquad \because E_2 \propto V_1, \text{ the supply voltage.}$$

Example 2.19. The power input to a 3-phase, 50 Hz induction motor is 50 kW. The total stator loss is 800 W. Find the total mechanical power developed, if the rotor emf makes 90 complete cycles per minute.

Solution :

Power input to motor, P_1 = 50 kW or 50,000 W

Total stator losses = 800 W

Power input to rotor, P_2 = Power output from stator = Input to motor – stator losses = 50,000 – 800 = 49,200 W

$$\text{Rotor emf frequency, } f' = \frac{90}{60} = 1.5 \text{ Hz}$$

$$\text{Slip, } s = \frac{f'}{f} = \frac{1.5}{50} = 0.03$$

Total mechanical power developed, $P_{mech} = P_2(1 - s) = 49{,}200\ (1 - 0.03) = 47{,}724$ W or 47.724 kW **Ans.**

Example 2.20. The core loss in a 3-phase induction motor is 100 W and equals the mechanical loss. Stator copper loss is 150 W when developing 2,000 W at the shaft output. What is the efficiency of the machine ? Assume the slip as 4%.

Solution :

Shaft output, P_{sh} = 2,000 W

Mechanical power developed, $P_{mech} = P_{sh}$ + mechanical loss = 2,000 + 100 = 2,100 W

$$\text{Power input to rotor, } P_2 = \frac{P_{mech}}{1-s} = \frac{2{,}100}{1-0.04} = 2{,}187.5 \text{ W}$$

Power input to motor, P_{in} = Power input to stator = Power input to rotor, P_2 + stator losses = 2,187.5 + 100 + 150 = 2,437.5 W

$$\text{Efficiency of machine, } \eta = \frac{P_{sh}}{P_{in}} \times 100 = \frac{2{,}000}{2{,}437.5} \times 100 = 82.05\% \quad \textbf{Ans.}$$

Example 2.21. The power input to the rotor of a 440 V, 50 Hz, 3-phase, 6-pole induction motor is 50 kW. The rotor emf makes 120 cycles per minute. Friction and windage losses are 2 kW. Calculate (*i*) slip (*ii*) rotor speed (*iii*) rotor copper losses (*iv*) mechanical power developed (*v*) output power (*vi*) output torque.

Solution :

Supply frequency, f = 50 Hz

$$\text{Rotor emf frequency, } f' = \frac{120}{60} = 2 \text{ Hz}$$

$$(i) \text{ Slip, } s = \frac{f}{f'} = \frac{2}{50} = 0.04 \quad \textbf{Ans.}$$

$$(ii) \text{ Rotor speed, N} = N_s(1-s) = \frac{120 \times f}{p}(1-s) = \frac{120 \times 50}{6}(1-0.04) = 960 \text{ rpm} \quad \textbf{Ans.}$$

Power input to rotor, P_2 = 50 kW

(*iii*) Rotor copper losses = $sP_2 = 0.04 \times 50 = 2$ kW **Ans.**

(*iv*) Mechanical power developed, $P_{mech} = P_2(1 - s) = 50(1 - 0.04) = 48$ kW **Ans.**

(*v*) Output power, $P_{out} = P_{mech}$ – friction and windage losses = 48 – 2 = 46 kW **Ans.**

$$(vi) \text{ Output torque, T} = \frac{P_{out} \times 60}{2\pi N} = \frac{46 \times 1{,}000 \times 60}{2\pi \times 960} = 457.57 \text{ Nm} \quad \textbf{Ans.}$$

Example 2.22. A 3-phase, 400 V, 50 Hz, induction motor takes a power input of 35 kW at its full-load speed of 980 rpm. The total stator losses are 1 kW and the friction and windage losses are 1.5 kW. Calculate slip, rotor ohmic losses, shaft power, shaft torque and efficiency. [*M.D. Univ. May 2006*]

Solution : The number of poles even in number and to give synchronous speed nearest higher than actual speed of motor (*i.e.* 980 rpm) that the motor must have is 6. $\because P = \frac{120 f}{N} = \frac{120 \times 50}{980} = 6$

$$\text{Synchronous speed of motor, } N_s = \frac{120 f}{P} = \frac{120 \times 50}{6} = 1{,}000 \text{ rpm}$$

$$\text{Slip, } s = \frac{N_s - N}{N_s} = \frac{1{,}000 - 980}{1{,}000} = 0.02 \text{ or } 2\% \textbf{ Ans.}$$

Power input to motor, P_1 = 35 kW

Total stator losses = 1 kW

Power input to rotor, P_2 = Power output from stator

= Input to motor – stator losses = 35 – 1 = 34 kW

Rotor ohmic losses = $sP_2 = 0.02 \times 34 = 0.68$ kW or 680 W **Ans.**

Mechanical power developed, $P_{mech} = P_2 (1 - s) = 34 (1 - 0.02) = 33.32$ kW

Shaft power = Motor output = P_{mech} – friction and windage losses

= 33.32 – 1.5 = 31.82 kW **Ans.**

$$\text{Shaft torque, } T_{sh} = \frac{P_{out}}{2\pi N/60} = \frac{P_{out} \times 60}{2\pi N} = \frac{31.82 \times 1{,}000 \times 60}{2\pi \times 980} = 310 \text{ Nm } \textbf{Ans.}$$

$$\text{Motor efficiency, } \eta = \frac{\text{Motor output}}{\text{Motor input}} \times 100 = \frac{31.82}{35} \times 100 = 90.91\% \textbf{ Ans.}$$

Example 2.23. A 4-pole, 400 V, 3-phase, 50 Hz squirrel cage induction motor runs at 1,450 rpm at 0.85 pf lagging developing 11 kW power. The stator losses are 1,100 W and mechanical losses are 400 W. Determine (*i*) Slip (*ii*) Rotor copper loss (*iii*) Rotor frequency (*iv*) Line current (*v*) Efficiency.

Solution : Synchronous speed of motor, $N_s = \frac{120 f}{P} = \frac{120 \times 50}{4} = 1{,}500$ rpm

$$(i) \text{ Slip, } s = \frac{N_s - N}{N_s} = \frac{1{,}500 - 1{,}450}{1{,}500} = 0.0333 \text{ or } 3.33\% \quad \textbf{Ans.}$$

Motor output = 11 kW

Rotor output, P_{mech} = Motor output + mechanical losses = 11,000 + 400 = 11,400 W

$$(ii) \text{ Rotor copper loss} = \frac{P_{mech}}{1 - s} \times s = \frac{11{,}400}{1 - 0.0333} \times 0.0333 = 393 \text{ W} \quad \textbf{Ans.}$$

(iii) Rotor frequency, $f' = sf = 0.0333 \times 50 = 1.665$ Hz **Ans.**

$$\text{Power input to rotor, } P_2 = \frac{P_{mech}}{1 - s} = \frac{11{,}400}{1 - 0.0333} = 11{,}793 \text{ W}$$

Power input to stator, $P_1 = P_2$ + stator losses = 11,793 + 1,100 = 12,893 W

$$(iv) \text{ Line current, } I_1 = \frac{P_1}{\sqrt{3}\, V_1 \cos\phi} = \frac{12{,}893}{\sqrt{3} \times 400 \times 0.85} = 21.89 \text{ A } \textbf{Ans.}$$

$$(v) \text{ Efficiency, } \eta = \frac{\text{Motor output}}{\text{Motor input}} \times 100 = \frac{11{,}000}{12{,}893} \times 100 = 85.32\% \textbf{ Ans.}$$

Example 2.24. A 50 hp, 6-pole, 50 Hz, slip-ring induction motor runs at 960 rpm on full-load with a rotor current of 40 A. Allowing 300 W for copper loss in the short-circuiting gear and 1,200 W for mechanical losses, find the resistance R_2 per phase of the 3-phase rotor winding.

Solution : Motor output, P_{out} = 50 × 735.5 = 36,775 watts

Rotor output = Motor output + short-circuiting gear loss + mechanical losses

= 36,775 + 300 + 1,200 = 38,275 watts

$$\text{Synchronous speed, } N_s = \frac{120\,f}{P} = \frac{120 \times 50}{6} = 1{,}000 \text{ rpm}$$

$$\text{Slip, } s = \frac{N_s - N}{N_s} = \frac{1{,}000 - 960}{1{,}000} = 0.04$$

$$\text{Rotor copper losses} = 3I^2 R_2 = \frac{\text{Rotor output}}{1-s} \times s = \frac{38{,}275}{1 - 0.04} \times 0.04 = 1{,}594.8 \text{ W}$$

$$\text{or Rotor resistance per phase, } R_2 = \frac{1{,}594.8}{3 \times I^2} = \frac{1{,}594.8}{3 \times 40^2} = 0.3322\ \Omega \textbf{ Ans.}$$

Example 2.25. The power input to a 6-pole, 50 Hz, 3-phase induction motor is 700 W at no-load and 10 kW at full-load. The no-load copper losses may be assumed negligible while the full-load stator and rotor copper losses are 295 W and 310 W respectively. Find the full-load speed, shaft torque and efficiency of the motor assuming rotational and core losses to be equal.

Solution : Full-load input, P_{in} = 10 kW or 10,000 watts

Power input on no load = Core loss + rotational losses   ∵ no-load copper losses are negligible

or 700 W = 2 × core loss (or rotational loss)

∵ both the rotational loss and core loss are equal,

$$\text{or Rotational loss} = \text{Core loss} = \frac{700}{2} = 350 \text{ W}$$

Stator core loss = Core loss = 350 W   assuming rotor core loss to be negligible

Power input to rotor, P_2 = P_{in} – stator core loss – stator copper loss

= 10,000 – 350 – 295 = 9,355 watts

Rotor output power = P_2 – rotor copper loss = 9,355 – 310 = 9,045 W

$$\text{Full-load speed, } N = N_s \times \frac{\text{Rotor output power}}{\text{Power input to rotor}}$$

$$= \frac{120 \times 50}{6} \times \frac{9{,}045}{9{,}355} = 967 \text{ rpm } \textbf{Ans.}$$

Motor output power = Rotor output power – mechanical loss

= 9,045 – 350 = 8,695 watts

$$\text{Shaft torque, } T_{sh} = \frac{\text{Motor output power in watts}}{\frac{2\pi N}{60}} = \frac{8{,}695}{2\pi \times \frac{967}{60}} = 85.865 \text{ Nm } \textbf{Ans.}$$

$$\text{Motor efficiency, } \eta = \frac{\text{Output}}{\text{Input}} \times 100 = \frac{8{,}695}{10{,}000} \times 100 = 86.95\% \textbf{ Ans.}$$

Example 2.26. A 440 V, 6-pole, 50 Hz, 3-phase delta-connected induction motor running on full load develops a useful torque of 160 Nm when the rotor emf makes 120 complete cycles/min. Determine the shaft power output. If the mechanical torque lost in friction and for core loss is 10 Nm, estimate:

(*i*) Copper loss in rotor windings

(*ii*) Input to the motor

(*iii*) Efficiency

The total stator loss is 1 kW.

Solution : Rotor frequency, $f' =$ Number of complete cycles per second

$$= \frac{120}{60} = 2 \text{ cycles per second or } 2 \text{ Hz}$$

$$\text{Slip, } s = \frac{f'}{f} = \frac{2}{50} = 0.04$$

$$\text{Synchronous speed, } N_s = \frac{120f}{P} = \frac{120 \times 50}{6} = 1{,}000 \text{ rpm}$$

Useful torque, $T_{sh} = 160$ Nm

Mechanical torque lost in friction and core loss,

$$T_f = 10 \text{ Nm}$$

Torque exerted on the rotor by the rotating flux

$$T = T_{sh} + T_f = 160 + 10 = 170 \text{ Nm}$$

$$\text{Power transferred from the stator to rotor, } P_2 = T \times \frac{2\pi N_s}{60} = 170 \times \frac{2\pi}{60} \times 1{,}000 = 17{,}800 \text{ W or } 17.8 \text{ kW}$$

(*i*) Copper loss in rotor windings $= s \times P_2 = 0.04 \times 17.8 = 0.712$ kW **Ans.**

(*ii*) Input to motor, $P_{in} = P_2 +$ stator losses $= 17.8 + 1.0 = 18.8$ kW **Ans.**

$$\text{Motor output or shaft power output} = T_{sh} \times \frac{2\pi N}{60} = 160 \times \frac{2\pi}{60} \times 1{,}000\,(1 - 0.04)$$

$= 16{,}085$ W or 16.085 kW **Ans.**

(*iii*) Motor efficiency, $\eta = \frac{\text{Motor output}}{\text{Motor input}} \times 100 = \frac{16.085}{18.8} \times 100 = 85.56\%$ **Ans.**

2.24. EQUIVALENT CIRCUIT OF AN INDUCTION MOTOR

The energy is transferred from primary (stator) winding to secondary (rotor) winding entirely by induction, therefore, induction motor is essentially a transformer. At standstill, the induction motor is actually a static transformer having its secondary (rotor) winding short-circuited.

When the motor operates at slip s, the frequency of rotor currents is s times the frequency of the stator currents, therefore, the revolving field produced by the rotor currents revolves with respect to rotor itself at speed

$$\frac{120sf}{P} = \frac{120f}{P} \times s = N_s \times s,$$

Mechanical speed of the motor, $N = N_s\,(1 - s)$. The speed of the revolving field of the rotor with respect to stator or space is obtained by combining the rotational speed of rotor field with respect to rotor with mechanical speed of the rotor. Hence speed of rotor revolving field with respect to stationary stator or space $= s\,N_s + N_s\,(1 - s) = N_s$.

Hence from the point of view of stator, the induction motor still can be considered as static transformer, even when its rotor is rotating, and it is possible to represent the performance of an induction motor by a transformer phasor diagram. Actually the rotor field does not exist alone but combines with the revolving field of stator to produce a resultant field, just as in the transformer, the resultant field is produced by the combination of primary and secondary ampere-turns.

In the transformer the load on secondary is electrical and in an induction motor the load is mechanical which can be replaced by an equivalent electrical load of resistance R'_L given by

$$R'_L = \frac{R_2}{K^2}\left(\frac{1}{s} - 1\right) \qquad \text{...(Refer to Art. 2.25),}$$

where R_2 is the resistance per phase of rotor and K is the turn-ratio of the secondary (rotor) to primary (stator).

The equivalent circuit for a polyphase induction motor is shown in Fig. 2.20 where V is applied voltage per phase; R_1 and X_1 are stator resistance and leakage reactance per phase respectively; R_2 and X_2 are rotor resistance and standstill leakage reactance per phase respectively; K is the turn-ratio of secondary to primary ; R_0 and X_0 are the no-load resistance and reactance per phase respectively, with the rotor being driven at synchronous speed.

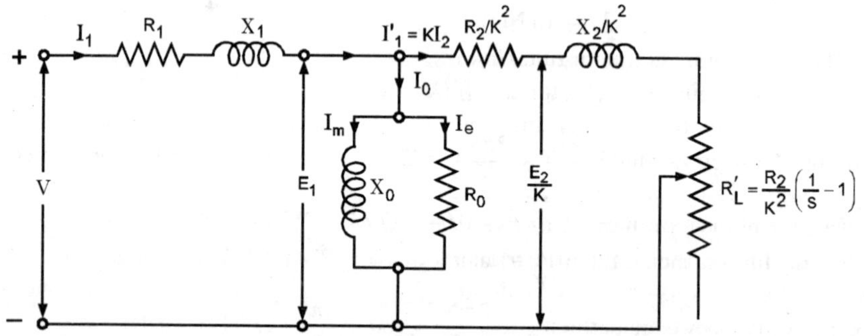

Fig. 2.20

At no load, the stator impedance drop, being very small can be neglected; $\frac{V}{R_0}$ gives the energy component of no-load current, which supplies core losses and stator copper losses (not windage and friction losses) equal to $\frac{V^2}{R_0}$ and $\frac{V}{X_0}$ gives the magnetising component of no-load current.

E_1 is the emf induced per phase in the stator, and is equal to applied voltage per phase minus impedance drop in primary (stator) winding. E_2 is the emf induced in the rotor, if it is assumed that the mechanical load is replaced by equivalent electrical load, as shown in Fig. 2.20.

The equivalent circuit can be simplified by transferring no-load current components to the supply terminals, as shown in Fig. 2.21.

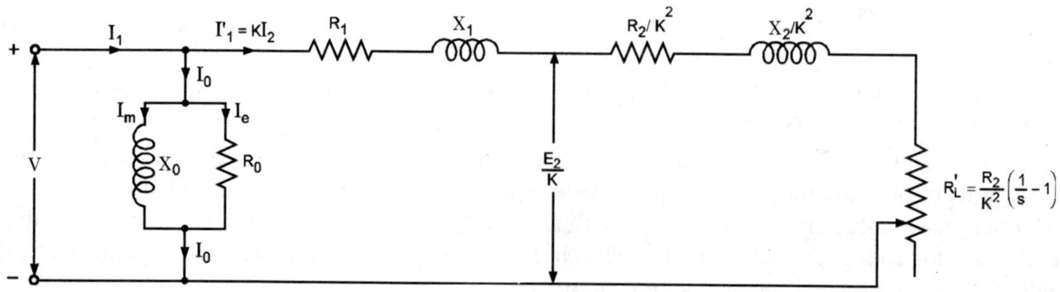

Fig. 2.21

2.24.1. Comparison Between 3-Phase Induction Motor and Transformer

The important differences between a 3-phase induction motor and transformer are as given below :

1. Because of distributed windings in an induction motor, the ratio of stator and rotor currents is not equal to the ratio of turns per phase in the rotor and stator windings.
2. In an induction motor the magnetic leakage and leakage reactances of rotor and stator are higher than those in a transformer.
3. The friction and windage losses are also present in induction motor, so the efficiency is lower than that of a transformer.
4. No-load current in an induction motor varies from about 30 to 50 per cent of full-load current whereas in transformer it varies from 2 to 5 per cent of full-load primary current. This is because of the fact that the mutual flux in a transformer completes its path through low-reluctance core, whereas the mutual flux in an induction motor has to cross the air gap between stator and rotor. Thus the reluctance offered to mutual flux in an induction motor is much higher than that in a transformer and, therefore, more magnetizing force (mmf) is required to develop the required mutual flux in case of induction motor than that in a transformer. It means that the magnetising current is large for the induction motor as compared to that for a transformer. Energy component of exciting or no-load current is also much higher in case of an induction motor than that in a transformer because of presence of windage and friction losses in induction motor. Thus no-load current of an induction motor is much higher than that of an equivalent transformer.

2.25. PHASOR DIAGRAM OF INDUCTION MOTOR

On the basis of equivalent circuit, shown in Fig. 2.20 for polyphase induction motor, the phasor diagram drawn is shown in Fig. 2.22. The diagram is exactly similar to that of a transformer except that in case of induction motor the magnetizing current, therefore, no-load current is very large on account of air gap (of high reluctance).

The revolving flux in the air gap links with stator and rotor windings and induces emfs in them. At any load, when the motor is operating at slip s, the induced emf in the rotor winding $E_2' = sE_2$ where E_2 is the standstill induced emf per phase in rotor winding.

This induced emf E_2' causes current I_2 through the rotor winding and is equal to phasor sum of voltage drops $I_2 R_2$ and $I_2 sX_2$ in the rotor winding.

$$i.e.\quad \mathbf{E'_2} = s\mathbf{E_2} = \mathbf{I_2}\,(R_2 + js\,X_2)$$

$$\text{or } s\,E_2 = I_2\sqrt{(R_2)^2 + (s\,X_2)^2}$$

$$\text{or } I_2 = \frac{s\,E_2}{\sqrt{(R_2)^2 + (s\,X_2)^2}}$$

$$= \frac{E_2}{\sqrt{\left(\frac{R_2}{s}\right)^2 + (X_2)^2}}$$

From the above equation, the rotor having true resistance R_2, variable leakage reactance sX_2 and variable induced emf sE_2 appears as a rotor having fixed reactance X_2, constant induced emf E_2 and variable resistance $\frac{R_2}{s}$.

$$\text{Total power supplied to rotor} = 3 I_2^2 \frac{R_2}{s}$$

$$\text{Power lost in rotor} = 3 I_2^2 R_2$$

$$\text{Mechanical power developed} = 3 I_2^2 \left(\frac{R_2}{s} - R_2\right)$$

$$= 3 I_2^2 R_2 \left(\frac{1}{s} - 1\right) = 3 I_2^2 R_L$$

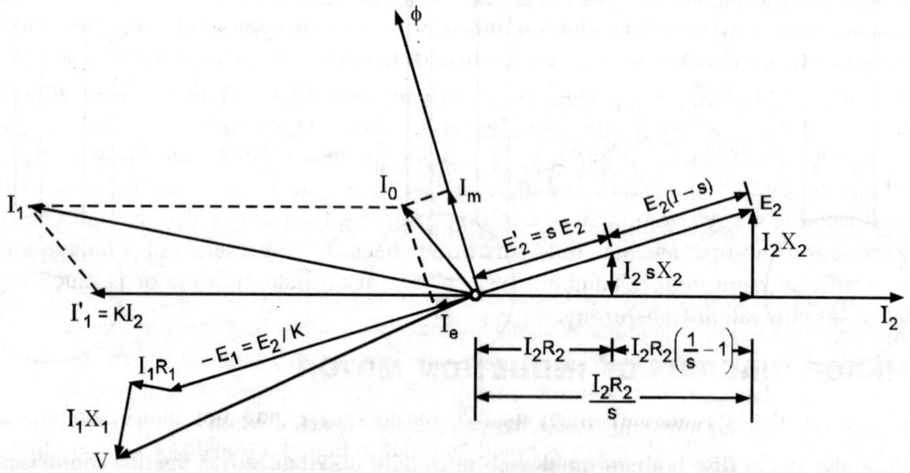

Fig. 2.22. ***Phasor Diagram For a Polyphase Induction Motor***

Hence mechanical power developed can be replaced by a pure resistive load of equivalent resistance R_L given by the relation $R_L = R_2\left(\frac{1}{s} - 1\right)$ *and induction motor may be considered* just as static transformer, in which stator winding acts as primary, rotor winding acts as secondary and load resistance $R_L = R_2\left(\frac{1}{s} - 1\right)$ equivalent to mechanical load acts as an electrical load connected across the terminals of the secondary winding.

In phasor diagram shown in Fig. 2.22, $\frac{I_2 R_2}{s}$ is the total resistive drop, I_2X_2 is the total leakage reactance drop where X_2 is the rotor standstill reactance per phase. The phasor sum of voltage drops in rotor *i.e.* of $\frac{I_2 R_2}{s}$ and $I_2 X_2$ gives emf E_2 induced in rotor. The difference of emf E_2 and actual emf sE_2 induced in the rotor *i.e.* $E_2 (1 - s)$ is due to slip and is known *as speed emf*, which would be induced in the rotor by its cutting the air-gap flux at actual rotor speed. Although under running conditions no such emf exists, but it is necessary to consider it in the operation of replacing the mechanical load by electrical one.

As in transformer, an emf E_1 equal to $\frac{E_2}{K}$ is induced in the stator. The air gap flux ϕ leads the induced emfs E_1 and E_2 by 90°. I_m and I_e are the magnetising and energy components of no-load current I_0. Stator current I_1 is the phasor sum of I_1' and I_0 where I_1' is the counter-balancing secondary (rotor) current and is equal to K I_2. I_1R_1 and I_1X_1 are the resistive and leakage reactance drops in the stator winding, V is the applied voltage per phase and is equal to the phasor sum of reversed induced emf E_1 and impedance drop in stator winding.

2.26. INDUCTION MOTOR TESTING

Various tests, which are performed on a polyphase induction motor, for determination of its efficiency and other operating characteristics are (*i*) stator resistance test (*ii*) no-load test (*iii*) blocked rotor test and (*iv*) heat run test.

As the circuit model of an induction motor is similar to that of a transformer, the parameters of the model can be had by means of no-load test as in the case of a transformer—*no-load test* (corresponding to the open-circuit test on the transformer) and the blocked rotor test (corresponding to the short-circuit test on the transformer.)

The stator core loss and friction and windage losses of an induction motor can be determined from no-load test, stator resistance can be determined from stator resistance test and rotor resistance can be determined from blocked rotor test and so the stator and rotor copper losses can be computed for any load and so the efficiency.

Heat run test is performed to determine the temperature rise of different parts of the motor while it is running at rated speed and that is why, it is sometimes called the *temperature rise test.*

If the motor is a 3-phase induction motor, it is customary to assume that the stator winding is star-connected*.

1. Stator Resistance Test. The stator winding resistance is measured between any two terminals, using direct current. This gives the resistance of two phases in series, which must be divided by 2 to obtain stator winding resistance per phase. For greater accuracy, the resistance between three pairs of stator terminals is measured. The average of these three values divided by 2 gives the resistance per phase. Since the effective ac resistance is higher than the dc resistance, the latter is usually multiplied by a figure of about 1.25 to obtain the former.

2. No-Load Test. This test is performed to determine no-load current I_0, no-load power factor cos ϕ_0, windage and friction losses P_{wf}, no-load core loss P_i, no-load power input P_o, and no-load resistance R_0 and reactance X_0. Moreover, it also reveals any mechanical fault, noise etc.

The no-load test is performed with different values of applied voltage below and above rated voltage, while the motor is running light (without load).

The power input for all three phases P_0, voltage V_0 (line to line value) and current I_0 are measured by wattmeter W, voltmeter V and ammeter A connected in the circuit Fig. (2.23). If the motor is of wound rotor type, the external resistor should be cut out completely. Since motor is running without load, the power factor would be less than 0.5 (about 0.05 to 0.15), hence total power input will be equal to the difference of the two readings of the wattmeter.

*This may not be actual case, but the result would be exactly the same, if delta connections were assumed.

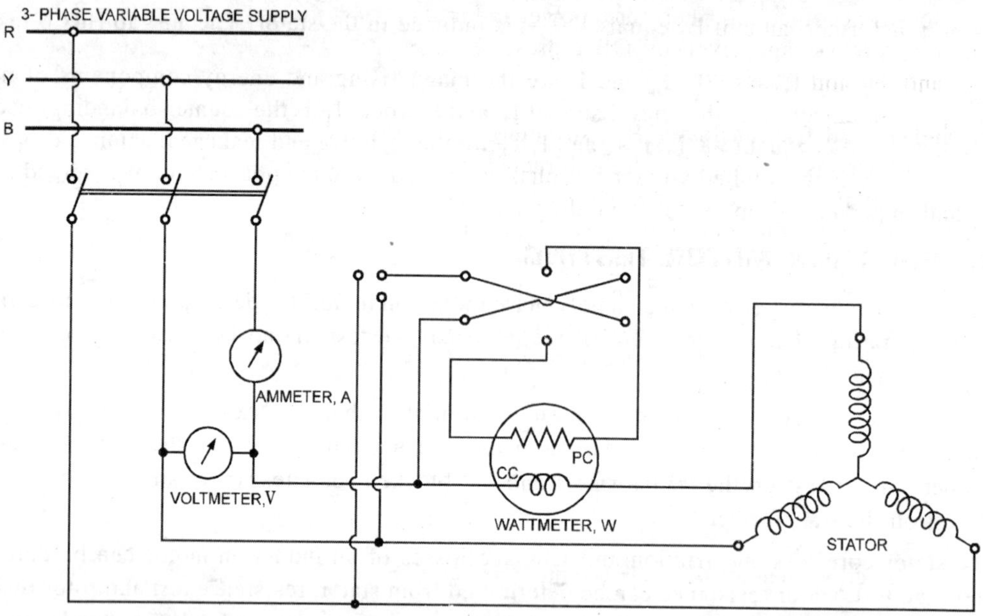

Fig. 2.23. ***Circuit Diagram For No-Load Test on a 3-Phase Induction Motor***

At no-load, power input is equal to the core loss P_i, stator copper loss P_{c1} and windage and frictional losses P_{wf}. Core loss occurs only in the stator as the slip is extremely small (of the order of 0.001) and so the frequency of rotor current is as low as 0.05 Hz. The magnitude of no-load current in an induction motor is about 30 to 40 % of full-load current because of the air gap. So the stator copper loss at no-load needs to be accounted for. This can be estimated by determining stator resistance, as mentioned above.

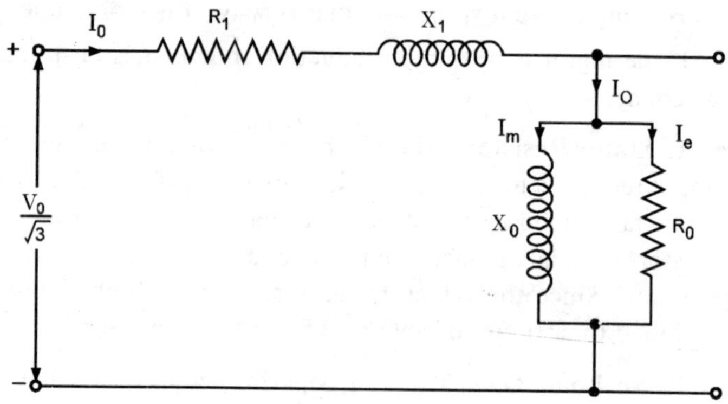

Fig. 2.24. ***Equivalent Circuit of a Polyphase Induction Motor at No-Load***

The mechanical power developed corresponds to windage and friction losses P_{wf} only and the rotor circuit can, therefore, be considered to be open. This is also evident from the magnitude of rotor circuit resistance $\frac{R_2/K^2}{s}$. This resistance becomes very high because the slip at no-load is extremely small and so the rotor circuit is practically open at no load. Thus the equivalent circuit of the motor at no load will be as illustrated in Fig. 2.24.

To determine windage and friction losses, the curve drawn between input power in watts and applied voltage in volts is extended to intersect vertical axis at any point (say B in Fig. 2.25). Point B corresponds to power input when applied voltage is zero. Since when applied voltage is zero, the stator copper losses and core losses become zero, therefore, power input at no load and zero voltage given by OB represents the windage and friction losses.

If windage and friction losses P_{wf}, so determined, and stator copper losses ($3\ I_0^2\ R_1$) are subtracted from power input at no load, P_0, core losses P_i can be determined. Knowing the total core losses P_i, no-load current I_0 and applied voltage V_0 (line-to-line value), the values of magnetising component and energy component of no-load current (I_m and I_e respectively), no-load resistance R_0 and no-load reactance X_0 can be determined from the following relations.

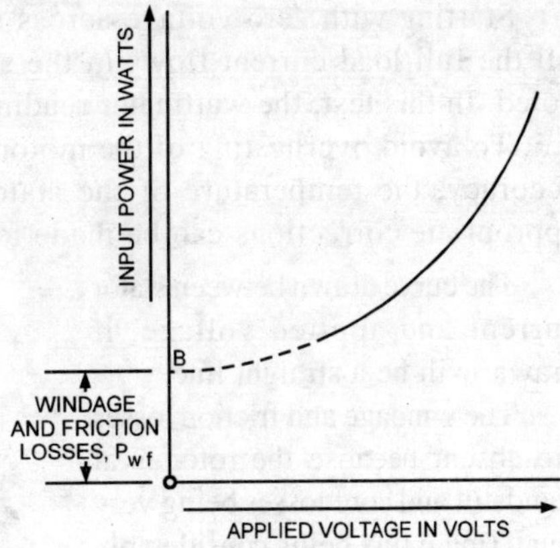

Fig. 2.25

$$\text{No-load power factor, } \cos\phi_0 = \frac{P_i}{\sqrt{3}V_0 I_0} \quad ...(2.36)$$

$$\text{Energy component of no-load current, } I_e = I_0 \cos\phi_0 = I_0 \times \frac{P_i}{\sqrt{3}\ V_0 I_0} = \frac{P_i}{\sqrt{3}\ V_0} \quad ...(2.37)$$

$$\text{Magnetising component of no-load current, } I_m = \sqrt{I_0^2 - I_e^2} \quad ...(2.38)$$

$$\text{No-load impedance, } Z_0 = \frac{V_0/\sqrt{3}}{I_0} \quad ...(2.39)$$

$$\text{No-load resistance, } R_0 = \frac{V_0/\sqrt{3}}{I_e} = \frac{V_0/\sqrt{3}}{P_i/\sqrt{3}\ V_0} = \frac{V_0^2}{P_i} \quad ...(2.40)$$

$$\text{No-load reactance, } X_0 = \sqrt{Z_0^2 - R_0^2} \quad ...(2.41)$$

3. Voltage-Ratio Test. This test can only be performed on a wound rotor motor by exciting the stator winding at rated voltage and frequency (the rotor circuit being kept open circuited and standstill). The ratio of rotor to stator voltage can be measured by means of a voltmeter. It is to be noted that emf appearing at the slip rings (induced emf in rotor) is of the supply frequency because the rotor is at standstill.

4. Blocked Rotor Test. This test is performed to determine the short-circuit current I_{sc} with normal applied voltage to stator; power factor on short circuit, total equivalent resistance and reactance of the motor as referred to stator (R_{01} and X_{01}).

In this test the rotor is held firmly (rotor windings are short circuited at slip rings in case of wound rotor motor) and stator is connected across supply of variable voltage. This test is just equivalent to short-circuit test on transformer. The connection diagram remains the same as in case of no-load test. Instruments must usually be changed to record the high power and current indications.

Though this test does not reveal any mechanical defect but is of importance because the short-circuit current I_{sc} and power factor $\cos\phi_s$ alongwith the no-load current I_0 and power factor $\cos\phi_0$ enables the circle diagram to be drawn, (as will be discussed in Art. 2.28). In addition, the copper losses computed from this test are necessary for determination of efficiency.

Starting with zero voltage across the stator, the applied voltage is gradually increased in steps till the full-load current flows in the stator. The readings of voltmeter, ammeter and wattmeter are noted. In this test, the wattmeter readings are additive because the power factor is much greater than 0.5. To avoid overheating of the motor, the readings must be taken as quick as possible. For greater accuracy, the temperature of the stator winding should also be noted during the test so that the appropriate corrections can be made to the recommended temperature of 75° C.

The curve drawn between stator current and applied voltage, if drawn, will be a straight line.

The windage and friction losses are absent because the rotor is at standstill and core losses being very small (main flux being considerably reduced under short-circuit condition at the low short-circuit test voltage) can also be ignored for practical purposes. Thus power input during this test can be considered to be entirely stator and rotor copper losses and no-load resistance R_0 of the equivalent circuit representing core losses can also be omitted. But the magnetizing reactance X_0, being much lower in an induction motor compared to a transformer cannot be ignored. The equivalent circuit of an induction motor under blocked rotor condition will, therefore, be as illustrated in Fig. 2.26.

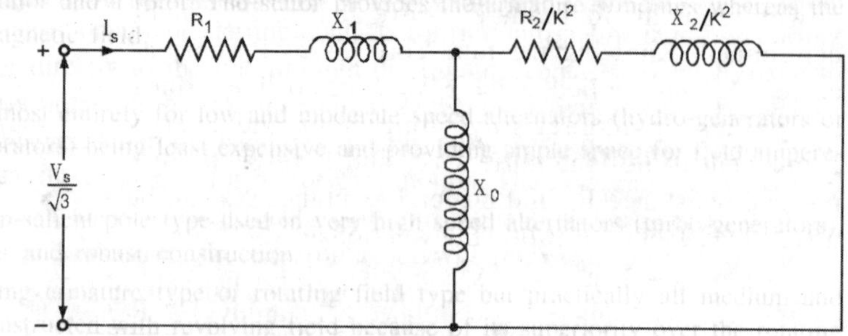

Fig. 2.26. ***Equivalent Circuit of a Polyphase Induction Motor Under Blocked Rotor Condition***

If V_s is the applied voltage (line-to-line value) and causes current I_s in the stator winding and P_s is the total input power at short-circuit then short-circuit current I_{sc} with normal voltage V (line-to-line) applied across the motor is given by

$$I_{sc} = I_s \frac{V}{V_s} \qquad ...(2.42)$$

$$\text{and power factor, } \cos \phi_s = \frac{P_s}{\sqrt{3} V_s I_s} \qquad ...(2.43)$$

Since input on short circuit meets with stator and rotor copper losses, as justified above, so

$$\text{Input on short-circuit, } P_s = 3 I_s^2 R_{01}$$

$$\text{or motor equivalent resistance per phase, as referred to stator, } R_{01} = \frac{P_s}{3 I_s^2} \qquad ...(2.44)$$

$$\text{Motor equivalent impedance per phase, as referred to stator, } Z_{01} = \frac{V_s / \sqrt{3}}{I_s} \qquad ...(2.45)$$

$$\text{Motor equivalent reactance per phase, as referred to stator, } X_{01} = \sqrt{Z_{01}^2 - R_{01}^2} \qquad ...(2.46)$$

These values constitute the series equivalent of the blocked rotor test (Fig. 2.27)

Usually the stator resistance per phase, X_1 is assumed to be equal to rotor reactance per phase, as referred to stator $X_2' = \left(\frac{X_2}{K^2}\right)$ and so

$$X_1 = X_2' = \frac{X_{01}}{2} \quad ...(2.47)$$

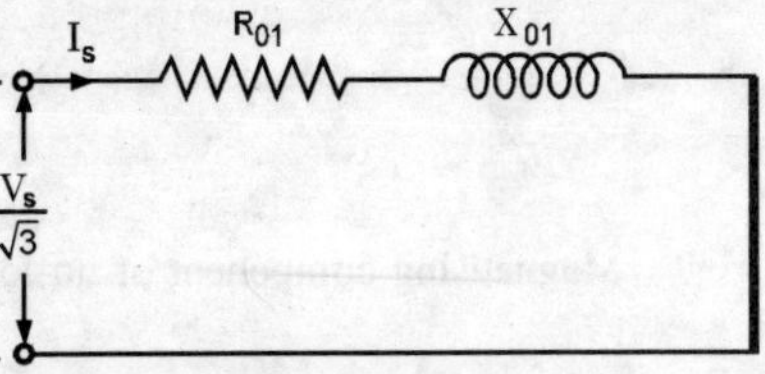

Fig. 2.27

In case of wound rotor motors, the stator and rotor resistances are separated by dividing R_{01} in the ratio of dc resistances of the stator and rotor windings. In case of squirrel cage induction motors, the rotor resistance per phase, as referred to stator, can be determined by subtracting R_1 from R_{01}.

$$\text{Thus } R_2' = R_{01} - R_1 \text{(determined from stator resistance test)} \quad ...(2.48)$$

5. Heat Run Test. The life of the insulation of electrical equipment depends upon the temperature attained during operation. The objective of this test is to find out the actual maximum temperature attained while the machine is operating under certain load conditions.

For motors with continuous rating and for motors for periodic duty and for continuous duty with intermittent load, this test is continued till equilibrium is reached. In the case of intermittent duty cycle motors, the duration of one cycle is kept 10 minutes.

The temperature is measured both while the motor is operating and after it is shut down. The motor should be stopped as soon as the motor attains the thermal equilibrium.

Example 2.27. A 400 volt, 3-phase delta-connected induction motor gave the following results on no load and short circuit (line values).

No-load test	**400 V ;**	**3.0 A;**	**645 watts**
Short-circuit test	**200 V;**	**12.0 A;**	**1,660 watts**

The friction and windage losses amount to 183 watts. Determine energy and magnetising components of no-load current, no-load power factor; no-load resistance R_0 and reactance X_0, equivalent resistance and reactance per phase as referred to primary; power factor on short circuit and short-circuit current with normal applied voltage of 400 volts across the stator. Stator resistance may be assumed to be 5 Ω.

Solution : Applied voltage per phase, V = 400 V $\quad\because$ motor is delta connected

No-load line current, $I_{L_0} = 3$ A

No-load phase current, $I_0 = \frac{I_{L0}}{\sqrt{3}} = \frac{3}{\sqrt{3}} = \sqrt{3}$ A

Input on no-load, $P_0 = 645$ watts

Stator copper loss, $P_{c1} = 3 I_0^2 R_1 = 3(\sqrt{3})^2 \times 5 = 45$ watts

Windage and friction loss, $P_{wf} = 183$ watts

Total stator core loss, $P_i = P_0 - P_{c1} - P_{wf} = 645 - 45 - 183 = 417$ watts

No-load power factor, $\cos\phi_0 = \frac{P_i}{3 VI_0} = \frac{417}{3 \times 400 \times \sqrt{3}} = 0.2$ **Ans.**

Energy component of no-load line current, $I_{Le} = I_{L_0} \cos\phi_0 = 3 \times 0.2 = 0.6$ A **Ans.**

Magnetising component of no-load line current, $I_{Lm} = \sqrt{(I_{L0})^2 - (I_{Le})^2} = \sqrt{(3)^2 - (0.6)^2} = 2.94$ A **Ans.**

Energy component of no-load phase current, $I_e = I_0 \cos\phi_0 = 1.732 \times 0.2 = 0.3464$ A

$$\text{No-load resistance, } R_0 = \frac{V}{I_e} = \frac{400}{0.3464} = 1{,}155\ \Omega \textbf{ Ans.}$$

$$\text{Magnetising component of no-load phase current, } I_m = \frac{2.94}{\sqrt{3}} = 1.7 \text{ A}$$

$$\text{No-load leakage reactance, } X_0 = \frac{V}{I_m} = \frac{400}{1.7} = 236\ \Omega \textbf{ Ans.}$$

$$\text{On short circuit, voltage applied per phase, } V_s = 200 \text{ V}$$

$$\text{On short circuit, stator line current, } I_{Ls} = 12.0 \text{ A}$$

$$\text{On short circuit, stator phase current, } I_s = \frac{12.0}{\sqrt{3}} = 6.928 \text{ A}$$

$$\text{Input on short circuit, } P_s = 1{,}660 \text{ watts}$$

$$\text{Power factor on short circuit, } \cos\phi_s = \frac{P_s}{3\,VI_s} = \frac{1{,}660}{3 \times 200 \times 6.928} = 0.4 \textbf{ Ans.}$$

$$\text{Equivalent impedance, } Z_{01} = \frac{V_s}{I_s} = \frac{200}{6.928} = 28.87\ \Omega/\text{ phase} \textbf{ Ans.}$$

$$\text{Equivalent resistance, } R_{01} = \frac{P_s}{3I_s^2} = \frac{1{,}660}{3\,(6.928)^2} = 11.53\ \Omega/\text{phase} \textbf{ Ans.}$$

$$\text{Equivalent reactance, } X_{01} = \sqrt{Z_{01}^2 - R_{01}^2} = \sqrt{(28.87)^2 - (11.53)^2}$$

$$= 26.5\ \Omega/\text{phase} \textbf{ Ans.}$$

Short-circuit current with normal voltage of 400 V applied to the stator,

$$I_{sc} = I_s \times \frac{V}{V_s} = 12 \times \frac{400}{200} = 24 \text{ A} \textbf{ Ans.}$$

Example 2.28. A 110 V, 3-phase, star-connected induction motor takes 25 A at a line voltage of 30 V with rotor locked. With this line voltage, power input to the motor is 440 W and core loss is 40 W. The dc resistance between a pair of stator terminals is 0.1 Ω. If the ratio of ac to dc resistance is 1.6. find the equivalent leakage reactance/phase of the motor and the stator and rotor resistance per phase.

Solution : Short-circuit voltage (line-to-line), V_s = 30 V

$$\text{Short-circuit current, } I_s = 25 \text{ A}$$

$$\text{Power input to motor on short circuit, } P_s = 440 \text{ W}$$

$$\text{Core loss during short-circuit, } P_i = 40 \text{ W}$$

$$\text{DC resistance per phase, } R_{dc} = \frac{0.1}{2} = 0.05\ \Omega$$

$$\text{Stator ac resistance per phase, } R_1 = 1.6\, R_{dc} = 1.6 \times 0.05 = 0.08\ \Omega \textbf{ Ans.}$$

$$\text{Stator and rotor copper losses, } P_c = P_s - P_i = 440 - 40 = 400 \text{ W}$$

Motor equivalent resistance per phase as referred to stator,

$$R_{01} = \frac{P_c}{3I_s^2} = \frac{400}{3 \times 25^2} = 0.213\ \Omega$$

Rotor equivalent resistance per phase as referred to stator,

$$R_2' = R_{01} - R_1 = 0.213 - 0.08 = 0.133\ \Omega\ \textbf{Ans.}$$

Motor equivalent impedance per phase as referred to stator,

$$Z_{01} = \frac{V_s/\sqrt{3}}{I_s} = \frac{30/\sqrt{3}}{25} = 0.693\ \Omega$$

Motor equivalent leakage reactance per phase as referred to stator,

$$X_{01} = \sqrt{Z_{01}^2 - R_{01}^2} = \sqrt{(0.693)^2 - (0.213)^2} = 0.66\ \Omega\ \textbf{Ans.}$$

2.27. MEASUREMENT OF SLIP

The various methods of measurement of slip are given below:

1. Actual Measurement of Rotor Speed. In this method of determination of slip, the actual speed of the rotor is measured and then subtracted from the synchronous speed to obtain slip

$$i.e.\ \text{slip, } s = N_s - N$$

This method of determination of slip is not accurate, since slip is very small as compared to synchronous or actual rotor speed and is the difference of two nearly equal quantities and, therefore, small percentage error in measurement of rotor speed will cause large percentage error in slip.

2. By Measure of Rotor Frequency. Since the rotor frequency f', being the product of slip and supply frequency, is very low, it can be accurately measured by means of a low value centre-reading dc moving coil voltmeter connected across the slip rings in case of wound rotor motor. The current in the voltmeter follows the variations of rotor current and so the pointer oscillates about its mean zero position. The number of complete cycles (movement from zero to a maximum to right, back to zero and on a maximum to the left and then back to zero) made by the pointer can be easily counted. The slip can be determined from the following relation

$$s = \frac{\text{Rotor frequency}}{\text{Supply frequency}} = \frac{f'}{f} = \frac{\text{Number of oscillations, counted over a period of } t \text{ seconds}}{50 \times t} \quad ...(2.49)$$

With a squirrel cage motor the above method cannot be applied. If, however, a large flat search coil having many turns is placed centrally against the end plate on the non-driving end of the motor, it is sometimes possible to pick up sufficient voltage by induction from the leakage fluxes to obtain a reading. A larger 50 Hz voltage will also be induced, but this is too rapid to affect the instrument. Commercial slip indicators have been developed employing such a search coil in conjunction with low-pass filter-amplifier to eliminate the fundamental frequency and then comparing f' and f by means of a bridge circuit or digital counter.

It is sometimes possible to obtain an indication by connecting a dc millivoltmeter between the ends of the shaft.

3. Stroboscopic Method. This is another method for measurement of slip. In this method a disc known as *stroboscopic disc*, with the geometrical pattern on it, is rigidly attached to the shaft of the motor, whose slip is to be determined, as shown in Fig. 2.28. The number of black sectors as well as white sectors, each are equal to the number of poles for which the motor is wound. For example for a 6-pole motor there will be 12 sectors, 6 black and 6 white. A neon glow lamp supplied from the motor line either through a transformer or directly is arranged to illuminate the stroboscopic disc. The neon lamp employed has the characteristics that it ceases to glow when voltage across it falls below a certain limit *i.e.* such a lamp glows twice in a cycle. If the disc were to rotate at synchronous speed, it would appear to be stationary. Since, in actual practice, its speed is slightly less than the synchronous speed, it appears to rotate slowly backwards. This fact is explained as follows.

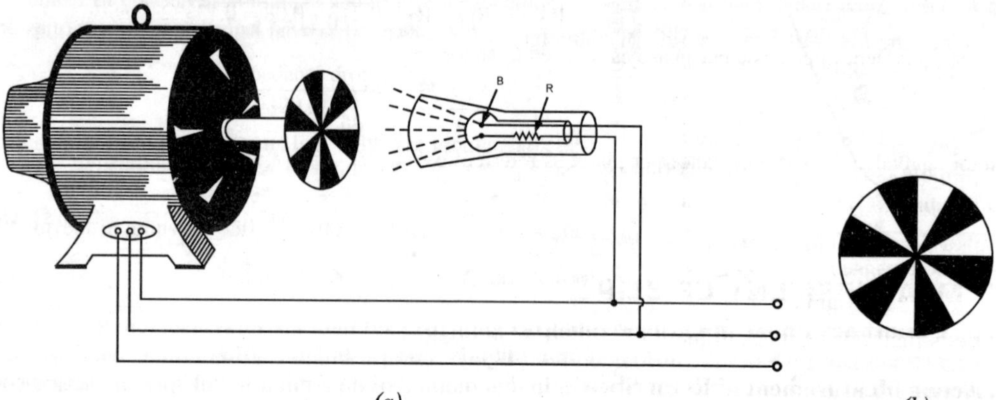

(a) *(b)*

Fig. 2.28

In one half-cycle the rotor of the motor would advance one pole if there were no slip and during this period each black sector (or white sector) would advance to the position just occupied by the adjacent black sector (or white sector) that preceded it. During the period of advancement the sectors are but faintly visible, for the illumination from the lamp is practically zero. Hence the disc would appear stationary. Owing to the fact that each conductor on the rotor does not advance one pole each half cycle, the sectors will not reach the position of the next adjacent sector of same colour but will fall short of this distance, owing to the slip. The sectors on the disc will not appear stationary but will seem to be rotating slowly backward.

The complete apparent revolutions of the disc are counted in a given time and slip is determined from the following relation.

$$\text{Slip, } s = \frac{\text{Apparent revolutions counted in the given time}}{\text{Time in seconds}} \times \frac{\text{pairs of poles}}{\text{supply frequency}} \quad ...(2.50)$$

2.28. CIRCLE DIAGRAM

It is known that in a series circuit consisting of constant reactance but variable resistance, the locus of the current is a circle of diameter $\frac{V}{X}$.

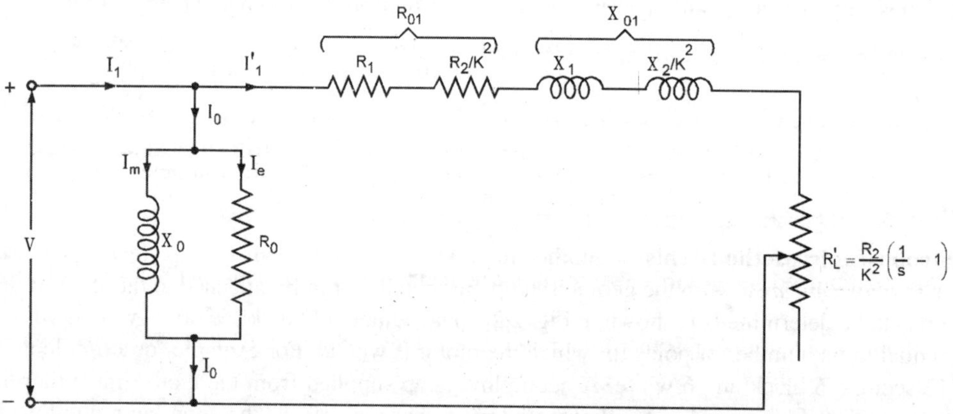

Fig 2.29. ***Approximate Equivalent Circuit of An Induction Motor***

In a circuit shown in Fig. 2.29 the current I_0 to the shunt circuit is constant, in the circuit to the right $X_{01} = X_1 + \frac{X_2}{K^2}$, the equivalent reactance referred to primary (stator) is constant; $R_{01} = R_1 + \frac{R_2}{K^2}$, the equivalent resistance referred to primary is also constant but load resistance R'_L varies with the load on the motor, Hence the locus of I_1' is a circle of diameter $\frac{V}{X_{01}}$ and having coordinate $\left(\frac{V}{2X_{01}}, 0\right)$ with respect to O'.

Since the total current I_1 drawn by the motor from the supply main is phasor sum of constant current I_0 and variable current I_1', the locus of I_1 is also a circle. The same circular arc O'HAG acts as locus of I_1 referred to point O, which acts as locus of I'_1 referred to point O'.

Thus operating characteristics of an induction motor can be computed by use of a circle diagram easily and conveniently.

2.28.1. Construction of the Circle Diagram

Datas for the construction of this diagram are obtained from no-load and short-circuit tests. [Refer to Art. 2.26]

From no-load test, line voltage V, line current I_0 and total power input are measured and no-load power factor $\cos \phi_0$ is determined. [Refer to Art 2.26 (2)].

From short-circuit test, short-circuit current I_{sc} corresponding to normal voltage applied to stator, short-circuit phase angle ϕ_s are determined.

The voltage phasor V is taken along Y-axis, *i.e.* in phasor diagram (Fig. 2.30) OV represents the normal voltage applied to the stator. Phasor OO ' is drawn at an angle ϕ_0 from OV (lagging) representing no-load line current I_0 in magnitude as well as in direction. Line OF is drawn perpendicular to voltage phasor OV.

Phasor OA be drawn at an angle ϕ_s from phasor OV (lagging) and equal to I_{sc} in magnitude. Thus phasor OA represents the line current that would exist were the rated line voltage V impressed across the stator when motor blocked and phasor O'A represents the corresponding rotor current I_1' as referred to primary (stator).

Obviously the points O′ and A lie on the circle. To determine the centre of the required circle, line O′G is drawn parallel to OF, O′A is joined and line BC is drawn perpendicular to line O′A bisecting the line O′A at point B and intersecting line O′G at C. Now with C as center and O′C as radius the semicircle O′HP′AG is drawn.

Since voltage phasor is drawn vertically, therefore, all vertical distances represent the active components of currents and power to different scales. For example O′J represents the active component of no-load current, which supplies core loss, friction and windage loss and small amount of stator copper losses. O′J also represents power input at no-load to some other scale. Similarly vertical component AF of OA represents the active component of short-circuit current I_{sc} and also represents power input on short circuit to some other scale.

Thus AF represents the power input when rotor is blocked. The power input when rotor is blocked meets core losses, and stator and rotor copper losses. Line AE represents the stator and rotor copper losses when EF = O′J is subtracted from AF.

Now the line AE is divided by D into two segments such that $\frac{AD}{DE} = \frac{\text{Rotor copper losses}}{\text{Stator copper losses}}$. Line joining points O′ and D is drawn. Line O′D is known as *torque line or rotor input line.*

In case of wound rotor motor, the line AE is divided directly into two segments such that

$$\frac{AD}{DE} = \frac{R_2/K^2}{R_1} = \frac{R_2'}{R_1}$$

In case of a squirrel cage motor, stator resistance R_1 is determined and DE is made equal to stator copper losses $[3I'^2_{sc} R_1]$, where I_{sc}' is the short-circuit current per phase.

In circle diagram (Fig. 2.30) line O′A represents the output line or mechanical power developed line.

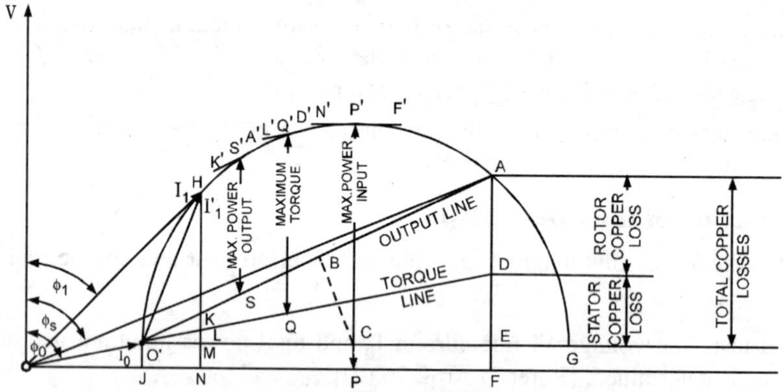

Fig. 2.30. ***Circle Diagram For An Induction Motor***

If a line N' P' F' is drawn parallel to line NPF and tangential to the semicircle, then the vertical intercept P'P between this tangent line N'P'F' and horizontal line NPF represents the maximum power input to the motor.

If a line L′Q′D′ is drawn parallel to torque line LQD and tangential to the semicircle, then the vertical intercept Q′Q between the tangential line L′Q′D′ and torque line LQD represents the maximum torque.

If a line K′S′A′ is drawn parallel to output line O' A and tangential to the semicircle, then the vertical intercept S' S between the tangential line K' S' A' and output line O' A represents the maximum power output.

At any load current I_1 (represented by OH), the secondary current referred to primary is I_1' (represented by O' H) being equal to phasor difference of I_1 and I_0 *i.e.* $(\mathbf{I_1} - \mathbf{I_0})$, HN represents the energy component of load current I_1. Now

$$\text{Motor input} = \sqrt{3}\ V \cdot NH$$

$$\text{Constant losses} = \sqrt{3}\ V \cdot NM$$

$$\text{Stator copper losses} = \sqrt{3}\ V \cdot ML$$

$$\text{Rotor input power} = \sqrt{3}\ V \cdot LH$$

$$\text{Rotor copper losses} = \sqrt{3}\ V \cdot LK$$

$$\text{Motor output} = \sqrt{3}\ V \cdot HK$$

$$\text{Efficiency} = \frac{\text{Output}}{\text{input}} = \frac{\sqrt{3}\ V \cdot KH}{\sqrt{3}\ V \cdot NH} = \frac{KH}{NH}$$

$$\text{Slip, } s = \frac{\text{Rotor copper losses}}{\text{Rotor input}} = \frac{\sqrt{3}\ V \cdot LK}{\sqrt{3}\ V \cdot LH} = \frac{LK}{LH}$$

$$\text{Power factor, } \cos\phi_1 = \frac{NH}{OH}$$

Example 2.29. Make use of the circle diagram or otherwise, prove that the maximum output of an induction motor per phase = $\dfrac{V_1 (I_{sc} - I_0)}{2(1 + \cos\phi_{sc})}$ where V_1 = source voltage per phase, I_{sc} = source current during blocked rotor test, I_0 = no-load currrent and ϕ_{sc} = angle of lag during blocked rotor test between source voltage and current.

Solution : Maximum power output of an induction motor per phase is given as

$$P_{gmax} = \frac{V_1^2}{2(R_{01} + Z_{01})} \text{ watts}$$

where V_1 is the source voltage per phase and R_{01} and Z_{01} are the motor equivalent resistance and reactance respectively, in ohms, as referred to the stator.

Neglecting the phase angle difference caused by the exciting current during blocked rotor test

$$\text{Cos } \phi_{sc} = \frac{R_{01}}{Z_{01}}$$

$$\text{and } I_{sc} - I_0 = \frac{V_1}{Z_{01}}$$

$$\text{So } P_{gmax} = \frac{V_1 \times \dfrac{V_1}{Z_{01}}}{2\left[\dfrac{R_{01}}{Z_{01}} + 1\right]} = \frac{V_1 (I_{sc} - I_o)}{2[\cos\phi_{sc} + 1]} \quad \text{proved.}$$

Example 2.30. Draw the circle diagram of a 20 hp, 400 V, 50 Hz, 3-phase star-connected induction motor from the following test data (line values)

No. load	**400 V**	**9A**	**pf 0.2**
Blocked rotor	**200 V**	**50 A**	**pf 0.4**

From the circle diagram find (*i*) line current and power factor at full load. (*ii*) maximum power output.

Solution : No-load current, $I_0 = 9$ A

No-load power factor, $\cos\phi_0 = 0.2$

or no-load phase angle, $\phi_0 = \text{Cos}^{-1}\, 0.2 = 78.5°$

Short-circuit current with normal voltage applied across the stator,

$$I_{sc} = 50 \times \frac{400}{200} = 100 \text{ A}$$

Short-circuit power factor, $\cos\phi_s = 0.4$

or $\phi_s = \text{Cos}^{-1}\, 0.4 = 66.5°$

From the above datas, the circle diagram (Fig. 2.31) is constructed as follows:

The current scale is assumed 10.0 A per cm

OV is drawn vertical to represent the input voltage. OO′ is drawn lagging behind phasor OV by an angle 78.5° to represent no-load current of 9 A. Phasor OA is drawn lagging behind phasor OV by an angle 66.5° representing short-circuit I_{sc} with full applied voltage (*i.e.* 100 A)

A horizontal line O′G is drawn through O′ and O′A are joined. The line O′A is bisected at right angle by line BC. Then point C on line O′G is the centre of the required semicircle. This semicircle drawn through points O′ and A with centre C is the locus of the current phasor for all loading conditions from no-load to short-circuit.

To determine the power scale, a perpendicular AF through point A is drawn. AF represents total input on short circuit or when rotor blocked.

Input on short-circuit, $P_{sc} = \sqrt{3}\ V I_{sc} \cos \phi_s = \sqrt{3} \times 400 \times 100 \times 0.4 = 27{,}720$ watts

and AF = 4 cm

So power scale = $\dfrac{27{,}720}{4} = 6{,}930$ watts / cm

AE represents total copper loss in the rotor and stator. Assuming copper losses being equally divided between rotor and stator losses, AE is bisected at point D.

Now motor output = 20 hp = 20 × 735.5 = 14,710 watts, which will be represented by $\dfrac{14{,}710}{6{,}930} = 2.12$ cm. Line FA is extended to point T so that AT = 2.12 cm. From point T line TH is drawn parallel to output line O' A cutting the semicircle at point H. Point H is joined with origin O and perpendicular HN is drawn.

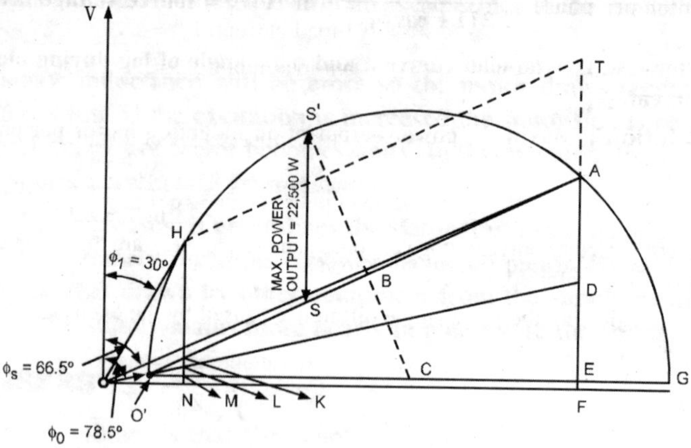

Fig. 2.31

Now from circle diagram

(*i*) Line current on full load = OH = 3 cm = 3 × 10 = 30 A **Ans.**

$$\text{Full-load power factor, } \cos \phi_1 = \frac{NH}{OH} = \frac{2.6}{3.0} = 0.866 \textbf{ Ans.}$$

(*ii*) Maximum power output is obtained by producing CB intersecting semicircle at S' and then dropping perpendicular from point S' on horizontal line OE meeting output line at S. Now SS' represents the maximum power output,

Maximum power output = SS' × power scale = 3.25 × 6,930 = 22,500 watts **Ans.**

Example 2.31. A 20 hp, 400 V, 50 Hz, three-phase star-connected induction motor has the following test data:

No-load test	**400 V**	**9 A**	**Cos φ = 0.2**
Blocked rotor test	**200 V**	**50 A**	**Cos φ = 0.4**

Draw a circle diagram and determine (*a*) line current (*b*) power factor (*c*) slip and (*d*) efficiency at full-load.

The stator and rotor copper losses are divided equally in the blocked rotor test.

Solution : The datas are the same as given in Example 2.30. So

(*a*) Line current at full-load, I_1 = 30 A **Ans.**

(*b*) Power factor at full-load, $\cos \phi_1$ = 0.866 **Ans.**

$$(c) \text{ Slip} = \frac{KL}{LH} = \frac{0.13}{2.3} = 0.056 \text{ or } 5.6\% \textbf{ Ans.}$$

$$(d) \text{ Efficiency, } \eta = \frac{KH}{NH} \times 100 = \frac{2.12}{2.6} \times 100 = 81.54\% \textbf{ Ans.}$$

Example 2.32. Draw the circle diagram for a 3-phase, 6-pole, 50 Hz, 400-V, star-connected induction motor from the following datas (line values) :

No load test	**400 V: 9A;**	**1,250 watts**
Short circuit test	**200 V; 50 A;**	**6,930 watts**

The stator loss at standstill is 55% of total copper losses and full-load current is 32.0 A. From circle diagram determine (*i*) pf, slip, output, efficiency, speed and torque at full load (*ii*) starting torque (*iii*) maximum power

factor (*iv*) maximum power output (*v*) maximum power input (*vi*) maximum torque in synchronous watts and slip for maximum torque.

Solution : Voltage applied, V = 400 V

No-load current, I_0 = 9.0 A

No-load input, P_0 = 1,250 W

$$\text{No load power factor, } \cos\phi_0 = \frac{P_0}{\sqrt{3}VI_0} = \frac{1,250}{\sqrt{3}\times 400 \times 9} = 0.2$$

$$\text{or No-load phase angle, } \phi_0 = \text{Cos}^{-1}(0.2) = 78^\circ\ 28'$$

$$\text{Short-circuit current with normal voltage, } I_{sc} = 50 \times \frac{400}{200} = 100 \text{ A}$$

$$\text{Short circuit input with normal voltage, } P_{sc} = \left(\frac{100}{50}\right)^2 \times 6,930 = 27,720 \text{ watts}$$

$$\text{Short-circuit pf , } \cos\phi_s = \frac{27,720}{\sqrt{3}\times 400 \times 100} = 0.4;\ \phi_s = \cos^{-1} 0.4 = 66^\circ\ 25'$$

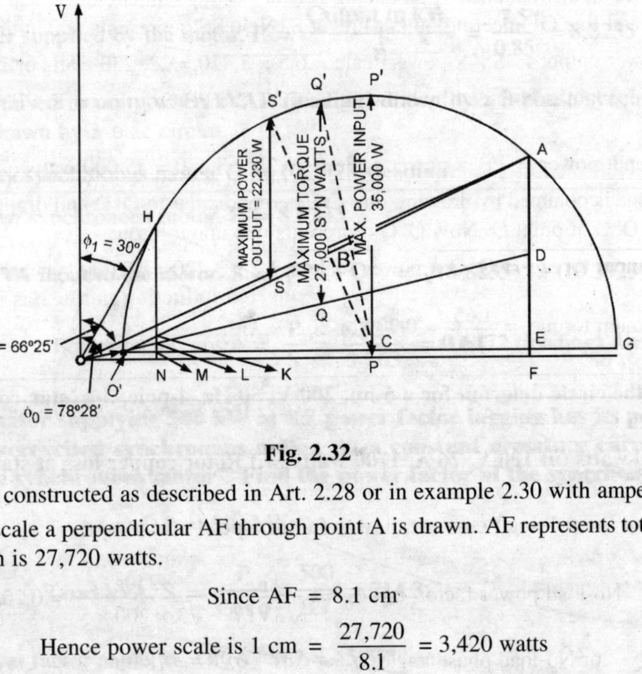

Fig. 2.32

The circle diagram is constructed as described in Art. 2.28 or in example 2.30 with ampere-scale 1 cm = 5 A.

To determine power scale a perpendicular AF through point A is drawn. AF represents total input on short circuit with normal voltage which is 27,720 watts.

Since AF = 8.1 cm

$$\text{Hence power scale is 1 cm} = \frac{27,720}{8.1} = 3,420 \text{ watts}$$

Since stator copper loss is 55% of total copper loss

∴ Torque line O' D is obtained by taking ED = 0.55 EA

(*i*) Since full-load current is 32.0 A, hence it is represented by a length of $\frac{32}{5} = 6.4$ cm.

With O as centre and 6.4 cm as radius an arc is drawn intersecting semicircle at point H. This point represents full–load conditions. Point O and H are joined. From point H, perpendicular HN is drawn to line OF intersecting lines O' A, O' D, O' E and O' F at points K, L, M and N respectively.

At full load

$$(a)\ \text{pf} = \frac{HN}{OH} = \frac{5.55}{6.4} = 0.866\ \textbf{Ans.}$$

also pf = Cos ∠VOH = cos 30° = 0.866 **Ans.**

(*b*) Slip $= \frac{\text{KL}}{\text{LH}} = \frac{0.30}{4.8}$ = 0.0625 or 6.25% **Ans.**

(*c*) Output = KH × power scale = 4.5 × 3,420 = 15,390 watts or 15.39 kW **Ans.**

(*d*) Efficiency $= \frac{\text{KH}}{\text{NH}} \times 100 = \frac{4.5}{5.55} \times 100$ = 81% **Ans.**

(e) Speed at full load $= \text{N}_s(1-s) = \frac{120f}{\text{P}}(1-s) = \frac{120 \times 50}{6}(1-0.625)$ = 937.5 rpm **Ans.**

Torque at full load T_f = LH × power scale = 4.8 × 3,420 = 16,416 synchronous watts **Ans.**

(*ii*) Starting torque T_{st} = AD × power scale = 3.50 × 3,420 = 11,970 synchronous watts **Ans.**

(*iii*) Maximum power factor is obtained by drawing a line from point O and tangential to semicircle. In this problem full–load current line OH is tangential (which is not necessary in every case) to the semicircle, therefore, maximum power factor,

$$\text{Cos}\,\phi_{max} = \text{Cos } 30° = 0.866 \textbf{ Ans.}$$

(*iv*) Maximum power output is obtained by producing CB intersecting semicircle at S' and then dropping perpendicular from point S′ on horizontal line OF meeting output line at S, Now SS' represents maximum power output.

Maximum power output = SS′ × power scale = 6.5 × 3,420 = 22, 230 watts or 22.23 kW **Ans.**

(*v*) Maximum power input is obtained by drawing vertical line PP' from point P. PP' represents maximum input power.

∴ Maximum input power = PP′ × power scale = 10.25 × 3,420 = 35,000 watts or 35 kW **Ans.**

(*vi*) Maximum torque is obtained by drawing line CQ′ perpendicular to O′ D and dropping a vertical line from Q' intersecting torque line O′ D at point Q. Now Q′ Q represents maximum torque.

∴ Maximum torque = Q′ Q × power scale = 7.2 × 3,420 = 27,000 synchronous watts **Ans.**

Slip for maximum torque $= \frac{1.65}{7.9}$ = 0.209 or 20.9% **Ans.**

Example 2.33. Draw the circle diagram for a 5 hp, 200 V, 50 Hz, 4-pole, 3-φ, star-connected induction motor from the following data :

(*i*) 200 V, 5 A, 350 watts (*ii*) 100 V, 26 A, 1700 watts (*iii*) Rotor copper loss at standstill = Half of the total copper loss.

Estimate therefrom the full load current, power factor, speed and torque.

Solution :

No-load power factor, $\cos\phi_0 = \frac{\text{P}_0}{\sqrt{3}\text{VI}_0} = \frac{350}{\sqrt{3} \times 200 \times 5} = 0.202$

or No-load phase angle, $\phi_0 = \text{Cos}^{-1}\, 0.202 = 78.35°$

Short-circuit current with normal voltage, $\text{I}_{sc} = 26 \times \frac{200}{100} = 52 \text{ A}$

Short-circuit input with normal voltage, $\text{P}_{sc} = 1{,}700 \times \left(\frac{200}{100}\right)^2 = 6{,}800$ watts

Short-circuit power factor, $\cos\phi_s = \frac{6{,}800}{\sqrt{3} \times 200 \times 52} = 0.3775$

or $\phi_s = \text{Cos}^{-1}\, 0.3775 = 67.82°$

The circle diagram is constructed as described in Art. 2.28 or example 2.30 with ampere scale 1 cm = 5A.

To determine the power scale a perpendicular AF through point A on O' G is drawn. AF represents total input on short circuit with normal voltage applied, which is 6,800 watts.

$$\text{Since AF} = 3.9 \text{ cm}$$

$$\text{Hence power scale 1 cm} = \frac{6{,}800}{3.9} = 1{,}744 \text{ watts}$$

Since rotor copper losses at standstill = Half of the total copper losses, therefore, line AE is bisected at point D to separate the stator and rotor copper losses.

Now power input = 5 hp = 5 × 735.5 = 3677.5 watts which will be represented by $\frac{3{,}677.5}{1{,}744}$ = 2.1 cm on the circle diagram. Line FA is extended to point T so that AT = 2.1 cm. From point T line TH is drawn parallel to output line O'A cutting the circle at point H. Point H is joined with origin O and perpendicular HN is drawn.

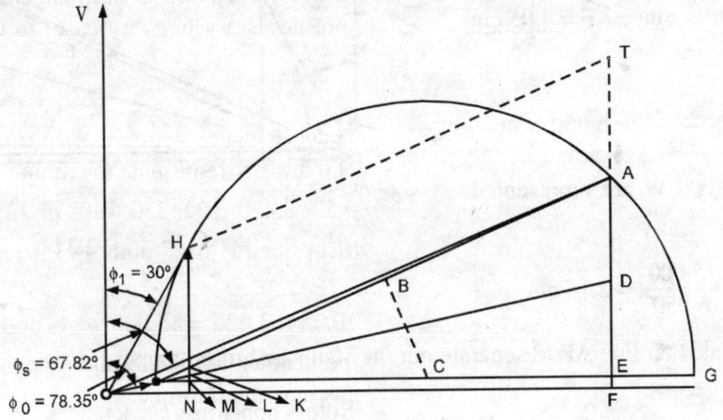

Fig. 2.33

Now from circle diagram (Fig. 2.33)

$$\text{Full load current, } I_f = \text{OH} = 3 \text{ cm} = 3 \times 5 = 15 \text{ A } \textbf{Ans.}$$

$$\text{Full load power factor, } \cos\phi_f = \frac{\text{NH}}{\text{OH}} = \frac{2.6}{3} = 0.866 \textbf{ Ans.}$$

$$\text{Full load speed, } N_f = \frac{\text{KH}}{\text{LH}} \times N_s = \frac{2.2}{2.3} \times \frac{120 \times 50}{4} = 1{,}435 \text{ rpm } \textbf{Ans.}$$

$$\text{Full load torque} = \text{LH} \times \text{power scale} = 2.3 \times 1{,}744 = 4{,}010 \text{ synchronous watts } \textbf{Ans.}$$

Example 2.34. A 400 V, 3-phase, 50 Hz, star-connected induction motor has the following test results :

No-load test	**400 V**	**8.5 A**	**1,100 W**
Blocked rotor test	**180 V**	**45 A**	**5,799 W**

Calculate the line current and power factor when operating at 4% slip. The stator resistance per phase is 0.5 Ω.

Solution :

$$\text{No-load power factor, } \cos\phi_0 = \frac{1{,}100}{\sqrt{3} \times 400 \times 8.5} = 0.187$$

$$\text{and } \phi_0 = \text{Cos}^{-1}\, 0.187 = 79.2^\circ$$

$$\text{Short-circuit current with normal voltage, } I_{sc} = \frac{45 \times 400}{180} = 100 \text{ A}$$

Short-circuit input with normal voltage, $P_{sc} = 5{,}700 \times \left(\frac{400}{180}\right)^2 = 28{,}150$ W

Short-circuit power factor, $\cos\phi_s = \frac{5{,}700}{\sqrt{3} \times 180 \times 45} = 0.4063$

and $\phi_s = \text{Cos}^{-1}\, 0.4063 = 66°$

The circle diagram is constructed as described in Art. 2.28 or in example 2.30 with ampere scale 1 cm = 10 A.

Stator copper losses on short circuit = $3\, I_{sc}^2 R_1 = 3 \times 100^2 \times 0.5 = 15{,}000$ W

To determine power scale a perpendicular AF through point A on O'G is drawn. AF represents total input on short circuit with normal voltage applied, which is 28,150 W. Since AF = 4.05 cm

Hence power scale,

$$1 \text{ cm} = \frac{28{,}150}{4.05} = 6{,}950 \text{ watts}$$

Stator losses of 15,000 W are represented by ED

$$\text{So ED} = \frac{15{,}000}{6{,}950} = 2.16 \text{ cm}$$

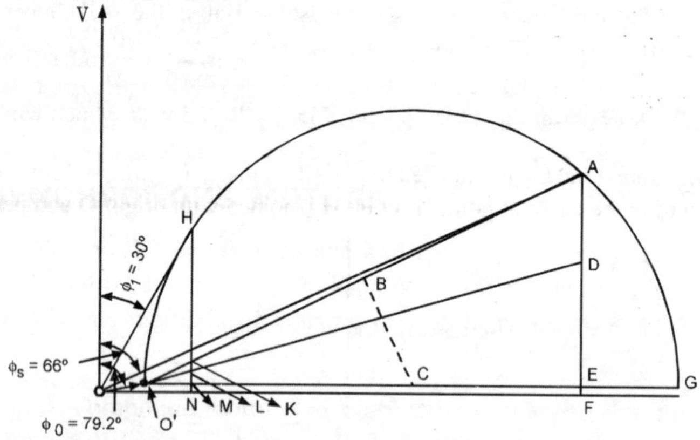

Fig. 2.34

Thus point D is located on line AF to separate out the stator and rotor copper losses.

$$\text{Slip, } s = \frac{\text{Rotor copper loss}}{\text{Rotor input}} = \frac{KL}{LH} = 0.04$$

Hence by trial and error point H is so located on the circle that $\frac{LH}{KL} = 25$.

Now line current, I_L = OH = 10 × 2.42 = 24.2 A **Ans.**

Power factor, $\cos\phi_1 = \text{Cos } 30° = 0.866$ **Ans.**

Example 2.35. Draw the circle diagram of a 3-phase mesh-connected, 30 hp, 500 V, 4-pole, 50 Hz cage-type induction motor. The figures given below give the measurements of line current and voltage and readings of two wattmeters connected to measure the input power.

No-load	**500 V**	**8.3 A**	**+ 2.85 kW**	**– 1.35 kW**
Short-circuit (blocked rotor)	**100 V**	**32 A**	**– 0.75 kW**	**+ 2.35 kW**

Find from the diagram for full-load (*i*) line current (*ii*) power factor (*iii*) efficiency and (*iv*) maximum output.

Solution : No-load power input, $P_0 = 2.85 - 1.35 = 1.5$ kW or 1,500 W

No-load power factor, $\cos\phi_0 = \frac{P_0}{\sqrt{3} V I_0} = \frac{1{,}500}{\sqrt{3} \times 500 \times 8.3} = 0.2087$

and $\phi_0 = \text{Cos}^{-1}\, 0.2087 = 78°$

Short-circuit line current with normal voltage applied,

$$I_{sc} = I_s \times \frac{V}{V_s} = 32 \times \frac{500}{100} = 160 \text{ A}$$

Short-circuit input power with normal voltage,

$$P_{sc} = P_s \times \left(\frac{V}{V_s}\right)^2 = (2.35 - 0.75) \times \left(\frac{500}{100}\right)^2 = 40 \text{ kW}$$

$$\text{Short-circuit power factor, } \cos \phi_s = \frac{P_s}{\sqrt{3}\, VI_s} = \frac{400 \times 1{,}000}{\sqrt{3} \times 500 \times 160} = 0.28868$$

$$\text{and } \phi_s = \text{Cos}^{-1}\ 0.28868 = 73.22°$$

The circle diagram is constructed, as described in Art. 2.28 or in example 2.30, with ampere scale 1 cm = 16 A (line value).

To determine power scale a perpendicular AF through point A on O' G is drawn. AF represents total power input on short circuit with normal voltage applied which is 40 kW or 40,000 W

$$\text{Hence power scale} = \frac{40{,}000}{AF} = \frac{40{,}000}{2.89} = 13{,}800 \text{ watts}$$

$$\text{Alternatively the power scale,} = \sqrt{3} \times \text{line voltage} \times \text{ampere per cm}$$

$$= \sqrt{3} \times 500 \times 16 = 13{,}850 \text{ watts/cm}$$

$$\text{Motor output} = 30 \times 735.5 = 22{,}065 \text{ watts,}$$

which will be represented by $\frac{22{,}065}{13{,}850}$ = 1.6 cm. Line FA is extended to point T so that AT = 1.6 cm. From point T draw line TH parallel to output line O' A cutting the semicircle at H. Point H is joined with origin O and perpendicular HN is drawn.

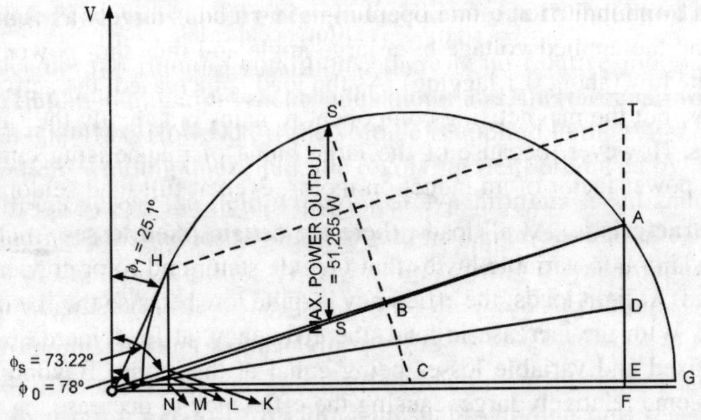

Fig. 2.35

Now from circle diagram (Fig. 2.35)

(*i*) Line current on full load = OH = 2 × 16 = 32 A **Ans.**

(*ii*) Full-load power factor, $\cos \phi_1 = \frac{NH}{OH} = \frac{1.81}{2} = 0.905$ **Ans.**

(*iii*) Motor efficiency, $\eta = \frac{KH}{NH} \times 100 = \frac{1.75}{2} \times 100 = 87.5\%$ **Ans.**

(*iv*) Maximum power is obtained by producing CB intersecting circle at S' and then dropping perpendicular from point S' on the horizontal line OF meeting output line at S. Now SS' represents the maximum power output.

So maximum power output = 3.7 × 13,850 = 51,245 watts **Ans.**

2.29. PERFORMANCE CURVES OF INDUCTION MOTORS

Steady-state operating characteristics of an induction motor, gives graphically the variation of speed (or slip), power factor, stator current and efficiency with the variation in power output from no load to full load. The performance curves of a 75 kW, 2200 V, 3-phase, 50 Hz squirrel cage induction motor are shown in Fig. 2.36.

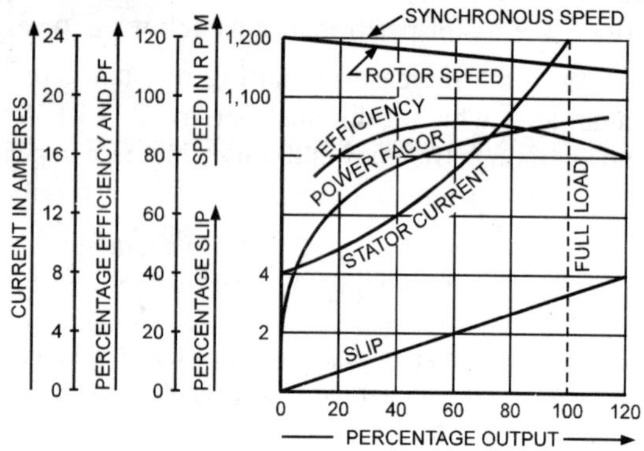

Fig. 2.36. ***Performance Curves of a 75 kw, 2,200 V, 3-phase, 50 Hz Squirrel Cage Motor***

1. Speed-Output Characteristic. The three-phase, squirrel cage induction motor like a dc shunt motor, operates at substantially constant speed from no load to full load. As the rotor cannot attain the speed of the rotating magnetic field (*i.e.* synchronous speed), it must at all times operate with a certain amount of slip. Because of the extremely low impedance of the rotor, only a slight decrease in speed is enough to cause a large increase in rotor current to develop the torque necessary to drive the load. The per cent slip at no load is less than 1 per cent while at full load it is usually 2 to 5 per cent. This small percentage change in slip (or speed) from no load to full load likewise indicates why a squirrel cage induction motor is considered a constant speed motor.

2. Power Factor Characteristic. Because of the air gap between the stator and rotor of an induction motor, the reluctance of the magnetic circuit is high. As a result an induction motor draws a large magnetising current (I_m) to produce the required flux in the air gap.

The current drawn by an induction motor operating at no load is largely a magnetising current, so no-load current lags behind the applied voltage by a large angle and thus the power factor of an induction motor at no load is quite low (about 0.1 lagging). As the load is increased, the active or power component of the current increases, but the magnetising component remains practically the same. Consequently, the power factor improves. However, because of the large value of magnetising current, which is present regardless of load, the power factor of an induction motor, even at full load seldom exceeds 90 per cent.

3. Efficiency Characteristic. At all loads, there are certain fixed losses (such as core loss, friction and windage loss) in addition to variable losses that include stator and rotor copper losses varying nearly as the square of the load. At light loads, the efficiency is quite low because the fixed losses are a relatively large part of the input. With the increase in load, the efficiency, at first, increases rapidly and attains a maximum value, the fixed and variable losses being equal at this point. Beyond this point, the copper losses (I^2 R losses) become relatively large, causing the efficiency to decrease.

The maximum efficiency occurs at about 80 to 95% of rated output, the higher values being applicable to larger motors.

4. Stator Current Characteristic. The no-load current is about 30 to 40 per cent of the rated current, the higher values being applicable to smaller motors. With increase in load, slip increases in a straight line characteristic, the rotor current will likewise increase in practically a direct proportion and cause the stator current to increase accordingly.

2.30. FACTORS GOVERNING PERFORMANCE OF INDUCTION MOTORS

The operating (or performance) characteristics of an induction motor are governed by its rotor resistance, air-gap length and shape of both stator and rotor slots and teeth.

One of the most important of the points given above is **rotor resistance.** Its value is determined primarily by the material used in construction and the area of cross section of the end rings in case of a squirrel cage induction motor. The two most widely used materials are copper and aluminium, the latter having a resistivity that is about 1.65 times as much as that of the former. An *increase in rotor resistance* affects the motor performance as below :

With the increase in starting resistance the starting torque increases until it attains the maximum value, after which it diminishes (Art. 2.11.2). An increase in rotor resistance reduces the starting current and operating efficiency and causes increase in per cent regulation. The maximum torque remains unaffected because it depends upon the rotor reactance only (Art. 2.11.1).

The **air gap** is one of the two main sources of the low power factor at which the induction motor operates. In induction motors, for constant supply voltage, the air-gap flux remains almost constant. Increase in air-gap length causes increase in magnetizing current necessary to set up the air-gap flux. This reduces the no-load power factor as well as full-load power factor of the induction motor. So in order to reduce the magnetizing current it is obvious to have air gap between stator and rotor as small as practicable (it cannot be too small because extremely small value of air gap presents mechanical difficulties) Too much reduction of the air gap may cause increase in noise, pulsation losses and magnetic pull and reduce the overload capacity.

The shapes of the rotor teeth and slots affect the reactance, which, in turn, affects the starting current and the maximum torque. When the rotor slots are deep and narrow, as opposed to shallow and wide, the rotor reactance is high because slot leakage flux is directly proportional to the slot depth. Thus induction motors with deeper slots have more leakage reactance and so less value of stating current, maximum torque and slip corresponding to maximum torque.

Further the slots may be open, semi-closed or closed. For a given slot mmf reluctance offered by (*i*) open slots is more (*ii*) semi-closed slots is less and (*iii*) closed slots is still less. Consequently the open slots have less leakage reactance than semi-closed slots, whereas the closed slots have more leakage reactance.

Open slots in an induction motor have the effect of increasing the air-gap length (1.1 to 1.2 times the actual air-gap length) as a result of which more magnetizing current is required and the operating power factor of the motor becomes poor. But an induction motor with open slots has less leakage reactance and, therefore, more starting torque and more maximum torque. Likewise an induction motor with semi-closed or closed slots needs less magnetizing current and, therefore, better power factor but reduced starting torque and maximum torque due to more leakage reactance.

The slots in the rotor are not always made parallel to the shaft, but are given a slight twist or *skew*. The motor noise and vibration, cogging and synchronous cusps can be reduced or even entirely eliminated by skewing either the stator or rotor slots. The practice usually followed in India is to skew the rotor slots. In order to eliminate the effect of any harmonic, the rotor bars should be skewed through an angle so that the bar lies under alternate harmonic poles of the same polarity (*i.e.* bars must be skewed through two pitches). The skewing of rotor slots also results in a smoother torque curve for different positions of the rotor and causes increase in effective transformation ratio between stator and rotor, increase in rotor resistance due to increased length of bars, increase in motor impedance at a given slip and also increase in slip for a given torque.

With certain relationship between the number of poles and of stator and rotor slots in cage rotor motors, peculiar behaviour may be observed when the motor is started. The motor may refuse to start or crawl at subnormal speed. So to avoid cogging and crawling of induction motor at start (*i*) the number of rotor slots should never be equal to stator slots but must either be larger or smaller (*ii*) the difference between the stator slots and rotor slots should not be equal to P, 2P, or 5 P to avoid synchronous cusps (*iii*) the difference between the number of stator slots and rotor slots should not be equal to 1,2, (P ± 1) or

(P ± 2) to avoid noise and vibration and (*iv*) the difference between the number of stator slots and rotor slots should not be equal to 3 P for 3-phase machines in order to avoid magnetic locking.

Harmonic torques can also be reduced by chording, integral slot winding and increasing air-gap length in addition to by skewing of rotor slots. Chorded windings with integral number of slots per pole per phase weakens the stator winding mmf harmonics and so eliminates the harmonic induction torques. Windings with fractional number of slots per pole per phase create asymmetrical mmf distribution around the air-gap and produces noise in the motor. So fractional slot windings are not employed for induction motor stators and only integral slot windings are employed. Though by increasing the air-gap length harmonic torques can be reduced but it will cause increase in no-load current and so in poor power factor.

2.31. HIGH TORQUE CAGE MOTORS (DEEP BAR AND DOUBLE CAGE MOTORS)

The great simplicity and robustness of the cage motor are very valuable characteristics of the motor from the stand point of cost and maintenance. The main drawback of a squirrel-cage motor is its poor starting torque, because of its low rotor resistance. If the rotor were made of high resistance (to obtain a good starting torque with reasonable starting current) its rotor copper loss will increase and, therefore, efficiency will be reduced and full-load slip (speed regulation) will increase. This also is not desirable. The difficulty with a cage rotor is that its cage is permanently short circuited, so there is no possibility of introducing external resistance temporarily in the rotor circuit at the start. Many efforts have been made to build squirrel cage motors with improved starting characteristics without sacrificing significantly its excellent running characteristics. Increased starting torque with short-circuited rotors can be obtained by using two types of rotors (either deep bar rotors or double cage rotors) which employ change of rotor circuit impedance with change in rotor frequency to have high starting torque and also high efficiency under normal running conditions.

2.31.1. Deep Bar Cage Rotor Motor

In deep bar motors use is made of skin effect with rotor bars caused by the slot leakage fluxes.

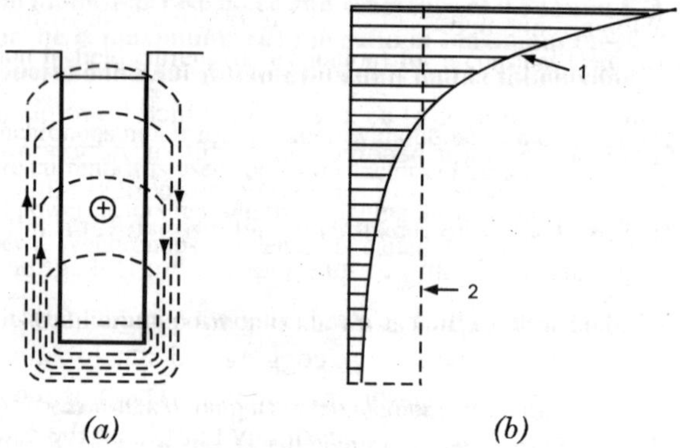

Fig. 2.37

Let us first consider the phenomena at starting. At starting the rotor frequency is equal to the supply frequency. The slot leakage flux paths of the rotor for these conditions are depicted in Fig. 2.37 (*a*). A glance at this figure shows that bar portions of different heights are linked by different number of leakage flux lines; the lower parts with greatest number of lines, the upper parts with the least number. For this reason the maximum leakage emf is induced in the lower parts of the bar, the minimum emf in the upper parts of the bar. The result is that the current in the rotor bars is forced to flow in the top layers of the bars. The current density is distributed along the conductor height as illustrated by curve 1 in Fig. 2.37 (*b*). This is called the *skin effect*. With the usual conductor height of 10-12 mm, the skin effect is almost unnoticeable. In deep bar motors, where the bars are 20-25 mm high (the depth-to-width ratio being 10 or 12 to 1), the effect is very strong and appreciably changes the rotor parameters.

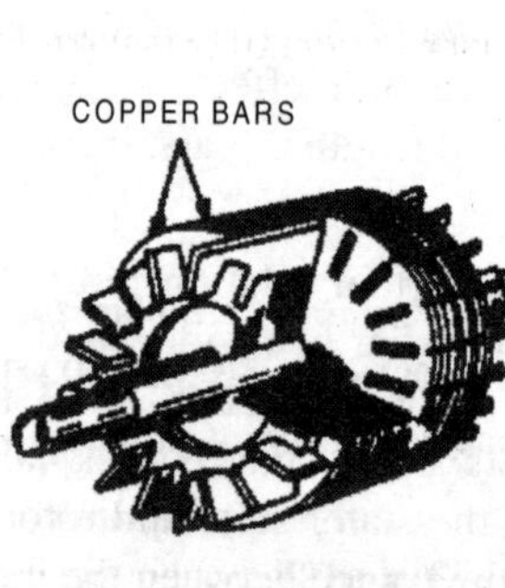

Fig. 2.37 (*c*). *Cutaway View of Deep Bar Rotor*

Non-uniform current distribution causes greater ohmic loss meaning thereby that the effective bar resistance (ac resistance) becomes much more than its dc resistance. As a result, the starting torque increases and the starting current decreases. As the motor accelerates, the frequency of rotor currents decreases to values corresponding to steady state operating conditions. The skin effect gradually diminishes, and the current becomes more uniformly distributed over the conductor cross section, as shown by dotted line 2 in Fig. 2.37 (*b*). The resistance of the rotor gradually decreases, and the inductive reactance related to the circuit frequency increases. At normal speed the deep bar motor is a plain squirrel-cage motor with the usual resistance, but with a somewhat higher inductive reactance, owing to which the power factor, efficiency and the overload capacity of the motor are somewhat reduced.

Cutaway view of deep bar rotor is given in Fig. 2.37 (*c*). The torque-speed characteristic of a deep bar rotor compared to a normal-bar design with low and high rotor resistances is illustrated in Fig. 2.38.

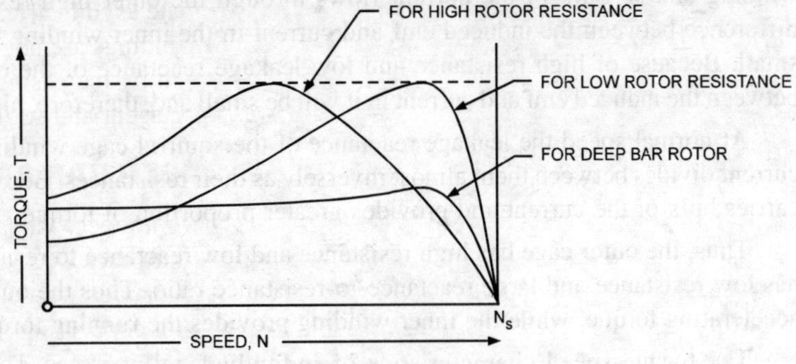

Fig. 2.38

2.31.2. Double Cage Rotor Motor

A rotor design, which though expensive gives still better starting and running characteristics than the deep-bar design, is the double cage rotor. The double squirrel cage motor is designed to provide a high starting torque with a low starting current. The rotor is so designed that the motor operates with the advantages of a high-resistance rotor circuit during starting and a low-resistance rotor circuit under running conditions. The starting torque of the double squirrel cage motor ranges from 200 to 250 per cent of full-load torque with a starting current of 400 to 600 per cent of full-load value. It is classed as a high-torque low starting current motor.

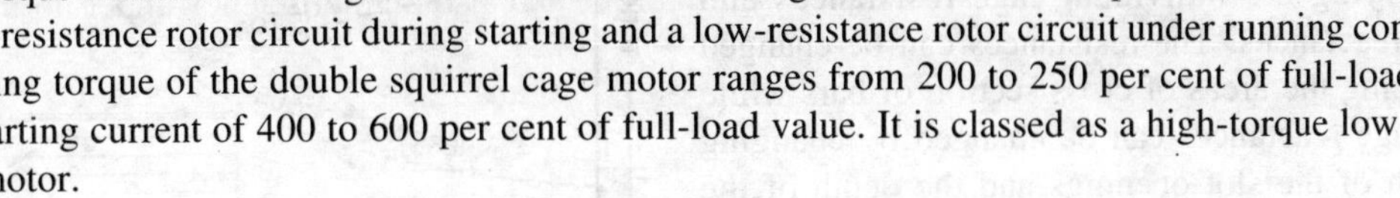

Fig. 2.39

The rotor of a double squirrel-cage motor carries two squirrel cage windings embedded in two rows of slots. The outer slots contain a high resistance and low leakage reactance winding and the inner slots contain a low resistance and high leakage reactance winding.

The outer winding bars are usually of manganese brass while the inner winding bars are of red copper. The inner winding has greater cross section than that of outer winding in order to have resistance one fourth or one fifth of outer winding. The end rings may be separate or combined.

At starting, the rotor current has the same frequency as the supply current, therefore, leakage reactance of the inner winding is high. As a result (*i*) the impedance of the outer winding is less than that of the inner winding and so bulk of the current flows through the outer high resistance winding and (*ii*) the phase difference between the induced emf and current in the inner winding will be large and the torque will be small. Because of high resistance and low leakage reactance of the outer winding, the phase difference between the induced emf and current in it will be small and, therefore, high starting torque will be developed.

At normal speed the leakage reactance of the squirrel cage windings is of minor importance and the current divides between them almost inversely as their resistances. Because of low resistance inner winding carries bulk of the current and provides greater proportion of torque.

Thus, the outer cage has high resistance and low reactance to resistance ratio, whereas the inner cage has low resistance and large reactance-to-resistance ratio. Thus the outer cage produces high starting and accelerating torque, while the inner winding provides the running torque at good efficiency.

The torque-speed characteristics of individual cages as well as resultant are shown in Fig. 2.40. It may be assumed to a first approximation, the two cages develop two separate torques, as shown in the figure. The resultant torque-speed characteristic can be modified and any desired characteristic can be obtained by modifying the individual cage resistances and leakage reactances. The resistances can be changed by changing the areas of cross section of bars while the leakage reactances can be changed by changing the width of the slot openings and the depth of the inner cage.

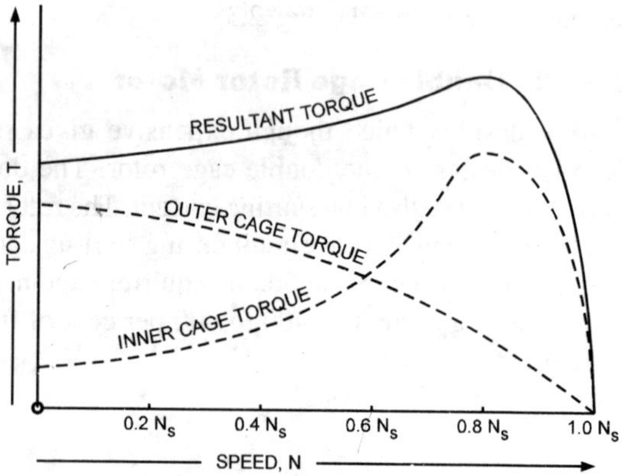

Fig. 2.40

The pull-out torque of a double cage motor is smaller than that of a plain squirrel cage motor because two cages produce their maximum torques at two different speeds. Also, the additional reactance of the inner cage reduces the full-load power factor while the high resistance of the outer cage increases the full-load copper loss thereby reducing the operating efficiency.

The cost of a double cage motor is 20 to 30 per cent higher than that of plain squirrel cage motor.

A deep-bar rotor construction is employed for higher starting torque applications and double cage construction for still higher starting requirements. For large-size motors with stringent starting torque requirements, the most expensive wound rotor construction is employed.

Example 2.36. If the standstill impedance of the outer cage of a double cage machine is 0.3 + *j* 0.4 ohm, and of the inner cage is 0.1 + *j* 1.5 ohms, compare the relative currents and torques of the two cages (*a*) at standstill (*b*) at a slip of 5 per cent.

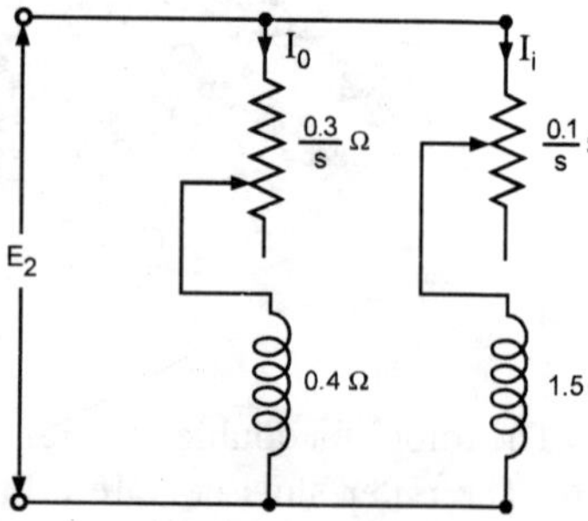

Fig. 2.41

Solution : The equivalent circuit diagram of the double squirrel cage motor is shown in Fig. 2.41.

(*a*) *At standstill*

The slip, $s = 1$

$$\text{Outer cage impedance, } \mathbf{Z}_0 = \frac{0.3}{1} + j0.4$$

$$\text{or } Z_0 = \sqrt{(0.3)^2 + (0.4)^2} = 0.5\ \Omega$$

$$\text{Outer cage current, } I_0 = \frac{E_2}{0.5} = 2\ E_2 \text{ amperes}$$

$$\text{Power input to outer cage, } P_0 = I_0^2\ R_0 = (2E_2)^2 \times 0.3 = 1.2\ E_2^2 \text{ watts}$$

$$\text{Inner cage impedance, } \mathbf{Z}_i = \frac{0.1}{1} + j\,1.5$$

$$\text{or } Z_i = \sqrt{(0.1)^2 + (1.5)^2} = 1.503\ \Omega$$

$$\text{Inner cage current, } I_i = \frac{E_2}{Z_i} = \frac{E_2}{1.503} \text{ amperes}$$

$$\frac{\text{Outer cage current}}{\text{Inner cage current}} = \frac{2\ E_2}{\dfrac{E_2}{1.503}} = 3.006 \quad \textbf{Ans.}$$

$$\text{Power input to inner cage, } P_i = (I_i)^2\ R_i = \left(\frac{E_2}{1.503}\right)^2 \times 0.1 = 0.0443 E_2^2$$

$$\frac{\text{Torque of outer cage}}{\text{Torque of inner cage}} = \frac{\text{Power input to outer cage}}{\text{Power input to inner cage}}$$

$$\text{or } \frac{T_0}{T_i} = \frac{1.2\ E_2^2}{0.0443\ E_2^2} = 27.12 \text{ or } T_0 : T_i :: 27.12 : 1 \quad \textbf{Ans.}$$

(b) At a slip of 5 per cent :

$$\text{The slip, } s = 0.05$$

$$\text{Outer cage impedance, } \mathbf{Z}_0 = \frac{0.3}{0.05} + j\,0.4 = 6 + j\,0.4\ \Omega$$

$$\text{or } Z_0 = \sqrt{(6)^2 + (0.4)^2} = 6.013\ \Omega$$

$$\text{Outer cage current, } I_0 = \frac{E_2}{Z_0} = \frac{E_2}{6.013} \text{ amperes}$$

$$\text{Power input to outer cage, } P_0 = I_0^2\ \frac{R_0}{s} = \left(\frac{E_2}{6.013}\right)^2 \times \frac{0.3}{0.05} = 0.166\ E_2^2 \text{ watts}$$

$$\text{Inner cage impedance, } \mathbf{Z}_i = \frac{0.1}{0.05} + 1.5 = 2 + j\,1.5\ \Omega$$

$$\text{or } Z_i = \sqrt{(2)^2 + (1.5)^2} = 2.5\ \Omega$$

$$\text{Inner cage current, } I_i = \frac{E_2}{Z_i} = \frac{E_2}{2.5} = 0.4\ E_2 \text{ amperes}$$

$$\frac{\text{Outer cage current}}{\text{Inner cage current}} = \frac{E_2/6.013}{0.4\ E_2} = \frac{1}{6.013 \times 0.4} = 0.416 \quad \textbf{Ans.}$$

$$\text{Power input to inner cage, } P_i = I_i^2 \frac{R_i}{s} = (0.4\,E_2)^2 \times \frac{0.1}{0.05} = 0.32\,E_2^2 \text{ watts}$$

$$\frac{\text{Torque of outer cage}}{\text{Torque of inner cage}} = \frac{\text{Power input to outer cage}}{\text{Power input to inner cage}}$$

$$\text{or } \frac{T_0}{T_i} = \frac{0.166\,E_2^2}{0.32\,E_2^2} = 0.52$$

So $T_0 : T_i :: 0.52 : 1$ **Ans.**

Example 2.37. In a double cage induction motor, if the outer cage has an impedance at standstill of (2 + *j* 1.2) ohm, determine the slip at which the two cages develop equal torques if the inner cage has an impedance of (0.5 + *j* 3.5) ohm at standstill.

Solution : Let s be the slip at which two cages develop equal torques

$$\text{Outer cage impedance, } Z_0 = \sqrt{\left(\frac{2}{s}\right)^2 + (1.2)^2}$$

$$\text{Outer cage current, } I_0 = \frac{E_2}{Z_0} = \frac{E_2}{\sqrt{\left(\frac{2}{s}\right)^2 + (1.2)^2}}$$

$$\text{Power supplied to the outer cage, } P_0 = I_0^2 \frac{R_0}{s} = \frac{E_2^2}{\left(\frac{2}{s}\right)^2 + (1.2)^2} \times \frac{2}{s}$$

$$\text{Similarly power supplied to the inner cage, } P_i = I_1^2 \frac{R_i}{s} = \frac{E_2^2}{\left(\frac{0.5}{s}\right)^2 + (3.5)^2} \times \frac{0.5}{s}$$

Since torque developed, T α power supplied

$$\text{or } \frac{T_0}{T_i} = \frac{P_0}{P_i} = \frac{\frac{0.25}{s^2} + 12.25}{\frac{4}{s^2} + 1.44} \times 4$$

$$\because T_0 = T_i$$

$$\therefore \left[\frac{0.25}{s^2} + 12.25\right] \times 4 = \frac{4}{s^2} + 1.44$$

$$\text{or } s = \sqrt{\frac{3}{47.56}} = 0.251 \text{ or } 25.1\ \%$$ **Ans.**

2.32. ADVANTAGES, DISADVANTAGES AND APPLICATIONS OF INDUCTION MOTORS

Squirrel cage induction motors are comparatively *cheaper* than wound rotor induction motors and are *simple* and *rugged* in construction.

A cage rotor needs considerably less conductor material than a wound rotor, so copper loss in cage rotor is low and *efficiency* of a cage motor is a *little more* than that of a wound rotor motor.

Cage motor is more rugged in construction and needs no slip-rings, brushes etc., therefore, its *maintenance cost is low*.

Cage motor is *explosion proof*, because the absence of slip-rings and brushes eliminates risk of sparking.

Cage motor can be *cooled better* because of its bare end rings. Also there is more space for rotor fans.

A squirrel cage induction motor has very small length of overhang, so it has low rotor overhang leakage flux. This causes standstill reactance X_2 low for a cage rotor than wound rotor. So a cage motor has *more pull-out torque, greater maximum power output* and *better operating power factor* as compared to a wound rotor motor.

Other advantages of squirrel cage motors are (*i*) nearly constant speed, high overload capacity and simple starting arrangement which may be made automatic.

The disadvantages of squirrel cage induction motors are enumerated below :

(*i*) Since in cage rotors, rotor bars are permanently short circuited, it is not possible to insert any external resistance in the rotor circuit for the purpose of increasing starting torque. Consequently its *starting torque is low* with *large starting current :* — starting torque varying between 1.5 and 2 times the full-load torque with starting current 5 to 9 times the full-load current. *Power factor* is also *low* at start in case of cage motors.

(*ii*) No possibility of speed control.

(*iii*) Squirrel cage induction motors are very sensitive to fluctuations in supply voltage.

(*iv*) The total energy lost during starting of cage motor is much more than with the slip-ring induction motor and this fact is very important where frequent starting is required.

Squirrel cage induction motor is suitable for constant speed industrial drives of small power where speed control is not required and where starting torque requirements are of medium or low value, such as for printing machinery, flour mills and other shaft drives of small power.

When the squirrel cage motors are used for crane or hoist work where large starting torque is of more importance than high efficiency owing to the intermittent nature of load, the rotor resistance is increased by employing the end rings, and sometimes the bars as well, of high resistance metal such as German silver.

Advantages of wound rotor (or slip-ring) induction motors are given below :

(*i*) *High starting torque* with low starting current. By inserting an external resistance of proper value (so that rotor resistance R_2 becomes equal to its standstill reactance X_2) in each phase of the rotor circuit, maximum torque can be obtained at start.

(*ii*) *Adjustable speed* as speed can be controlled easily in case of a wound rotor motor simply by adding external resistors in the rotor circuit. This is the reason that wound rotor induction motors are considered to be variable speed motors.

Other advantages are high overload capacity, nearly constant speed, smooth acceleration under heavy loads, no abnormal heating during starting, adoptability to automatic starting and good running characteristics after the external resistances are taken out of the rotor circuit.

The disadvantages of wound rotor induction motors are :

(*i*) Initial and maintenance costs are more than those of a squirrel cage motor, because a slip-ring induction motor needs slip rings, brushes, short-circuiting devices etc.

(*ii*) Speed regulation is poor when operated with external resistances in the rotor circuit.

(iii) Lower efficiency and power factor as compared with cage rotor motor.

(iv) Low power factor at light loads.

(v) Sensitivity to fluctuations in supply voltage.

Wound rotor (or slip-ring) induction motors are used for loads requiring severe starting conditions or for loads requiring speed control such as for driving line shafts, lifts, pumps, generators, winding machines, cranes, hoists, elevators, compressors, small electric excavators, printing presses, turn tables, strokers, large ventilating fans, crushers etc.

2.33. COMPARISON OF SQUIRREL CAGE AND SLIP-RING (PHASE WOUND) INDUCTION MOTORS

S.No.	*Characteristics*	*Squirrel Cage Induction Motor*	*Slip-Ring Induction Motor*
1.	Construction	Simple and rugged	Needs slip rings, brushes, short-circuiting device etc.
2.	Overhang	Less	Large
3.	Space factor in slots	Better	Poor
4.	Cost (initial as well as maintenance)	Less	More
5.	Copper losses	Small	More
6.	Efficiency	High (only for machines, not designed for high starting torque)	Low
7.	Power factor	Better	Poor
8.	Cooling	Cooling better because of its bare end rings and availability of more space for rotor fans	Not quite efficient
9.	Speed regulation	Better	Poor when operated with external resistances in the rotor circuit
10.	Starting	Simple	The motor needs slip rings, brushgear, short-circuiting device and starting resistors etc.
11.	Starting torque	Starting torque is low with large starting current	Possibility of increasing starting torque by insertion of external resistances in the rotor circuit
12.	Power factor at start	Poor	Can be improved
13.	Speed control	No possibility	Possible by insertion of external resistor in the rotor circuit
14.	Protection against explosion	Explosion proof	Not explosion proof

2.34. EFFECTS OF OPERATING CONDITIONS

(i) *Effect of Loading.* With the increase in load, speed falls and consequently torque increases. A point will reach when the maximum torque will be developed. A further increase in load causes further drop in speed, consequently the driving torque decreases and the rotor stops ultimately.

(ii) *Effect of Unbalance Supply Voltage.* If the 3-phase supply to the motor is not balanced, the rotating magnetic field will move at a non-uniform rate and in a non-uniform strength. This will cause more unbalanced currents in the stator windings and, therefore, produce an unequal heating.

(*iii*) *Effect of Break in One Phase.* If one phase of a polyphase induction motor breaks, may be due to any reason, when the motor is in operaton, the motor will continue to operate single phase, provided that the load does not exceed 57.7% of the normal rating, with about the same temperature rise as when carrying rated load as a 3-phase induction motor. It will not, however, start, single phase but if brought up to speed by some external means, such as pulling on the belt, it will continue to operate satisfactorily on load below 57.7% of rated one.

A break in one phase of the rotor of a wound rotor induction motor may prevent it from starting but if it is brought up to speed before the circuit is opened, it will operate under reduced load but usually with considerable vibrations.

(*iv*) *Effects of Variation in Line Frequency and Line Voltage.* It is usual for manufacturers to guarantee satisfactory operation of motors with only slight altered characteristics with voltage or frequency variations amounting to not more 10% above or below normal, provided, however, that when both voltage and frequency vary at the same time the combined variation shall not exceed 10% above or below.

Since the speed of an induction motor depends upon the speed of the revolving field, which in turn depends upon the line frequency, hence any variation in line frequency will cause corresponding changes in motor speed. But with modern power systems there is seldom any marked change in frequency, hence no trouble need be anticipated on this account. It should be understood, however, that the motors designed for a certain commercial frequency should not be operated upon a system of decidedly different frequency, otherwise aside from the resulting change in speed there would be changes in the torque which the motor would develop and in the operating temperature of its various parts and thus motor operation will not be satisfactory.

Variation in	Synchronous speed	%age slip	Full-load speed	Starting current	Full-load current	Starting torque and maximum running torque
Supply voltage, V	Constant	$\alpha \frac{1}{V^2}$	Decreases with the decrease in V	α V	Increases with decrease in V	αV^2
Supply frequency, f	αf	Constant	αf	$\alpha \frac{1}{f}$	Increases slightly with decrease in f	$\alpha \frac{1}{f^2}$

With voltages below normal the torque developed is reduced in proportion to the square of the voltage; therefore, with a below-normal supply voltage the motor may not be able to start its customary load or may become stalled on a small overload. If the motor is operated with its rated kW load at reduced voltage, the strength of the stator field will be weaker comparatively, resulting in decrease in speed as well as in core losses but increase in stator and rotor currents and the operating temperature may become abnormal owing to increased copper losses. If, on the other hand, the motor is operated on line voltage above normal, the stator field strength will be increased resulting in increase in speed as well as core losses but decrease in input current and, therefore, reduction in copper losses. The tendency in either case is to overheat some part of the machine even though the temperature as a whole does not become abnormal.

The variation in supply frequency and supply voltage will not affect only the speed and torque of the motor but also the starting current, full-load current, starting torque, maximum torque (as given below) and operating temperature of various parts of the machine and thus make their operation unsatisfactory.

2.35. INDUCTION GENERATOR

If a polyphase induction motor while connected to a constant voltage and frequency mains is mechanically coupled to a prime mover and is driven by it at a speed higher than synchronous speed, the rotor overtakes the rotating magnetic field and the rotor conductors cut the magnetic flux in a reverse direction to motor rotation. As a result, the emfs and currents in the rotor will reverse their direction. Consequently the reaction force between rotating field and rotor current also changes its direction and opposes the rotation. This all has been illustrated in Fig. 2.42. To maintain this rotation mechanical energy is required to be delivered to the rotor; the magnetising current remaining the same, since the rotating field conditions are the same in an induction generator and motor. The power of the machine under such conditions is negative *i.e.* the machine delivers electrical energy to the mains instead of taking it from the mains. The slip of an induction machine operating as a generator is negative* and the direction of its emf now coincides with the active component of current instead of being opposed to it, as is the case in a motor. It is to be noted that this latter circumstance may serve as a characteristic feature of a machine passing from motor to generator action.

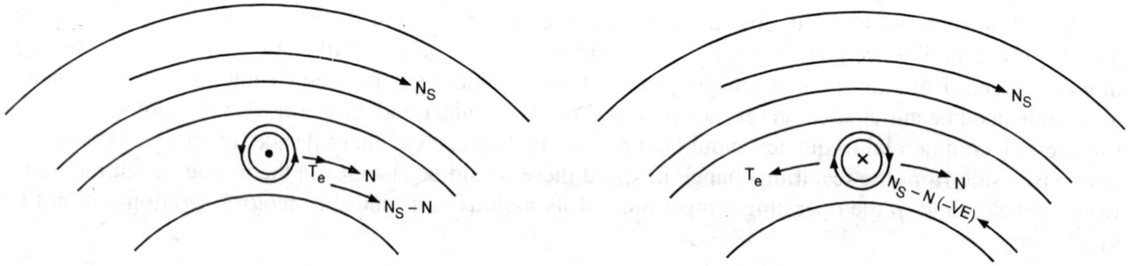

(a) Motoring Action (s Positive) *(b) Generating Action (s Negative)*

Fig. 2.42. ***Motoring and Generating Action***

At super-synchronous speeds, the conditions of an induction generator are electrically similar to those at subsynchronous speeds for induction motor. In this connection it is to be noted that the equivalent circuit and all the characteristic equations of the induction motor also apply to the induction generator when the slip s is introduced with a negative sign.

With the increase in negative slip, the inductive reactance of the rotor increases and as a result the phase shift between the emf and current in the rotor also increases. On the other hand, the rotor emf increases in proportion to the slip, and since the braking torque of an induction generator is expressed in the same way as the torque of a motor

$$T \propto \phi_2 I_2 \cos \phi_2$$

it follows that in the same way as for a motor with slip corresponding to the condition $s = R_2/X_2$ there will be a maximum torque above which the operation of the machine will be unstable. The maximum value of torque which the induction generator can develop, thus, limits its range of operation. The torque-slip characteristics of the machine as a generator resemble to those of the motor (Fig. 2.43).

An induction generator supplies an active current to the mains, but at the same time, it receives from the mains a lagging reactive current (magnetising current), in the same way as a motor. That is why, it requires a source of reactive energy to obtain this current. Such a source may be a synchronous generator operating in parallel with an induction generator and feeding into a common mains. Therefore, an induction generator cannot operate independently, and, operating in parallel and feeding into the mains, it reduces the power factor through its reactive current. However, the necessary reactive current may be supplied to an induction generator by static capacitors connected in parallel with it, as illustrated in Fig. 2.44. The terminal voltage

* In generator operation, the actual rotor speed N exceeds the synchronous speed N_s so the slip s, by equation $s = N_s - N$, becomes negative.

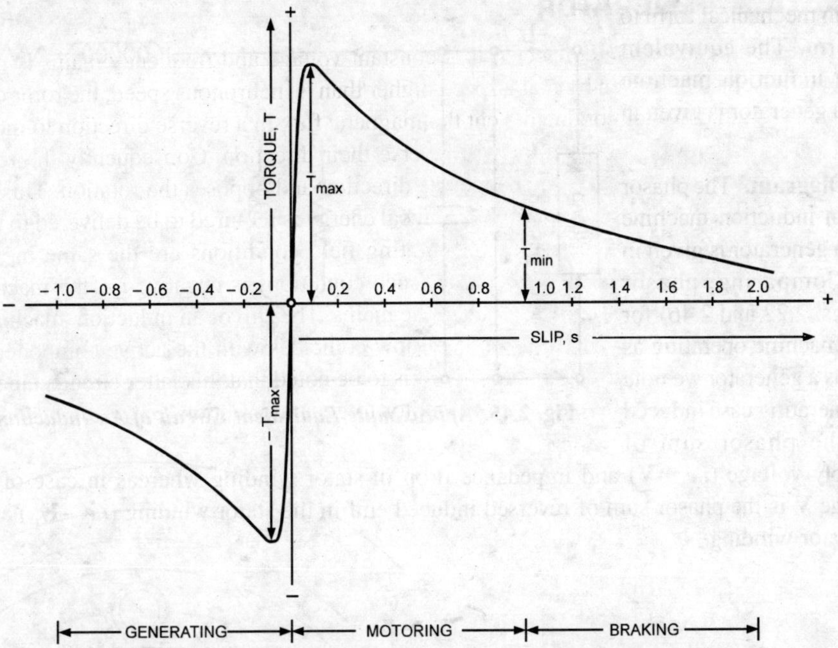

Fig. 2.43. ***Torque-Slip Characteristics of Induction Machine***

increases with the capacitance. If the capacitance is insufficient, the generator will not build up voltage. Hence capacitor bank must be large enough to supply the reactive power normally drawn by the machine as a motor. In such an arrangement (shown in Fig. 2.44), when the induction generator is started, self excitation of the machine occurs at the expense of the residual magnetism in it.

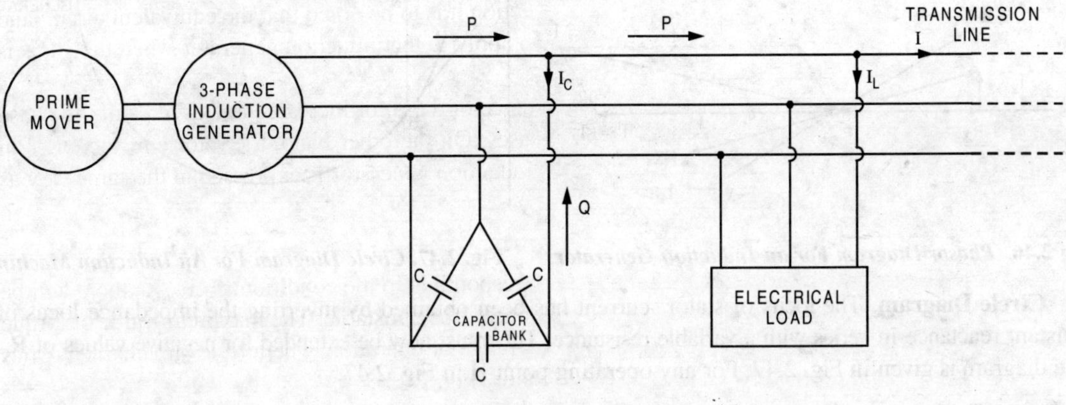

Fig. 2.44.

Equivalent Circuit. In the equivalent circuit shown in Fig. 2.29, the induction motor is depicted as a transformer delivering power to an equivalent load resistance $R'_2\left(\frac{1}{s}-1\right)$ which represents the power converted into mechanical work. For negative value of *s* this load resistance is also negative being equal to $R'_2\left(\frac{1}{-s}-1\right)$ *i.e.* $-R'_2\left(\frac{1+s}{s}\right)$, that is, it no longer absorbs power but is really a source or in other words there is a transfer of energy from mechanical form to electrical form. The equivalent circuit of an induction machine operating as a generator is given in Fig. 2.45.

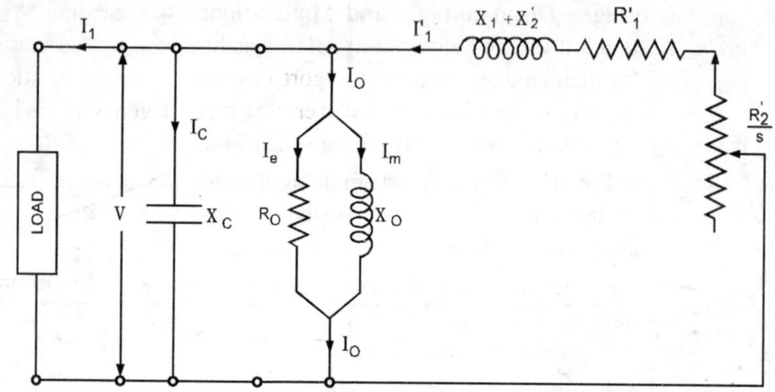

Fig. 2.45. *Approximate Equivalent Circuit of An Induction Generator*

Phasor Diagram. The phasor diagram of an induction machine operating as a generator is given in Fig. 2.46. Comparing phasor diagrams (Figs. 2.22 and 2.46) for an induction machine operating as a motor and as a generator we note that in the generating case induced emf $\mathbf{E}_1$ is the phasor sum of reversed supply voltage (*i.e.* –**V**) and impedance drop in stator winding whereas in case of motoring the applied voltage **V** is the phasor sum of reversed induced emf in the stator winding (*i.e.* $-\mathbf{E}_1$) and impedance drop in the stator winding.

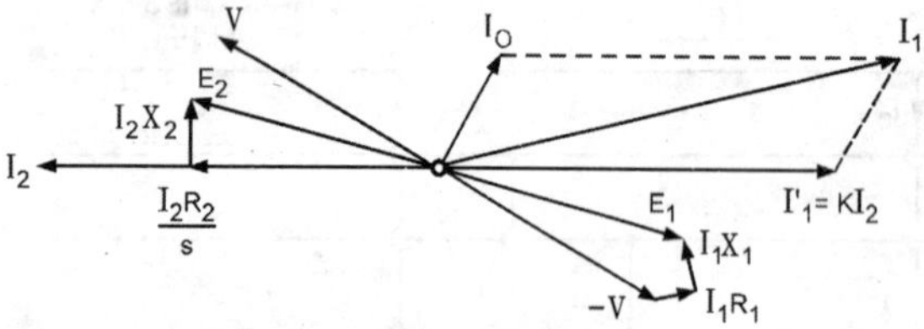

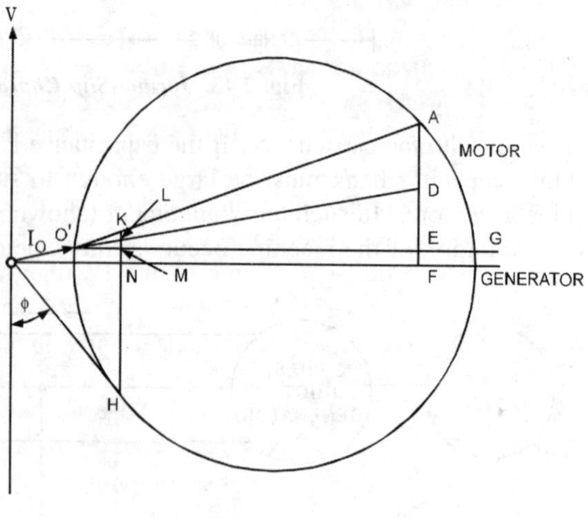

Fig 2.46. *Phasor Diagram For an Induction Generator*

Fig. 2.47. *Circle Diagram For An Induction Machine*

Circle Diagram. The locus of stator current has been obtained by inverting the impedance locus of a constant reactance in series with a variable resistance. This must now be extended for negative values of R_2/s. The diagram is given in Fig. 2.47. For any operating point H in Fig. 2.47.

$$\text{Total power delivered to generator} = \sqrt{3}\ V.\ KH$$

$$\text{Rotor copper loss} = \sqrt{3}\ V.\ KL$$

$$\text{Stator copper loss} = \sqrt{3}\ V.\ LM$$

$$\text{Stray losses} = \sqrt{3}\ V.\ MN$$

$$\text{Output} = \sqrt{3}\ V.\ NH$$

$$\text{Slips, } s = KL/LH$$

$$\text{Power factor, } \cos\phi = \frac{NH}{OH}$$

Advantages, Disadvantages and Applications. The distinct advantage of the induction generator lies in the fact that it does not hunt or drop out of synchronism, it is simple and rugged in construction, cheaper in cost, easy in maintenance, and when short circuited it delivers little or no sustained power,* because its excitation quickly becomes zero. However, this type of generator is little used to supply commercial power because of the distinct superiority of the synchronous generators. Its main drawbacks are :

1. As already mentioned, it cannot be operated independently and the method of providing self excitation by connecting suitable capacitors across its terminals is of limited practical importance in view of the magnitude of capacitors required.
2. The other drawback is that it can deliver only leading current whereas most commercial loads require lagging current. Considerable kVA capacity synchronous apparatus is required to supply the total quadrature (lagging) current.
3. Induction generators are not helpful in system stability than the addition of parallel reactors with a rating equal to the generator magnetising reactance.
4. Dangerously high voltages may occur with induction generators over long transmission lines if the synchronous machines at the far end become disconnected and the line capacitance excites the induction machines. Motors with power factor correction capacitors may generate if the capacitors remain connected to the running machine when for some reason the supply is cutoff.

It is highly important to prevent shock loads from exceeding the maximum generating torque of the machine or causing the overcurrent trips to function during periods of regeneration.

The induction generator is useful when the prime mover does not run at constant speed such as in hydroelectric power stations having a variable low-head water supply. Where the generator and turbine run submerged in a river to utilise tidal power, induction generators might be considered. The utilisation of wind power is another suitable application for this type of generator since the stator frequency depends on that of paralleled synchronous machines, not on the rotor speed.

The induction generator is very useful for installation in small power stations where it can be operated in parallel and feeding into a common mains without attendent. It does not require synchronising and requires no separate exciter but floats on the line delivering power when the water supply is sufficient to drive it above synchronous speed, otherwise running idle as a motor.

The induction generator is very useful for braking purposes in railway work. If the induction motor be left connected across the line on a downgrade; any tendency of the train to drive it above synchronous speed will cause generator action. In addition to providing effective braking the energy is fed back to the ac mains and thus main generating station is relieved from some of its load. The machine requires no complicated control apparatus when used for regenerative braking, such as is required in case of dc motors operating under similar conditions.

When used on such applications as hoists, the polyphase induction machine supplies very effective over–speed brake since a comparatively small increase in speed causes full generator action to take place exerting a strong braking action, besides furnishing reverse power to the supply line. Since the retarding torque ceases at synchronous speed, the braking from synchronous speed to standstill must be accomplished by some other means.

2.35.1. Self Excited or Isolated Induction Generator

An isolated induction generator supplying a load is shown in Fig. 2.48. The capacitor bank connected across the stator terminals of induction generator provides the lagging reactive power of both the induction generator as well as load as shown in Fig. 2.48. The terminal voltage increases with the increase in capacitance

* An induction generator delivers an instantaneous 3-phase short-circuit current equal to terminal voltage divided by its standstill reactance, but its rate of decay is much faster than that of a synchronous generator of the same rating, and its sustained short-circuit current is zero.

C. If the capacitance is insufficient, the generator will not build up voltage. Hence the capacitor bank must be large enough to supply the reactive power normally drawn by the machine as motor. In such an arrangement (shown in Fig. 2.48), when the induction generator is started, self excitation of the machine occurs at the expense of residual magnetism in it.

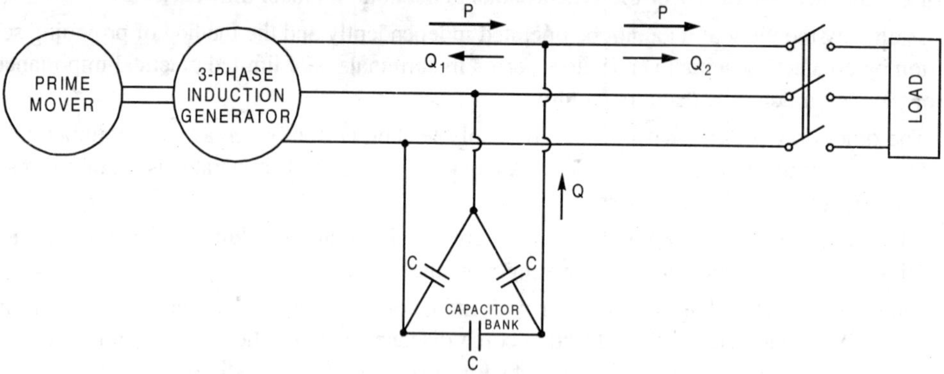

Fig. 2.48

Capacitance reactive power, Q = Reactive power required by 3-phase induction generator, Q_1 + Reactive power required by the load, Q_2

In a self-excited induction generator, stator terminal voltage depends upon its magnetization curve as shown in Fig. 2.49. The reactive current of a capacitor is dependent on the voltage across it terminals *i.e.* capacitor voltage $V_C = \omega CI$. In Fig. 2.50, load lines of capacitor banks having different capacitances $C_1 > C_2 > C_3$ are drawn.

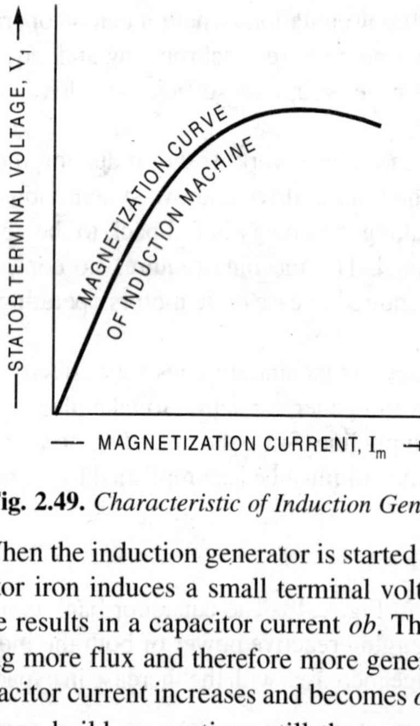

Fig. 2.49. *Characteristic of Induction Generator*

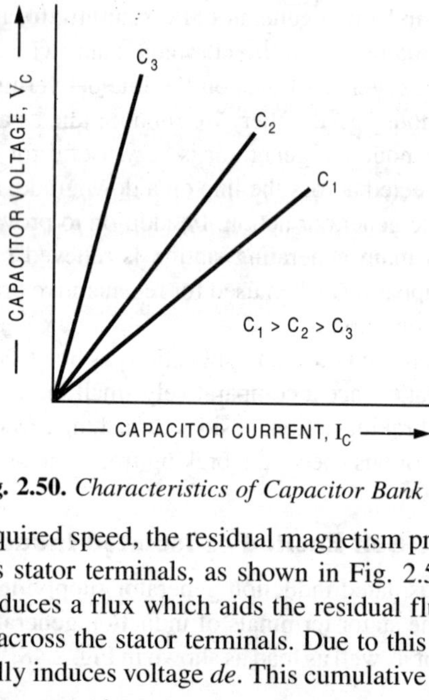

Fig. 2.50. *Characteristics of Capacitor Bank*

When the induction generator is started and run at the required speed, the residual magnetism present in the rotor iron induces a small terminal voltage *oa* across its stator terminals, as shown in Fig. 2.51. This voltage results in a capacitor current *ob*. This current ob produces a flux which aids the residual flux, thus creating more flux and therefore more generated voltage *bc* across the stator terminals. Due to this voltage *bc* capacitor current increases and becomes *od* which eventually induces voltage *de*. This cumulative process of voltage build up continues till the capacitor load line intersects the saturation curve of the induction generator at point *f*, thus giving a no-load induced emf *gf* for magnetizing current I_{m_1}. The noteworthy point is that the induction generator builds up voltage across its stator terminals like a dc shunt generator.

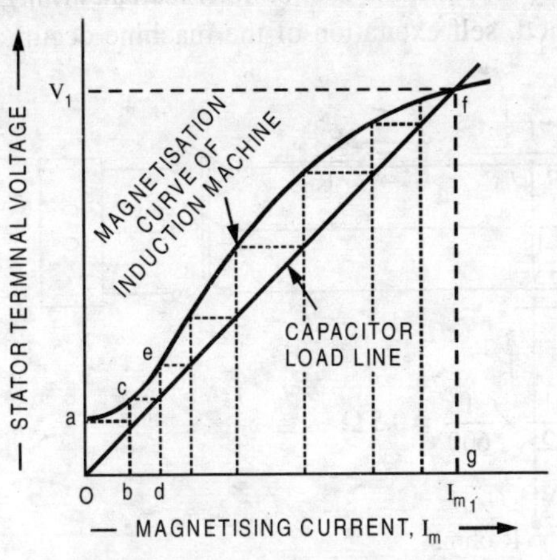

Fig. 2.51. ***Building-up Voltage in Induction Generator at No Load***

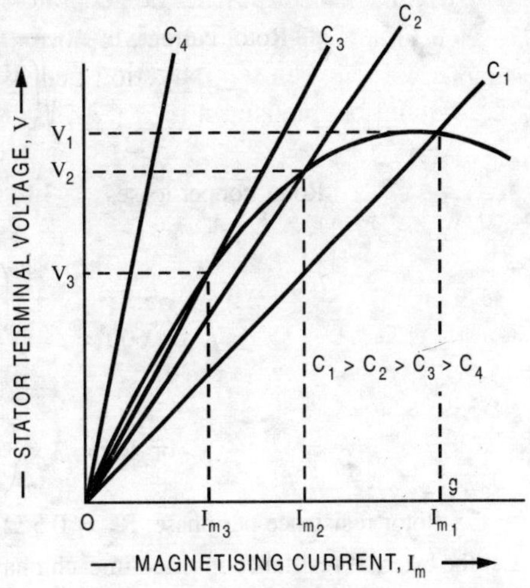

Fig. 2.52. ***No-load Voltages For Different Capacitances***

In case of absence of residual flux in the rotor iron of induction generator there would be no voltage build up. This problem can, however, be solved by operating the induction machine as a 3-phase induction motor for sometime to induce residual magnetism.

The voltage build up depends upon the value of capacitor as shown in Fig. 2.52. Higher the value of capacitor, more is the voltage build up. In case the capacitor load line does not intersect the magnetization curve of induction generator, there would be no voltage build up, as in case of capacitor C_4.

Induction generator is quite suitable for generation of electrical energy in coastal areas, where wind energy is available in abundance.

Example 2.38. An induction generator having a synchronous speed of 750 rpm is driven at 800 rpm and with the rotor short-circuited, the stator output is 9 kW. Determine *(a)* the rotor copper loss *(b)* the rotor resistance per phase, given that the voltage across slip rings at standstill equals 260 V. Neglect stator resistance and rotor reactance. *(c)* By the addition of external resistance in the rotor circuit, the stator output is reduced to 4 kW. Determine the value of the external resistance inserted per phase.

Solution : Alternator speed, N = 800 rpm

Synchronous speed, N_s = 750 rpm

$$\text{Slip, } s = \frac{N_s - N}{N_s} = \frac{750 - 800}{750} = -\frac{1}{15}$$

Stator output = 9 kW = 9,000 watts

Rotor output = Stator output + stator copper losses + stray losses

Since stator resistance is negligible, stator copper losses are negligible and also neglecting stray losses

Rotor output = Stator output = 9,000 watts

$$\therefore \text{Slip, } s = \frac{\text{Rotor copper losses}}{\text{Rotor output}}$$

$$\therefore (a) \text{ Rotor copper losses} = \text{Rotor output} \times \text{slip} = 9{,}000 \times \frac{1}{15} = 600 \text{ watts } \textbf{Ans.}$$

Induced emf per phase in the rotor at standstill,

$$E_2 = \frac{260}{\sqrt{3}} \text{ V} \qquad \text{assuming rotor star-connected}$$

$$\text{Rotor current, } I_2 = \frac{E_2}{\sqrt{\left(\frac{R_2}{s}\right)^2 + (X_2)^2}} = \frac{E_2}{R_2/s} \qquad \because \text{ rotor reactance is negligible}$$

$$\text{Rotor copper losses} = 3\, I_2^2\, R_2 = 3 \times \frac{E_2^2}{R_2^2/s^2} \times R_2 = \frac{3E_2^2\, s^2}{R_2}$$

$$\text{or } 600 = \frac{3\times\left(\frac{260}{\sqrt{3}}\right)^2 \times \frac{1}{225}}{R_2}$$

$$\text{or } R_2 = 3 \times \left(\frac{260}{\sqrt{3}}\right)^2 \times \frac{1}{225} \times \frac{1}{600} = 0.5\ \Omega$$

∴ Rotor resistance per phase, $R_2 = 0.5\ \Omega$ **Ans.**

(*c*) Let the external resistance inserted in each phase of rotor be R ohms

$$\text{Rotor copper losses} = \frac{3E_2^2 s^2}{R_2 + R} = \frac{3\times\left(\frac{260}{\sqrt{3}}\right)^2 \times \left(\frac{1}{15}\right)^2}{R_2 + R} = \frac{300}{R_2 + R} \text{ watts} \qquad ...(i)$$

$$\text{Also rotor copper losses} = \text{Slip} \times \text{rotor output} = \frac{1}{15} \times 4{,}000 = 266.66 \text{ watts} \qquad ...(ii)$$

∵ rotor output = stator output, stator losses being negligible

Equating Eqs. (*i*) and (*ii*) we have

$$\frac{300}{R_2 + R} = 266.66$$

$$\text{or } R_2 + R = \frac{300}{266.66} = 1.125\ \Omega$$

or external resistance to be inserted in each phase,

$$R = 1.125 - R_2 = 1.125 - 0.5 = 0.625\ \Omega$$

HIGHLIGHTS

1. An induction motor is simply an electric transformer whose magnetic circuit is separated by an air gap into two relatively movable portions, one carrying the primary winding and the other the secondary winding.

It essentially consists of (*i*) a stator and (*ii*) a rotor.

Rotors are of two types (*i*) squirrel-cage rotor and (*ii*) phase-wound or wound rotor.

2. In a polyphase induction motor, slip-frequency currents are induced in the rotor windings as the rotor slips past the synchronously rotating stator flux wave. These rotor currents in turn produce a flux wave which rotates in synchronism with the stator flux wave, the torque is produced from the interaction of these two flux waves. For increased load on the motor, the rotor speed falls, resulting in larger slip and increased induced rotor currents, and greater torque.

Examination of the flux-mmf interactions in a polyphase induction motor shows that, electrically, the machine is a generalized transformer. The synchronously rotating air-gap flux wave in the induction machine is the counterpart of the mutual core flux in the transformer. The rotating field induces emfs of stator frequency in stator windings and of slip frequency in the rotor windings for all rotor speeds other than synchronous speed. Thus, the induction machine transforms voltages and at the same time changes frequency. When viewed from the stator, all rotor electrical and magnetic phenomena are transformed to stator frequency. The rotor mmf reacts on the stator windings in the same way that the mmf of secondary current in a transformer reacts on the primary.

3. *Working principle.* When a 3-ϕ supply is provided to a 3-phase wound stator of an induction motor, a rotating magnetic field is established rotating at synchronous speed. At start, stationary rotor conductors cut across the revolving magnetic field and an emf is induced in them by the electromagnetic phenomenon. Current flows through the rotor conductors, as they form a closed path, and so rotor field is developed. By interaction of stator magnetic field and rotor field, torque develops and causes the rotation of rotor in the same direction as that of revolving magnetic field.
4. The direction of rotation of a 3-phase induction motor can be reversed by interchanging the connections to the supply of any two terminals.
5. *Synchronous speed,* $N_s = \frac{120\,f}{P}$, f is supply frequency and P is the number of poles on stator.
6. **Slip.** The rotor never attains the synchronous speed because at that speed there would be no relative speed between rotor conductors and stator revolving field and induction phenomenon is not possible. Its speed is always less than synchronous speed.

   The difference between the speed of the stator field, known as synchronous speed (N_s), and the actual speed of the rotor (N) is known as the slip and is denoted by s

$$s = \frac{N_s - N}{N_s}.$$

7. Both the stator and rotor fields rotate synchronously, which means that they are stationary w.r.t. each other at all possible rotor speeds.
8. Frequency of rotor current or emf, $f' = sf$.
9. Rotor emf at start, $E_2 = E_1 \times \frac{N_2}{N_1}$.
10. Rotor current at start, $I_2 = \frac{E_2}{\sqrt{R_2^2 + X_2^2}}$.

    Rotor current at slip s, $I_2' = \frac{sE_2}{\sqrt{R_2^2 + (sX_2)^2}}$.
11. Starting torque, $T_{st} = \frac{KR_2E^2}{R_2^2 + X_2^2} = \frac{K_1R_2}{R_2^2 + X_2^2}$, if applied voltage is constant.

    Torque under running condition, $T = \frac{K_1 sR_2 E_2^2}{R_2^2 + s^2 X_2^2} = \frac{K_2 sR_2}{R_2^2 + s^2 X_2^2}$, if applied voltage is constant.
12. Condition for maximum torque under running condition

$$R_2 = s\,X_2.$$

13. Slip corresponding to maximum torque, $s_{\max T} = \frac{R_2}{X_2}$.
14. Condition for maximum starting torque, $R_2 = X_2$.
15. Maximum torque, $T_{max} \propto \frac{1}{X_2}$.
16. One of the salient facts affecting induction motor applications is that the slip at which maximum torque occurs can be controlled by varying the rotor resistance. A high rotor resistance gives optimum starting conditions but poor running performance. A low rotor resistance, on the other hand, may result in unsatisfactory starting conditions.
17. Starting torque of an induction motor is proportional to the square of applied voltage.
18. When the induction motor is running with slip s, the torque acting on the rotor,

$$T \propto s \times (\text{supply voltage})^2 \propto s\,V^2.$$

19. A change of frequency will give rise to a change of flux level unless the induced emf is changed in the same ratio. An unbalance in this ratio will cause either an excessive flux and saturation, or diminished flux and reduced torque per ampere of current.

With the lowering of frequency, keeping $\frac{V}{f}$ constant, the shape of torque-slip (or torque-speed) curves do not change, but displaced horizontally.

20. $$\frac{T_f}{T_{max}} = \frac{2}{\frac{s_{max\,T}}{s_f} + \frac{s_f}{s_{max\,T}}}.$$

21. $$\frac{T_{st}}{T_{max}} = \frac{2R_2/X_2}{\left(\frac{R_2}{X_2}\right)^2 + 1} = \frac{2a}{a^2+1}.$$

22. The rotor current of a 3-phase induction motor reduces slightly with the slight increase in speed but with large increase in speed the current drops rapidly (almost nearly) until at synchronous speed the rotor emf and rotor current are both zero. This is because at high speed, the slip s is low and so the rotor reactance (sX_2) becomes small as compared to rotor resistance R_2 and the rotor current becomes proportional to rotor induced emf which is proportional to emf.

23. The value of breakdown torque varies with the design of motor but ranges from 200 to 300% of full-load torque in standard squirrel cage motors.

24. Gross torque, $T_g = \frac{P_{mech}}{2\pi N/60} = 9.55\,\frac{P_{mech}}{N}$ Nm

Shaft torque, $T_{sh} = \frac{P_{output}}{2\pi N/60} = 9.55\,\frac{P_{output}}{N}$ Nm where $P_{output} = P_{mech}$ – copper losses

Power Relations

$$\text{Rotor copper loss} = s \times \text{power input to rotor}$$

$$\frac{\text{Rotor copper loss}}{\text{Gross rotor output}} = \frac{s}{1-s}$$

$$\frac{\text{Gross rotor output}}{\text{Rotor input}} = \frac{N}{N_s} = 1 - s$$

Power input to rotor : gross rotor output power : rotor copper loss

i.e. $P_2 : P_{mech} : P_{cu} :: 1 : (1-s) : s$

25. Torque developed by rotor is given by

$$T = \frac{3 \times 60}{2\pi N_s} \cdot \frac{R_2 s E_2^2}{R_2^2 + s^2 X_2^2}.$$

26. Equivalent electrical load of resistance R_L' is given as

$$R_L' = \frac{R_2}{K^2}\left(\frac{1}{s} - 1\right).$$

27. No load test is performed to determine no-load current I_0, no-load pf $\cos\phi_0$, windage and friction losses $P_{\omega f}$, no-load core loss P_i, no-load power input P_0, and no-load resistance R_0 and reactance X_0.
Blocked rotor test is performed for determination of short-circuit current I_{sc} with normal voltage applied to stator, power factor on short-circuit, total equivalent resistance and reactance of the motor as referred to stator (R_{01} and X_{01}).

28. The operating characteristics of an induction motor can be computed by using a circle diagram easily and conveniently. The locus of stator load current I' is a circle of diameter $\frac{V}{X_{01}}$ and having coordinate $\left(\frac{V}{2X_{01}}, 0\right)$ with respect to O'.

29. Increased starting torque with short-circuited rotors can be obtained by using two types of rotors (either deep bar rotors or double cage rotors).

In deep-bar rotors use is made of skin effect with rotor bars caused by the slot leakage fluxes.

The rotor of a double cage motor carries two squirrel cage windings embedded in two rows of slots. The outer slots contain a high resistance and low leakage reactance winding and the inner slots contain a low resistance and high leakage reactance winding.

Double cage rotor motor is classed as a high-torque, low-starting current motor and costs approximately 20 to 30 per cent higher than that of plain squirrel cage motor.

30. For applications requiring a substantially constant speed without excessive severe starting conditions, the squirrel cage induction motor usually is unrivaled because of its ruggedness, simplicity, and relatively low cost. Its only disadvantage is its relatively low pf (about 0.85 to 0.90 at full load and considerably lower at light loads and for lower-speed motors). A wound-rotor motor can be used for very severe starting conditions or when speed control by rotor resistance is required. A wound-rotor motor is more expensive than a squirrel cage motor.

31. Induction generator is often called asynchronous generator, because its speed is not in synchronism with line frequency. It needs a source of reactive energy to provide magnetising (lagging reactive) current to it, so it cannot operate independently. Induction generator is very useful for braking purpose in railway work and in providing effective overspeed brake when an induction motor is used on such applications in hoists.

EXERCISES

1. Describe with neat sketches the construction and principle of operation of a 3-phase cage type induction motor. **[M.D. Univ. Electrical Machines-II, May 2009]**

2. Bring out clearly, with the help of neat sketches the differences between the 3-phase slip-ring induction motor and the three- phase squirrel cage induction motor.

3. Sketch a layout of squirrel cage induction motor. Label all the parts and material used for each part. Explain why rotor of an induction motor can never attain synchronous speed.

4. Give the description of the construction of a squirrel cage induction motor. Also explain the principle of operation. **[M.D. Univ. December 2005]**

5. Explain the principle of operation of a 3-phase induction motor. **[M.D. Univ. May 2004]**

6. Explain how torque is produced in a 3-phase induction motor. **[M.D. Univ. December 2004]**

7. Explain why the rotor of an induction motor rotates. Is it possible for the rotor of an induction motor to run at synchronous speed ? Give reasons. **[M.D. Univ. December 2006, May 2008]**

8. Define slip. Why cannot an induction motor run at synchronous speed ? **[M.D. Univ. Electrical Machine-II, December 2010]**

9. Discuss principle of operation of 3-phase induction motor. What is the frequency of rotor current? Discuss.

10. Explain the terms slip, slip frequency, wound rotor and cage rotor in induction motor.

11. Derive an expression for the frequency of rotor current is an induction motor. **[M.D. Univ. Electrical Machines-II, December 2010]**

12. Develop an expression to show that the torque developed in a 3-ϕ induction motor depends upon the supply voltage, rotor reactance and slip. Also obtain the conditions of maximum torque.

13. Drive condition for maximum torque and also the expression for maximum torque for a 3-phase induction motor. **[M.D. Univ. May 2007]**

14. Explain the principle of working of polyphase induction motor. What is slip in an induction motor ? Explain why slip is never zero in an induction motor? Also draw torque-slip characteristics. **[M.D. Univ. December 2007]**

15. Explain the torque-slip characteristics of an induction motor. **[M.D. Univ. December 2006]**

16. Explain the torque-slip characteristics of an induction motor. Why does an induction motor run at a speed less than the synchronous speed. **[M.D. Univ. December 2005]**

17. Show how increase in rotor circuit resistance will increase the starting torque of an induction motor. State the maximum limit of such increased resistances.

18. Draw the torque-slip characteristic of 3-phase induction motor and show with suitable derivations that maximum torque is independent of rotor resistance.

19. Sketch the torque-slip characteristic of a 3-phase induction motor, indicating therein the starting torque, maximum torque and operating region. How do starting and maximum torques vary with rotor resistance ?

[M.D. Univ. Electrical Machines-II, May 2009]

20. Discuss the effect of variation of rotor resistance and supply voltage on the torque-slip characteristics.

21. Show that in induction motor, the rotor always run in the direction of stator field.
Sketch the torque-speed characteristic of an induction motor working at normal voltage and frequency. Indicate starting torque, pull-out torque, and normal operating region of induction motor on characteristic itself.

22. With single line diagram, describe power flow in a three-phase induction motor. Also explain why it is advantageous to operate induction motor with a slip as small as possible.

23. Neglecting stator impedance, show that the torque T at slip s can be expressed as $T = \dfrac{2T_m}{\left(\dfrac{s}{s_m} + \dfrac{s_m}{s}\right)}$

where T_m is the maximum torque occurring at slip s_m.

24. What are the terms air-gap power, internal mechanical power developed and shaft power? How are these terms related with each other?

25. Enumerate various components of power loss in an induction motor and name the parts wherein these occur. Give typical values of power factor and efficiency under full load condition for a 100 kW, 1480 rpm motor.

26. Show and explain through power flow diagram, how electrical input is converted into mechanical power output in an induction motor; and hence show that

$$P_g : \text{rotor copper loss} : P_m = 1 : s : 1 - s \quad \text{where } s \text{ is slip.}$$

27. Derive, from the fundamental, the expression for maximum torque in the case of 3-φ induction motor and show that while maximum torque is independent of rotor resistance but slip at which maximum torque occurs depends upon rotor resistance.

28. Derive an expression for the torque developed in an induction motor. **[M.D. Univ. December 2006, May 2008]**

29. Explain the similarities between the transformer and induction motor equivalent circuits.

[M.D. Univ. December 2004]

30. Draw the equivalent circuit of a 3-phase induction motor. What do the various elements on it represent of induction motor ? Explain the steps in deriving the expression for torque from the equivalent circuit. Draw the torque-slip curve of the motor.

31. Develop the equivalent circuit for a 3-phase induction motor and explain how the mechanical power developed is taken care in the equivalent circuit.

32. Draw and explain no-load and load phasor diagrams of a 3-phase induction motor and from these derive the equivalent circuit model of induction motor.

33. Explain the basic principle of induction motor. Draw its equivalent circuit. Derive its torque equation.

[M.D. Univ. May 2009]

34. Explain why the no-load test on 3-φ induction motor is conducted at rated voltage whereas only a small voltage is applied to stator winding during blocked rotor test.

35. Describe the no-load test and blocked rotor test on an induction motor.

36. Explain how the parameters of a 3-phase induction motor can be obtained from the test results.

37. Explain the procedure of drawing circle diagram of an induction motor. What information can be drawn from the circle diagram ? Discuss.

38. For high starting torque application, which type of rotor constructions are preferably adopted in induction motor ? Explain above rotor constructions with its suitable torque-slip characteristic.

39. Explain the working of double cage induction motor.

40. Explain how the desirable features of high-starting torque and low-operating slip are obtained in a double-cage polyphase induction motor.

41. Discuss the relative merits and demerits of single cage and double cage induction motors.

42. Compare deep bar induction motor with a double cage induction motor.

43. How is it possible to obtain a high starting torque and good running performance with a double cage induction motor ? Explain with the help of torque-slip curves. Compare a double cage induction motor with a deep bar induction motor.

44. Distinguish between 3-phase slip-ring and squirrel cage induction motors, from the points of view of starting torque, pull-out torque, efficiency and power factor.

45. Describe the constructional features of wound and cage rotor induction motors and hence discuss and compare salient features of their performance characteristics, merits and demerits.

46. Differentiate the constructional details of squirrel cage and slip ring (wound rotor) 3-phase induction motors and also give their relative merits and demerits.

47. Explain the principle of operation of an induction generator. What are its limitations ?

[M.D. Univ. Electrical Machines-II, May 2009; December 2010]

48. Describe an induction generator. How does it differ from a synchronous generator ?

49. The performance of the induction generator at constant voltage and constant frequency requires a prime mover with rising speed characteristic.' Justify the statement.

50. How is torque developed in a 3-phase induction motor ? Why cannot it attain synchronous speed of its own. Discuss what happens if it is made to run at speed higher than synchronous.

51. How can an induction generator feed power into a constant voltage constant frequency bus ? Give circuit configuration only.

52. Write short notes on the following :

(*i*) Principle of operation of 3-phase induction motor.

(*ii*) Testing of 3-phase induction motor.

(*iii*) Equivalent circuit of a 3-phase induction motor.

(*iv*) Circle diagram of 3-phase induction motor. **[M.D.U. May 2009]**

(*v*) Double cage and deep bar motors. **[M.D.U. May 2009]**

(*vi*) Deep bar motor. **[M.D.U. December 2010]**

(*vii*) 3-phase induction generator.

SHORT ANSWER TYPE QUESTIONS WITH ANSWERS

Q. 1. What is meant by an induction motor ?

Ans. An induction motor derives its name from the fact that the current in the rotor conductors is induced by the motion of rotor conductors relative to the magnetic field developed by the stator currents.

Q. 2. What are the two types of 3-phase induction motors ? Which type is generally preferred ?

Ans. Two types of 3-phase induction motors are (*i*) squirrel cage and (*ii*) wound rotor or slip-ring induction motors. Squirrel cage induction motor is generally preferred.

Q. 3. Why are induction motors called ''asynchronous'' ?

Ans. Because the rotor does not turn in synchronism with the rotating field developed by the stator currents.

Q. 4. Why closed type slots are often used for small induction motors ?

Ans. Use of closed slots makes the reluctance more uniform, reduces exciting current and gives better power factor.

Q. 5. Why the air gap between stator core and rotor of an induction motor is made very small ?

Ans. The air gap is made as small as possible so as to produce required flux with a minimum exciting current and give leakage reactance as small as possible. This results in improved power factor.

Q. 6. Why the slots on the rotor of an induction motor are usually skewed ?

Ans. The slots on the rotor of an induction motor are usually skewed in order to obtain a uniform torque, reduce the magnetic locking of the stator and rotor, and reduce the magnetic humming noise while running.

Q. 7. Explain why the number of poles of the stator and of the rotor of an ac motor should be equal.

Ans. The flux set up by the stator windings is a rotating flux with a definite number of poles; operation is only possible when the number of poles is same on both (stator and rotor).

Q. 8. Which type of induction motor develops higher starting torque ?

Ans. Slip-ring induction motor.

Q. 9. What is the necessity of short-circuited rotor conductors in a squirrel cage induction motor ?

Ans. If the rotor conductors in a squirrel cage motor are not short circuited, the rotor winding will be open circuited and when the stator is connected to 3-ϕ ac supply, the rotating magnetic field will be set up in the air gap and an emf will be induced in the rotor but no current will flow through the rotor conductors and, therefore, no torque will be developed and the rotor will remain stationary.

Q. 10. How can the direction of rotation of the 3-phase induction motor be reversed ?

Ans. The direction of rotation of a 3-phase induction motor may be reversed by interchanging the connection to the supply of any two leads of the motor.

Q. 11. Give the expression of speed in terms of poles and frequency of supply.

Ans. Speed, $N = \dfrac{120 \times \text{frequency of supply}}{\text{Number of poles}}$.

Q. 12. Define slip in 3-phase induction motor. What is its value at starting and at synchronous speed?

Ans. The difference between the speed of the stator field and the actual speed of the rotor expressed as a fraction of synchronous speed is known as slip of an induction motor.

i.e. Fraction slip, $s = \dfrac{N_s - N}{N_s}$

At starting, slip, $s = 1.0$ **Ans.**

At synchronous speed, $s = 0$ **Ans.**

Q. 13. State the operating conditions of the motor when the slip is either positive greater than one or negative.

Ans. When the slip is positive greater than 1, the induction motor acts as a brake and it absorbs kinetic energy from the still revolving field causing its speed to fall. When the slip is negative (*i.e.* when the speed is more than synchronous speed), the induction machine operates as an induction generator and feeds power back to the supply line.

Q. 14. What is the speed of rotor mmf of a 3-phase induction motor w.r.t. its stator mmf ?

Ans. zero.

Q. 15. In case of a 3-phase induction motor why is slip always positive ?

Ans. The induction motor cannot run at synchronous speed because at synchronous speed, there will not be relative speed between the revolving field and the rotor conductors and so the basic phenomenon will not occur to develop torque. It means that a 3-phase induction motor always run at a speed lesser than that of the field. The slip being the difference of synchronous speed, N_s and actual speed N ($N_s > N$) is, therefore, always positive.

Q. 16. Explain why rotor of an induction motor can never attain synchronous speed.

Ans. An induction motor cannot run at synchronous speed. If it were possible, by some means, for the rotor to attain synchronous speed, the rotor would then be standstill with respect to rotating flux. Then no emf would be induced in the rotor, no rotor current would flow, and therefore, there would be no torque developed.

Q. 17. Show that in induction motor, the rotor always run in the direction of stator field.

Ans. According to Lenz's law we know that the direction of induced emf would be in such a direction that it would try to oppose the very cause for which it is due. Because the cause producing the induced currents in the rotor conductors is the relative speed between the rotating magnetic field and stationary rotor conductors, therefore, they circulate in such a way that a torque is produced in the rotor tending to cause it to follow the rotating magnetic field and thus reducing the relative speed.

Q. 18. How does the slip vary with load ?

Ans. With the increase in load, the motor slows down until the relative motion between the rotor and the rotating magnetic field is just sufficient to result in the development of torque required for a particular load. Thus slip increases with the increase in load.

Q. 19. If the full-load speed of an induction motor is 960 rpm, what will be the synchronous speed ? Give an intuitive answer.

Ans. 1,000 rpm.

Q. 20. At what slip the torque developed in an induction motor will be maximum ?

Ans. Torque developed in an induction motor will be maximum at a slip given by the ratio of rotor resistance to rotor standstill reactance. *i.e.* $s = \frac{R_2}{X_2}$.

Q. 21. What is the condition for maximum torque at starting in a 3-phase induction motor ?

Ans. Torque developed in an induction motor at start will be maximum when rotor resistance equals rotor standstill reactance *i.e.* $R_2 = X_2$.

Q. 22. Is the maximum torque of a 3-phase induction motor dependent on the rotor resistance ?

Ans. No, it is independent of rotor resistance.

Q. 23. What is the relationship of developed torque of a 3-phase induction motor with the supply voltage?

Ans. $T \propto sV^2$.

Q. 24. When the applied rated voltage per phase is reduced to one-half, what will be the starting torque of a squirrel cage induction motor in terms of its starting torque with full voltage ?

Ans. One-fourth of starting torque with full rated voltage.

Q. 25. How the starting torque and maximum torque vary with rotor circuit resistance ?

Ans. As the rotor resistance R_2 is increased from zero, power factor is improved but rotor impedance is increased and so the rotor current is reduced. At first, the effect of improved power factor predominates over the effect of decrease in rotor current due to increase in impedance and the starting torque, therefore, increases. But after a certain value of rotor resistance, the effect of increased impedance predominates over the effect of improved power factor and so the torque starts decreasing.

Maximum torque is independent of rotor circuit resistance.

Q. 26. Why is that the *V/f* ratio is kept constant while controlling the speed of a 3-phase induction motor by varying the supply frequency ?

Ans. In order to maintain flux level constant.

Q. 27. Speed of a 3-phase induction motor is varied by varying the supply frequency while *V/f* is kept constant. How will the maximum torque vary ?

Ans. The magnitude of maximum torque does not change but the speed at which it occurs changes in the same ratio as the supply frequency.

Q. 28. What are the different losses taking place in a 3-phase induction motor ?

Ans. Stator copper loss; stator iron loss; rotor copper loss, friction and windage losses and stray load loss.

Q. 29. Why are the iron losses in the rotor of an induction motor normally neglected ?

Ans. Because of very low frequency of emf induced in rotor.

Q. 30. How is the mechanical power output represented in the circuit model of an induction motor ?

Ans. Mechanical power output per phase is represented by a pure resistive load of equivalent resistance R_L given by the relation $R_L = R_2 \left(\frac{1}{s} - 1\right)$ where R_2 is the rotor circuit resistance per phase and *s* is the slip.

Q. 31. The induction motor can be called a generalised transformer. Explain why ?

Ans. Since the two fields set up by stator (primary) and rotor (secondary) are stationary relative to each other, so the polyphase induction motor can be considered equivalent to a transformer having an air gap, separating the iron portions of the magnetic circuit carrying the primary and secondary windings. Thus the induction motor can be called a generalized transformer.

Q. 32. The reactance of the rotor of the induction motor varies greatly between starting and running conditions. Explain why ?

Ans. The reactance of the rotor of an induction motor is given as

$$X_2' = 2\pi f' L_2$$

where f' is the frequency of rotor current and is equal to slip multiplied by supply frequency f

$$i.e.\ f' = sf$$

So $X_2' = 2\pi s f L_2 = sX_2$ where X_2 is standstill reactance of the rotor.

Since the slip s varies greatly from starting to normal running condition, rotor reactance varies greatly between starting and running conditions.

Q. 33. An induction motor can be represented by an equivalent transformer. Are their pu magnetising currents of the same order ?

Ans. No.

Q. 34. List out the tests necessary to draw the circle diagram of a 3-phase induction motor.

Ans. No-load and blocked rotor tests.

Q. 35. What parameter of the equivalent circuit should change the diameter of the circle diagram and have it effect the pull-out torque ?

Ans. Total standstill leakage reactance per phase referred to stator changes the diameter of the circle diagram and increase in standstill leakage reactance reduces the pull-out torque.

Q. 36. Explain the reasons for lower power factor of low-speed, 3-phase induction motors as compared to that of high-speed motors.

Ans. Because of larger number of poles in low-speed 3-phase induction motors as compared to high-speed motors, magnetising current is more due to increase in leakages, which increases with the increase in number of poles. So the power factor of low-speed induction motors is poor in comparison to that for high-speed induction motors.

Q. 37. Why the power factor of a squirrel cage induction motor is low at light loads ?

Ans. The current drawn by an induction motor operating at light load is largely a magnetising current so current at light loads lags behind the applied voltage by a large angle and thus the power factor of an induction motor at light loads is quite low.

Q. 38. The magnetising current in a transformer is 2 to 5 % whereas in 3-phase induction motor, it is about 25-40 % of rated current. Mention the reason.

Ans. Owing to the air gap in an induction motor, the magnetising current is large for the induction motor as compared to that for a transformer. The friction and windage losses are also present in the induction motor. So no-load current of an induction motor is much higher than that of an equivalent transformer.

Q. 39. Why the power factor of an induction motor is low at starting ?

Ans. The rotor frequency and rotor reactance are high under starting conditions and, therefore, rotor currents lag the rotor emf by a large angle. This results in low power factor at starting.

Q. 40. Why the efficiency of a three phase induction motor is less than that of a transformer ?

Ans. The large magnetising current, due to air gap and friction and windage losses in the induction motor are the factors which make the efficiency of the induction motor less than that of the transformer.

Q. 41. The starting torque of a squirrel cage induction motor cannot be altered, when the applied voltage is constant, why ?

Ans. The starting torque of a squirrel cage induction motor cannot be increased as there is no provision for inserting resistance in the rotor circuit.

Q. 42. How will you improve the starting torque of a 3-phase induction motor ?

Ans. The starting torque of a 3-phase induction motor can be improved by inserting external resistance in each phase of the rotor circuit, which is possible only in case of wound rotor motors.

Q. 43. What are the main advantages of a cage motor ?

Ans. The main advantages of a cage motor are simple and rugged construction, low initial as well as maintenance cost, better efficiency and power factor, better cooling, more pull-out torque, greater maximum power output, nearly constant speed and simple starting arrangement.

Q. 44. What are the main disadvantages of a cage motor ?

Ans. The main drawbacks of a cage motor are low starting torque with large starting current and no possibility of a speed control.

Q. 45. What are the advantages and disadvantages of wound rotor over squirrel cage rotor in case of induction motor ?

Ans. In wound rotor external resistance can be inserted and, therefore,

(*i*) high starting torque can be obtained by inserting an external resistance of proper value in each phase of the rotor circuit and

(*ii*) adjustable speed can be obtained simply by adding external resistors in the rotor circuit.

The drawback of wound rotor over squirrel cage rotor are initial and maintenance costs are more because wound rotor needs slip rings, brushes, short-circuiting devices etc.

Q. 46. What is meant by single phasing ?

Ans. The term 'single phasing' refers to operation with one stator terminal accidently disconnected (as by a blown fuse or a faulty contact on the run side of a star-delta switch).

Q. 47. Why the induction generator is often called an asynchronous generator ?

Ans. Because the speed of an induction generator is not in synchronism with line frequency, so it is often called an asynchronous generator.

Q. 48. Why an induction generator cannot be operated independently ?

Ans. An induction generator needs a source of reactive energy to provide magnetising (lagging reactive) current to it, so it cannot operate independently.

Q. 49. Can the operating principle of a self excited induction generator be used for braking the induction motor ?

Ans. Yes, In this case the induction machine receives its excitation from the bank of capacitors and operates as an induction generator.

Q. 50. What is the application for induction generator ?

Ans. Induction generator is very useful for braking purpose in railway work and in providing effective over-speed brake when an induction motor is used on such applications as hoists.

PROBLEMS

1. A 12-pole, 3-phase alternator driven at a speed of 500 rpm supplies power to an 8-pole, 3-phase induction motor. If the slip of the motor is 0.03 pu, calculate the speed. [**Ans.** 727.5 rpm]

2. A 50 Hz, 4-pole, 3-phase induction motor has rotor current frequency of 2 Hz. Obtain slip and speed of motor. [**Ans.** 4%, 1,440 rpm]

3. A 3-ϕ induction motor is wound for 4 poles and is supplied from 50 Hz system. Calculate:
(*i*) synchronous speed, (*ii*) rotor speed when slip is 4%, and (*iii*) rotor frequency when rotor runs at 600 rpm. [**Ans.** (*i*) 1,500 rpm (*ii*) 1,440 rpm (*iii*) 30 Hz]

4. A 12-pole, 3-phase alternator is coupled to an engine running at 500 rpm. It supplied a 3-phase induction motor having a full-load speed of 1,440 rpm. Find the percentage slip, frequency of rotor current and number of poles of the motor. [**Ans.** 4%, 2 Hz, 4]

5. A 4-pole, 3-phase induction motor operates from a supply whose frequency is 50 Hz, calculate (*i*) the speed at which the magnetic field of the stator is rotating (*ii*) the speed of the rotor when the slip is 0.04 (*iii*) the frequency of rotor currents when the slip is 0.03 (*iv*) the frequency of the rotor currents at standstill. [**Ans.** (*i*) 1,500 rpm (*ii*) 1,440 rpm (*iii*) 1.5 Hz (*iv*) 50 Hz]

6. A 3-ϕ, 6-pole, 50 Hz, induction motor has a slip of 1% at no-load and 3% at full load. Determine synchronous speed, no-load speed, full-load speed, frequency of rotor current at standstill and frequency of rotor current at full load. **[Ans.** 1,000 rpm, 990 rpm, 970 rpm, 50 Hz, 1.5 Hz]

7. A 3-phase, 50 Hz induction motor has 6 poles and operates with a slip of 5% at a certain load. Determine
   (*i*) the speed of the rotor with respect to the stator.
   (*ii*) the frequency of rotor current.
   (*iii*) the speed of the rotor magnetic field with respect to rotor.
   (*iv*) the speed of the rotor magnetic field with respect to stator and
   (*v*) the speed of the rotor magnetic field with respect to the stator magnetic field.
   [Ans. (*i*) 950 rpm (*ii*) 2.5 Hz (*iii*) 50 rpm (*iv*) 1,000 rpm (*v*) zero]

8. A 3-phase, 50 Hz induction motor has a full-load speed of 960 rpm. Calculate the speed of rotor field with respect to rotor structure, with respect to stator structure and with respect to stator field.
   [Ans. 40 rpm, 1,000 rpm; zero]

9. A 5 hp, 230 V, 50 Hz induction motor has a rated full-load speed of 950 rpm. The induced voltage per phase of rotor at standstill is 100 V. Calculate:
   (*i*) Number of poles and percentage full-load slip.
   (*ii*) Rotor induced voltage and its frequency at full load. **[Ans.** (*i*) 6, 5% (*ii*) 5 V, 2.5 Hz]

10. The star-connected rotor of an induction motor has a standstill impedance of (0.4 + *j*4) ohm per phase and the rheostat impedance per phase is (6 + *j*2) ohm.
    The motor has an induced emf of 80 V between slip rings at standstill when connected to its normal supply voltage. Find (*i*) rotor current at standstill with the rheostat in the circuit (*ii*) when the slip rings are short circuited and motor is running with a slip of 3%.
    [Ans. (*i*) 5.265 A lagging behind induced emf by an angle 43.15°
    (*ii*) 3.32 A lagging behind induced emf by an angle of 16.7°]

11. The rotor resistance and reactance per phase of a 4-pole, 50 Hz, 3-phase induction motor are 0.025 Ω and 0.12 Ω respectively. Make simplifying assumption, state them and find (*i*) speed at maximum torque (*ii*) find value of additional rotor resistance per phase required to give three-fourth of maximum torque at starting.
    [Ans. 1,187.5 rpm; 0.0292 Ω]

12. A 3-phase, 50 Hz, 8-pole induction motor has a full-load slip 4%. The rotor resistance is 0.001 Ω per phase and standstill reactance 0.005 Ω per phase. Find the ratio of maximum to full-load torque and the speed at which the maximum torque occurs. **[Ans.** 2.6 ; 600 rpm]

13. A 746 kW, 3-phase, 50 Hz, 16-pole induction motor has a rotor impedance of (0.02 + *j* 0.15) Ω at standstill. Full-load torque is obtained at 360 rpm. Calculate (*i*) the ratio of maximum to full-load torque (*ii*) the speed for maximum torque and (*iii*) the rotor resistance to be added to get maximum starting torque.
    [Ans. (*i*) 1.817 (*ii*) 325 rpm (*iii*) 0.13 Ω]

14. A 3-phase induction motor has starting torque of 100 % and a maximum torque of 200 % of the full-load torque. Find slip at maximum torque. **[Ans.** 26.8%]

15. The starting and maximum torques of a 3-phase induction motor are 1.5 times and 2.5 times its full-load torque. Determine the percentage change in rotor circuit resistance to obtain a full-load slip of 0.03. Neglect stator impedance. **[Ans.** 56.9%]

16. The power input to the rotor of a 400 V, 50 Hz, 6-pole, 3-phase induction motor is 80 kW. The rotor emf is observed to make 100 complete alternations per minute. Calculate slip, mechanical power developed and rotor copper loss. **[Ans.** 0.0166; 78.667 kW, 1.333 kW]

17. Draw the slip-torque curve of 3-ϕ induction motor and show the region of stable operation. The stator loss of 3-ϕ induction motor is 2 kW. When the input is 90 kW, what will be rotor mechanical power developed and rotor copper loss, if the motor is running at slip of 4 % ? **[Ans.** 84.48 kW; 3.52 kW]

18. The power input to the rotor of a 440 V, 50 Hz, 3 -phase, 6-pole induction motor is 60 kW. It is observed that the rotor emf makes 90 complete cycles per minute. Calculate (*i*) the slip (*ii*) the rotor speed (*iii*) rotor copper loss and (*iv*) mechanical power developed. **[Ans.** (*i*) 0.03 (*ii*) 970 rpm (*iii*) 1.8 kW (*iv*) 58.2 kW]

19. A 6- pole, 3-phase induction motor develops 30 hp including 2 hp mechanical losses at a speed of 950 rpm on 550 V, 50 Hz mains. The power factor is 0.88 lagging. Calculate (*i*) slip (*ii*) rotor copper loss (*iii*) total input if stator losses are 2 kW (*iv*) efficiency and (*v*) line current.

[Ans. (*i*) 0.05 (*ii*) 1,161 W (*iii*) 25,226 W (*iv*) 81.64% (*v*) 30 A]

20. The power input to the rotor of a 400 V, 50 Hz, 6-pole, 3-phase induction motor is 80 kW. The rotor electromotive force is observed to make 100 complete alterations per minute. Calculate (*i*) the slip (*ii*) the rotor speed (*iii*) mechanical power developed (*iv*) rotor copper loss per phase (*v*) the rotor resistance per phase if the rotor current is 65 A. **[Ans.** (*i*) 3.33% (*ii*) 966.7 rpm (*iii*) 77.333 kW (*iv*) 889 watts (*v*) 0.21 Ω]

21. A 3-phase, star-connected , 220 V (line-to-line), 50 Hz, 4-pole induction motor has the following constants in ohm per phase referred to stator :
$r_1 = 0.29$, $r_2 = 0.14$, $x_1 = 0.5$, $x_2 = 0.21$ and magnetising reactance $x_{mag} = 13.25$ ohms. The core loss may be assumed to be constant at 400 watts. For a slip of 2% compute (*i*) speed (*ii*) output torque and (*iii*) stator current of the motor. Neglect frictional and windage losses. **[Ans.** (*i*) 1,470 rpm (*ii*) 40 Nm (*iii*) 21.5 A]

22. The following data refer to a 12-pole, 420 V, 50 Hz, 3-phase, mesh connected induction motor: $R_1 = 2.95$; $X_1 = 6.82$; $R_2 = 2.08$, $X'_2 = 4.11$ ohms per phase. On no-load the line value of magnetising current is 6.7 A and total core loss is 269 W. Determine power factor, input current, equivalent rotor current and torque at a slip of 3% from the approximate equivalent circuit. **[Ans.** 0.78 (lag); 13.1 A; 9.95A; 393.3 Nm]

23. A 3-phase, 400 V induction motor gave the following test readings:

No-load	:	400 V,	1250 W,	9 A
Short-circuit	:	150 V,	4 kW,	38 A

Draw the circle diagram. If the normal rating is 20.27 hp (metric), find from the circle diagram, the full-load values of current, power factor and slip. **[Ans.** 30 A, 0.866 (lagging). 6.7%]

24. Draw the circle diagram for 5 hp (3.73 kW), 200 V, 50 Hz, 4-pole, 3-ϕ star-connected induction motor from the following test data :

No-load	Blocked rotor
Line voltage 200 V, line current 5 A, total input 350 W.	Line voltage 100 V, line current 26 A, total input 1,700 W.

Estimate from the diagram for full load condition, the line current, power factor and also the maximum torque in terms of the full-load torque. The rotor copper loss at standstill is half the total copper loss.

[Ans. 15 A, 0.83, 180%]

25. Draw the circle diagram of a 10 hp, 200 V, 50 Hz, 3-ϕ slip ring induction motor with a star-connected stator and rotor, a winding ratio of unity, a stator resistance of 0.38 Ω/phase and a rotor resistance of 0.24 Ω per phase. The following are test readings:

No-load	:	200 V,	7.7 A,	pf 0.195
Short-circuit	:	100V,	47.6 A,	pf 0.454

Find (*a*) the starting torque (*b*) maximum torque in synchronous watts (*c*) the maximum power factor (*d*) the slip for maximum torque (*e*) the maximum output.

[Ans. (*a*) 5,600 synchronous watts (*b*) 12,500 watts (*c*) 0.88 (*d*) 0.195 (*e*) 10.4 kW]

26. An induction motor has a double cage rotor with equivalent impedance at standstill of (1.0 + *j* 1.0) and (0.2 + *j* 4.0) ohms. Find the relative values of torque given by each cage (*a*) at starting and (*b*) at slip of 5%.

[Ans. (*a*) 40 : 1 (*b*) 0.4 : 1]

27. Two impedances at standstill of the inner and outer windings of a double cage motor are (0.015 + *j*0.5) Ω and (0.05 + *j*0.11) Ω respectively. Calculate the ratio of torque due to the two windings at standstill. **[Ans.** 57 : 1]

❑❑❑

3

CHAPTER

Starting and Speed Control of Three Phase Induction Motors

3.1. STARTING OF SQUIRREL CAGE INDUCTION MOTORS

As a general rule, a motor is subjected to the most severe service during its starting and accelerating period; it is during this interval of time that an extremely high current tends to flow unless steps are taken to keep its value down to a reasonable level. In the case of an induction motor, at start when the rotor is at standstill, the squirrel-cage rotor is just like a short-circuited secondary. Therefore, the current in the rotor circuit will be very high and consequently the stator also will draw a high current from the supply lines if full-line voltage were applied at start. Of course, this high current is momentarily and falls off rapidly as the motor picks up speed. The magnitude of this current depends on the electrical design of the motor and is independent of the mechanical load. The duration of starting current, however, depends upon the time required for acceleration, which in turn depends on the nature of the driven load.

It is true that the starting of a squirrel cage induction motor because of its rugged construction does not impose electrical stresses that are as severe as those experienced by commutator type motors; nevertheless, it is important that the starting current be minimised to such an extent that the line-voltage drop does not affect the operation of other equipment connected to the same distribution network, Modern well-designed induction motors will usually draw about 5 to 7 times the rated full-load current at the starting instant if rated voltage is applied to it. One obvious way to reduce the starting current is to impress the lower voltage across the stator winding at starting instant.

With the reduction in voltage applied to the stator of an induction motor the short-circuit current will be reduced in the same proportion. The main flux, which is determined by the applied voltage, and, therefore, the magnetising current will fall in the same proportion if the magnetic circuit is not saturated. Further, since the core losses are approximately proportional to the square of the flux density and, therefore, the voltage, the energy component of the no-load current will be reduced in proportion to the reduced voltage.

One serious objection to the practice of employing reduced voltage for starting of squirrel cage induction motor is the large reduction in starting torque, because starting torque is proportional to the square of impressed voltage [Refer to Eq. (2.15)]. Whether or not such a large reduction in starting torque is permissible will depend, of course, upon the application.

There are several methods of starting squirrel cage induction motors. The most common methods are described below.

1. Direct-on -Line Starting or Full Voltage Starting. Although there is no limitation on the size of the motor that may be started by this method, it should be understood that objectionable line voltage drops will usually occur, especially if large motors are started frequently. Whether or not normal voltage is employed will, therefore, depend upon the following factors :

(*i*) the size and design of the motor

(*ii*) the kind of application

(*ii*) the location of the motor in the distribution system and

(*iv*) the capacity of the power system and the rules governing such installations as established by power supply companies.

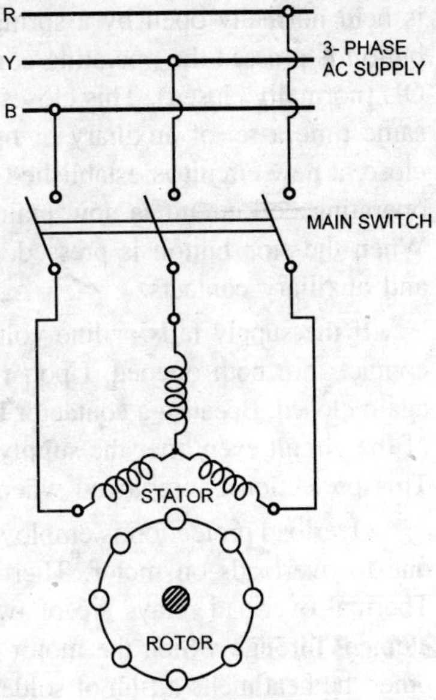

Fig. 3.1

Squirrel cage motors of capacity up to 1.5 kW, double cage motors and squirrel cage motors of large capacity having a large rotor resistance are started by this method. This is the most economical method of starting of induction motors.

As the name suggest, this method involves direct switching of polyphase squirrel cage induction motor to the supply mains, as illustrated in Fig. 3.1. In this method of starting of cage motors the rate of temperature rise is high and the motor may get damaged if the starting period is large, which may be due to (*i*) excessive load (*ii*) insufficient rotor resistance or (*iii*) excessive voltage drop in the supply lines. The starting period usually lasts a few seconds because starting torque for induction motor is about twice the full–load torque.

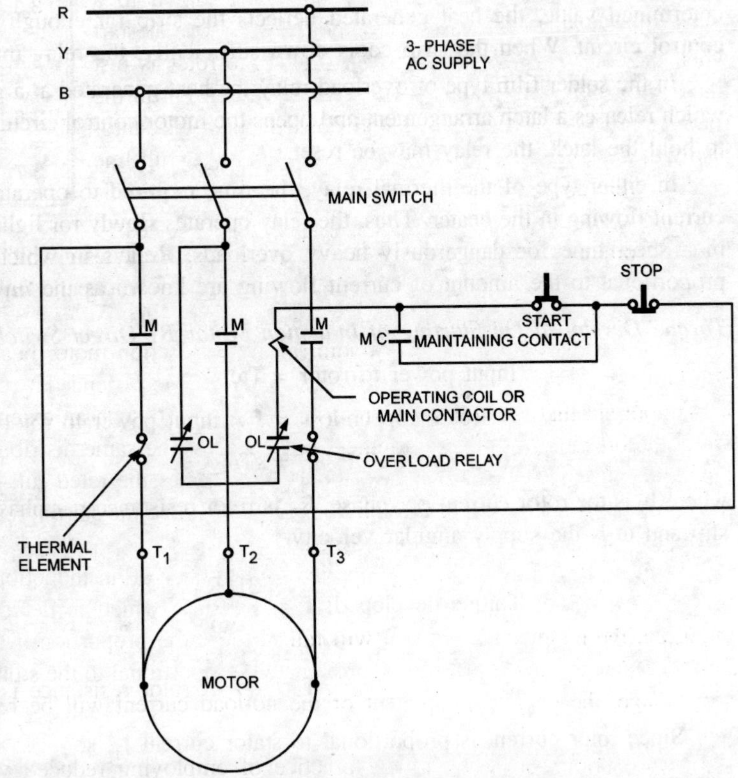

Fig. 3.2. Wiring Diagram of a Direct-on-Line Starter With Protective Devices

The push button type direct-on-line starter, which is very common in use, is shown in Fig. 3.2. It is simple, inexpensive and easy to install and maintain. It consists of a set of 'start' and 'stop' push buttons, a contactor (an electromagnet) with its associated contacts and usually an overload and under-voltage protection devices. The start button is a momentary contact switch that is held normally open by a spring. The stop button is held normally closed by a spring. When the start button is pressed the operating coil (or main contactor) gets energized through the overload relay contacts OL (normally closed). This closes the three main contacts M that connects the motor to the supply. At the same time a set of auxiliary or maintaining contacts MC are closed. When the maintaining contacts MC close, a new circuit is established through stop button, maintaining contacts MC and operating coil. Since operating coil circuit is now maintained by the auxiliary contacts MC, the start button may be released. When the stop button is pressed, the operating coil gets de-energized, thereby opening all main contacts and auxiliary contacts.

If the supply fails or line voltage drops below a certain value, the main contacts and the maintaining contacts are both opened. Upon return of the supply, the contactor cannot close until the start button is again closed. Because a contactor that is controlled by a three-wire control circuit maintains the interruption of the circuit even after the supply is restored, it is said to provide *under-voltage protection* for the motor. This protection is employed when it is desired to prevent the unexpected starting of a motor.

Overload protection is employed for protecting the motor and control apparatus from excessive heating due to overloads on motor. Thermal overload relays are commonly used for motor overload protection. Thermal overload relays are of two types. Both types are operated from the heat generated in a thermal element through which the motor current flows. In one type, the heat bends a bimetallic strip, and in the other the heat melts a film of solder. Both act to open the motor control circuit and, therefore, to disconnect the motor from the source of supply.

A bimetallic strip is made of two different metals whose surface has been welded together. One of the metal expands rapidly as compared to the other, when heated. When heat is applied, the strip is caused to deflect or curl up due to unequal expansion of the two metals. When the motor current attains a pre-determined value, the heat generated deflects the strip far enough to trip a latch that opens the motor control circuit. When the strip cools down sufficiently, the relay may be reset and the motor re-started.

In the solder-film type of overload relay the heat generated at a given overload current melts the film, which releases a latch arrangement and opens the motor control circuit. When the solder has cooled enough to hold the latch, the relay may be reset.

In either type of the thermal relay, the time required to operate is determined by the magnitude of current flowing in the heater. Thus, the relay operates slowly for light overloads but disconnects the motor in a short time for dangerously heavy overloads. Relays in which the time of operation is inversely proportional to the amount of current flowing are known as the *inverse time relays.*

Torque Developed On Starting of Induction Motor By Direct Switching.

$$\text{Input power to rotor} = T\omega$$

$$\text{Rotor copper loss} = s \times \text{input power to rotor}$$

$$\text{or } 3\,I^2 R_2 = sT\omega$$

where I_2 is the rotor current per phase, R_2 is rotor resistance per phase, T is the torque developed, s is the slip and ω is the supply angular velocity.

$$\text{or Torque developed, } T = \frac{3I^2 R_2}{s\omega}$$

$$\textit{i.e. } T \propto \frac{I_2^2}{s} \text{ if rotor resistance is constant}$$

Since rotor current is proportional to stator current I_1, so

$$T \propto \frac{I_1^2}{s}$$

$$\text{or } T = K \frac{I_1^2}{s} \quad \text{where K is any constant}$$

At starting slip, $s = 1$

So starting torque, $T_{st} = K (I_{st})^2$ where I_{st} is the starting current

$$\text{Full-load torque, } T_f = \frac{KI_f^2}{s_f}$$

where I_f is the full-load motor current and s_f is the full-load slip

$$\text{So } \frac{T_{st}}{T_f} = \frac{KI_{st}^2}{KI_f^2/s_f} = \left(\frac{I_{st}}{I_f}\right)^2 s_f$$

$$\text{or } T_{st} = T_f \times \left(\frac{I_{st}}{I_f}\right)^2 s_f$$

When the motor is connected directly across the supply mains, the starting current is equal to the short-circuit current. I_{sc}

$$\therefore \text{ Starting torque, } T_{st} = T_f \left(\frac{I_{sc}}{I_f}\right)^2 s_f \qquad ...(3.1)$$

2. Primary Resistor (or Reactor) Starting. This method of starting of a polyphase cage motor is very simple and provides smooth acceleration of the motor. In this method of starting of a 3-phase squirrel cage induction motor reduced voltage is obtained by means of resistors (or reactors) that are connected in series with each stator lead, as shown in Fig. 3.3, during the starting period. The voltage drop in resistors (or reactors) causes a reduced voltage across the motor terminals. As the motor picks up the speed, the resistors (or reactors) are cut out in steps and finally short circuited when the motor attains the operating speed.

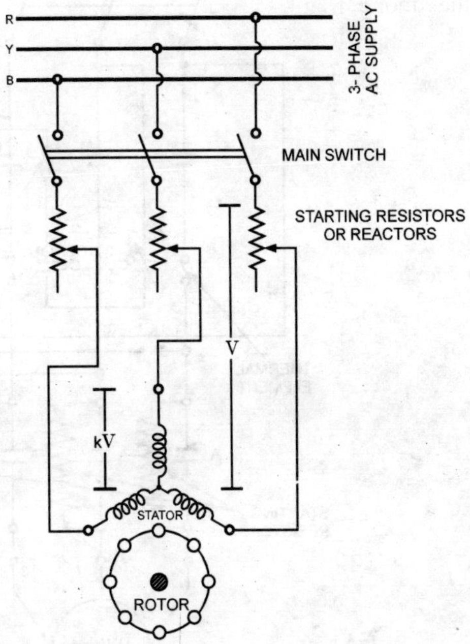

Fig. 3.3

The primary resistor method is quite similar to the dc shunt motor starter with its series armature resistor, because both provide a voltage drop between the voltage supply and the motor terminals. There are, however, some very important differences. In a dc shunt motor the starting torque is directly proportional to the starting armature current whereas in case of a polyphase induction motor the torque varies as the square of the motor terminal voltage *i.e.*, there is a far greater reduction of starting torque for an equivalent reduction of starting current.

Although the initial cost of reactors is high in comparison to that of resistors, reactor starting is preferred because this method incurs small power losses and is more effective* in reducing the voltage applied to the stator at starting.

---

* The power factor of an induction motor is quite low and if reactors are used for reducing voltage across the motor terminals, the voltage drop across the reactors will be nearly in phase with the supply voltage and so the net reduction in the applied voltage to the stator will be more. On the other hand in case of resistance starting, the voltage drop across the resistors will be nearly in quadrature with the applied voltage and so the net reduction in the applied voltage to the stator terminals will be less.

The advantages and disadvantages of this method of starting of cage motors are given below:

Advantages. 1. Smooth acceleration.

2. High power factor during start.
3. Less expensive than auto-transformer starter in lower output ratings.
4. Closed transition starting.
5. Available with as many as 7 accelerating points.

Disadvantages 1. Resistors give off heat.

2. Low torque efficiency.
3. Starting duration usually exceeds 5 seconds, so needs expensive resistors.
4. Starting voltage is difficult to adjust to meet varying conditions.

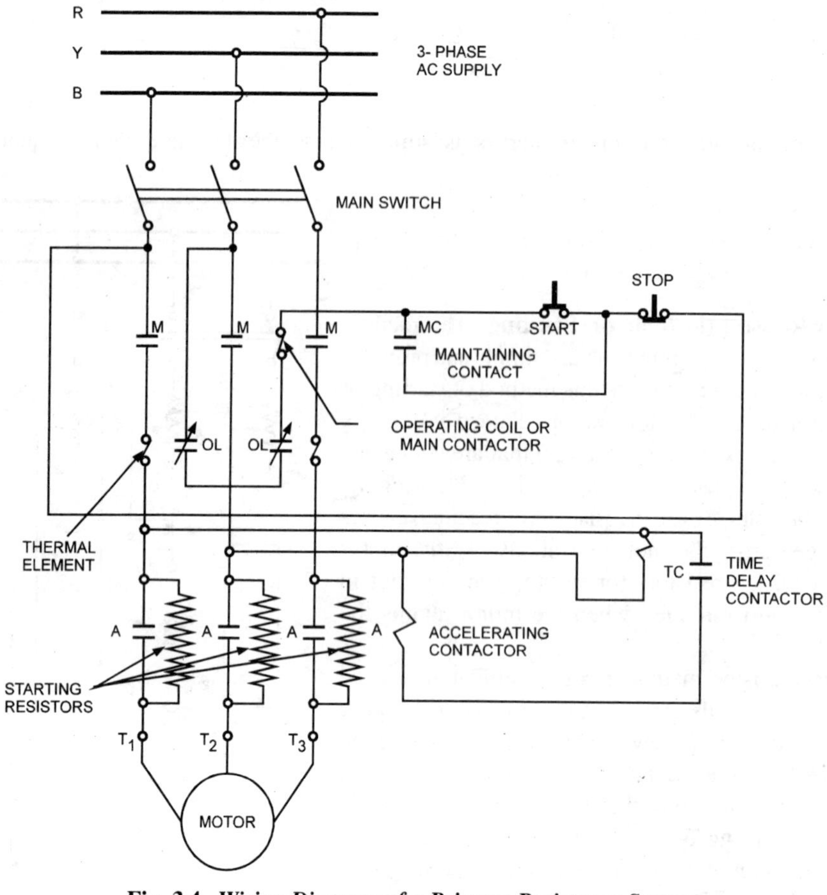

Fig. 3.4. ***Wiring Diagram of a Primary Resistance Starter***

Circuit diagram of a magnetically operated primary resistance starter is illustrated in Fig. 3.4. It consists of three line resistors, two sets of three line contacts, two contactors, overload protection, a set of start and stop buttons and several auxiliary contacts. When the start button is pressed, the main contactor is energized through the overload relay contacts OL (normally closed), which in turn closes the three main contacts M and one auxiliary contact MC. Thus the motor is connected to the line through the starting resistors. At the same time, a mechanical escapement timer is started. After a definite time has elapsed during which the motor has speeded up, the time delay contacts TC close. This energizes the accelerating contactor A, which in turn closes the three contacts A and short circuiting the starting resistors applies full voltage to the motor.

Overload and under-voltage protections are provided in the same manner as in the direct-on-line starter.

In the starter described above, the starting resistors are cut out in one step. Starters are also available, in which arrangement is to cut out starting resistances in several steps and thus to give smooth acceleration with less line disturbance.

If the normal supply phase voltage = V volts and by using line resistance starter the voltage is reduced to kV volts, then starting current is also reduced in the same ratio *i.e.* starting current $I_{st} = k\, I_{sc}$ where I_{sc} is the short-circuit current.

$$\text{Starting torque, } T_{st} = T_f \times \left(\frac{I_{st}}{I_f}\right)^2 s_f = T_f \times k^2 \left(\frac{I_{sc}}{I_f}\right)^2 s_f \qquad ...(3.2)$$

$$= k^2 \times \text{Torque obtained by switching the motor directly.}$$

3. Auto-Transformer Starting. In auto-transformer starting method the reduced voltage is obtained by taking tappings at suitable points from a three phase auto-transformer, as shown in Fig. 3.5. The auto-transformers are generally tapped at the 50, 60 and 80 per cent points, so that adjustment at these voltages may be made for proper starting torque requirements. Since the contacts frequently break large values of current, arcing is sometimes quenched effectively by having them assembled to operate in an oil bath.

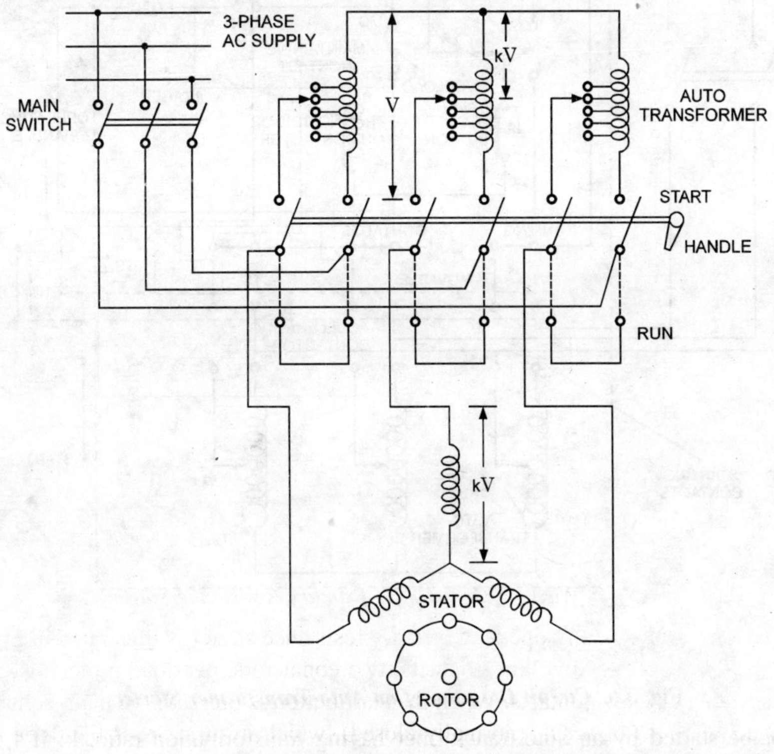

Fig. 3.5. ***Auto-Transformer Starter***

Auto-transformer starters may be either manually or magnetically operated. The wiring diagram of an auto-transformer starter is shown in Fig. 3.6.

The manual auto-transformer starter is essentially a multi-pole double-throw switch. It consists of three sets of contacts starting, running (both stationary) and the movable contacts (the contacts attached to the operating handle). The operating handle is spring loaded lever mounted on the outside of a steel cabinet to stand vertically in the "off" position. When the operating handle is moved to the start position, the movable contacts are moved against the starting contacts. This energizes the star–connected auto-transformer and impresses reduced voltage (50 or 60 or 80 per cent according to the setting on auto-transformer) across the motor. After the motor has accelerated to about full speed, the operating handle is moved to the run position. This instantly opens the starting contacts (or disconnects the auto-transformer from the line) and connects the motor directly to the line through the running contacts. A latch, fixed to the mechanism, is made to drop into a notch so that the operator is prevented from throwing the handle accidently to the run position first, however, when the handle is quickly pushed from the start to the 'run' position, the latch is kicked up to make the lever free for its forward motion. The operating handle is held in the run position by an electromagnet or under voltage (latch) coil until the stop button is pressed. If the supply fails or supply voltage drops to low value, the electromagnet will release and trip the holding mechanism. Overload protection is provided by thermal overload relays.

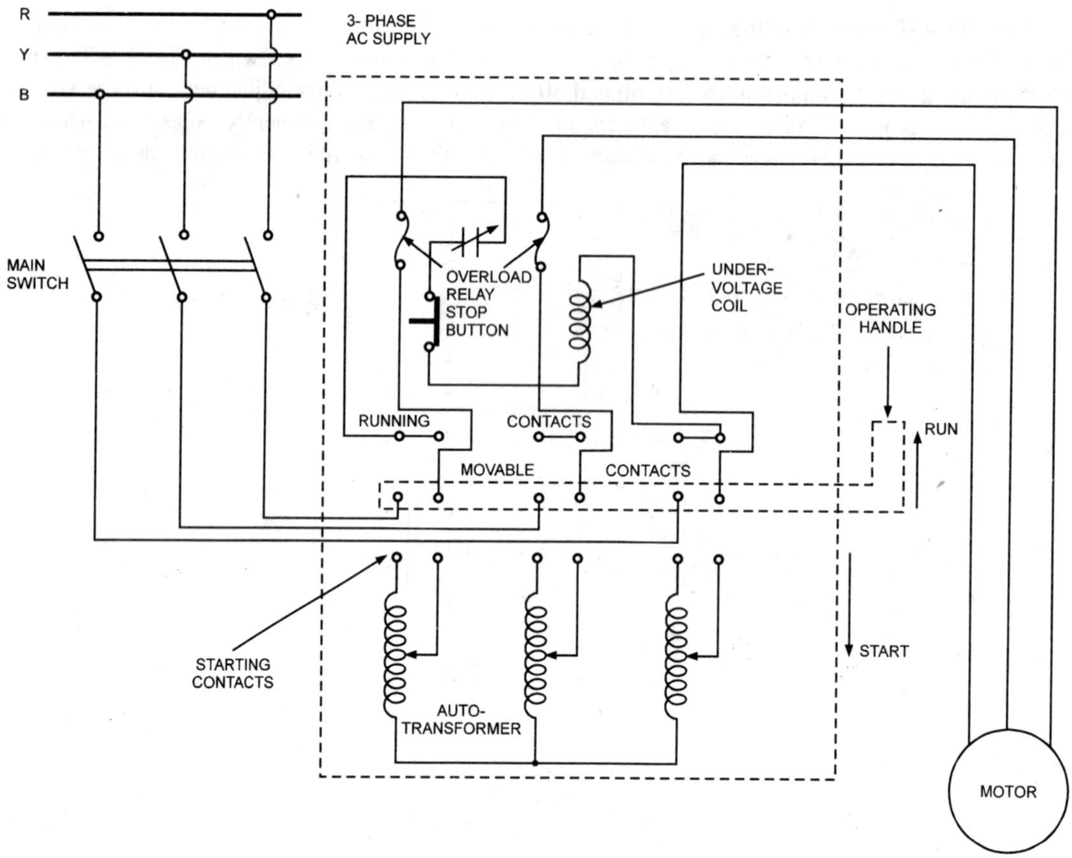

Fig. 3.6. ***Circuit Diagram of an Auto-Transformer Starter***

Let the motor be started by an auto-transformer having transformation ratio, k. If I_{sc} is the starting current when normal voltage is applied, and applied voltage to stator winding at starting = kV

then motor input current, $I_{st} = kI_{sc}$ ....(3.3)

$$\text{Supply current} = \text{Primary current of auto-transformer}$$
$$= k \times \text{secondary current of auto-transformer} = k^2 I_{sc}$$

$$\text{Starting torque, } T_{st} = T_f \left(\frac{I_{st}}{I_f}\right)^2 \times s_f = T_f k^2 \left(\frac{I_{sc}}{I_f}\right)^2 \times s_f \qquad ...(3.4)$$
$$= k^2 \times \text{Torque obtained by direct switching.}$$

Hence the line current and the starting torque are reduced in the square ratio.

The advantages of this method of starting of cage induction motors lies in the fact that the voltage is reduced by transformation and not by dropping the voltage in the resistors (or reactors) and, therefore, the current and power drawn from the supply mains are also reduced in comparison to primary resistor (or reactor) starting. The internal losses of the starter itself are small during long starting periods. Other advantages of this method are (*i*) availability of highest torque per ampere of supply current (*ii*) adjustment of starting voltage by selection of proper tap on the auto-transformer (*iii*) suitability for long starting periods (*iv*) closed transition starting and (*v*) motor current larger than supply current.

The drawbacks of such starters are low power factor and higher cost in case of lower output rating motors.

This method can be employed for starting of star-connected as well as delta-connected motors.

For starting of large cage motors (of output rating exceeding 20 kW) this method of starting is often used.

4. Star-Delta Starting. This method of starting of cage induction motors is based upon the principle that with 3 windings connected in star, the voltage across each winding is $1/\sqrt{3}$ *i.e.* 57.7 % of the line to line voltage whereas the same winding connected in delta will have full line-to line voltage across each.

The star-delta starter connects the three stator windings in star across the rated supply voltage at the starting instant. After the motor attains speed the same windings, through a change-over switch, are re-connected in delta across the same supply voltage.

The basic diagram of connection is shown in Fig. 3.7. An actual starter incorporates under-voltage and overvoltage coils, as shown in Fig. 3.8. The starter is also provided with a mechanical interlocking device to prevent the handle from being put in the 'Run' position first.

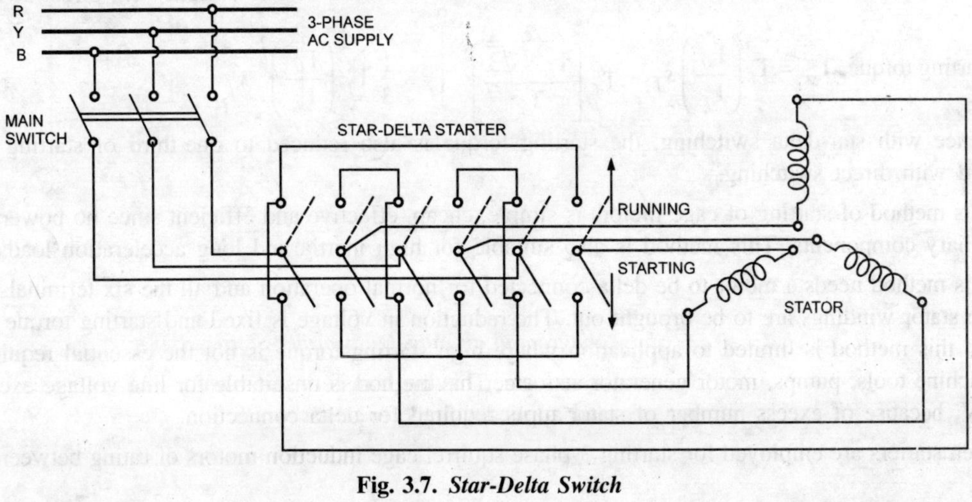

Fig. 3.7. ***Star-Delta Switch***

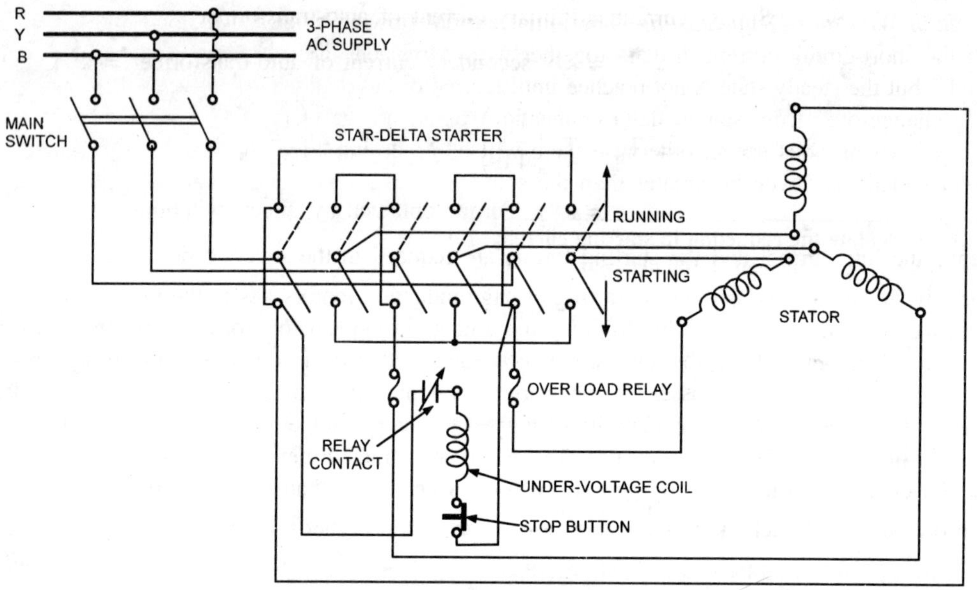

Fig. 3.8. ***Star-Delta Starter***

Since at starting instant, the stator windings are connected in star, voltage across each phase winding is reduced to $\frac{1}{\sqrt{3}}$ of line voltage and, therefore, starting current per phase becomes equal to $I_{sc}/\sqrt{3}$.

Starting line current by connecting the stator windings in star at the starting instant

$$= \text{Starting motor current per phase} = I_{sc}/\sqrt{3}.$$

Starting line current by direct switching with stator windings connected in delta = $\sqrt{3}\, I_{sc}$.

$$\therefore \quad \frac{\text{Line current with star-delta starting}}{\text{Line current with direct switching}} = \frac{I_{sc}/\sqrt{3}}{\sqrt{3}\, I_{sc}} = \frac{1}{3}$$

Hence by star-delta starting line current is reduced to one-third of line current with direct switching.

$$\text{Starting torque, } T_{st} = T_f \left(\frac{I_{st}}{I_f}\right)^2 s_f = T_f \left[\frac{I_{sc}/\sqrt{3}}{I_f}\right]^2 s_f = \frac{1}{3} T_f \left(\frac{I_{sc}}{I_f}\right)^2 s_f \qquad ...(3.5)$$

Hence with star-delta switching, the starting torque is also reduced to one-third of starting torque obtained with direct switching.

This method of starting of cage motors is simple, cheap, effective and efficient since no power is lost in auxiliary components. This method is also suitable for high inertia and long acceleration loads.

This method needs a motor to be delta-connected for normal operation and all the six terminals of the 3-phase stator windings are to be brought out. The reduction in voltage is fixed and starting torque is also low. So this method is limited to application where high starting torque is not the essential requirement *e.g.* machine tools, pumps, motor generator sets etc. This method is unsuitable for line voltage exceeding 3,000 V, because of excess number of stator turns required for delta connection.

Such starters are employed for starting 3-phase squirrel cage induction motors of rating between 4 and 20 kW.

Precaution With Star-Delta Starting. The initial current flowing when the motor is started in star is 57.7 % of the short-circuit current in delta together with a transient in each phase. The transient currents decay rapidly but the steady state is not reached until the motor has attained 70 per cent of its synchronous speed. The change-over from star to delta connection should not be made until the motor attains about 90 per cent of synchronous speed, otherwise there will be a current surge considerably greater than full-load current which may even be greater than the standstill current with star-connection.

Example 3.1. Calculate the reduction in starting current and starting torque when the supply voltage to a cage rotor motor is 80 per cent instead of 100 per cent.

Solution : Starting current with normal voltage $= I_{sc}$

Starting current with 80 % of normal supply voltage $= 0.8\, I_{sc}$

$$\text{Reduction in starting current} = \frac{I_{sc} - 0.8\, I_{sc}}{I_{sc}} \times 100 = 20\ \%\ \textbf{Ans.}$$

$$\text{Starting torque with normal voltage} = T_f \left(\frac{I_{sc}}{I_f}\right)^2 s_f$$

where I_f, s_f, T_f, and I_{sc} are the full-load current, full-load slip, full-load torque and short-circuit current with normal voltage applied to the stator respectively.

Starting torque with 80 % of normal supply voltage applied to the stator

$$= T_f \left(\frac{0.8\, I_{sc}}{I_f}\right)^2 s_f = 0.64 \text{ times of starting torque with normal supply voltage.}$$

Reduction in starting torque $= (1 - 0.64) \times 100 = 36\ \%$ **Ans.**

Example 3.2. A 3-phase induction motor has a ratio of maximum torque to full-load torque as 2.5 : 1. Determine the ratio of actual starting torque to full-load torque for star-delta starting. Given $R_2 = 0.4\ \Omega$ and $X_2 = 4\ \Omega$.

Solution : Ratio of maximum torque to full-load torque, according to Eq. (2.18), is given as

$$\frac{T_{max}}{T_f} = \frac{R_2^2 + s_f^2 X_2^2}{2\, s_f\, R_2\, X_2}$$

where R_2 is rotor resistance per phase, X_2 is standstill rotor reactance per phase and s_f is the full-load slip.

$$\text{or } 2.5 = \frac{0.4^2 + 4^2 s_f^2}{2 s_f \times 0.4 \times 4}$$

$$\text{or } s_f^2 - 0.5 s f + 0.01 = 0$$

$$\text{or } s_f = 0.02087 \text{ rejecting higher value}$$

$$\text{Full-load rotor current per phase, } I_f = \frac{s_f\, E_2}{\sqrt{R_2^2 + s_f^2 X_2^2}} \qquad \text{[see Eq. (2.10)]}$$

$$\text{Short-circuit rotor current per phase, } I_{sc} = \frac{E_2}{\sqrt{R_2^2 + X_2^2}} \qquad \because \text{ at standstill slip } s \text{ is } 1$$

$$\text{So } \frac{I_{sc}}{I_f} = \frac{\sqrt{R_2^2 + s_f^2 X_2^2}}{s_f \sqrt{R_2^2 + X_2^2}} = \frac{\sqrt{0.4^2 + 0.02087^2 \times 4^2}}{0.02087\sqrt{0.4^2 + 4^2}} = 4.87$$

Starting torque with star-delta starter, $T_{st} = \frac{1}{3} T_f \left(\frac{I_{sc}}{I_f}\right)^2 s_f = \frac{1}{3} \times T_f \times (4.87)^2 \times 0.02087 = 0.165\ T_f$

So the ratio of actual starting torque to full-load torque for star-delta starting = 0.165 **Ans.**

Example 3.3. A 12 kW, 3-φ, 6-pole, 50 Hz, 400 V, delta-connected induction motor runs at 960 rpm on full-load. If it takes 85 A on direct starting, find the ratio of the starting torque to full-load torque with a star-delta starter. Full-load efficiency and power factor are 88 % and 0.85 respectively.

Solution : Full-load line current drawn by a 3-phase induction motor (delta-connected)

$$= \frac{\text{Output in kW} \times 1{,}000}{\sqrt{3} \times \text{line voltage} \times \text{full-load efficiency} \times \text{full-load power factor}}$$

$$= \frac{12 \times 1{,}000}{\sqrt{3} \times 400 \times 0.88 \times 0.85} = 23.16 \text{ A}$$

$$\text{Full-load current per phase, } I_f = \frac{23.16}{\sqrt{3}} = 13.37 \text{ A}$$

$$\text{Short-circuit current per phase, } I_{sc} = \frac{85}{\sqrt{3}} \text{ A}$$

$$\text{Synchronous speed, } N_s = \frac{120 f}{P} = \frac{120 \times 50}{6} = 1{,}000 \text{ rpm}$$

$$\text{Full-load speed } f, N_f = 960 \text{ rpm}$$

$$\text{Full-load slip, } s_f = \frac{N_s - N_f}{N_s} = \frac{1{,}000 - 960}{1{,}000} = 0.04$$

$$\frac{\text{Starting torque}}{\text{Full-load torque}} = \frac{T_{st}}{T_f} = \frac{1}{3}\left(\frac{I_{sc}}{I_f}\right)^2 \times s_f = \frac{1}{3}\left(\frac{85/\sqrt{3}}{13.37}\right)^2 \times 0.04 = 0.18 \textbf{ Ans.}$$

Example 3.4. A 3-φ Δ-connected squirrel cage induction motor, when connected directly to 400 V, 3-φ, 50 Hz supply takes a starting current of 100 A in each stator phase. Calculate line current for direct on-line starting, and line and phase currents for Y-Δ starting.

Solution : Starting current (phase value), I_{SC} = 100 A

Line current for direct-on-line starting, $I_{st} = \sqrt{3} \times 100 = 173.2$ A **Ans.**

Line current for Y-Δ starting = One-third of line current with direct switching

$$= \frac{1}{3} \times 173.2 = 57.73 \text{ A } \textbf{Ans.}$$

Phase current for Y-Δ starting = Line current for Y-Δ starting = 57.73 A **Ans.**

Example 3.5. Determine approximately the starting torque of a 3-phase induction motor in terms of full-load torque when started by (*i*) star-delta starter (*ii*) an auto-transformer starter with 50 per cent tapping. The short-circuit current of motor is 5 times the full-load current and the full-load slip is 5 per cent.

Solution : Short-circuit current, $I_{sc} = 5\ I_f$

Full-load slip, s = 5 per cent or 0.05

(*i*) Starting torque with star-delta starter $= \frac{1}{3}\left(\frac{I_{sc}}{I_f}\right)^2 \times s_f\, T_f = \frac{1}{3}(5)^2 \times 0.05 = 0.4167\, T_f$

or 41.67% of full-load torque **Ans.**

(*ii*) Starting torque with an auto-transformer starter with 50 % tapping (*i.e.* k = 0.5)

$$= k^2 \left(\frac{I_{sc}}{I_f}\right)^2 s_f\, T_f = (0.5)^2 \times (5)^2 \times 0.05\, T_f = 0.3125\, T_f$$

or 31.25 % of full-load torque. **Ans.**

Example 3.6. A 50 kVA, 400 V, 3-phase, 50 Hz squirrel cage induction motor has full-load slip of 5 %. Its standstill impedance is 0.866 Ω/ phase. It is started using a tapped auto-transformer. If the maximum allowable supply current at the time of starting is 100 A, calculate the tap position and the ratio of starting torque to full-load.

Solution : Full-load current drawn by a 3-phase induction motor,

$$I_f = \frac{\text{Output in kVA} \times 1{,}000}{\sqrt{3} \times \text{line voltage}} \quad \text{neglecting losses}$$

$$= \frac{50 \times 1{,}000}{\sqrt{3} \times 400} = 72.2 \text{ A}$$

Short-circuit current, $I_{sc} = \frac{\text{Supply voltage (line value)}/\sqrt{3}}{\text{Standstill impedance per phase}} = \frac{400/\sqrt{3}}{0.866} = 266.7$ A

Supply current at start, I_{st} = 100 A

Tap position of auto-transformer, $k = \sqrt{\frac{I_{st}}{I_{sc}}} = \sqrt{\frac{100}{266.7}} = 0.6124$ or 61.24 % **Ans.**

Full-load slip, s = 5 % or 0.05

Ratio of starting torque to full-load torque,

$$\frac{T_{st}}{T_f} = k^2 \left[\frac{I_{sc}}{I_f}\right]^2 s_f = 0.6124^2 \left[\frac{266.7}{72.2}\right]^2 \times 0.05 = 0.256 \textbf{ Ans.}$$

Note : 3-phase induction motor has been assumed to be star-connected.

Example 3.7. A cage type induction motor when started by means of a star-delta starter takes 180 % of full-load line current and develops 35 % of full-load torque at starting. Calculate the starting torque and current in terms of full-load values, if an auto-transformer with 80 % tappings were employed.

Solution : Supply line current at start, I_{sc} = 3 times the current drawn by motor with star-delta starter

= 3 × 180 % of full-load current

= 540 % of full-load current or 5.4 I_f

Starting torque with star-delta starter, $T_{st} = \frac{1}{3}\left[\frac{I_{sc}}{I_f}\right]^2 s_f = 35\%$ of T_f or $0.35\, T_f$

$$\text{or} \quad \frac{T_{st}}{T_f} = \frac{1}{3}\left[\frac{I_s}{I_f}\right]^2 s_f = 0.35$$

$$\text{or} \quad \frac{1}{3} \times (5.4)^2 \times s_f = 0.35$$

$$\text{or full-load slip, } s_f = \frac{0.35 \times 3}{(5.4)^2} = 0.036$$

Auto-transformer transformation ratio, k = 80 % or 0.8

$$\text{Starting current (line value)} = k^2 I_{sc} = (0.8)^2 \times 5.4 \times I_f$$

$$= 3.456\, I_f \text{ or } 345.6 \text{ \% of full-load current} \quad \textbf{Ans.}$$

$$\text{Starting torque, } T_{st} = k^2 \left[\frac{I_{sc}}{I_f}\right]^2 s_f\, T_f$$

$$= (0.8)^2 \times (5.4)^2 \times 0.036\, T_f$$

$$= 0.672\, T_f \text{ or } 67.2 \text{ \% of full-load torque} \quad \textbf{Ans.}$$

Example 3.8. The full-load slip of a 400-volts, 3-phase cage induction motor is 3.5 % and with locked rotor, full load current is circulated when 92 volts is applied between lines. Find the necessary tapping on an auto-transformer to limit the starting current to twice the full-load current of the motor. Determine also the starting torque in terms of the full-load torque.

Solution : Short-circuit current with full voltage (rated voltage) applied,

$$I_{sc} = \frac{400}{92} I_f = \frac{100}{23} I_f \text{ amperes}$$

$$\text{Limiting line current, } I_{st} = 2\, I_f \qquad \text{...(given)}$$

$$\text{Tapping required on auto transformer, } k = \sqrt{\frac{I_{st}}{I_{sc}}} = \sqrt{\frac{2\, I_f}{\frac{100}{23} I_f}} = \sqrt{0.46} = 0.678 \text{ or } 67.8 \text{ \%} \quad \textbf{Ans.}$$

$$\text{Starting torque, } T_{st} = k^2 \left(\frac{I_{sc}}{I_f}\right)^2 s_f\, T_f = 0.46 \times \left(\frac{100}{23}\right)^2 \times 0.035 \times T_f = 0.304\, T_f$$

i.e. 30.4 % of full-load torque **Ans.**

Example 3.9. A 20 hp, 400 V, 950 rpm, three phase, 50 Hz, 6-pole cage induction motor draws 6 times the full-load current at standstill from a 400 V mains and develops 1.8 times full-load running torque. The full-load current is 30 A.

(*i*) What voltage must be applied to produce full-load torque at starting ?

(*ii*) What current will this voltage produce ?

(*iii*) If this voltage is obtained by an auto-transformer, what will be the line current ?

(*iv*) If the starting current in the line is limited to full-load current by means of an auto-transformer, what will be the starting torque as a percentage full-load torque?

Assume full-load slip as 5% and neglect magnetising current and the stator impedance drops.

Solution : Short-circuit current, $I_{sc} = 6\, I_f = 6 \times 30 = 180$ A

Starting torque developed with direct switching (400 V applied to the stator),

$$T_{st} = 1.8\, T_f$$

(*i*) Let the voltage applied to develop starting torque T_{st} equal to full-load torque be V′ volts.

$$\because \text{ Starting torque } \alpha \text{ (applied voltage)}^2$$

$$\therefore \left(\frac{V'}{V}\right)^2 = \frac{T'_{st}}{T_{st}} = \frac{T_f}{1.8\, T_f} = \frac{1}{1.8}$$

$$\text{or } V' = \sqrt{\frac{1}{1.8}} \times V = \frac{400}{\sqrt{1.8}} = 298.14 \text{ V } \textbf{Ans.}$$

(*ii*) Current produced at start by V′ (298.14 V) $= I_{st} \times \frac{V'}{V} = 180 \times \frac{298.14}{400} = 134.163$ A **Ans.** $\because I_{st} \propto V$

$\because$ starting current drawn by motor is proportional to applied voltage

$$\text{Transformation ratio, } k = \frac{V'}{V} = \frac{298.14}{400} = 0.7454$$

(*iii*) Line current with auto-transformer of ratio k of $0.7454 = k^2 I_{sc} = (0.7454)^2 \times 180 = 100$ A **Ans.**

(*iv*) Auto transformer transformation ratio to produce starting line current equal to full-load current

$$k = \sqrt{\frac{I_{st}}{I_{sc}}} = \sqrt{\frac{I_f}{6\, I_f}} = 0.408.$$

$$\text{Starting torque} = k^2 \left[\frac{I_{sc}}{I_f}\right]^2 s_f\ T_f = 0.408^2\, [6]^2 \times 0.05\, T_f = 0.3\, T_f$$

i.e. 30 per cent of full-load torque **Ans.**

3.2. STARTING OF SLIP-RING INDUCTION MOTORS

Though the methods described in the preceding section can also be employed for starting of 3-phase slip-ring induction motors but this is not done in general because then the advantages of wound rotor cannot be realized fully. *Rotor resistance starter* cannot be employed in case of squirrel cage type motors because external resistance cannot be introduced in the cage rotor.

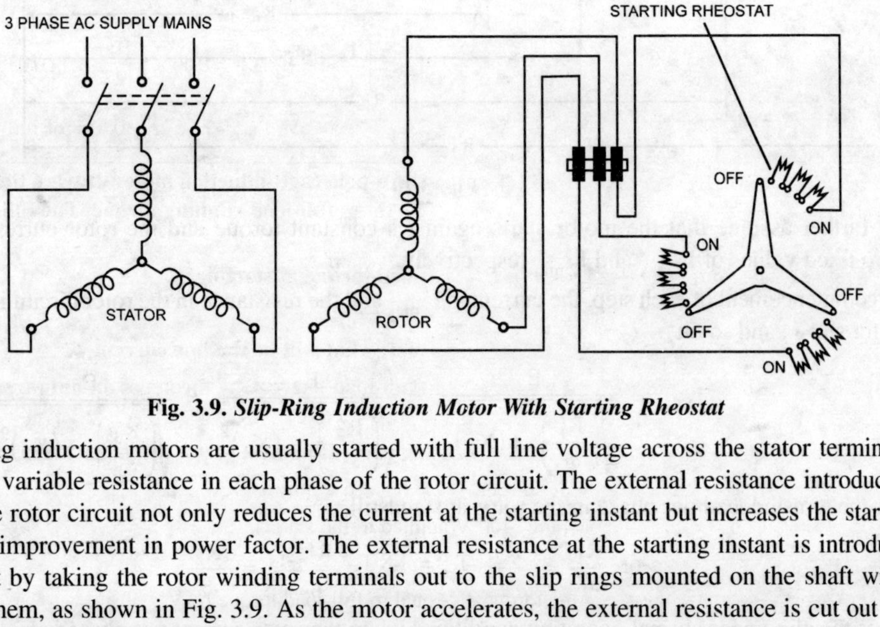

Fig. 3.9. ***Slip-Ring Induction Motor With Starting Rheostat***

Slip-ring induction motors are usually started with full line voltage across the stator terminals and by introducing variable resistance in each phase of the rotor circuit. The external resistance introduced in each phase of the rotor circuit not only reduces the current at the starting instant but increases the starting torque also due to improvement in power factor. The external resistance at the starting instant is introduced in the rotor circuit by taking the rotor winding terminals out to the slip rings mounted on the shaft with brushes resting on them, as shown in Fig. 3.9. As the motor accelerates, the external resistance is cut out in steps so that the available electromagnetic torque remains maximum during the accelerating period. Ultimately when the motor attains the normal speed, the rotor winding is short circuited automatically through the slip rings by means of a metal collar which is pushed along the shaft under the slip rings touching them all there by eliminating the resistance of brush contacts on the slip rings. The brushes are then lifted to reduce the wear and frictional losses.

For small and medium size motors face-plate starters are employed but for large size motors liquid starters are employed.

A starter is also provided with no-voltage (or under-voltage) and overcurrent protective devices. It is also provided with some type of interlocking to ensure proper sequential operation of the line contactor and starter. This interlocking arrangement prevents the closing of stator contactor unless the starter is 'all in'.

Since slip-ring induction motor develops high starting torque due to introduction of external resistance in the rotor circuit at the starting instant with reasonable starting current, such motors can be started under load also.

3.2.1. Calculations of Resistance Steps For 3-phase Slip-ring Induction Motor Starters

Consider one phase of a slip-ring rotor with resistance R_2 and standstill reactance X_2. Let $r_1, r_2 ... r_{n-1}$ be the resistances of the different steps in each phase and $R'_1, R'_2, R'_3 ... R'_{n-1}$ and R'_n be the total resistance per phase in the rotor circuit on the 1*st*, 2*nd*, 3*rd*...(*n* – 1)*th* and *nth* stud respectively (Fig. 3.10).

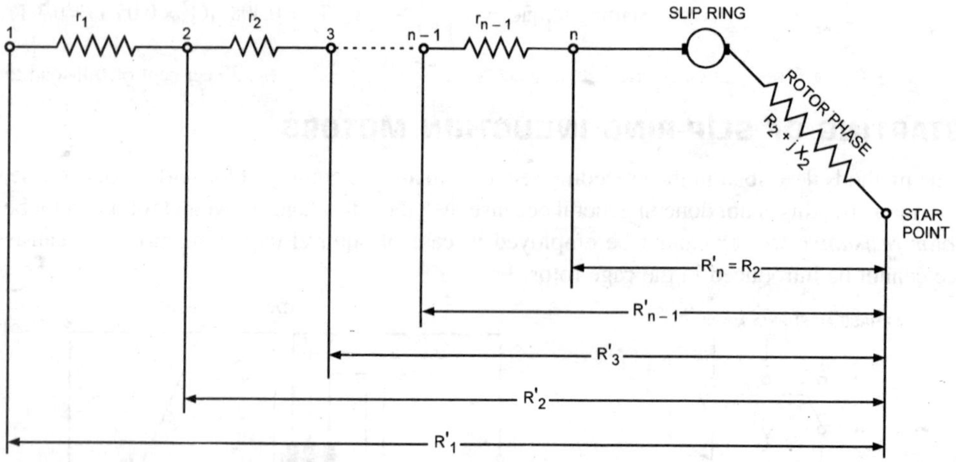

Fig. 3.10

Let us further assume that the motor starts against a constant torque and the rotor current fluctuates between two fixed values of $I_{2\,max}$ and $I_{2\,min}$ respectively.

At the commencement of each step, the current is $I_{2\,max}$ and the resistance in the rotor circuit is R'_1 for slip $s_1 = 1$, R_2' for slip s_2 and so on

$$I_{2max} = \frac{E_2}{\sqrt{\left(\frac{R_1'}{s_1}\right)^2 + (X_2)^2}} = \frac{E_2}{\sqrt{\left(\frac{R_2'}{s_2}\right)^2 + (X_2)^2}} = \frac{E_2}{\sqrt{\left(\frac{R_3'}{s_3}\right)^2 + (X_2)^2}} = ...$$

where E_2 is the emf induced per phase in the rotor at standstill

$$\text{or} \quad \frac{R_1'}{s_1} = \frac{R_2'}{s_2} = \frac{R_3'}{s_3} = ... = \frac{R'_{n-1}}{s_{n-1}} = \frac{R_2}{s_{max}} \quad ...(3.6)$$

where s_{max} is the slip under normal operating conditions when the rotor current per phase is $I_{2\,max}$.

When the motor is switched on, it is on 1st stud, the resistance in the rotor circuit per phase is R_1', the slip s_1 is unity and rotor current per phase is the maximum permissible value $I_{2\,max}$. As the motor gains speed, the current decreases till it becomes $I_{2\,min}$ at slip s_2,

$$\text{Hence } I_{2\,min} = \frac{E_2}{\sqrt{\left(\frac{R_1'}{s_2}\right)^2 + (X_2)^2}}$$

Similarly for other studs also

$$I_{2\,min} = \frac{E_2}{\sqrt{\left(\frac{R_2'}{s_3}\right)^2 + (X_2)^2}} = \frac{E_2}{\sqrt{\left(\frac{R_3'}{s_4}\right)^2 + (X_2)^2}} = \ldots$$

$$\text{So that } \frac{R_1'}{s_2} = \frac{R_2'}{s_3} = \frac{R_3'}{s_4} = \ldots \frac{R_{n-1}'}{s_{max}} \qquad \ldots(3.7)$$

Combining Eqs. (3.6) and (3.7) we get

$$\frac{s_2}{s_1} = \frac{s_3}{s_2} = \frac{s_4}{s_3} = \ldots \frac{s_{max}}{s_{n-1}} = \frac{R_2'}{R_1'} = \frac{R_3'}{R_2'} = \frac{R_4'}{R_3'} \ldots \frac{R_2}{R'_{n-1}} = \text{K (say)}$$

From Eq. (3.6) total resistance in the rotor circuit per phase on first stud is

$$R_1' = R_2 \times \frac{s_1}{s_{max}} = \frac{R_2}{s_{max}} \qquad \ldots(3.8) \quad \text{since } s_1 = 1$$

$$\text{while thereafter } R_2' = K R_1';\ R_3' = KR_2' = K^2 R_1'$$

$$\text{and } R_2 = KR'_{n-1} = K^{n-1} R_1' = K^{n-1} \frac{R_2}{s_{max}}$$

$$\text{or } K = (s_{max})^{\frac{1}{n-1}} \qquad \ldots(3.9)$$

where $(n - 1)$ is number of sections in the starter

The resistances of various sections are determined as follows:

$$r_1 = R_1' - R_2' = R_1'(1 - K)$$

$$r_2 = R_2' - R_3' = R_1\,(K - K^2) = Kr_1$$

$$r_3 = R_3' - R_4' = R_1'\,(K^2 - K^3) = Kr_2 = K^2 r_1 \text{ etc.}$$

Example 3.10. Design the five sections of a 6-stud rotor starter for a 3-phase wound rotor induction motor. The slip at full-load is 2 per cent and the starting current is 1.5 times the full-load current. The resistance of rotor is 0.02 Ω per phase.

Solution : Full-load slip, s_f = 2 % or 0.02

$$\text{Rotor current at full-load is given as } I_{2f} = \frac{s_f E_2}{\sqrt{R_2^2 + s_f^2 X}} \simeq \frac{s_f E_2}{R_2}$$

∵ when slip s is very low, the value of term sX_2 is very small and can be neglected *i.e.* slip s directly varies with rotor current

So slip at 1.5 times full-load current, s_{max} = 1.5 × 0.02 = 0.03

The total resistance in the rotor circuit per phase on first stud,

$$R_1' = \frac{R_2}{s_{max}} = \frac{0.02}{0.03} = 0.6667\ \Omega$$

$$\text{and } K = (s_{max})^{\frac{1}{n-1}} = (0.03)^{\frac{1}{5}} = 0.496$$

The resistances of the various sections of the starter are, therefore,

$$r_1 = (1 - K)\, R_1' = (1 - 0.496) \times 0.6667 = 0.336\ \Omega$$

$$r_2 = K\, r_1 = 0.496 \times 0.336 = 0.1667\ \Omega$$

$$r_3 = K^2\, r_1 = (0.496)^2 \times 0.336 = 0.0826\ \Omega$$

$$r_4 = K^3\, r_1 = (0.496)^3 \times 0.336 = 0.041\ \Omega$$

$$\text{and } r_5 = K^4\, r_1 = (0.496)^4 \times 0.336 = 0.0203\ \Omega$$

3.3. SPEED CONTROL OF 3-PHASE INDUCTION MOTORS

The problem of speed control of electrical motors in general and of induction motors in particular is of great practical importance.

In a number of industries motors must satisfy very strict speed characteristic requirements, both with respect to the range and smoothness of control and also with respect to economical operation. From the view point of speed control characteristics, induction motors are inferior to dc motors. The speed of a dc shunt motor can be adjusted between wide range with good efficiency and speed regulation, but in induction motors speed cannot be varied without loosing efficiency and good speed regulation.

The speed of an induction motor is given by the expression $N = \frac{120f}{P}(1 - s)$. Thus there are three factors viz, supply frequency *f*, number of poles P and slip s on which the speed of an induction motor depends. Hence to change the speed of an induction motor it is essential to change at least one of the above three factors.

Methods of speed control are distinguished according to the main action on the motor : (*i*) from the stator side, and (*ii*) from the rotor side.

Various methods of speed control from stator side are (*a*) variation of supply frequency (*b*) variation of applied voltage and (c) by varying the resistance or reactance connected in between the supply and stator terminal. From the rotor side the speed may be controlled (*a*) by changing the resistance in the rotor circuit and (*b*) by introducing into the rotor circuit an additional emf of the same frequency as the fundamental emf of the rotor.

For the latter method of speed control of induction motors, an additional electrical machine or several such machines are required. A set consisting of a regulated induction motor and one or more additional electrical machines connected to it electrically or mechanically is called *cascade.* Commutator machines are commonly used as the additional machines.

3.3.1. Speed Control By Variation of Supply Voltage (or Stator Voltage Control)

This is a slip-control method with constant frequency variable supply voltage. In this method of speed control of induction motors, the voltage applied to the stator is varied for varying the speed.

By reference to Eq. (2.14) it is seen that, for a given slip and constant internal parameters (rotor resistance and standstill reactance), the torque developed by an induction motor is directly proportional to the square of induced emf in rotor, E_2, which is approximately proportional to the impressed voltage, V. Thus torque developed is proportional to the square of supply voltage, V. By reference to Art. 2.10.3, it is seen that slip at maximum torque is independent of supply voltage. The variation in supply voltage does not cause any change in synchronous speed. Torque-speed characteristics for a particular induction motor for rated impressed voltage,

$\frac{1}{\sqrt{2}}$ rated voltage and one-half the rated value are given in Fig. 2.17. It readily can be seen from the torque-speed characteristics that for a given load, the speed of motor cannot be varied over a wide range:

From Eq. (2.12), rotor current is given as

$$I_2 = \frac{E_2}{\sqrt{\left(\frac{R_2}{s}\right)^2 + (X_2)^2}}$$

and rotor current with reference to stator is given as

$$I_2' = \frac{V}{\sqrt{\left(\frac{R_2'}{s}\right)^2 + (X_2')^2}}$$

where R_2' and X_2' are the rotor resistance and standstill reactance per phase as referred to stator, s is slip and V is supply voltage per phase.

Around the rated speed, slip of the induction motor is so small that $\frac{R_2'}{s}$ is much greater than X_2' and so X_2' can be neglected in comparison to $\frac{R_2'}{s}$ and rotor current referred to stator is given

$$I_2' = \frac{sV}{R_2'} \quad ...(3.10)$$

In Art 2.13, we have seen that torque developed by an induction motor, $T \propto sV^2$

From the above relations it is revealed that for an induction motor operating at full-load slip, if the slip is to be doubled for constant load torque, then the supply voltage must be reduced by a factor of $\frac{1}{\sqrt{2}}$ and consequently the rotor current I_2' will become $\sqrt{2}$ times of the full-load value. The motor, therefore, tends to get overheated. It is, therefore, necessary to take care, while using this method of speed control, that the motor is not damaged due to excessive current at the reduced speeds.

This method of speed control is simple, low in first cost and has low maintenance cost but it has limited use because (*i*) the operation at voltages exceeding rated voltage is restricted by magnetic saturation, (*ii*) a large change in voltage is required for a relatively small change in speed (*iii*) the developed torque reduces greatly with the reduction in supply voltage and the motor will come to standstill if the load torque exceeds the pull-out torque of the motor corresponding to reduced supply voltage and (*iv*) the range of speed control is very limited in the downward direction *i.e.*, from rated speed to lower speeds.

In general, this method of speed control is used only on loads where the torque required drops off considerably as the speed is reduced such as with small squirrel cage motors driving fans.

The variable voltage may be obtained by means of either saturable reactors, variac, or tap-changing transformers.

3.3.2. Stator Resistance Control

This method is also stator voltage control. In this method of speed control resistances (or reactances) are connected in series with each stator lead, as shown in Fig. 3.3. The voltage drop in series resistors (or reactors) cause reduction in the voltage across the motor stator terminals. Thus, by varying the resistance (or reactance) connected in between the supply and motor stator, the voltage across the stator terminals can be varied.

Stator resistance control method involves wastage of energy in controlling resistances.

3.3.3. Speed Control By Variation of Supply Frequency (or Frequency Control)

This method of speed control provides wide speed-control range with gradual variation of the speed throughout this range. The major difficulty with this method is how to get the variable frequency supply. The auxiliary equipment required for this purpose results in a high first cost, increased maintenance and lowering of the overall efficiency. That is why, this method is not employed for general-purpose speed control applications. Inspite of the fact that this scheme is complicated, there are certain applications in which its wide, continuously variable, speed range and good speed regulation makes its use highly desirable.

If an induction motor is to be operated at different frequencies with practically constant values of efficiency, power factor, overload capacity and a constant absolute slip, then, with the iron unsaturated, it is essential that the supply voltage V be varied with the change in frequency according to the equation

$$\frac{V'}{V} = \frac{f'}{f}\sqrt{\frac{T'}{T}} \qquad ...(3.11)$$

where V′ and T′ are the voltage and torque corresponding to the frequency f' and V and T to the frequency f.

For constant torque *i.e.* T' = T, we have

$$\frac{V'}{V} = \frac{f'}{f} \qquad ...(3.12)$$

i.e., the voltage applied to the stator must vary in proportion to the frequency.

This kind of variation can occur for example, if the supply generator is subjected to the speed variations, either due to momentary overloads or because of a noticeable speed regulation. Both output voltage and frequency vary as the speed if no automatic correction is provided. On some large marine drives, the propellor motors are induction type and are speed controlled from such a local supply which is provided by synchronous generators coupled to variable speed turbines. Even in such cases the range of speed variation is limited, because the efficiency of the prime movers falls rapidly with the change in speed from that for what they are designed.

From the emf equation, it can be understood that a change of frequency will result in a change of flux level unless the induced emf is changed in the same ratio. An imbalance in this ratio will cause either an excessive flux and saturation, or reduced flux and reduced torque per ampere of current. Excessive flux created due to reduction in frequency made for the purpose of reducing speed at constant voltage will cause increase in iron losses and increased iron losses alongwith reduced ventilation caused by the lower speed can result in excessive heating.

Constant flux per pole is approximated if the applied voltage per hertz $\frac{V}{f}$ is maintained constant, because $V \simeq E_1$, induced emf in stator, However, an idealized variable-frequency control scheme might aim to ensure that any particular torque should be obtained at the same flux $\frac{E_1}{f}$, as when operating at normal voltage and frequency. For any specified flux level torque developed is proportional to

$$I_2' \cos \phi_2 \;\alpha\; \frac{sE_1}{Z_2'} \cdot \frac{R_2'}{Z_2'} \;\alpha\; \frac{N_s - N}{N_s} . E_1 . \frac{R_2'}{(Z_2')^2}$$

Thus with this criterion of controlling $\frac{E_1}{N_s}$ to a constant value for particular torque, the speed difference $(N_s - N)$ is also fixed for a given torque. Since $(N_s - N)$ determines the slip frequency f, the emf induced in rotor, rotor impedance and rotor current, so these quantities remain unchanged. From this it follows, as illustrated in Fig. 3.11, that as the frequency is reduced from a value corresponding to N_s say, the torque-speed characteristic curves are of the same shape but displaced horizontally. Under these conditions the maximum developed torque remains approximately constant and the output power varies approximately in direct proportion to the synchronous speed.

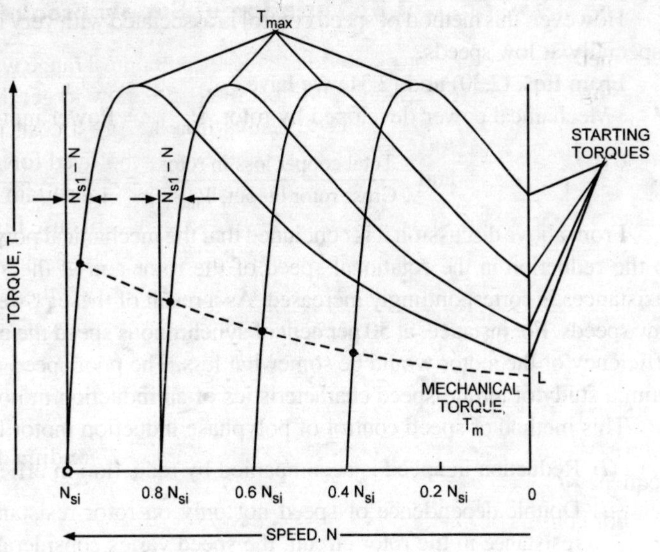

Fig. 3.11

With rapid growth made in power electronics and circuits combined with ruggedness and higher efficiency of an induction motor, motors with this type of drive are replacing dc motors in several applications. Variable frequency supply may be had from various types of frequency conversion equipments such as adjustable frequency generators, rotating frequency changers etc. However, dc link invertors are presently most popular in practice for providing variable frequency variable voltage supply to induction motors for smooth and wide range speed control.

3.3.4. Speed Control By Variation of Rotor Resistance (or Rotor Resistance Control)

As the name implies, this type of control is only possible with wound rotor induction motors *i.e.* this method cannot be applied to squirrel cage motor.

Wound rotor motors are usually started by connecting starting resistances in the secondary circuit (Refer to Art. 3.2), which are shorted out as the motor speeds up. If the ohmic values of these resistors are properly chosen and if these resistors are designed for continuous operation, they can serve dual purpose — starting and speed control. This method of speed control has characteristics similar to those of dc shunt motor speed control by means of resistance in series with the armature.

From Eq. (2.14) torque developed by an induction motor is given as $T = \dfrac{ks\, R_2\, E_2^2}{R_2^2 + s^2 X_2^2}$

When the speed is very near to synchronous speed N_s *i.e.*, when slip s is very low the value of the term sX_2 is very small and can be neglected as compared to R_2 and torque developed becomes proportional to $\dfrac{s}{R_2}$ $\left(i.e.\ T \propto \dfrac{s}{R_2}\right)$. So it is obvious that for a given torque, slip s can be increased (or speed can be reduced) by increasing the rotor resistance. It is also obvious that by this method speeds only below rated speed can be had. This method of speed control is stepped. However, the larger the number of steps of the external resistances, the smoother is the speed control.

Secondary resistance control is simple and low in initial as well as in maintenance cost and it is possible to have a large starting torque with low starting current and large pull-out torques at small values of slip, as obvious from Fig. 2.15 (*b*). The circuit arrangement is the same as shown in Fig. 3.9.

However, this method of speed control is associated with very large power losses in the external resistances, especially at low speeds.

From Eqs. (2.30) and (2.31) we have

Mechanical power developed by rotor, P_{mech} = Power input to rotor $(1 - s)$

$$\text{and} \quad \frac{\text{Total copper loss in rotor}}{\text{Gross rotor output, } P_{mech}} = \frac{s}{1-s} = \frac{(N_s - N)/N_s}{1-s} = \frac{N_s - N}{N_s(1-s)} = \frac{N_s - N}{N} \quad ...(3.13)$$

From above discussion it is concluded that the mechanical power developed is reduced in direct proportion to the reduction in the rotational speed of the rotor and at the same time power losses in the rotor circuit resistances is correspondingly increased. As a result of these power losses, the efficiency is largely reduced at low speeds. For instance, at 50 per cent of synchronous speed the efficiency of rotor is only 50 per cent and the efficiency of the motor would be somewhat less. The poor speed regulation at low speeds can be appreciated from a study of torque-speed characteristics of an induction motor [Fig. 2.15 (*a*)].

This method of speed control of polyphase induction motor has the following disadvantages.

(*i*) Reduction in speed is accompanied by reduction in efficiency.

(*ii*) Double dependence of speed not only on rotor resistance R_2 but on load as well— with a large resistance in the rotor circuit, the speed varies considerably with variation in torque [Fig. 2.15 (*a*)].

(*iii*) The external rotor resistors are comparatively bulky and expensive as they have to dissipate a good deal of power without getting overheated.

This method of speed control as such is, therefore, not suitable for controlling speed at constant torque. But this method is widely used for loads where the torque required drops off considerably as the speed is reduced such as fan loads for which the power input drops noticeably as the speed is reduced which in turn reduces the rotor copper losses. Moreover, this method is not adopted for continuous speed control but is preferred for intermittent (short-time) operation. Such a method of speed control is widely used in practice, for low-power motors and in overhead cranes. It is sometimes used however, for speed regulation of rolling mills especially where they are provided with flywheels for reducing the load peaks in the circuit. Here the rheostat, called the *speed regulator,* is automatically switched on when the load increases, and as a result speed decreases and part of the load is compensated at the expense of kinetic energy of the flywheel. Conversely, when the load decreases, the rotor circuit resistance is reduced; the speed increases and the flywheel begins to store kinetic energy.

3.3.5. Slip Power Recovery Control

In this method, the speed of an induction motor is controlled by injecting a voltage of slip frequency in the secondary circuit. If injected voltage is in phase with the induced emf in the secondary circuit, it is equivalent to decreasing of resistance in the secondary circuit, therefore, slip decreases or speed increases. It is possible to achieve super-synchronous speeds by injecting a voltage of slip frequency in phase with the induced emf in the rotor circuit. If injected voltage is in phase opposition to the induced emf in the secondary circuit, it is equivalent to increasing of resistance in the secondary circuit, hence resulting in increase in slip or reduction in speed. Thus a large range of speed control is possible by this method by including frequency-converting equipment in the rotor circuit.

In this method the disadvantages of lower efficiency and poor speed regulation are overcome, but, since it requires one and usually more auxiliary machines to inject emf of slip frequency in the secondary circuit, therefore, it is costlier and used with motors of very large rating, such as for motors in steel mills.

Several methods have been developed to control speed of induction motors by injecting a voltage of slip frequency in the secondary circuit. The different methods vary in their requirements for auxiliary equipment; from the incorporation of the auxiliary equipment in the design of induction motor itself to a fairly complicated system of auxiliary rotating machines and adjustable ratio transformers. The basic creteria that such a system must ful fill are : (*i*) that the variable foreign voltage have a slip frequency *i.e.* frequency equal to *sf* and (*ii*) that the energy associated with the foreign voltage be recoverable. A comprehensive discussion of this method of speed control is beyond the scope of this book. The most common systems based on this principle viz the Kramer system and Scherbius system of speed control of polyphase induction motors will be discussed in this very chapter in the following sections.

3.4. SPEED CONTROL BY CONCATENATION

In this method of speed control, two motors are required, at least one of which must have a wound rotor. The speed is changed by changing the slip of one motor, which changes the frequency supplied to the other motor. In this method of speed control, instead of dissipating energy in the control rheostat, this energy can be supplied to another induction motor connected to the first one as a cascade, and converted into mechanical energy. Hence the power drawn from the power circuit can be utilised more effectively.

For normal operation of the cascade, motor I and II (Fig. 3.12) must be joined both electrically and mechanically. An electrical connection alone does not suffice, since motor I begins to operate from a power circuit with normal voltage and frequency, and runs at normal speed, whereas motor II receives a very low voltage with a very low frequency from the slip rings of motor I. Hence the motor does not develop any torque and is only an additional resistance introduced into the rotor circuit of motor I.

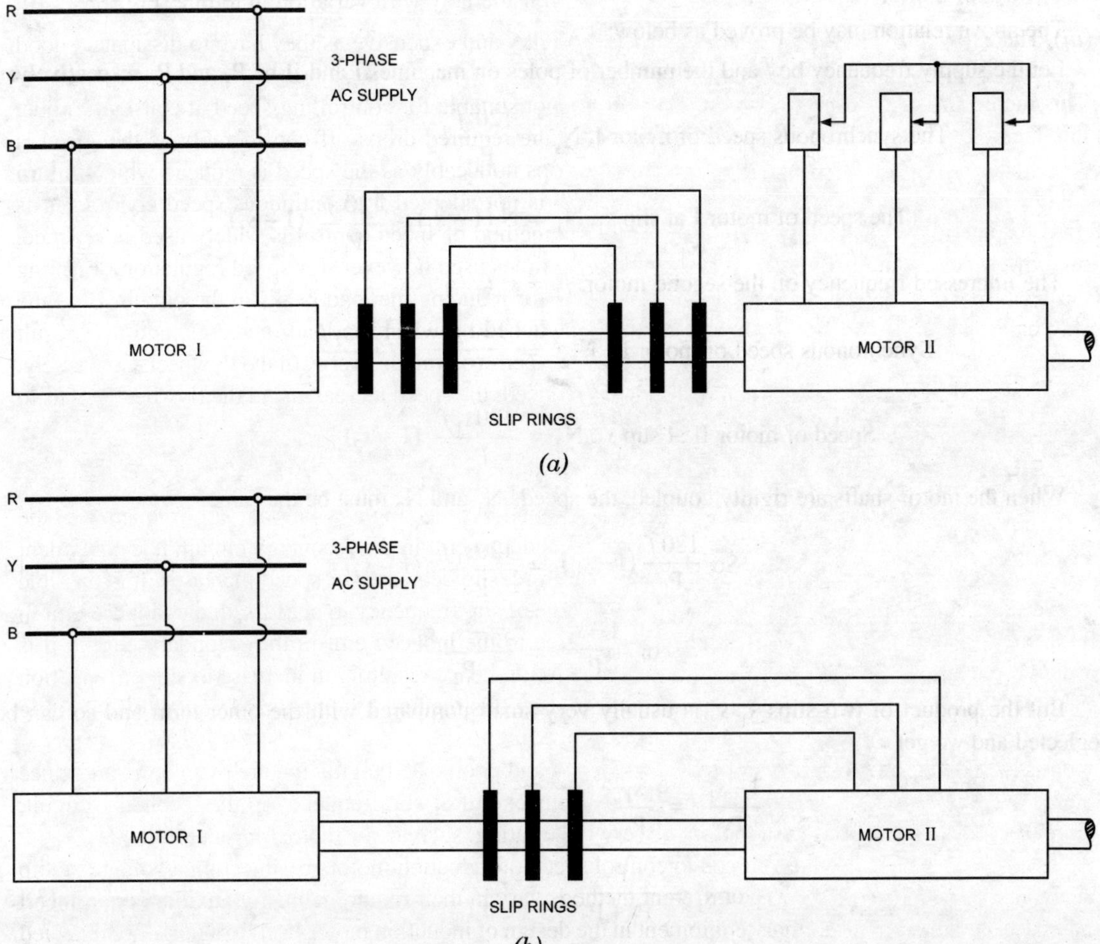

Fig. 3.12. ***Cascade Connections of Induction Motors***

Motors I and II should be so selected that their characteristics correspond to each other. Thus, for instance, if both motors are identical in design (conditions prevailing in railway traction), the stators of both the motors being high-voltage circuits and the rotor, low-voltage circuits, it is necessary to interconnect the rotors of both motors; to connect the rotor of motor I to the stator of motor II would not be rational. In special installations motor II may have a squirrel cage rotor; here the rotor of motor I is connected to the stator of motor II, as shown in Fig. 3.12 (*b*), but the characteristics of both circuits should be matched. Motor I should have a one-to-one ratio of transformation between stator and rotor so that at standstill with the external circuit of rotor open, the voltage across the rotor slip rings should be equal to line voltage. When two rotors are connected together, as shown in Fig. 3.12 (*a*), it is possible to dispense with slip rings, but they are generally retained to permit the use of either motor alone. The variable resistances are employed to start the unit and are short circuited during running condition. The speed of the combination will be determined by the sum or difference of the number of poles in the two machines. In practice it is customary, to connect the two machines in such a way that the phase rotation of the stator fields of both motors is in the same direction. The synchronous speed of the cascade set, in this case, is given as $N_s = \dfrac{120\,f}{P_1 + P_2}$

where f is supply frequency and P_1 and P_2 are the number of poles on machines I and II respectively.

The above relation may be proved as below:

Let the supply frequency be f and the number of poles on machines I and II be P_1 and P_2 respectively.

$$\text{The synchronous speed of motor I, } N_s = \frac{120\,f}{P_1}$$

$$\text{The speed of motor I at slip } s_1,\ N_1 = N_s(1 - s_1) = \frac{120\,f}{P_1}(1 - s_1)$$

$$\text{The impressed frequency on the second motor, } f_2 = s_1 f$$

$$\text{Synchronous speed of motor II, } N_{s_2} = \frac{120\,f_2}{P_2} = \frac{120 s_1 f}{P_2}$$

$$\text{Speed of motor II at slip } s_2,\ N_2 = \frac{120 s_1 f}{P_2}(1 - s_2)$$

When the motor shafts are rigidly coupled, the speeds N_1 and N_2 must be the same

$$\text{So } \frac{120 f}{P_1}(1 - s_1) = \frac{120 s_1 f}{P_2}(1 - s_2) \qquad ...(3.14)$$

$$\text{or } \frac{1 - s_1}{P_1} = \frac{s_1 - s_1 s_2}{P_2}$$

But the product of two slips $s_1\ s_2$ is usually very small compared with the other term and so can be neglected and we get

$$\frac{1 - s_1}{P_1} = \frac{s_1}{P_2}$$

$$\text{or } s_1 = \frac{P_2}{P_1 + P_2} \qquad ...(3.15)$$

$$\text{and concatenated speed of the set, } N = \frac{120\,f}{P_1}(1 - s_1) = \frac{120\,f}{P_1}\left(1 - \frac{P_2}{P_1 + P_2}\right) = \frac{120\,f}{P_1 + P_2} \qquad ...(3.16)$$

Now if the two motors are so connected electrically that the torques of the two motors are in opposite directions, then Eq. (3.14) becomes $\frac{120f}{P_1}(1-s_1) = -\frac{120 s_1 f}{P_2}(1-s_2)$

$$\text{or} \quad \frac{1-s_1}{P_1} = \frac{s_1 s_2 - s_1}{P_2}$$

Again neglecting $s_1 s_2$ we have

$$s_1 = \frac{P_2}{P_2 - P_1} \qquad ...(3.17)$$

$$\text{and concatenated speed of the set, N} = \frac{120f}{P_1}(1-s_1) = \frac{120f}{P_1}\left(1-\frac{P_2}{P_2-P_1}\right) = \frac{120f}{P_1 - P_2} \qquad ...(3.18)$$

If the two machines are operated individually, then

$$\text{Synchronous speed of machine I, } N_{s1} = \frac{120f}{P_1} \qquad ...(3.19)$$

$$\text{and synchronous speed of machines II, } N_{s2} = \frac{120f}{P_2} \qquad ...(3.20)$$

Thus four different speeds are possible by cascade connections of two induction motors. Since the shaft torque is largely reduced in differential cascade [Eq. (3.18)], so it is not employed in practice.

If the two motors have the same number of poles and are connected in cumulative cascade [Eq. (3.16)], they will operate at practically half speed.

The power distribution in a cascade is the same as in a normal motor. The power input to rotor of machine I is equal to the power supplied to the stator of motor 1 minus the copper and core losses. The power input to the rotor of machine 1 is further divided into two parts; one part proportional to the speed of the set (*i.e.*, $1 - s_1$) and is converted into mechanical power and the other proportional to s_1 *is* developed as electrical power at the slip frequency and supplied to machine II, which uses it for producing mechanical power and losses. Thus the mechanical outputs of the two machines are approximately in the ratio of $(1 - s_1) : s_1$ or $\left(1-\frac{P_2}{P_1+P_2}\right) : \frac{P_2}{P_1+P_2}$ or $P_1 : P_2$ *i.e.* in the ratio of the number of poles on the machines.

The set will not start, if connected in concatenation, because rotors will tend to rotate in opposite directions. Hence it must first be brought up to speed by an auxiliary motor or one motor alone be started and then second motor be connected.

It will be recognized that the concatenation method of speed control is similar to the series-parallel control of dc motors. In concatenation, at starting and for intermediate speeds, resistance is introduced in the rotor circuit of the second motor. When the motors are connected in parallel across the line, resistance is introduced in each rotor circuit and is gradually cut out. Full torque is available only when the motors are cascade connected.

A cascade of two induction motors has substantial disadvantages. Indeed, the magnetising current necessary to produce the fluxes in both motors is drawn from the supply mains through motor I. Therefore, no-load current of the set is appreciably greater than no-load current of one motor. On the other hand motor II is connected is series with the motor I and, therefore, the short-circuit current of the set is lesser than that of motor I, alone. The cascade characteristics are much inferior to the corresponding characteristics of a motor operating alone. A cascade of two induction motors are, therefore, used only in special cases, mainly on railways with 3-phase current.

Example 3.11. A 50 Hz, 440 V, 3-phase, 4-pole induction motor develops half the rated torque at 1,490 rpm. With the applied voltage magnitude remaining at the rated value, what should be its frequency if the motor has to develop the same torque at 1,600 rpm ? Neglect stator and rotor winding resistances and leakage reactances and iron losses.

Solution : Synchronous speed at supply frequency of 50 Hz, $N_s = \frac{120f}{P} = \frac{120 \times 50}{4} = 1{,}500$ rpm

Slip at a speed of 1,490 rpm (*i.e.* N = 1,490 rpm), $s = \frac{N_s - N}{N_s} = \frac{1{,}500 - 1{,}490}{1{,}500} = 0.00667$

Since torque developed by an induction motor, $T \propto sV^2$, slip s for constant torque and constant applied voltage remains unchanged.

New speed, $N' = 1{,}600$ rpm

Slip, $s = 0.00667$

New synchronous speed, $N_s' = \frac{N'}{1-s} = \frac{1{,}600}{1 - 0.00667} = 1{,}610.7$ rpm

New frequency, $f' = \frac{P \times N_s'}{120} = \frac{4 \times 1{,}610.7}{120} = 53.7$ Hz **Ans.**

Example 3.12. A 3-phase, 440 V, 1,000 rpm slip-ring induction motor is operating with 2% slip and taking a stator current of 50 A. Speed of the motor is reduced at constant torque to 500 rpm using stator voltage control. Calculate the new value of stator current.

Solution :

Supply voltage, V = 440 V

Slip, s = 2% or 0.02

Stator current, I_1 = 50 A

Synchronous speed, N_s = 1,000 rpm

Reduced speed, N' = 500 rpm

Slip at reduced speed, $s' = \frac{N_s - N'}{N_s} = \frac{1{,}000 - 500}{1{,}000} = 0.5$

Torque developed by an induction motor $T \propto sV^2$, so for constant torque

$$V \propto \sqrt{\frac{1}{s}}$$

or $V' = V \times \sqrt{\frac{s}{s'}} = 440\sqrt{\frac{0.02}{0.5}} = 88$ V

Stator current $I_1 \propto sV$

So new stator current $I_1' = I_1 \times \frac{s'V'}{sV} = \frac{50 \times 0.5 \times 88}{0.02 \times 440} = 250$ A **Ans.**

Example 3.13. An 8-pole, 50 Hz, 3-phase induction motor is running at 4% slip when delivering full load torque. It has standstill rotor resistance of 0.1 Ω and reactance of 0.6 Ω per phase. Calculate the speed of the motor if an additional resistance of 0.5 Ω per phase is inserted in the rotor circuit. The full-load torque remains constant.

Solution : Synchronous speed of motor, $N_s = \frac{120f}{P} = \frac{120 \times 50}{8} = 750$ rpm

Full load slip, $s_1 = 0.04$

The motor torque Eq. (2.14) is given by the equation

$$T = \frac{KsR_2E_2^2}{R_2^2 + s^2X_2^2}$$

Let the new slip at full load be s_2

$$\text{Now } T_1 = \frac{K s_1 R_2 E_2^2}{R_2^2 + s_1^2 X_2^2} = \frac{K \times 0.04 \times 0.1 \times E_2^2}{(0.1)^2 + (0.04 \times 0.6)^2} = \frac{KE_2^2}{2.644}$$

$$\text{and } T_2 = \frac{K s_2 (R_2 + R) E_2^2}{(R_2 + R)^2 + (s_2 X_2)^2} = \frac{K \times s_2 \times (0.1 + 0.5) E_2^2}{(0.1 + 0.5)^2 + (0.6 s_2)^2} = \frac{0.6 s_2 KE_2^2}{0.36 + 0.36 s_2^2}$$

Since torque remains the same *i.e.* $T_1 = T_2$

$$\therefore \quad \frac{0.6 s_2 KE_2^2}{0.36 + 0.36 s_2^2} = \frac{KE_2^2}{2.644}$$

$$\text{or } 0.6 (1 + s_2^2) = 2.644 s_2$$

$$\text{or } s_2^2 - 4.407 s_2 + 1 = 0$$

or $s_2 = 0.24$ neglecting higher value, being impractical

Speed of motor, $N_2 = N_s (1 - s) = 750 (1 - 0.24) = 570$ rpm **Ans.**

Example 3.14. A 4-pole induction motor and 6-pole induction motor are connected in cumulative cascade at 50 Hz supply. The frequency in the secondary circuit of the 6-pole motor is observed to be 1.0 Hz. Determine the slip in each machine and the combined speed of the set.

Solution : Concatenated speed of the set, $N_{sc} = \frac{120 f}{P_1 + P_2} = \frac{120 \times 50}{4 + 6} = 600$ rpm **Ans.**

The frequency in the secondary circuit of the 6-pole motor,

$$f'' = s'' f = 1.0 \text{ Hz}$$

$$\text{So slip, } s'' = \frac{f}{f''} = \frac{1.0}{50} = 0.02$$

Actual speed of the set, $N = N_{sc} (1 - s'') = 600 (1 - 0.02) = 588$ rpm **Ans.**

Synchronous speed of 4-pole induction motor, $N_s = \frac{120 \times 50}{4} = 1{,}500$ rpm

Slip for 4-pole induction motor, $s = \frac{N_s - N}{N_s} = \frac{1{,}500 - 588}{1{,}500} = 0.608$ or 60.8 % **Ans.**

The impressed frequency on the second motor, $f' = s f = 0.608 \times 50 = 30.4$ Hz

Synchronous speed of 6-pole induction motor with frequency f',

$$N_s' = \frac{120 \times f'}{P_2} = \frac{120 \times 30.4}{6} = 608 \text{ rpm}$$

Slip for 6-pole induction motor, $s = \frac{N_s' - N}{N_s'} = \frac{608 - 588}{608} = 0.033$ or 3.3 % **Ans.**

Example 3.15. Two 50 Hz, 3-phase induction motors having six and four poles respectively are cumulatively cascaded, the 6-pole motor being connected to the main supply. Determine the frequency of the rotor currents and slips referred to each stator field if the set has a slip of 2%.

Solution : Synchronous speed of the set, $N_{sc} = \frac{120 f}{P_1 + P_2} = \frac{120 \times 50}{6 + 4} = 600$ rpm

Actual speed of the set, $N = N_{sc} (1 - s') = 600 (1 - 0.02) = 588$ rpm

Synchronous speed of 6-pole motor stator field, $N_s = \frac{120 \times f}{P_1} = \frac{120 \times 50}{6} = 1{,}000$ rpm

Slip referred to the stator field, $s = \frac{N_s - N}{N_s} = \frac{1{,}000 - 588}{1{,}000} = 0.412$ or 41.2 % **Ans.**

The 6-pole motor rotor current frequency or the impressed frequency on 4-pole motor stator field,

$$f' = sf = 0.412 \times 50 = 20.6 \text{ Hz } \textbf{Ans.}$$

The synchronous speed of 4-pole induction motor with frequency f',

$$N_s' = \frac{120 f'}{P_2} = \frac{120 \times 20.6}{4} = 618 \text{ rpm}$$

The slip as referred to the 4-pole induction motor, $s'' = \frac{N_s' - N}{N_s'} = \frac{618 - 588}{618} = 0.0485$ or 4.85 % **Ans.**

Frequency of rotor current of 4-pole induction motor, $f'' = s''f = 0.0485 \times 20.6 = 1.0$ Hz **Ans.**

Example 3.16. A cascade set consists of two motors A and B with 4 poles and 6 poles respectively. The motor is connected to a 50 Hz supply. Find (*i*) the speed of the set (*ii*) the electric power transferred to motor B when the input to motor A is 25 kW. Neglect losses.

Solution : Synchronous speed of the set, $N_{sc} = \frac{120 f}{P_A + P_B} = \frac{120 \times 50}{4 + 6} = 600$ rpm

Because the outputs of the two motors are proportional to the number of their poles, so power transferred to motor B with 6 poles

= Power output of motor B neglecting losses

$$= P \times \frac{P_2}{P_1 + P_2} = 25 \times \frac{6}{4 + 6} = 15 \text{ kW } \textbf{Ans.}$$

Example 3.17. A 10 hp, 4-pole, 50 Hz, 220 V induction motor is connected in cascade with another 15 hp, 6-pole, 50 Hz, 220 V motor. What are the possible speeds obtainable with this combination and what is the maximum load in hp which may be delivered without overloading either machine. ? Also determine the ratio of mechanical power outputs of the two machines at this load.

Solution : Synchronous speed of machine I (6-pole machine), $N_{s1} = \frac{120 f}{P_1} = \frac{120 \times 50}{6} = 1{,}000$ rpm

Synchronous speed of machine II (4-pole machine), $N_{s2} = \frac{120 f}{P_2} = \frac{120 \times 50}{4} = 1{,}500$ rpm

Concantenated speed of the set when cumulatively connected, $N_{sc1} = \frac{120 f}{P_1 + P_2} = \frac{120 \times 50}{6 + 4} = 600$ rpm

Concantenated speed of the set when differentially connected, $N_{sc2} = \frac{120 f}{P_1 - P_2} = \frac{120 \times 50}{6 - 4} = 3{,}000$ rpm

So speeds available are 1,000, 1,500, 600 and 3,000 rpm **Ans.**

In cascade connection the rating of the main motor should be such that it covers the power output of the auxiliary motor also. So the maximum load which can be delivered is 15 hp **Ans.**

The ratio of the mechanical power outputs of the main and auxiliary induction motors are

$P_1 : P_2$ or 6 : 4 or 1.5 : 1 **Ans.**

3.5. KRAMER SYSTEM OF SPEED CONTROL

This system is based on the extraction of power from the rotor circuit of a polyphase wound rotor induction motor via the slip rings. This power at slip frequency is supplied to a synchronous convertor, and the resulting direct current is used to supply the armature of a separately excited dc motor directly coupled to the induction motor. By varying the excitation of the dc motor, its emf and hence the induction motor terminal voltage, are varied, thereby changing the motor speed. Variation in excitation of the synchronous convertor controls the power factor of the main induction motor.

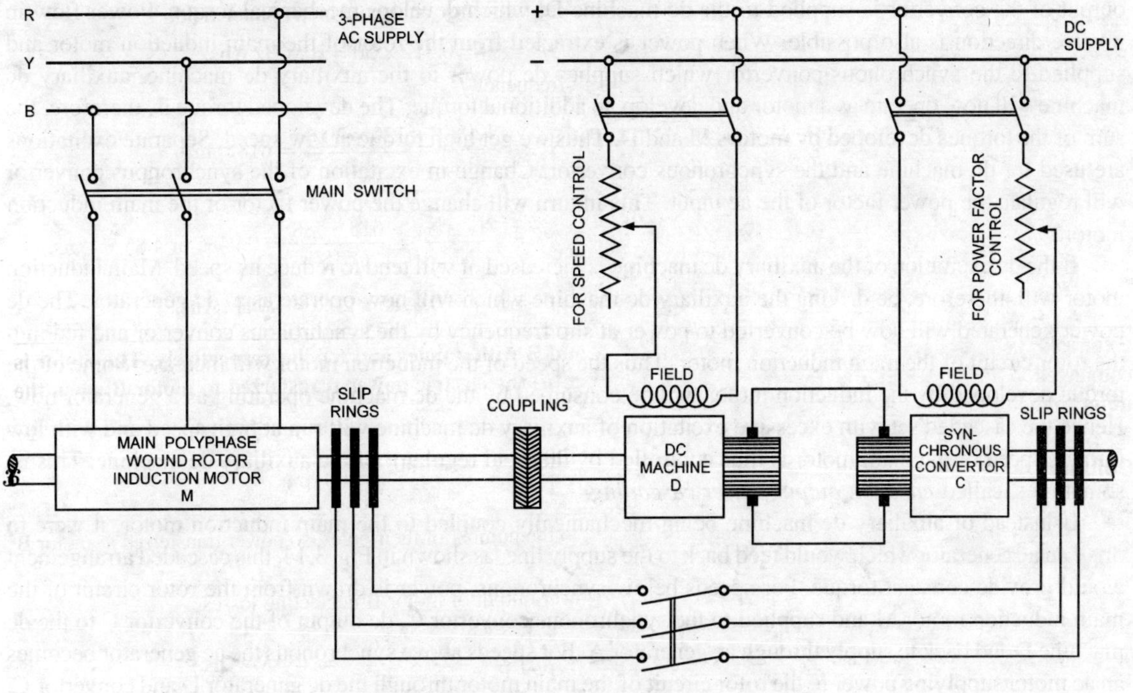

Fig. 3.13. ***Constant Output Power Kramer Drive***

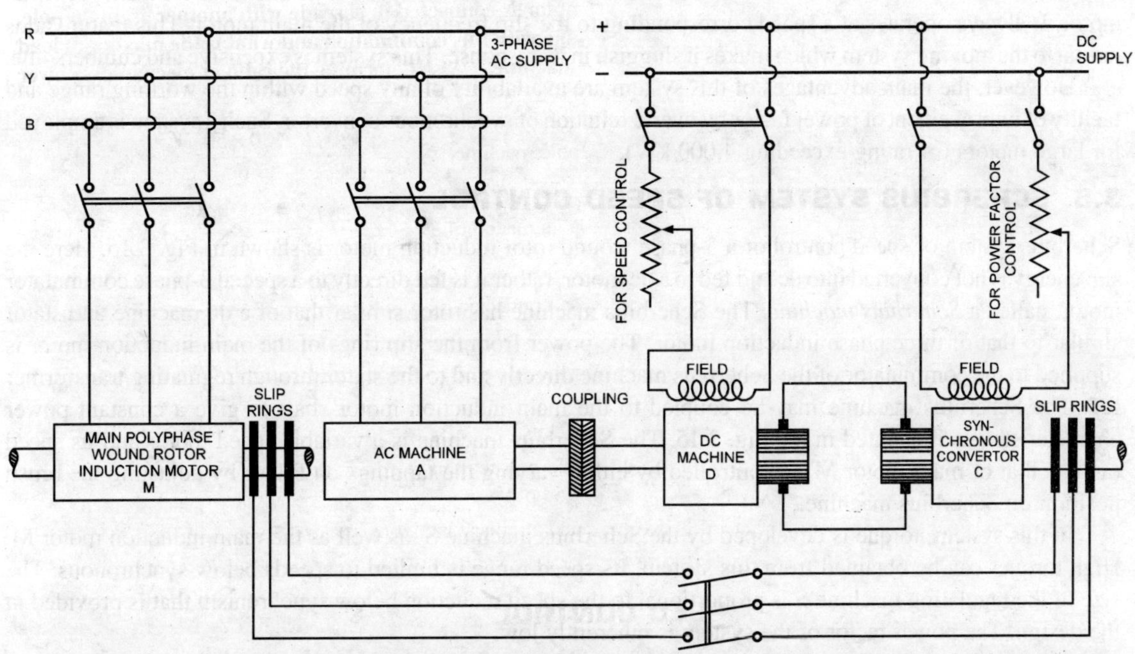

Fig. 3.14. ***Constant Torque Kramer Drive***

Constant output power Kramer drive is illustrated in Fig. 3.13. It develops no torque at synchronous speed. The main induction motor M is mechanically coupled to the auxiliary dc machine. Power from the rotor circuit of the main induction motor M is supplied to the slip rings of the synchronous convertor C and dc output of the convertor is supplied to the dc machine D, which develops mechanical torque. Power flow in reverse direction is also possible. When power is extracted from the rotor of the main induction motor and supplied to the synchronous convertor which supplies dc power to the auxiliary dc machine, auxiliary dc machine will now operate as a motor and develop an additional torque. The developed torque is, therefore, the sum of the torques developed by motors M and D. Thus we get high torque at low speed. Separate excitations are used for dc machine and the synchronous convertor. Change in excitation of the synchronous convertor will regulate the power factor of the ac input. This in turn will change the power factor of the main induction motor.

If the dc excitation of the auxiliary dc machine is increased, it will tend to reduce its speed. Main induction motor will, therefore, be driving the auxiliary dc machine which will now operate as a dc generator. The dc power generated will now be converted to power at slip frequency by the synchronous convertor and fed into the rotor circuit of the main induction motor. Thus the speed of the induction motor will increase. Some of the torque developed by the induction motor will be consumed by the dc machine operating as a generator now. Hence the cascaded set with excessive excitation of auxiliary dc machine will run at high speed and with low torque. Speed of the main motor is thus controlled by the field regulator of the auxiliary dc machine. This is, sometimes, called *constant output power cascading.*

If instead of auxiliary dc machine being mechanically coupled to the main induction motor, it were to drive an ac generator which would feed back to the supply line, as shown in Fig. 3.14, this cascaded arrangement would provide constant torque. For speeds below synchronous, power is drawn from the rotor circuit of the main induction motor M and supplied to the synchronous convertor C, dc output of the convertor C to the dc machine D and back to supply through ac generator A. For speeds above synchronous the ac generator becomes an ac motor supplying power to the rotor circuit of the main motor through the dc generator D and convertor C.

In both of the Kramer drives the convertor changes the speed with speed adjustments of the main induction motor. It always operates at a speed corresponding to the slip frequency of the main motor. This feature adds inertia to the moving system which makes it sluggish in its response. This system is expensive and cumbersome.

However, the main advantages of this system are availability of any speed within the working range and facility of improvement of power factor by over-excitation of synchronous convertor. Such a system is employed for large motors (of rating exceeding 4,000 kW).

3.6. SCHERBIUS SYSTEM OF SPEED CONTROL

Scherbius system of speed control of a 3-phase wound rotor induction motor is shown in Fig. 3.15. Here the slip energy is not converted into dc and fed to a dc motor, rather it is fed directly to a special 3-phase commutator motor, called a *Scherbius machine*. The Scherbius machine has rotor similar that of a dc machine and stator similar to that of three phase induction motor. The power from the slip rings of the main induction motor is supplied to the commutator of the Scherbius machine directly and to the stator through regulating transformer RT. The Scherbius machine may be coupled to the main induction motor shaft to give a constant power characteristic, as illustrated in the Fig. 3.15. The Scherbius machine is a variable speed motor and its speed (and so that of main motor M) is controlled by either varying the tappings on RT or by adjusting the brush position on Scherbius machine.

In this system, torque is developed by the Scherbius machine S as well as the main induction motor M. High torques can be obtained from this system. Its speed range is limited to speeds below synchronous. The size of the regulating machine S is proportional to the speed reduction below synchronism that is provided in the design. The power factor of the system is inherently low.

A Scherbius system for providing a constant-torque drive is shown in Fig. 3.16. Here the Scherbius motor is coupled to separate induction machine D, acting as an induction generator to feed the power back to the ac supply mains. For light loads the induction machine may act as a motor to supply rotational losses of the motor-generator set SD.

In case of Scherbius machine S whose stator is fed from the slip rings of the main induction motor M, it is not possible to obtain speed above synchronous. This is because as soon as synchronous speed is approached there is no rotor power to excite the Scherbius machine S. To overcome this, stator is fed from separate frequency convertor FC, coupled to the main motor and directly supplied from main supply from where main induction motor M is fed. This Scherbius machine will drive induction generator D to supply power back to the supply line as illustrated in Fig. 3.17.

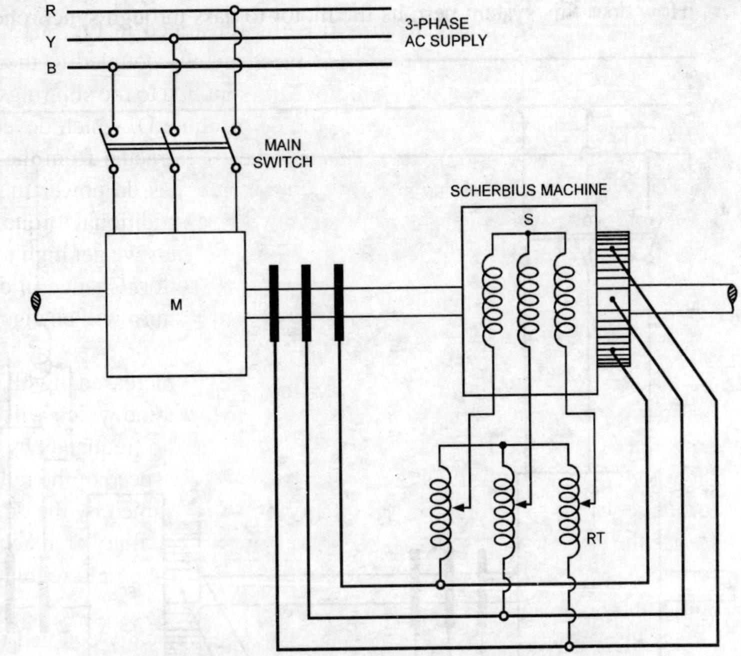

Fig. 3.15. ***Scherbius Constant Output Power Drive***

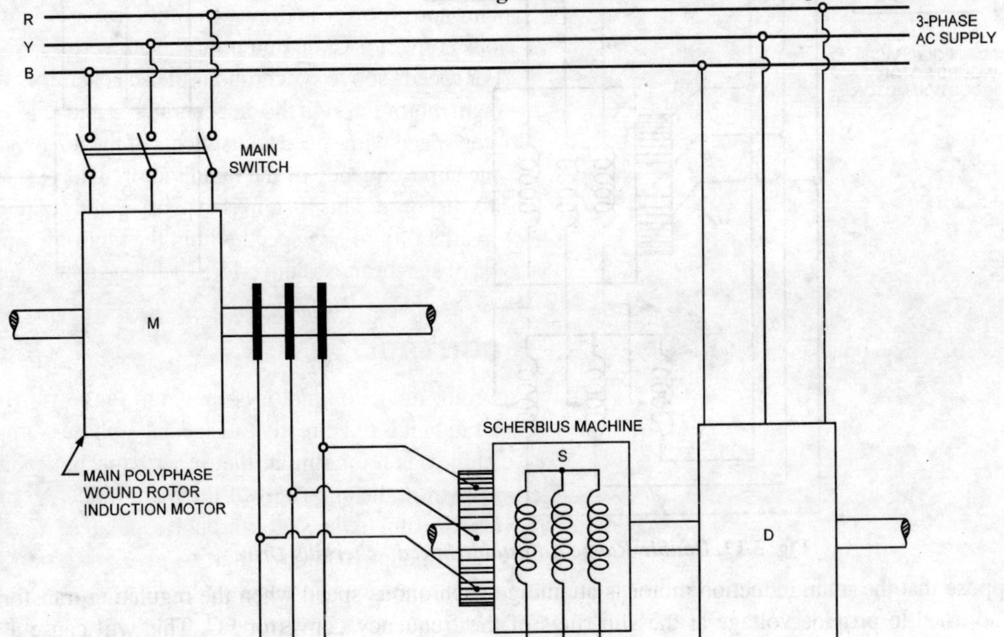

Fig. 3.16. ***Scherbius Constant Torque Drive***

How does this system permits the motor to pass through synchronous speed, it is explained below.

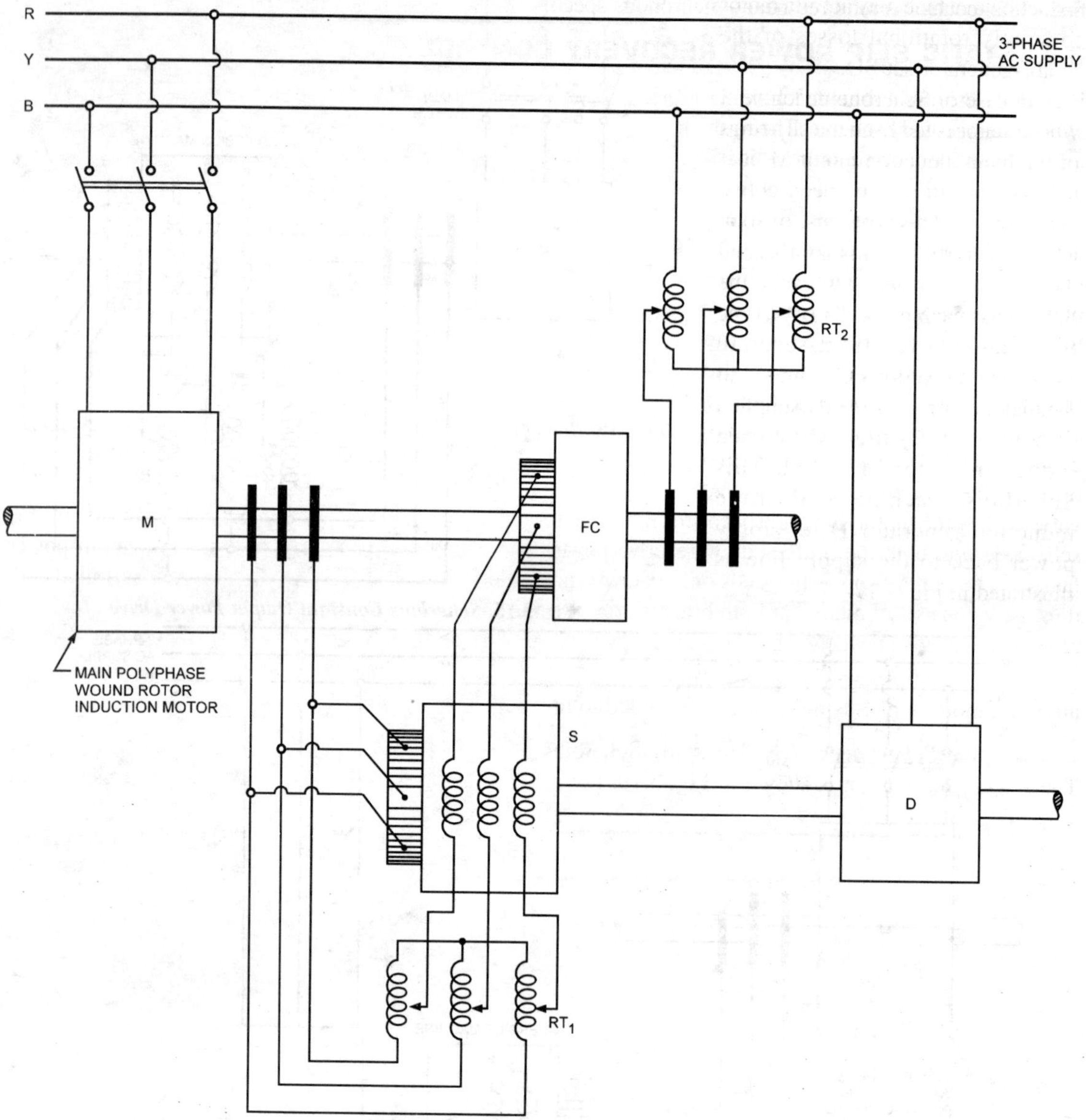

Fig. 3.17. ***Double Range Adjustable Speed Scherbius Drive***

Suppose that the main induction motor is attaining synchronous speed when the regulating transformer RT_2 is adjusted to provide voltage at the slip rings of the frequency convertor FC. This will cause a low voltage of slip frequency at the brushes of the frequency convertor. This voltage will then produce a magnetic field in the machine S which is rotating, causing slip frequency voltage to appear at the brushes of the machine S. This voltage forces a current through the rotor winding of the main induction motor, thus developing torque and accelerate the motor M. The frequency convertor is usually provided with brush shifting device so that the phase position of the voltages at its commutator can be altered. This feature provides the possibility of collecting a current that will develop torque in the main induction motor M but also provides power-factor correction of the main motor.

Though the Scherbius drive is costlier than Kramer system but it is superior to the Kramer drive when fine speed control is required near synchronous speed.

3.7. STATIC SLIP POWER RECOVERY CONTROL

Figure 3.18 shows a scheme known as *slip-power* recovery scheme. The rotor terminals are connected to the 3-phase ac supply mains through two fully controlled thyristor bridges. Bridge 1 acts as a rectifier (or converter) and bridge 2 acts as an inverter. Power output from rotor can be supplied back to the supply source. Since the frequency of rotor currents is slip frequency, this method is known as *slip-power recovery scheme*. By controlling the firing angles of the two bridges, the rotor power output can be varied. Thus the motor slip and speed (for the same torque) will also change. However, the drawback of this scheme is that both the bridges draw reactive power from the supply mains. Therefore, the overall power factor of the motor is poor. If speeds only below synchronous are desired, bridge 1 may be uncontrolled and, thus, may consists of diodes. If both bridges are controlled, the operation of the two bridges can be reversed also to get speeds above synchronous one.

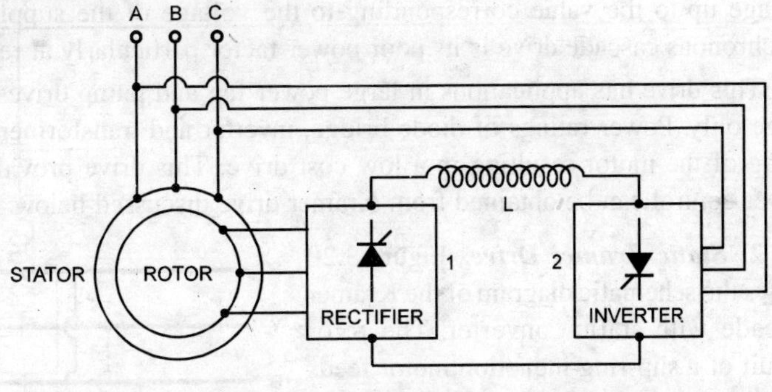

Fig. 3.18. ***Rotor Voltage Control of a 3-Phase Wound Rotor Induction Motor (Slip-Power Recovery Scheme)***

Infact, the slip-power is either returned to the supply network as in Scherbius scheme or used to drive an auxiliary motor which is mechanically coupled to the induction motor shaft, as in Kramer scheme.

1. ***Static Scherbius Drive.*** The static Scherbius drive also uses the principle of slip-power recovery. The schematic diagram is shown in Fig. 3.19. For achieving both sub-synchronous and supersynchronous

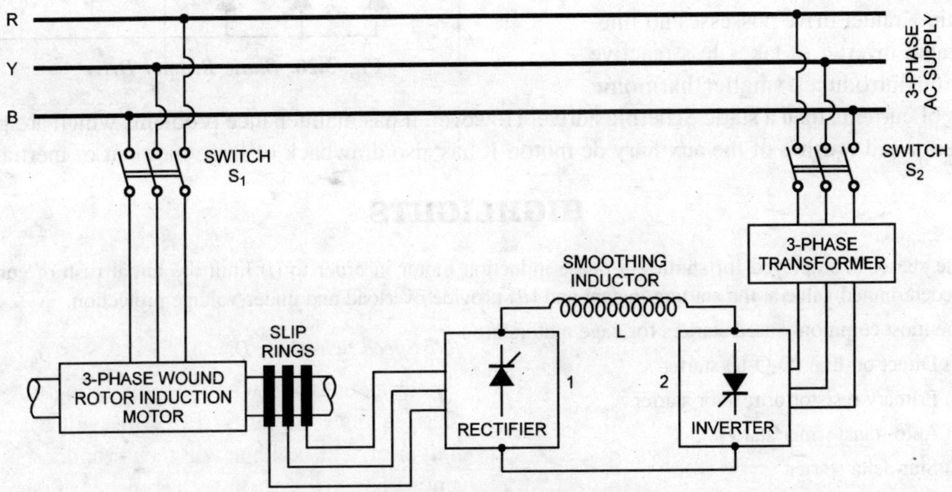

Fig. 3.19. ***Static Scherbius Drive***

speed control, converters 1 and 2 must be fully-contnrolled thyristor bridges, one functioning at slip frequency as a rectifier or inverter while the other operating at supply system frequency as an inverter or rectifier. The cost of converters is quite appreciable and a slip-frequency gating circuit is also required. Moreover, at speeds near synchronous, when the slip-frequency emfs are quite small for natural commutation, special connections for forced commutation methods are required. If converter 1 is taken uncontrolled one (diode bridge) the converter cascade and the control unit would become economical and simple, but only sub-synchronous speed control would then be available.

The 3-phase transformer between the supply source and inverter 2 is meant to bring the rotor circuit voltage up to the value corresponding to the voltage of the supply source. The main drawback of sub-synchronous cascade drive is its poor power factor particularly at reduced speeds.

This drive has applications in large power fan and pump drives which need speed control in a narrow range only. Power ratings of diode bridge, inverter and transformer is just maximum slip times the power rating of the motor resulting in a low cost drive. This drive provides a constant torque control. Constant power control can be obtained from Kramer drive discussed below.

2. *Static Kramer Drive.* Figure 3.20 shows the schematic diagram of the Kramer cascade with static converter. The rotor circuit of a slip-ring induction motor feeds the slip power, rectified by a diode bridge, to the armature of a separately excited dc motor mechanically coupled to the induction motor. Speed control is achieved by varying the field current of the dc motor. An emf proportional to the back emf of the dc motor may be considered to be injected into the rotor circuit of the induction motor to cause variation in the speed of the system.

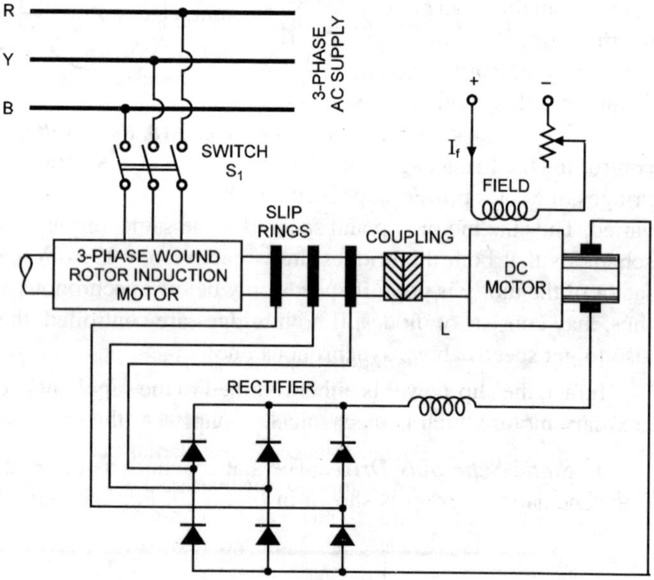

Fig. 3.20. ***Static Kramer Drive***

For achieving larger speed range, diode bridge will need replacement by a thyristor bridge. With thyristor bridge speed can be controlled up to standstill.

Static Kramer drive possesses no line-commutated inverter, it takes less reactive power and introduces smaller harmonic contents of currents than a static Scherbius drive. However, it has maintenance problems which are posed by commutator and brushes of the auxiliary dc motor. It has also drawback of large moment of inertia.

HIGHLIGHTS

1. The starter is employed for starting 3-phase induction motor in order to (*i*) limit the initial rush of current to a predetermined value at the starting instant and (*ii*) provide overload and under-voltage protection.
2. The most commonly used starters for cage motors are :

   (*a*) Direct-on-line (D-O-L) starter.

   (*b*) Primary resistor or reactor starter.

   (*c*) Auto-transformer starter.

   (*d*) Star-delta starter.

   (*a*) **Direct-on-line starting** is the cheapest method of starting of a 3-phase induction motor. Squirrel cage motors of capacity up to 1.5 kW capacity can be started by this method.

   In D-O-L starting

   Starting current, $I_{st} = I_{sc}$

and starting torque, $T_{st} = T_f \left(\frac{I_{sc}}{I_f}\right)^2 s_f$

(*b*) **Primary resistor or reactor starting method** is very simple and provides smooth acceleration of the motor. In this method reduced voltage is obtained by means of resistors (or reactors) connected in series with each stator lead. The drawbacks of this method are requirement of expensive resistors and low torque efficiency.

Starting current, $I_{st} = K I_{sc}$

Starting torque, $T_{st} = K^2 T_f \left(\frac{I_{sc}}{I_f}\right)^2 s_f = K^2 T_f$

(*c*) **Auto-transformer starting** is often employed for starting of large cage motors (of output rating exceeding 20 kW). In this method, the reduced voltage is obtained by taking tappings at suitable points from a 3-phase auto-transformer. The advantages of auto-transformer starting are (*i*) availability of highest torque per ampere of supply current (*ii*) suitability for long starting periods (*iii*) closed transition starting and (*iv*) motor current larger than supply current. The drawbacks of this method are poor power factor and higher cost in case of lower output rating motors.

Starting current, $I_{st} = K^2 I_{sc}$

Starting torque, $T_{st} = K^2 T_f \left(\frac{I_{sc}}{I_f}\right)^2 \times s_f = K^2 \times$ Torque obtained by direct switching

(*d*) **Star-delta starting method** is simple, cheap, effective and efficient method of starting cage motors as no power is lost in auxiliary equipment. The star-delta starter connects the three stator windings in star across the rated supply voltage at the starting instant. After attainment of normal speed, the same windings are re-connected in delta.

Star-delta starter reduces line current at starting and starting torque to one-third of those with direct switching.

3. Slip-ring induction motors are usually started with full-line voltage across the stator terminals and by introducing variable resistor in each phase of the rotor circuit. The external resistances introduced in each phase of the rotor circuit not only reduces the current at the starting instant but increases the starting torque also due to improvement in *pf*.

4. The problem of speed control of electrical motors in general and of induction motors in particular is of great practical importance.

The speed of a 3-phase induction motor can be controlled by any of the following methods.

1. From stator side : (*i*) By variation of supply frequency (*ii*) By variation of applied voltage and (*iii*) by changing the number of poles.

2. From rotor side : (*i*) By changing the resistance in the rotor circuit and (*ii*) By introducing into the rotor circuit an additional emf of the same frequency as the fundamental emf of the rotor.

(*i*) Speed control by variation of supply frequency. This method of speed control provides wide speed-control range with gradual variation of the speed throughout this range. The major difficulty with this method is how to get the variable frequency supply.

(*ii*) Speed control by variation of supply voltage. This method of speed control is simple, low in first cost and has low maintenance cost but it has limited use because (*i*) the operation at voltages exceeding rated voltage is restricted by magnetic saturation, (*ii*) a large change in voltage is required for a relatively small change in speed (*iii*) the developed torque reduces greatly with the reduction in supply voltage and the motor will come to standstill if the load torque exceeds the pull-out torque of the motor corresponding to reduced supply voltage and (*iv*) the range of speed control is very limited in the downward direction *i.e.*, from rated speed to lower speeds.

In general, this method of speed control is used only on loads where the torque required drops off considerably as the speed is reduced such as with small squirrel cage motors driving fans.

***(iii)* Speed control by variation of rotor resistance.** This type of control is only possible with wound rotor induction motors. This method of speed control is simple and low in initial as well as maintenance cost and it is possible to have a large starting torque with low starting current and large pull-out torques at small values of slip.

***(iv)* In secondary foreign voltage control.** The speed of an induction motor is controlled by injecting a voltage of slip frequency in the secondary circuit. It is costlier method and is used with motors of very large rating, such as for motors in steel mills.

***(v)* Speed control by concatenation.** In this method of speed control, two motors are required, at least one of which must have a wound rotor.

The synchronous speed of the cascade set is given as

$$N_s = \frac{120f}{P_1 \pm P_2}$$

+ sign for connections providing torques of the two motors in the same direction.

– sign for connections providing torques of the two motors in the opposite directions.

EXERCISES

1. Why starters are necessary for starting induction motor ? Name different starting methods for 3-phase induction motor and explain any one of them. **[M.D. Univ. Electrical Machines-II, December 2010]**
2. Explain why starter is needed for starting a large capacity induction motor.
3. Why starter is required to start a 3-phase induction motor ? Describe star-delta method of starting a squirrel cage induction motor.
4. Draw the star-delta starter for an induction motor with necessary protective devices.
5. Name various methods of starting the polyphase induction motors and describe auto-transformer method of starting, in detail.
6. Why a starter is necessary to start an induction motor ? Mention the various methods of starting. Explain in detail auto-transformer method of starting a squirrel cage induction motor.
7. Compare the star/delta and auto-transformer methods of starting of 3-phase squirrel cage induction motors on the basis of

   (i) starting torque and starting current *(ii)* adoptability to motor rating, terminal arrangements and *(iii)* starting condition.

   Discuss also the limitation of these methods.
8. What are different methods of starting 3-ϕ induction motor ? Describe, with the help of neat diagram, the method of starting slip-ring induction motor and squirrel cage induction motor.
9. Describe briefly the various methods of speed control of induction motor.

   [M.D. Univ. Electrical Machines-II, May 2009]
10. Discuss any two methods of speed control of 3-phase squirrel cage induction motors.
11. How is the speed of a 3-phase induction motor controlled by its stator voltage control ? Show that the region of speed control by this method is limited by the slip at which maximum torque develops.
12. Explain the method of speed control of 3-phase induction motor by varying the rotor resistance.

    [M.D. Univ. Electrical Machines-II, December 2010]
13. Describe speed control of a 3-phase slip-ring induction motor by injecting voltage in the rotor circuit.
14. Explain 'cascade arrangement' for controlling speed of three phase induction motors.
15. Describe a Scherbius scheme for a speed control of 3-ϕ slip-ring induction motor. What are its merits and demerits compared to other methods of speed control of the same machine.
16. Explain with circuit the Scherbius system of speed control of 3-phase induction motor.
17. Discuss a static slip-power recovery scheme for speed control of slip-ring induction motor.

18. Write short notes on the following :
   (*i*) Auto-transformer starting of 3-phase induction motor
   (*ii*) Star-delta starter.
   (*iii*) Rotor resistance control.
   (*iv*) Slip power recovery control.

SHORT ANSWER TYPE QUESTIONS WITH ANSWERS

Q. 1. Why do you require starters for starting of 3-phase induction motors ?
Ans. The starter is required to limit the initial rush of current to a predetermined value at the time of starting of a 3-phase induction motor.
Q. 2. What are the various types of starters used for squirrel cage motors ?
Ans. Starters used for starting of a squirrel cage induction motor are
(*a*) Direct-on-line starter (*b*) Line resistance starter
(*c*) Star-delta starter and (*d*) Auto-transformer starter.
Q. 3. Which is the cheapest method of starting a 3-phase induction motor ?
Ans. Direct-on-line starting.
Q. 4. How many terminals do you expect to find on the terminal box of a squirrel cage induction motor to be started by star-delta starter ?
Ans six.
Q. 5. In what ratio line current and starting torque are reduced with star-delta starting ?
Ans. Line current and starting torque are reduced to one-third with star-delta starting.
Q. 6. What is the serious objection to the practice of employing reduced voltage for the starting of a squirrel cage induction motor ?
Ans. Large reduction in staring torque because the starting torque varies as the square of voltage applied to the stator.
Q. 7. What type of protection is provided in the starter meant for 3-phase induction motors ?
Ans. Overload and under-voltage protection.
Q. 8. A 3-phase induction motor is to be started first using an auto-transformer with 80 per cent tapping and next direct-on-line. What will be the ratio of the starting torques ?
Ans. Starting torque with auto-transformer with 80% tapping/starting torque on direct on line = 0.64.
Q. 9. What is the parameter other than the ratio of rotor current limits which decides the number of resistance sections of a slip-ring induction motor starter ?
Ans. Slip at full load.
Q. 10. Is it possible to add an external resistance in the rotor circuit of a 3-phase cage induction motor ? Give reason for your answer.
Ans. No; because the rotor winding is permanently short-circuited.
Q. 11. Why is the rotor rheostat starter unsuited for a cage motor ?
Ans. Because there is no provision for inserting resistance in the rotor circuit of a cage motor.
Q. 12. Rotor resistance starting is preferred to reduced voltage starting of a wound rotor induction motor. Why ?
Ans. Because rotor resistance starting of a wound rotor induction motor not only reduces the starting current at the starting instant but increases the starting torque also due to improvement in power factor whereas the reduced voltage starting reduces only the starting current at the starting instant.
Q. 13. For which type of loads, induction motor with voltage controller is suitable and why ?
Ans. Induction motor with voltage controller is suitable only for loads where the torque required drops off considerably with the drop in speed such as for driving fans because a large change in voltage is required for a relatively small change in speed and developed torque reduces greatly with the reduction in supply voltage.
Q. 14. On what factors does the speed of an induction motor depend ?
Ans. The speed of an induction motor depends on supply frequency, slip and number of poles on stator.

Q. 15. If frequency variation is used to control the speed of an induction motor, what else must be changed ?

Ans. The voltage applied to the stator must be varied in direct proportion to the frequency so as to maintain flux level constant.

Q. 16. While controlling induction motor speed, how super-synchronous speed is achieved ?

Ans. Super-synchronous speed can be achieved, while controlling the speed of an induction motor, by injecting a slip frequency voltage in phase with emf induced in the rotor.

Q. 17. What will be the effect on the torque developed by an induction motor when the applied voltage is reduced to half, supply frequency remaining unchanged ?

Ans. Torque developed by an induction motor with half-rated voltage, supply frequency remaining unchanged, will be reduced to one-fourth because $T \alpha V^2$ (*i.e.* square of the supply voltage).

Q. 18. What will be the effect on the torque developed by an induction motor when both the applied voltage and frequency are reduced to half ?

Ans. The torque developed by an induction motor will be reduced to one half of normal value when both the applied voltage and frequency are reduced to half because $T \propto \frac{V^2}{f}$.

PROBLEMS

1. Determine approximately the starting torque of an induction motor in terms of its full-load torque when started by means of (*i*) star-delta starter and (*ii*) by an auto-transformer starter with 50 % taps. The motor draws 5 times full-load current when switched directly on line and has a full-load slip of 4 %.
[**Ans.** One-third and one-fourth of full-load torque]
2. Find the ratio of starting to full-load current in a 10 kW (output), 400 V, 3 phase induction motor with star/delta starter, given that full-load power factor is 0.85, the full-load efficiency is 0.88 and the blocked rotor current at 200 V is 40 A. Ignore magnetising current. [**Ans.** 1.38]
3. A squirrel cage induction motor takes 5 times full-load current and develops 2.5 times full-load torque when switched directly on the line. Determine the torque developed and the initial value of the current drawn from the supply mains when started through an auto-transformer with 60 % tapping.
[**Ans.** 1.8 times of full-load current; 81 % of full load torque]
4. A 3-phase squirrel cage induction motor has a ratio of maximum torque to full-load torque as 2.5 :1. Determine the ratio of actual starting torque to full-load torque for auto-transformer starting with a tapping of 75 per cent. The rotor resistance = 0.4 Ω and X_2 = 4 Ω. [**Ans.** 0.278]
5. The full-load efficiency and power factor of a 16 hp, 440 V, 3-phase induction motor are 85 % and 0.8 respectively. The blocked rotor line current is 45 A at 220 V. Calculate the ratio of starting to full-load current if the motor is provided with a star-delta starter. Neglect magnetising current. [**Ans.** 2.288]
6. Calculate the steps in a 5-section rotor starter for a 3-phase induction motor. The maximum starting current = full-load current. Full-load slip = 1.8 % with rings shorted. Rotor resistance per phase = 0.015 Ω.]
[**Ans** r_1 = 0.46 Ω, r_2 = 0.206 Ω, r_3 = 0.0923 Ω; r_4= 0.0413 Ω and r_5 = 0.0185 Ω]
7. The rotor of a 6 pole, 50 Hz slip-ring induction motor has a resistance of 0.2 Ω per phase and runs at 960 rpm on full load. Calculate the approximate resistance per phase of a rotor rheostat such that speed is reduced to 800 rpm for full-load torque. [**Ans.** 0.8 Ω/phase]
8. The voltage across two slip rings of a stationary star-connected 100 hp, 415 V induction motor is 273 V with full voltage across stator. If full-load slip is 2 per cent, find the resistance to be added in series with each slip-ring to reduce the speed by 20 %. Assume load torque to be constant and rotor reactance negligible. [**Ans.** 0.196 Ω]
9. Two 3-phase induction motors having 4 and 6 poles respectively operate in cascade connected to a supply of 400 V, 50 Hz. What speeds are obtainable ? [**Ans.** 1000 rpm; 1,500 rpm, 600 rpm and 3,000 rpm]
10. A cascade set consists of two motors A and B with 4 poles and 6 poles respectively. The motor A is connected to 50 Hz supply. Determine (*i*) the speed of the set (*ii*) the electric power transferred to motor B when the input to motor A is 25 kW. Losses may be neglected. [**Ans.** (*i*) 600 rpm (*ii*) 10 kW]

❑❑❑

SECTION B

SINGLE PHASE INDUCTION MOTORS

CONTENTS OF SECTION-B

SINGLE PHASE INDUCTION MOTORS : Double Revolving Field Theory; Cross Field Theory; Different Types of Single Phase Induction Motors; Circuit Model of Single Phase Induction Motor.

4

CHAPTER

Single Phase Induction Motors

4.1. INTRODUCTION

There are two basic reasons for the use of single-phase rather than 3-phase motors.

First, for reasons of economy, most houses, offices and also rural areas are supplied with single phase ac, as power requirements of individual load items are rather small.

The second factor is the economics of the motor and its branch circuit. Fixed loads requiring not more than 0.5 kW can generally be served most economically with single phase power and a single phase motor. Single phase motors are simple in construction, reliable, easy to repair and comparatively cheaper in cost and, therefore, find wide use in fans, refrigerators, vacuum cleaners, washing machines, other kitchen equipment, tools, blowers, centrifugal pumps, small farming appliances etc. To avoid the need for special branch circuits, portable tools are also usually single phase.

Because of above reasons motors of comparatively small ratings (mostly in fractional kW ratings) are manufactured in large number to operate on single phase ac at standard frequencies. An indication of the number of such motors can be had from the fact that the sum of total of all fractional kilowatt motors in use today far exceeds the total of integral kilowatt motors of all types. Single phase motors perform a great variety of useful services in the home, the office, the factory, in business establishments, on the farm, and many other places where electricity is available. They find interesting applications in automatic control devices of various types. Since the requirements of the numerous applications differ so

INSIDE THIS CHAPTER

widely, the motor manufacture industry has developed several types of such machines, each type having operating characteristics that meet definite demands. For example, one type operates satisfactory on dc or any frequency up to 50 Hz; another rotates at absolutely constant speed, regardless of the load; another develops considerable starting torque; and still another, although not capable of developing much starting torque is nevertheless extremely cheap to make and very rugged.

Single phase ac motors may be divided in three general classes namely (*i*) induction motors (*ii*) commutator motors and (*iii*) synchronous motors.

Induction motors are further classified as split-phase motors, shaded-pole motors and repulsion– start induction motors according to the method of producing starting torque. The commutator motors are the series motor, universal (ac/dc) motor, the repulsion-induction motor, with various modifications and combinations of these types being in use.

Single phase induction motors in very small sizes $\left(\frac{1}{400}\text{ kW to }\frac{1}{25}\text{ kW}\right)$ are used is toys, hair dryers, vending machines etc. The universal motor is widely used in portable tools, vacuum cleaners and kitchen equipment.

The main drawbacks of single phase motors are low overload capacity, low efficiency, low power factor and low output as compared to that of a 3-phase motor of the given frame size.

4.2. SINGLE PHASE INDUCTION MOTORS

If one line of a three phase induction motor is opened while the motor is running with moderate or light load, it is found that the motor continues to run though at slightly lower speed. This condition is known as single phase operation. Opening of one line (phase) leaves the equivalent of a single phase connected to the stator. In other words a 3-phase induction motor has become a single phase motor.

4.2.1. Construction

A single phase induction motor is similar to a 3-phase squirrel cage induction motor in physical appearance. The rotor of a single phase squirrel cage motor is essentially the same as that employed in 3-phase induction motors and needs no further description. There is uniform air gap between stator and rotor but no electrical

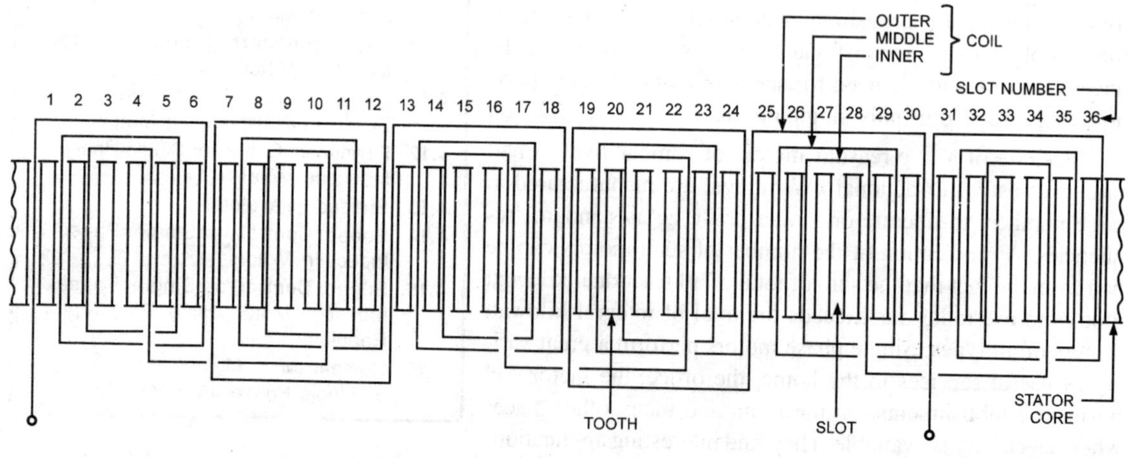

Fig. 4.1. ***Winding Diagram For a Single Phase 6-Pole Motor With 36 Slots (Main Winding Only)***

connection between them (stator and rotor). Except for shaded-pole types, the stator core is also very similar. A single phase motor can be wound for any even number of poles, two, four, and six being most common. Like three-phase machines, adjacent poles have opposite magnetic polarity and synchronous speed equation ($N_s = \frac{120f}{P}$) also applies.

The stator windings, however, are distinctively different in two aspects. First, single phase motors are usually provided with concentric coils, as illustrated in Fig. 4.1. In this diagram the coils appear to have only one turn of wire, but in practice each coil has many turns. With concentric coils, the number of turns per coil can be adjusted to provide an approximately sinusoidal distribution of mmf along the air gap and this is one of their advantages. Second, single phase squirrel cage motors normally have two stator windings. Except for shaded pole motors, the two windings are similar to that illustrated in Fig. 4.1, but one of them usually have few turns of much thin wire, and the two windings are in space quadrature with respect to each other. In motors that operate with both windings energized, the winding with the heaviest wire is known as the *main winding* and the other is called the *auxiliary winding.* If the motor runs with auxiliary winding open, these windings are usually referred as running and starting windings respectively. A motor that runs with auxiliary winding open, uses a centrifugal switch to cut out the starting (or auxiliary) winding after starting. In most of the motors the main winding is placed at the bottom of the slots and the starting winding on top but shifted 90° from the running winding.

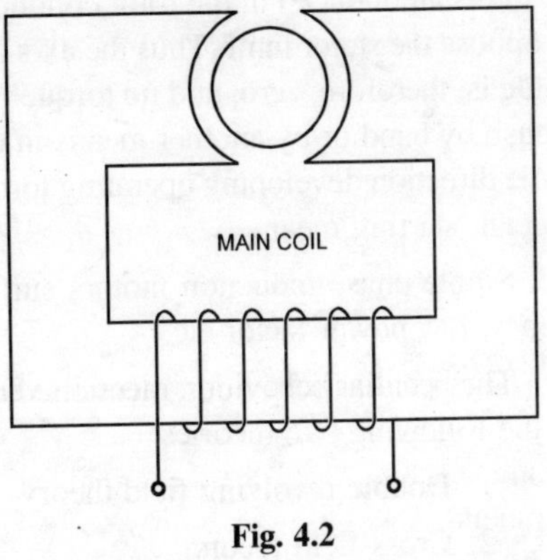

Fig. 4.2

If the motor has four or more poles, the main winding of a shaded pole motor is similar to the windings that have already been described, except that it generally has only one coil per pole. A 2-pole shaded-pole motor may be constructed similarly to the magnetic circuit given in Fig. 4.2.

An elementary single phase induction motor is shown in Fig. 4.3.

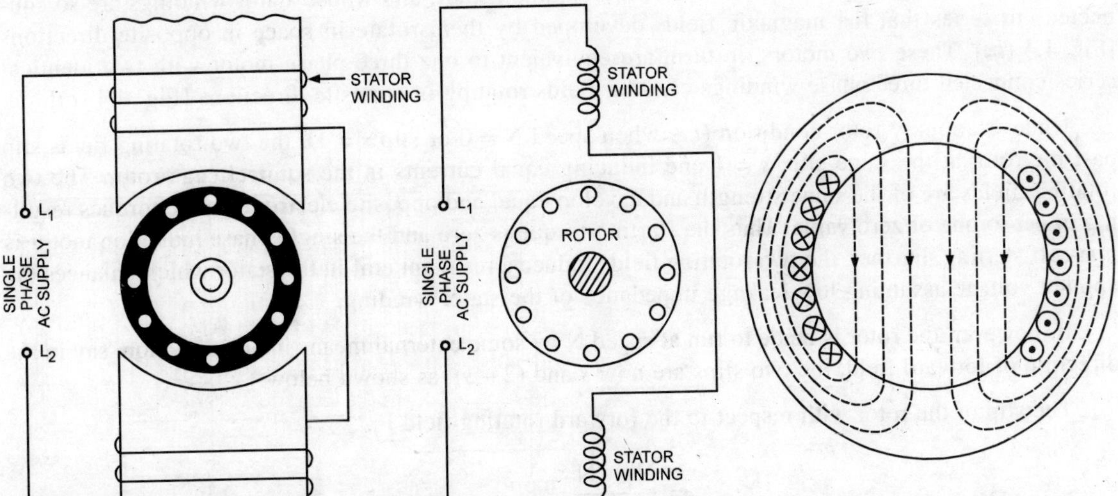

Fig. 4.3. ***Elementary Single Phase Induction Motor***

4.2.2. Action

When the stator winding of a single phase induction motor is connected to single phase ac supply, a magnetic field is developed, whose axis is always along the axis of stator coils. With alternating current in the fixed stator coil the mmf wave is stationary in space but pulsates in magnitude and varies sinusoidally with time. Currents are induced in the rotor conductors by transformer action, these currents being in such a direction as to oppose the stator mmf. Thus the axis of the rotor mmf wave coincides with that of the stator field, the torque angle is, therefore, zero, and no torque is developed at starting. However, if the rotor of such a motor is given a push by hand or by another means in either direction, it will pick up the speed and continue to rotate in the same direction developing operating torque. Thus a single phase induction motor is not self starting and needs special starting means.

Single phase induction motors suffer from several drawbacks such as low overload capacity, low efficiency, low power factor etc.

The peculiar behaviour, mentioned above, of a single phase induction motor may be explained by any one of the following two theories

1. Double revolving field theory
2. Cross-field theory.

Both theories have their own advantages and fields of application. They ultimately lead to the result that are approximately same.

4.3. DOUBLE REVOLVING FIELD THEORY

Though double revolving field theory is not simpler in application than the cross-field theory, but once this theory is understood, the fundamental concepts are more readily grasped. This theory is based on the idea that pulsating field produced in a single phase motor can be resolved into two components of half its amplitude and rotating in opposite directions with synchronous speed (Ferraris principle—Art. 1.2).

It may perhaps simplify the visual picture by placing these two rotating fluxes in two separate motors mounted on the same shaft. This is equivalent of replacing a single phase motor by two identical three phase motors whose rotors are fitted on a common shaft, and whose stator windings are so connected in series that the magnetic fields developed by them rotate in space in opposite directions [Fig. 4.4 (*b*)]. These two motors, in turn, are equivalent to one three phase motor with two identical series connected three phase windings creating fields rotating in opposite directions [Fig. 4.4 (*c*)].

Under stationary rotor condition (*i.e.* when speed N = 0 or slip $s = 1$), the two rotating fields slip past the rotor at the same slip, $s = 1$ and inducing equal currents in the squirrel cage rotor. The two rotating fields are of the same strength and develop equal and opposite electromagnetic torques resulting in net torque of zero value. Thus the starting torque is zero and the single phase induction motor is non-self starting. Further, the two rotating fields induce a resultant emf in the stator which balances the applied voltage assuming low leakage impedance of the stator winding.

If, however, the rotor is made to run at speed N by some external means in any direction, say in the direction of forward field, the two slips are now s and $(2 - s)$, as shown below:

The slip of the rotor with respect to the forward rotating field F_f,

$$s_f = \frac{N_s - N}{N_s} = s \qquad ...(4.1)$$

The slip of the rotor with respect to the backward rotating field F_b,

$$s_b = \frac{N_s-(-N)}{N_s} = \frac{2N_s-(N_s-N)}{N_s} = (2-s) \qquad \text{...(4.2)}$$

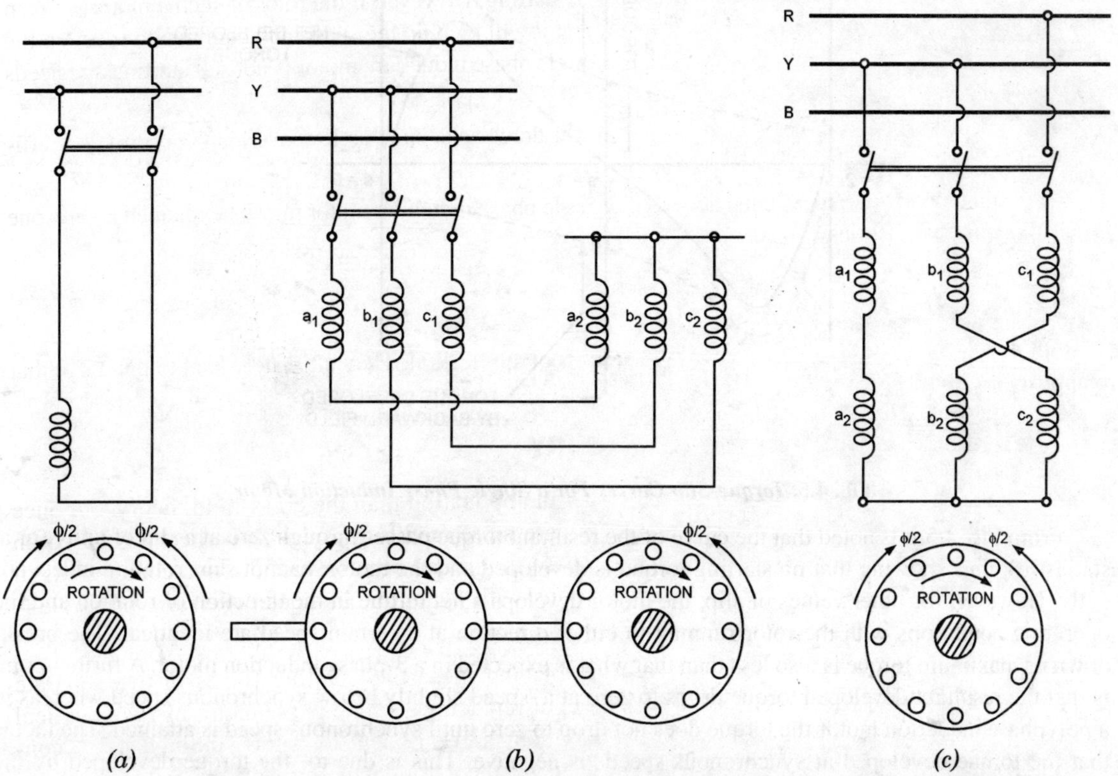

Fig. 4.4. ***Principle of Operation of a Single Phase Induction Motor***

For normal operation $(2-s) >> s$ and as a consequence the backward field rotor currents are much larger than at standstill and have a low power factor. The corresponding opposing rotor mmf, owing to stator impedance, causes the backward field to be greatly reduced in strength. On the other hand, the low-slip forward rotating field induces smaller currents of a high power factor in the rotor than at standstill. This leads to greatly strengthening of the forward field. Thus weakening of backward field and strengthening of forward field depends upon the slip (or speed of rotor) and the difference increases with the decrease in slip with respect to the forward field or with the increase in rotor speed in forward direction. In fact, at near about the synchronous speed, the forward field may be several times the backward field. As a result there is a net running torque.

Let us assume that the torque developed by forward field in the direction of rotor rotation is positive. Then the torque developed by the backward field will be negative and, evidently, be a braking torque.

The torque-slip curve of actual motor may be obtained by applying the principle of superposition to the hypothetical constituent motors. Their individual torque-slip curves will be of the form shown in

Fig. 4.5 by dotted line curves. The net or resultant torque at different slips is the algebraic sum of individual torques and is shown by the full-line curve in Fig. 4.5.

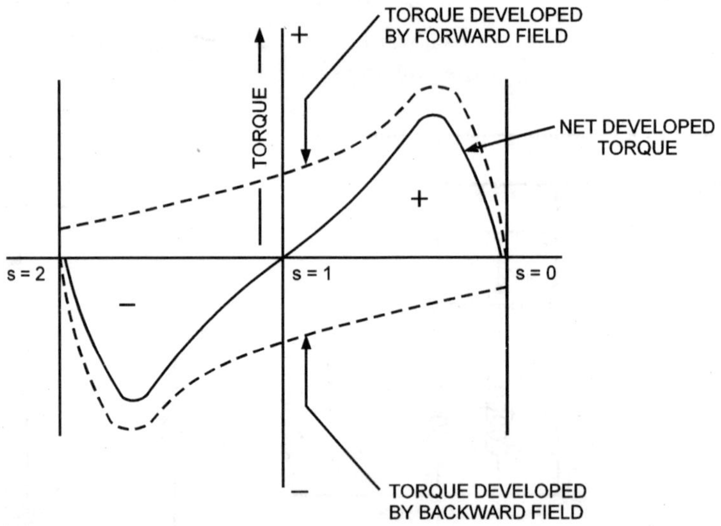

Fig. 4.5. ***Torque-Slip Curves For a Single Phase Induction Motor***

From Fig. 4.5 it is noted that the curve of the resultant torque passes through zero at a slip of unity (or at standstill), thus showing that no starting torque is developed and the motor cannot start rotating independently. However, for other values of slip, the motor develops a net torque in the direction of rotation and the operating conditions with the rotor running in either direction at a certain speed are identical. The breakdown or maximum torque is also less than that what is expected in a 3-phase induction motor. A further point is that the resultant developed torque drops to zero at a speed slightly below synchronous speed whereas in a polyphase induction motor the torque does not drop to zero until synchronous speed is attained. The fact is that the torque developed at synchronous speed is negative. This is due to the torque developed by the backward field.

The effect of increasing rotor resistance is rather different in a single phase induction motor than in a 3-phase induction motor. In the case of a 3-phase induction motor the value of the maximum torque is not affected by the change in rotor resistance, the only change being the value of slip at which maximum torque occurs. In a single phase machine, the increase in rotor resistance increases the effectiveness of the backward field, which reduces the breakdown torque, lowers the efficiency and increases the slip corresponding to maximum torque. Double squirrel cages and deep rotor bars are also impractical for the same reasons, and some other means must be adopted to obtain high starting torque.

The performance characteristics of a single phase induction motor, in general, are of the same nature as those for polyphase induction motor. However, the performance of a single phase induction motor, in every respect , is somewhat inferior to that of a 3-phase induction motor for the given frame size. This is due to the presence of backward rotating field. It has a lower breakdown torque at larger slip and increased power losses. Further, it has greater power (true as well as apparent) input because of their consumption in the backward rotating field. Even the copper losses occuring in the stator winding of a single phase induction motor are higher because all the current flows through a single winding. As a consequence, efficiency is lowered and temperature rise is increased. Further, the speed regulation of a single phase induction motor tends to be poorer than that for a polyphase motor.

The normal slip of a single phase induction motor under load conditions is rather greater than that of the corresponding 3-phase motor and, therefore, power factor of a single phase motor tends to be lower. For a given power and speed rating, a single phase motor must therefore, have a larger frame size than that of a 3-phase motor.

The forward field and the rotor's backward reaction field and also the backward field and the rotor's forward reaction field move in opposite directions with relative speeds of double the synchronous speed and develop second harmonic pulsating torques with zero average value. As a consequence single phase motors tend to be somewhat noisier than three-phase ones, which have no such pulsating torque.

4.4. CROSS-FIELD THEORY

Although the double revolving field theory of a single phase induction motor gives some idea of its method of operation, it does not give a conception of the reaction, that actually takes place in the machine. The mechanism of torque production is as follows:

Consider standstill conditions with the stator winding connected to a single phase ac supply. The stator current establishes a field along the vertical axis YY', similar to that shown in Fig. 4.6. During the half cycle that the stator current is flowing in the direction shown, north pole is established on the stator surface at A and south pole at C. During the next half cycle, the stator poles are reversed. Although the stator field strength is varying and reversing its polarity periodically, its action line is always along the vertical line YY'. Thus with alternating current in the stator winding, the stator mmf wave is stationary in space but pulsates in magnitude, the stator field strength alternating in polarity and varying sinusoidally with time.

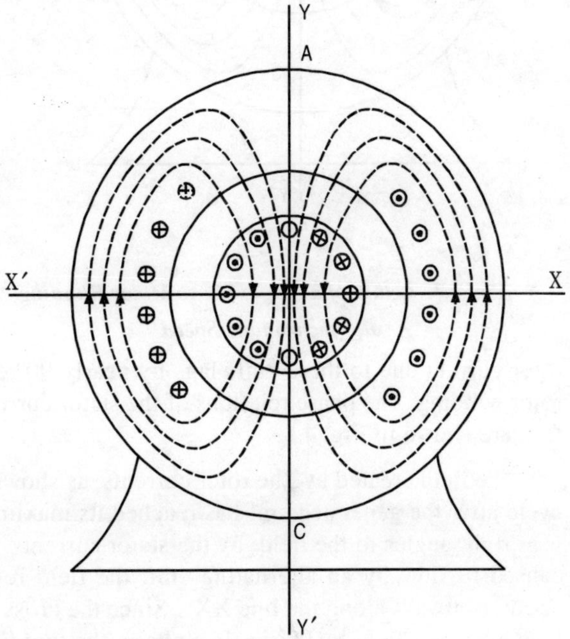

Fig. 4.6. ***Action of a Single Phase Motor at Standstill***

The rotor winding acts like a short-circuited secondary winding of a transformer and, therefore, carry induced currents. The directions of currents which would exist throughout one half cycle are shown in Fig. 4.6. Throughout the other half cycle the currents would be in respective opposite directions. The rotor currents establish poles on the rotor surface, but since these poles are always in direct line (along the vertical axis YY') with the stator poles no torque is developed in either direction. Therefore, no starting torque is developed to start the machine.

When, however, the single phase stator winding is connected to single phase ac supply and the rotor is made to rotate by some external means, the rotor conductors cut across the stator field, causing an emf known as *speed emf*, to be generated in them. In Fig. 4.7 for example, the stator field is downward and the rotor is being turned in clockwise direction, the direction of the generated emfs as determined by Fleming's right hand rule, will be outward in the lower half of the rotor and inward in the upper half of the rotor as indicated by the dots and crosses. During the next half cycle, the direction of the generated emfs will be reversed. The generated rotor emfs vary in phase with the stator current and flux. The

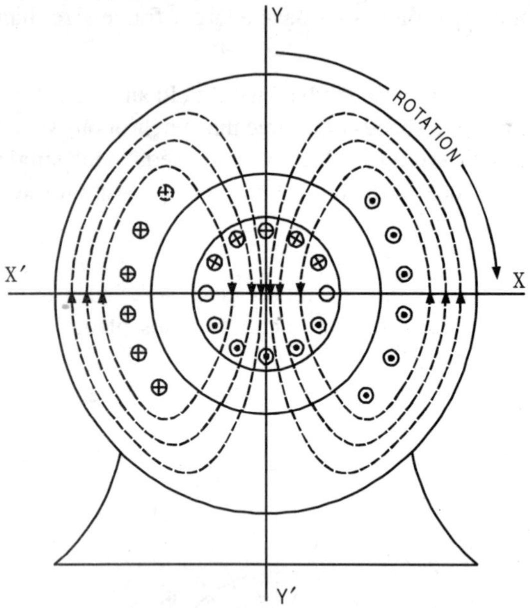

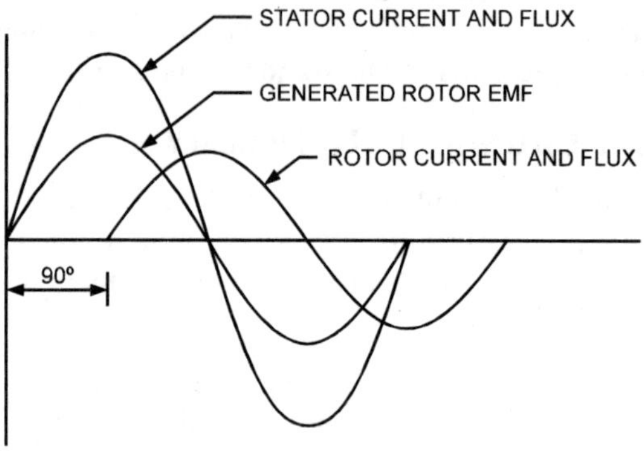

Fig. 4.7. ***Action of a Single Phase Motor Running at Synchronous Speed***

Fig. 4.8

rotor current due to these emfs lags by nearly 90° owing to low resistance and high inductance of the rotor winding. The phase relations of the stator current and flux, the rotor emf and the rotor current and flux are shown in Fig. 4.8.

The field created by the rotor currents, as shown in Fig. 4.8, becomes maximum nearly one-fourth cycle after the generated emf has reached its maximum value. Since the field created by rotor currents is at right angles to the field by the stator currents, it is known as *cross-field.* As the rotor currents are caused to flow by an alternating emf, the field resulting from these currents also alternates and its action is always along the line XX'. Since the cross-field acts at right angles to the stator field and also lags the stator field by 90° in time phase, the two fields combine to form a resultant rotating field that revolves at synchronous speed.

Since the cross-field is produced by a generator action, therefore, it is present only when the rotor is rotating and its strength is proportional to the speed of rotor. At synchronous speed the cross-field is nearly of same strength as the stator field. But since an induction motor must operate at some speed lower than synchronous speed, the cross-field is somewhat weaker than the stator field at actual operating speeds. Thus the rotating field produced in a single phase induction motor is not of constant strength as is the field of a polyphase motor but irregular. Therefore, the torque developed by a single phase induction motor is pulsating. This is the reason why many single phase motors are set on rubber or spring mounts to reduce the vibrations and noise which are inherent in such motors.

From the above discussion it may be concluded that at standstill there can be no cross-field, only the pulsating stator field, and therefore, the single phase induction motor cannot start itself *i.e.* the inherent starting torque of a single phase induction motor is zero. If, however, the rotor is made to run by some external means, then it will continue to develop torque as a single phase induction motor in the direction of rotation.

4.5. EQUIVALENT CIRCUIT (OR CIRCUIT MODEL) OF A SINGLE PHASE INDUCTION MOTOR

The equivalent circuit of a single phase induction motor can be developed on the basis of double revolving field theory. To develop the equivalent circuit it is first necessary to consider standstill or blocked rotor conditions. The motor with a blocked rotor merely acts like a transformer with its secondary short- circuited and its equivalent circuit will be as shown in Fig. 4.9 (*a*), E_m being emf induced in the stator. The motor may now be viewed from the point of view of the two revolving field theory. The two flux components induce in the stator winding the respective emf E_{mf} and E_{mb}. The phasor sum of E_{mf} and E_{mb} equals the applied voltage V (less the voltage drops in stator resistance R_1 and leakage reactance X_1). Since at standstill the two oppositely rotating fields are of same strength, the magnetizing and rotor impedances are divided into two equal halves connected in series, as illustrated in Fig. 4.9 (*b*). Thus, the equivalent circuit of a single phase induction motor at standstill, on the basis of the double revolving field theory will be that given in Fig. 4.9 (*b*).

When the rotor is running at speed N with respect to forward field, the slip is *s* with respect to forward field and (2 – *s*) with respect to backward rotating field and the equivalent circuit becomes as given in Fig. 4.9 (*c*). For simplification, if the core losses are neglected, the equivalent circuit is modified as given in Fig. 4.9 (*d*). The core losses, while making use of the equivalent circuit given in Fig. 4.9 (*d*), are handled as rotational losses and subtracted from the power converted into mechanical power. The amount of error thus introduced is relatively small.

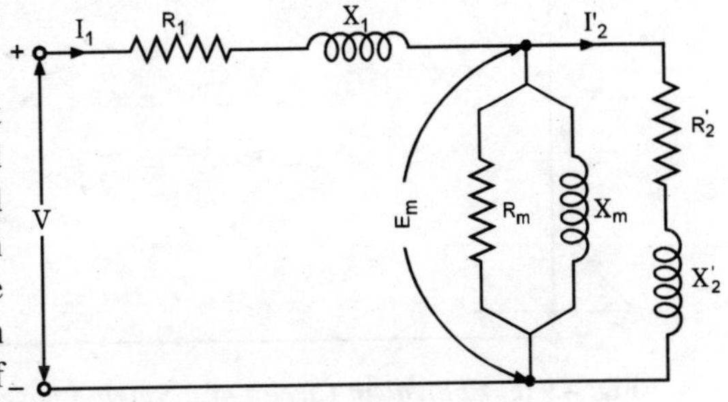

Fig. 4.9 (*a*) *Equivalent Circuit of a Single Phase Induction Motor at Standstill*

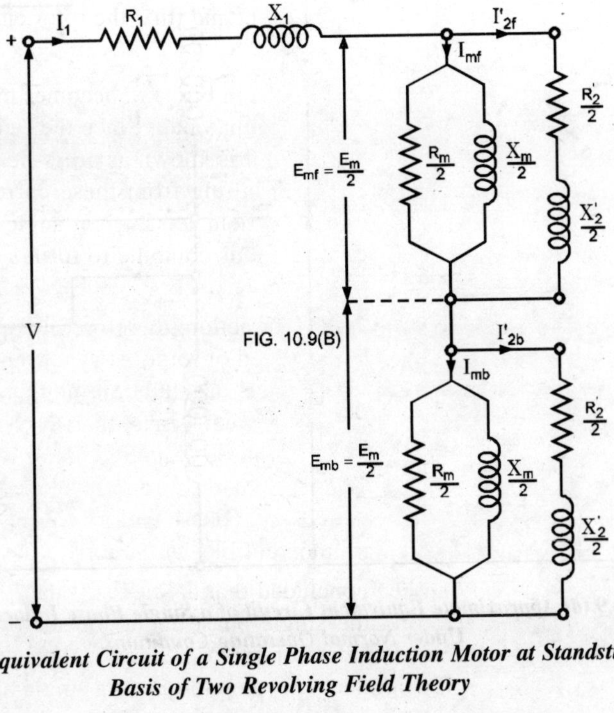

Fig. 4.9 (*b*) *Equivalent Circuit of a Single Phase Induction Motor at Standstill on The Basis of Two Revolving Field Theory*

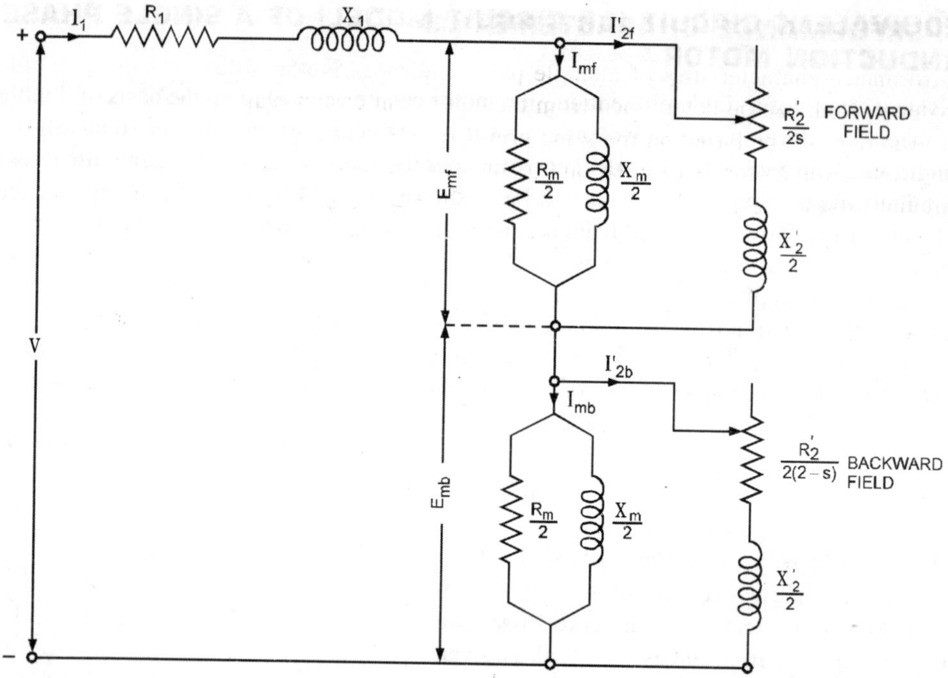

Fig. 4.9 ***(c) Equivalent Circuit of a Single Phase Induction Motor Under Normal Operating Conditions***

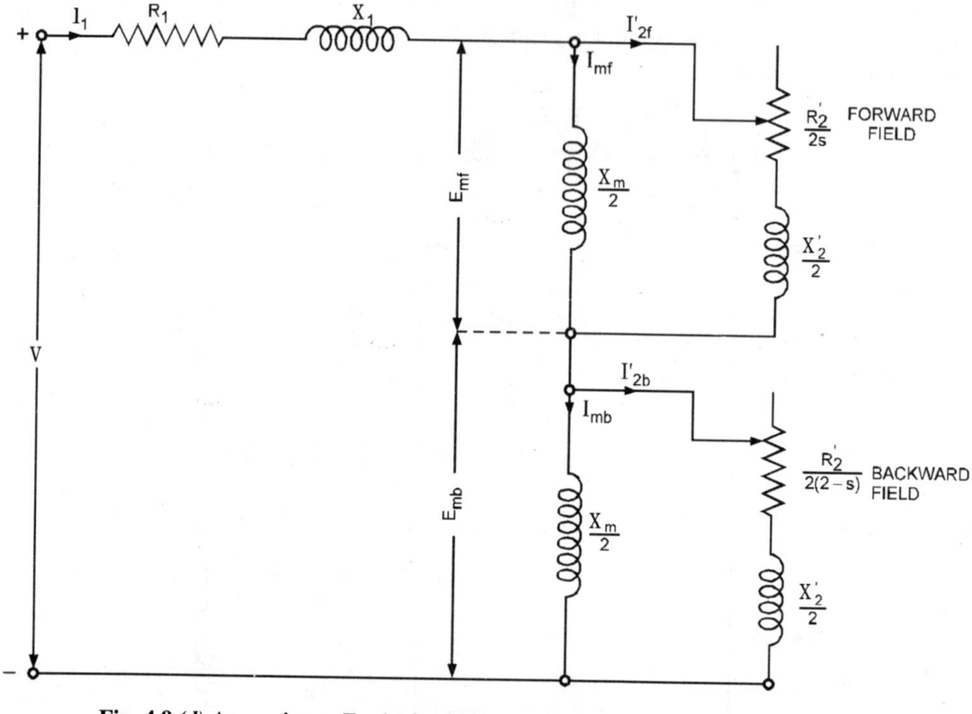

Fig. 4.9 ***(d) Approximate Equivalent Circuit of a Single Phase Induction Motor Under Normal Operating Conditions***

4.6. PERFORMANCE CHARACTERISTICS

The performance characteristics of a single phase induction motor, when running with only its main winding energized, can be determined from the equivalent circuit diagram for different values of slip. The calculations will be based on the same principles as were applied to the equivalent circuit for the polyphase induction motor. In calculation for single phase induction motor the core losses will be taken as rotational losses.

Impedance due to forward field $\mathbf{Z}_f = R_f + j\,X_f$

$$= \frac{R_2'}{2s} + j\frac{X_2'}{2} \text{ in parallel with } \frac{jX_m}{2} \quad ...(4.3)$$

Impedance due to backward field, $\mathbf{Z}_b = R_b + j\,X_b$

$$= \frac{R_2'}{2(2-s)} + j\frac{X_2'}{2} \text{ in parallel with } \frac{jX_m}{2} \quad ...(4.4)$$

Stator impedance $\mathbf{Z}_1 = R_1 + j\,X_1$

Total equivalent impedance $\mathbf{Z}_{eq} = \mathbf{Z}_1 + \mathbf{Z}_f + \mathbf{Z}_b$ ...(4.5)

$$\text{Motor current } I_1 = \frac{\mathbf{V}}{\mathbf{Z}_{eq}} = \frac{\mathbf{V}}{\mathbf{Z}_1 + \mathbf{Z}_f + \mathbf{Z}_b} \quad ...(4.6)$$

$$\text{Power factor} = \frac{R_{eq}}{\mathbf{Z}_{eq}} = \frac{R_1 + R_f + R_b}{\mathbf{Z}_1 + \mathbf{Z}_f + \mathbf{Z}_b} \quad ...(4.7)$$

$$\mathbf{E}_{mf} = \mathbf{I}_1\,\mathbf{Z}_f \quad ...(4.8)$$

$$\mathbf{E}_{mb} = \mathbf{I}_1\,\mathbf{Z}_b \quad ...(4.9)$$

$$I_{2f}' = \frac{E_{mf}}{\dfrac{R_2'}{2s} + j\dfrac{X_2'}{2}} = \frac{\mathbf{I}_1\,\mathbf{Z}_f}{\sqrt{\left(\dfrac{R_2'}{2s}\right)^2 + \left(\dfrac{X_2'}{2}\right)^2}} \quad ...(4.10)$$

$$I_{2b}' = \frac{E_{mb}}{\dfrac{R_2'}{2(2-s)} + j\dfrac{X_2'}{2}}$$

$$= \frac{\mathbf{I}_1\,\mathbf{Z}_b}{\sqrt{\left(\dfrac{R_2'}{2(2-s)}\right)^2 + \left(\dfrac{X_2'}{2}\right)^2}} \quad ...(4.11)$$

Air-gap power for forward field, $P_{gf} = (I_{2f}')^2\,\dfrac{R_2}{2s}$ watts ...(4.12)

Air-gap power for backward field, $P_{gb} = (I'_{2b})^2 \dfrac{R'_2}{2(2-s)}$ watts ...(4.13)

Mechanical power output for the forward field,

$$P_{mech\,f} = (1-s)\,P_{gf} = (I'_{2f})^2\,\frac{R'_2}{2}\cdot\frac{1-s}{s} \qquad ...(4.14)$$

Mechanical power output for the backward field,

$$P_{mech\,b} = [1-(2-s)\,P_{gb}]$$

$$= -(I'_{2b})^2\,\frac{R'_2}{2}\cdot\frac{1-s}{2-s} \qquad ...(4.15)$$

Mechanical power output, $P_{mech\ net} = P_{mech\,f} + P_{mech\,b}$

$$= (1-s)\,P_{gf} + [1-(2-s)\,P_{gb}]$$

$$= (1-s)\,[P_{gf} - P_{gb}] = P_{mech\,f} - P_{mech\,b} \qquad ...(4.16)$$

Torque developed for the forward field

$$= \frac{P_{gf}}{2\pi N_s/60} = \frac{9.55 P_{gf}}{N_s} \qquad ...(4.17)$$

Torque developed for the backward field

$$= \frac{9.55 P_{gb}}{N_s} \qquad ...(4.18)$$

$$\text{Net torque developed} = \frac{9.55}{N_s}\left(P_{gf} - P_{gb}\right) \qquad ...(4.19)$$

$$\text{Net power output} = P_{mech\ net} - \text{friction and windage losses} - \text{core losses} \qquad ...(4.20)$$

$$\text{Motor efficiency, } \eta = \frac{\text{Output power}}{\text{Input power}} \times 100 \qquad ...(4.21)$$

Power delivered to the backward field

$$= (2-s)\,P_{gb} = P_{gb} + (1-s)\,P_{gb} = P_{gb} - P_{mech\,b} \qquad ...(4.22)$$

From the above Eq. (4.22) it is obvious that power from the supply is delivered to the backward field in two ways. Power from the supply is delivered directly to the backward field across its air gap. The backward field also acts as a generator and must be driven by the forward field. Thus the power is delivered to the backward field from the power converted into mechanical power by the forward field. All of the power delivered to the backward field is converted into heat through the rotor copper losses caused by backward field.

Example 4.1. A 230 V, 4-pole, 50 Hz, 1-phase induction motor has an effective rotor resistance and leakage reactance of 0.5 Ω and 5 Ω respectively. It is running at a speed of 1,350 rpm. Determine :

(*i*) frequencies of forward and backward rotor current components

(*ii*) relative magnitudes of forward and backward fluxes.

Neglect magnetizing current and stator impedance.

Solution :

$$\text{Synchronous speed, } N_s = \frac{120 \times f}{P} = \frac{120 \times 50}{4} = 1{,}500 \text{ rpm}$$

$$\text{Speed, } N = 1{,}350 \text{ rpm}$$

$$\text{Slip, } s = \frac{N_s - N}{N_s} = \frac{1{,}500 - 1{,}350}{1{,}500} = 0.1$$

(*i*) Frequency of forward rotor currents, $f'_f = s\,f = 0.1 \times 50 = 5$ Hz **Ans.**

Frequency of backward rotor currents, $f'_b = (2 - s)\,f = (2 - 0.1) \times 50 = 95$ Hz **Ans.**

Impedance due to forward field, $\mathbf{Z}_f = R_f + j\,X_f$

$$= \frac{R'_2}{2s} + j\frac{X'}{2} = \frac{0.5}{2 \times 0.1} + j\frac{5}{2}$$

$$= (2.5 + j\,2.5)\,\Omega = 3.536\ \angle 45^\circ\ \Omega$$

Impedance due to backward field, $Z_b = R_b + j\,X_b = \dfrac{R'_2}{2\,(2-s)} + j\dfrac{X'}{2}$

$$= \frac{0.5}{2\,(2-0.1)} + j\frac{5}{2}$$

$$= (0.1316 + j\,2.5)\,\Omega = 2.5035\ \angle 86.99^\circ\ \Omega$$

$$\text{Now } \frac{\phi_f}{\phi_b} = \frac{E_f}{E_b} = \frac{I_1\,Z_f}{I_1\,Z_b} = \frac{Z_f}{Z_b} = \frac{3.536}{2.5035} = 1.4 \quad \textbf{Ans.}$$

Example 4.2. A 125-W, 4-pole, 110-V, 50-Hz single phase induction motor delivers rated output at a slip of 6%. The total copper loss at full load is 25 watts. Calculate the full-load efficiency and the rotor copper loss caused by the backward field. Rotational losses may be assumed to be 25 watts. Neglect stator copper loss.

Solution :

Output of motor = 125 watts

Rotational losses = 25 watts

Mechanical power developed by rotor, P_{mech} = Output + rotational losses

= 125+ 25=150 watts

Full-load slip, s_f = 6% or 0.06

$$\text{Power input to rotor} = P_{gf} - P_{gb} = \frac{P_{mech}}{1 - s_f} = \frac{150}{1 - 0.06} = 160 \text{ watts}$$

$$\text{i.e. } P_{gf} - P_{gb} = 160 \text{ watts} \qquad ...(i)$$

$$\text{and total copper losses} = s\,P_{gf} + (2 - s)\,P_{gb}$$

$$\text{or } 25 = 0.06\,P_{gf} + 1.94\,P_{gb}$$

$$\text{or } 0.06\,P_{gf} + 1.94\,P_{gb} = 25 \text{ watts} \qquad ...(ii)$$

Solving Eqs. (*i*) and (*ii*) we have rotor copper loss caused by backward field

$$1.94\ P_{gb} = \frac{25-(0.06\times 160)}{2}\times 1.94 = 14.94 \text{ watts } \textbf{Ans.}$$

Input power to motor = Output power of motor + rotational losses + rotor copper losses + stator copper losses + core losses

Neglecting stator copper losses and core losses

Input power to motor = 125 + 25 + 25 = 175 watts

$$\text{Full-load efficiency, } \eta = \frac{\text{Output}}{\text{Input}}\times 100 = \frac{125}{175}\times 100 = 71.43\% \ \textbf{Ans.}$$

Example 4.3. A 230 V, 50 Hz, 4-pole, class A, single phase induction motor has the following parameters at an operating temperature of 63° C :

$$r_{1m} = 2.51\ \Omega,\ r'_2 = 7.81\ \Omega,\ x_m = 150.88\ \Omega,\ x_{1m} = 4.62\ \Omega,\ x'_2 = 4.62\ \Omega$$

Determine the main winding current and power factor when the motor is running at a slip of 0.05 at the specified temperature of 63° C.

Solution : Resistance of stator main winding, $R_1 = 2.51\ \Omega$

Leakage reactance of stator main winding, $X_1 = 4.62\ \Omega$

Magnetizing reactance of stator main winding, $X_m = 150.88\ \Omega$

Rotor resistance, $R'_2 = 7.81\ \Omega$

Standstill rotor reactance, $X'_2 = 4.62\ \Omega$

Slip, $s = 0.05$

$(2-s) = 2 - 0.05 = 1.95$

Stator impedance, $\mathbf{Z}_1 = (2.51 + j\,4.62)\ \Omega$

$$\text{Forward field impedance, } \mathbf{Z}_f = \frac{\dfrac{jX_m}{2}\left[\dfrac{R'_2}{2s}+\dfrac{jX'_2}{2}\right]}{\dfrac{R'_2}{2s}+j\left(\dfrac{X_m+X'_2}{2}\right)}$$

$$= \frac{j\dfrac{150.88}{2}\left[\dfrac{7.81}{2\times 0.05}+j\dfrac{4.62}{2}\right]}{\dfrac{7.81}{2\times 0.05}+j\dfrac{150.88+4.62}{2}}$$

$$= \frac{j\,75.44\,(78.1+j\,2.31)}{78.1+j\,77.75}$$

$$= \frac{75.44\ \angle 90^\circ \times 78.134\ \angle 1.694^\circ}{110.203\ \angle 44.87^\circ}$$

$$= 53.487\ \angle 46.824^\circ \text{ ohms}$$

$$\text{Similarly backward field impedance, } \mathbf{Z}_b = \frac{\dfrac{jX_m}{2}\left[\dfrac{R_2'}{2(2-s)} + j\dfrac{X_2'}{2}\right]}{\dfrac{R_2'}{2(2-s)} + j\left(\dfrac{X_m + X_2'}{2}\right)}$$

$$= \frac{j\,75.44\left[\dfrac{7.81}{2\times 1.95} + j2.31\right]}{\dfrac{7.81}{2\times 1.95} + j\dfrac{150.88 + 4.62}{2}}$$

$$= \frac{j\,75.44\,(2.003 + j\,2.31)}{2.003 + j\,77.75}$$

$$= \frac{75.44\,\angle 90^\circ \times 3.0572\,\angle 4.91^\circ}{77.776\,\angle 88.52^\circ}$$

$$= 2.965\,\angle 50.58^\circ \text{ ohms}$$

$$\text{Total equivalent impedance, } \mathbf{Z}_{eq} = \mathbf{Z}_1 + \mathbf{Z}_f + \mathbf{Z}_b$$

$$= (2.51 + j\,4.62) + 53.487\,\angle 46.824^\circ + 2.965\,\angle 50.58^\circ$$

$$= (2.51 + j\,4.62) + 36.598 + j\,39 + 1.883 + j\,2.29$$

$$= 40.991 + j\,45.91 = 61.547\,\angle 48.24^\circ\ \Omega$$

$$\text{Stator main winding current, } I_1 = \frac{V}{Z_{eq}} = \frac{230}{61.547\,\angle 48.24^\circ} = 3.737 \text{ A } \textbf{Ans.}$$

$$\text{Power factor, } \cos\phi = \cos 48.24^\circ = 0.666 \text{ lag } \textbf{Ans.}$$

Example 4.4. The constants of a $\frac{1}{4}$ hp, 230 V, 4-pole, 60 Hz, single phase induction motor are as follows :

Stator resistance, $R_1 = 10.0\ \Omega$
Stator reactance, $X_1 = 12.8\ \Omega$
Rotor resistance referred to stator, $R'_2 = 11.65\ \Omega$
Rotor reactance referred to stator, $X_2' = 12.8\ \Omega$
Magnetising reactance, $X_m = 258.0\ \Omega$

The total load is such that the machine runs at 3% slip, when the applied voltage is at 210 V. The iron losses are 35.5 watts at 210 V

Calculate (*i*) input current, (*ii*) power developed (*iii*) shaft power (if mechanical losses are 7 watts) and (*iv*) efficiency.

Solution : Core loss current, $I_e = \dfrac{35.5}{210} = 0.169$ A

Since under running condition V_f is very high (almost 90 per cent of the applied voltage) and V_b is very low, so most of the core loss takes place in the 'forward motor' consisting of the common stator and forward running motor. So half value of core loss equivalent resistance,

$$\frac{R_m}{2} = \frac{210}{0.169} = 1{,}242\ \Omega$$

For forward Field [Fig 4.9 (*c*)]

$$\text{Conductance of core-loss branch} = \frac{1}{\frac{R_m}{2}} = \frac{1}{1{,}242} = 0.0008\ \text{S}$$

$$\text{Susceptance of magnetising branch} = \frac{-j}{\frac{X_m}{2}} = \frac{1}{\frac{258}{2}} = -j\,0.00775\ \text{S}$$

$$\text{Rotor admittance} = \frac{1}{\frac{R_2'}{2s} + \frac{jX_2'}{2}} = \frac{1}{\frac{11.65}{2 \times 0.03} + \frac{j12.8}{2}}$$

$$= \frac{1}{194.17 + j6.4} = (0.005145 - j\,0.00017)\ \text{S}$$

$$\text{Admittance}, \mathbf{Y}_f = (0.0008 - j\,0.00775 + 0.005145 - j\,0.00017)\ \text{S}$$

$$= (0.005945 - j\,0.00792)\ s$$

$$\text{Impedance}, \mathbf{Z}_f = \frac{1}{\mathbf{Y}_f} = \frac{1}{0.005945 - j\,0.00792}$$

$$= 60.62 + j\,80.759 = 100.979\ \angle 53.1^\circ\ \Omega$$

For backward field

$$\text{Rotor admittance} = \frac{1}{\frac{R_2'}{2(2-s)} + \frac{jX_2'}{2}} = \frac{1}{\frac{11.65}{2(2-0.03)} - \frac{j12.8}{2}}$$

$$= \frac{1}{2.957 + j6.4} = (0.0595 - j\,0.012876)\ \text{S}$$

$$\text{Admittance}, \mathbf{Y}_b = (0.0008 - j\,0.00775 + j\,0.0595 - j\,0.12876)\ \text{S}$$

$$= (0.0603 - j\,0.1365)\ \text{S}$$

$$\text{Impedance}, \mathbf{Z}_b = \frac{1}{\mathbf{Y}_b} = \frac{1}{0.0603 - j0.1365}$$

$$= 2.708 + j\,6.13 = 6.70\ \angle 66.17^\circ\ \Omega$$

$$\text{Equivalent impedance}, \mathbf{Z}_{eq} = \mathbf{Z}_1 + \mathbf{Z}_f + \mathbf{Z}_b$$

$$= (10 + j\,12.8) + (60.62 + j\,80.759) + j\,(2.708 + j\,6.13)$$

$$= 73.323 + j\,99.689 = 123.75\ \angle 53.66^\circ\ \Omega$$

$$\text{Input current}, I_1 = \frac{V}{Z_{eq}} = \frac{230}{123.75} = 1.8586\ \text{A}$$ **Ans.**

$$\text{Power input} = VI_1 \cos\phi = 230 \times 1.8586 \times \cos 53.66° = 253.3 \text{ W}$$

$$E_{mf} = I_1 \times Z_f = 1.8586 \times 100.979 = 187.68 \text{ V}$$

$$E_{mb} = I_1 \times Z_b = 1.8586 \times 6.70 = 12.45 \text{ V}$$

$$I'_{2f} = \frac{E_{mf}}{\sqrt{\left(\frac{R'_2}{2s}\right)^2 + \left(\frac{X'_2}{2}\right)^2}} = \frac{187.68}{\sqrt{(194.17)^2 + (6.4)^2}} = 0.966 \text{ A}$$

$$I'_{2b} = \frac{E_{mb}}{\sqrt{\left(\frac{R'_2}{2(2-s)}\right)^2 + \left(\frac{X'_2}{2}\right)^2}} = \frac{12.45}{\sqrt{(2.2957)^2 + (6.4)^2}} = 1.831 \text{ A}$$

$$\text{Air-gap power for forward field, } P_{gf} = (I'_{2f})^2 \times \frac{R'_2}{2s}$$

$$= (0.966)^2 \times \frac{11.65}{2 \times 0.03} = 181.2 \text{ W}$$

$$\text{Air-gap power for backward field, } P_{gb} = (I'_{2b})^2 \times \frac{R'_2}{2(2-s)} = (1.831)^2 \times \frac{11.65}{2 \times 1.97} = 9.9 \text{ W}$$

(*ii*) Power developed $P_{\text{mech net}} = (1 - s)(P_{gf} - P_{gb})$

$= (1 - 0.03)(181.2 - 9.9) = 166.2$ W **Ans.**

(*iii*) Shaft power $= P_{\text{mech net}}$ – mechanical losses $= 166.2 - 7 = 159.2$ W **Ans.**

(*iv*) Efficiency, $\eta = \frac{\text{Power output}}{\text{Power input}} \times 100 = \frac{159.2}{253.3} \times 100 = 62.85\%$ **Ans.**

4.7. DETERMINATION OF EQUIVALENT CIRCUIT PARAMETERS BY TESTS

The performance (efficiency, losses and power factor) of single phase induction motors is obtained by applying load to it and measuring output, input, speed and current, as such motors are usually of small ratings. However, to determine the constants of the machine it is necessary to perform various tests. The most common tests performed on single phase induction motors for determination of various constants are:

(*a*) **Measurement of Stator Winding Resistance.** The resistance of each stator winding is measured separately and in case of capacitor start motor the condenser is omitted.

(*b*) **Blocked Rotor Test.** This test is performed with main winding and starting winding separately and input voltage, amperes, and watts are measured in each case. This test is sometimes performed at reduced voltage (about 40 per cent of rated voltage) to avoid excessive short-circuit current and heating.

(*c*) **No-Load Test.** This test is performed at rated voltage, exciting the main winding only and input current, voltage and power, as well as the speed, are measured.

(*d*) **Determination of Friction and Windage Losses.** In this test the voltage applied to the motor is reduced until the motor will just run and the power input is measured. The power input, just sufficient to keep the motor running on no load, gives approximately the friction and windage losses.

When the rotor is blocked

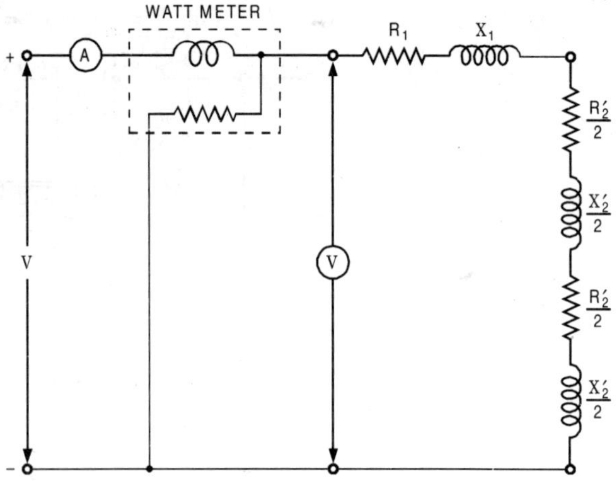

Fig. 4.10

Slip, s = 1.0. In the equivalent circuit $\frac{X_m}{2}$ can be neglected because $R'_2, X'_2 < X_m$. Further iron loss is also negligible as flux during blocked rotor test is small. The equivalent circuit on blocked rotor test with instruments is reduced to that shown in Fig. 4.10.

If the readings of the ammeter, voltmeter and wattmeter are I, V and P respectively,

$$\text{then } R'_2 + R_1 = \frac{P}{I^2} \qquad ...(4.23)$$

R_1 is measured separately with dc by ammeter-voltmeter method. This dc value is usually taken as the ac value because skin effect at normal supply frequency is negligible in small single phase induction motors.

$$\text{So } R'_2 = \frac{P}{I^2} - R_1 \qquad ...(4.24)$$

$$\text{Now equivalent impedance, } Z_{eq} = \sqrt{(R_1 + R'_2)^2 + (X_1 + X'_2)^2} = \frac{V}{I} \qquad ...(4.25)$$

$$\text{or } X_1 + X'_2 = \sqrt{Z_{eq}^2 - (R_1 + R'_2)^2} \qquad ...(4.26)$$

Assuming $X_1 = X'_2$ we have

$$X_1 = X'_2 = \frac{X_1 + X'_2}{2} \qquad ...(4.27)$$

In no-load test

Slip, s is very small, so $\frac{R'_2}{2s}$ in the equivalent circuit [Fig. 4.9 (*d*)] may be considered as infinity and the corresponding rotor circuit may be taken as open circuit. $\frac{X_m}{2}$ in the backward field circuit may also be omitted because $X_m > \frac{R'_2}{2}, X'_2$. Thus the equivalent circuit on no-load test with the measuring instruments becomes as given in Fig. 4.11. A fictitious resistance R' is added in the circuit shown in Fig. 4.11 to represent stray power loss*, which occur in addition to power losses in the resistances R_1 and $\frac{R'_2}{4}$. However, current through R' negligibly small, because R' is very high.

* Stray power losses are disregarded in the equivalent circuit of the motor.

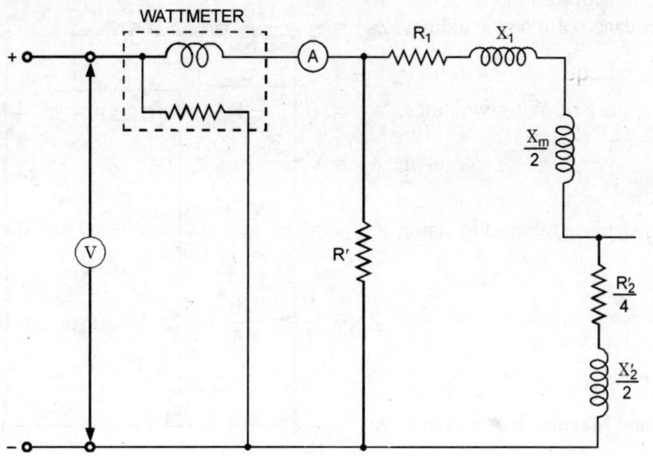

Fig. 4.11

If the readings of voltmeter, ammeter and wattmeter are V_0, I_0 and P_0 respectively, then

$$\text{Stray power loss, } P_s = P_0 - I_0^2\left[R_1 + \frac{R'_2}{4}\right] \qquad \text{...(4.28)}$$

The equivalent motor impedance, excluding R′, which carries negligibly small current in comparison to I, can be computed from the relation

$$Z_0 = \frac{V_0}{I_0} \qquad \text{...(4.29)}$$

$$\text{and } X_1 + \frac{X'_2}{2} + \frac{X_m}{2} = \sqrt{Z_0^2 - \left(R_1 + \frac{R'_2}{4}\right)^2} \qquad \text{...(4.30)}$$

Since X_1 and X'_2 are known from Eq. (4.27), so

$$X_m = 2\sqrt{Z_0^2 - \left(R_1 + \frac{R'_2}{4}\right)^2} - 2\left(X_1 + \frac{X'_2}{2}\right) \qquad \text{...(4.31)}$$

Example 4.5. A 220 V, 1-ϕ induction motor gave the following test results:

Blocked rotor test	:	**120 V**	**9.6 A**	**460 W**
No-load test	:	**220 V**	**4.6 A**	**125 W**

The stator winding resistance is 1.5 Ω and during blocked rotor test, the starting winding is open. Determine equivalent circuit parameters.

Solution : Stator winding resistance, $R_1 = 1.5\ \Omega$

Equivalent resistance of motor windings, $R_{eq} = R'_2 + R_1 = \frac{P}{I^2} = \frac{460}{(9.6)^2} = 5\ \Omega$ (From blocked rotor test)

or $R'_2 = 5 - 1.5 = 3.5\ \Omega$ **Ans.**

Equivalent impedance of motor windings, $Z_{eq} = \frac{V}{I} = \frac{120}{9.6} = 12.5\ \Omega$ (From blocked rotor test)

Equivalent reactance of motor windings, $X_{eq} = \sqrt{Z_{eq}^2 - R_{eq}^2} = \sqrt{(12.5)^2 - 5^2} = 11.46\ \Omega$

Assuming $X_1 = X'_2$

Rotor winding reactance referred to stator, $X'_2 = X_1 = \frac{X_{eq}}{2} = \frac{11.46}{2} = 5.73\ \Omega$ **Ans.**

$$Z_0 = \frac{V_0}{I_0} = \frac{220}{4.6} = 47.8\ \Omega \quad \text{(From no-load test)}$$

and Magnetising reactance, $X_m = \sqrt{Z_0^2 - \left(R_1 + \frac{R'_2}{4}\right)} - 2\left(X_1 + \frac{X'_2}{2}\right)$

$$= 2\sqrt{(47.8)^2 - \left(1.5 + \frac{3.5}{4}\right)^2} - 2\left(5.73 + \frac{5.73}{2}\right)$$

$= 84.3\ \Omega$ **Ans.**

4.8. STARTING METHODS AND TYPES OF SINGLE PHASE INDUCTION MOTORS

As already discussed the single phase induction motor is not self starting and, therefore, it is necessary to employ some means for making it self starting. Commercial single phase induction motors employ the principle of 'phase split', described below, for the purpose of starting.

4.8.1. Principle of Phase Split

It is known that when two windings, spaced 90 electrical degrees apart on the stator of a motor, are excited by two alternating emfs that are 90° displaced in time phase, a rotating magnetic field is produced. If two windings so placed are connected in parallel to a single phase source, the field produced will alternate but will not revolve since the two windings are equivalent to but one single phase winding. If, however, an impedance (resistance, inductance or capacitance) is connected in series with one of these windings, the currents may be made to differ in time phase. By proper selection of such an impedance the currents may be made to differ by as much as 90°*, thereby producing a rotating field very much like the field of a two phase motor. This is the principle of *phase splitting* and the single phase induction motors employing this principle for starting are known as *split-phase motors*. Split-phase motors are of the following types:

(*a*) Resistance start motor or simply split-phase motor

(*b*) Capacitor-start motor

(*c*) Permanent capacitor or single-value capacitor motor and

(*d*) Capacitor start-capacitor run or two value capacitor motor

* If the phase difference between the two currents is less than 90° and if the two mmfs are not equal, starting torque will be small but in many applications it is sufficient to start the motor.

4.8.2. Constructional Features

All of the single phase induction motors mentioned above, have one construction feature in common, *viz.*, the *auxiliary or starting winding*. The starting winding is provided so that, together with the main or working winding, the motor can stimulate a *two-phase motor* to develop torque at start. In the case of the permanent capacitor motor, the auxiliary winding remains in the circuit after starting, and thus closely approximates a two phase motor also when it is operating under load. Hence, the term '*split-phase*' may rightly be applied to all of the above mentioned single phase induction motors. In practice, however, it is the high resistance starting winding type that retains the name, while those that use a capacitor to split the single phase are called by the particular capacitor circuit required.

Except for the single-value capacitor motor, all the motors are provided with a cut-out for the auxiliary winding. Despite the fact that other circuits are now being designed and employed, the centrifugal switch remains an important element in the starting-winding circuit. One important advantage of a centrifugal switch is that it opens at a predetermined speed, and subsequently closes at a lower speed when the line switch is opened and the motor is brought to rest. It is thus not subject to fluctuations in line voltage, or to current variations for sudden change in load.

Centrifugal Switch and its Importance. It is mounted inside the motor. It consist, of a stationary part and a rotating part. The stationary part, consisting of one stationary and the other movable contact, is located in one of the housing or end bell, and does not rotate. The movable contact is held against the stationary one by spring action. It is thus normally closed (NC). Connections to the starting winding and line or capacitor (if employed) are made from this end bell. The rotating part of the switch has no electrical connection. It contains the actuating mechanism, the truely centrifugally operated component. It is mounted on the shaft and held close to it, usually by spring action, when the motor is at rest. When the motor is energized, the rotor speed increases, and, at approximately 75 per cent of rated speed, there is sufficient centrifugal force on the actuating mechanism to overcome its spring tension. It then makes contact with the actual switch, opening it, and holding it open, unless the speed drops to approximately 50 per cent of rated one.

Nevertheless it is the centrifugal switch that is the major cause of trouble and breakdown of motor in the operation of the single phase induction motor. Should failure of the centrifugal switch not be detected, the motor may get permanently damaged. In the case of a capacitor-start or a high resistance start motors, the auxiliary winding is not designed to remain energized for more than a few seconds. If the centrifugal switch remains open after the motor has been operating, little harm should occur if the motor is provided with overload protection or if an operator is present. The motor simply won't start on the next attempt, and will be disconnected from service. But in case of remotely located motors, such as in case of heating systems having motors for the oil pump, circulating pump, and one or more blowers, the motor may attempt to start many times during a day, and the excessive cycling may cause damage to the protective relay, if not the motor itself.

On the other hand, should the switch remain closed, the motor will start and run. The starting winding will get quickly overheated, and, if not soon detected, or cut out by a protective relay, this winding may cause damage to the main winding due to their proximity.

With a two-value capacitor motor (or capacitor-start, capacitor-run motor), the starting winding is designed to remain energized after starting, but one capacitor is not. Should the centrifugal switch remain closed, this starting capacitor may quickly get damaged. Also, when running, the higher capacitance in series with the inductive winding may actually reduce the total impedance of the winding or phase, thereby increasing its current above normal. The added capacitor will cause a change in the overall power factor, thereby negating one of the reasons for the motor's use. Should an auto-transformer be used to have the higher capacitance for starting, similar effects occur, although the capacitor may not get damaged.

From the above discussions the importance of centrifugal switch is obvious and it is never repaired. *It should always be replaced.*

In single phase induction motors, the main winding usually occupies only $\frac{2}{3}$ rd of a pole pitch and this makes it possible to achieve a winding factor of 0.827. Filling the other slots with the winding is not rational, since this leads to such a small winding factor (up to $\frac{2}{\pi}$ *i.e.* 0.638) that the copper of the stator winding will be poorly utilized. The remaining empty one-third of the stator slots can be used for the auxiliary winding, whose axis is shifted in space through 90° with respect to the axis of the main winding. A two-winding system of this type can be had from a star-connected three-phase winding by disconnecting one phase winding and using it as the starting winding, the remaining two series connected phase windings serving as the main winding.

In order to provide high resistance and low inductance, the starting winding is wound with fewer turns of fine wire and placed on the top of the stator slots. The main winding, on the other hand, is wound with thicker wire and placed at the bottom of the stator slots. The idea behind it is that the current through the starting winding is merely nearly in phase with the line voltage than is that through the main winding. This gives the effect of splitting the single phase applied to the motor terminals.

With the increase in speed of motor, the torque developed increases and above 85 per cent of synchronous speed, the torque developed by the main winding alone is actually greater than that developed by both windings, and it might be advantageous to open the auxiliary or starting winding circuit at this cross-over point . To allow for individual variations among motors and switches, however, the contacts are usually designed to open at 75 per cent of synchronous speed. This does not affect the operation, because the main winding alone usually develops approximately twice full-load torque at this speed.

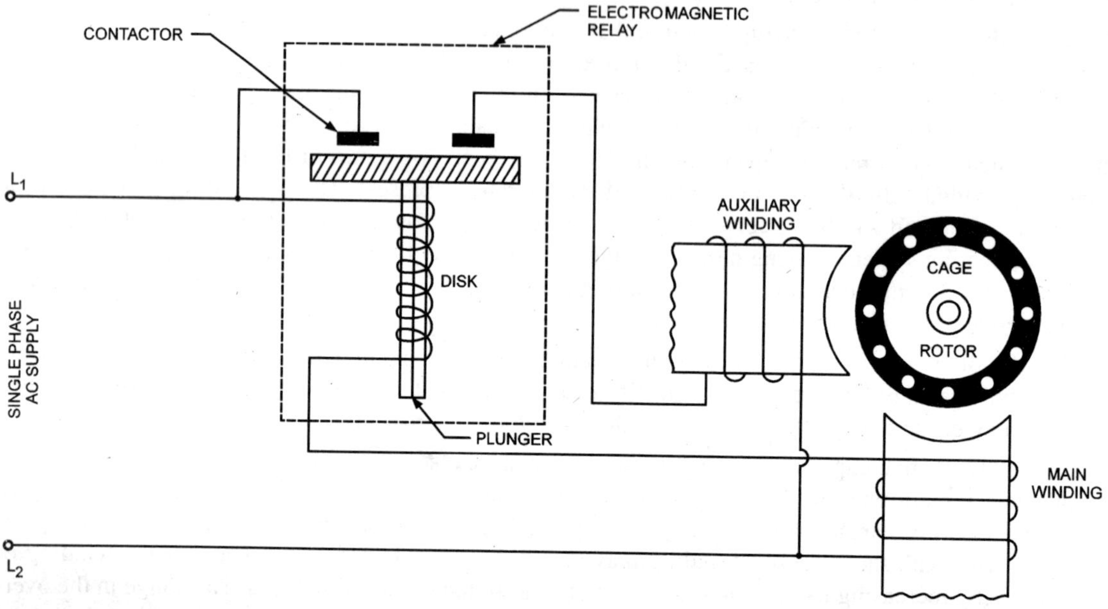

Fig. 4.12

The starting winding is not designed for continuous operation and care should be taken that it does not remain connected to the supply after it should have been disconnected by the switch. The series switch is usually centrifugally operated. In case of motors that are hermetically sealed, such as in refrigeration units, electromagnetic type relay is used.

Since split-phase motors are usually started directly from the line hence these motors draw large current (4 to 6 times the rated value) at start. This line current has two components, one of them (the larger of the two) passing through the main winding, while the other passes through the auxiliary winding. Advantage is taken of the high starting current in connection with the use of an electromagnetic type relay that performs the same function as the centrifugal switch. This relay, shown in Fig. 4.12, has a coil connected in series with the main winding, and a pair of contacts, normally open, in the auxiliary-winding circuit. At the start, when large current flows through the main winding, the contacts close so that the auxiliary winding circuit is energized. When the motor picks up the speed (75 or 80% of synchronous speed), the main-winding current drops to a value that is low enough to cause the contacts to open. Such relays are located outside the motor, where they can be serviced or replaced, a situation that would be impossible if an internally mounted centrifugal switch were used.

Reversal of Direction of Split—Phase Induction Motor. The direction of rotation of a split-phase induction motor can be reversed simply by reversing the line connections of either the main winding or the auxiliary winding. The noteworthy point is that reversal of a split -phase induction motor takes place only when it is at rest and the starting switch is in its normally closed de-energized position. If reversal is attemped under normal operating conditions, nothing will happen because under normal operating conditions, the starting switch is in open position and the motor is running as a pure single phase induction motor and developing torque in the direction of rotation.

4.9. RESISTANCE-START SINGLE PHASE INDUCTION MOTOR

It is a form of split-phase motor having a high resistance connected in series with the starting winding as illustrated in Fig. 4.13 (*a*). Because of low resistance and high inductive reactance of main winding and high resistance and low inductive reactance of starting winding, the starting winding current I_s does not lag so much as the main winding current I_m. At starting the main winding current I_m lags behind the applied voltage V by 65-75 degrees, whereas the starting winding current I_s lags by 35-45 degrees. Thus the phase displacement between main winding current and starting winding current is about 20-30 degrees as shown in phasor diagram [Fig. 4.13 (*b*)]. Since the currents in the two windings are different in magnitude and constants of two windings are different, therefore, the arrangement forms imperfect two phase motor, that produces a rotating

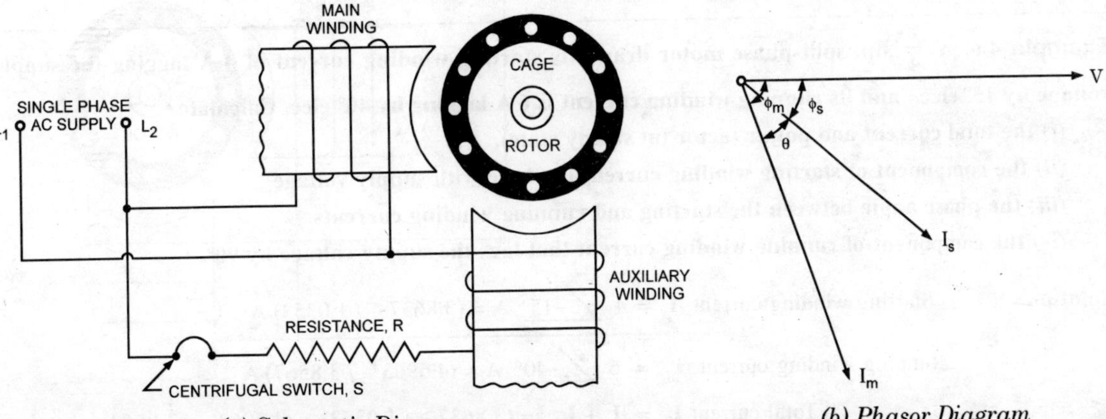

(a) Schematic Diagram *(b) Phasor Diagram*

Fig. 4.13

magnetic field that is not uniform in either time or space, but is sufficient for starting the motor. The best starting condition will be obtained if I_s and I_m differ in time by about 90º, but this condition is not possible in the motor of this type.

A typical torque-speed characteristic is shown in Fig. 4.13 (*c*). The starting torque is 1.5 times to twice the full-load starting torque and starting current is 6 to 8 times full-load current. The speed regulation of a resistance–start single phase induction motor is very good. The per cent slip in most fractional kilowatt split-phase motors is about 4-6 per cent. Such a motor may operate with a power factor of 0.55-0.65 and efficiency of 60-65 per cent.

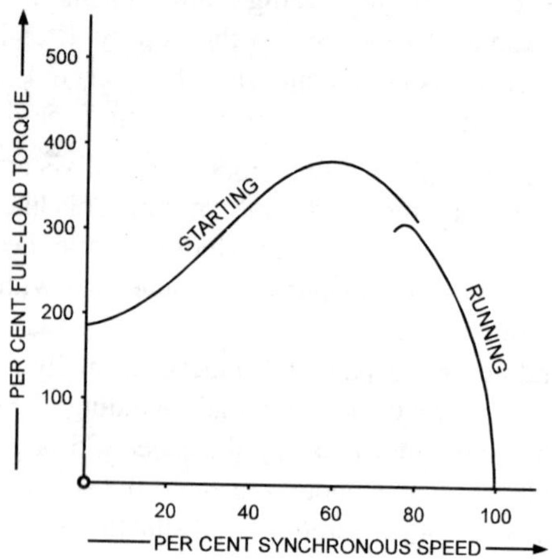

(*c*) *Torque-Speed Characteristic*

Fig. 4.13

Owing to their low cost, resistance start split–phase motors are most popular single phase motors. These motors are made in fractional kilowatt $\left(\frac{1}{20} \text{ to } \frac{1}{4} \text{ kW}\right)$ ratings with speed ranging from 2,875 to 700 rpm. Such motors find wide applications for low inertia loads, continuous operating loads and applications requiring moderate starting torque such as for driving washing machines, fans, blowers, centrifugal pumps, domestic refrigerators, duplicating machines, office appliances, food processing machines, wood working tools, grinders, oil burners etc. Because of low starting torque these motors are seldom employed in sizes larger than $\frac{1}{4}$ kW.

4.9.1. Two-Speed, Two-Winding Motors-Dual Voltage Motors

Resistance–start split-phase motors can be constructed as that two speeds are available. This is done by providing two sets of windings, one for each speed.

Split-phase motors are also convertible for use with either a 115 V or 230 V source of supply merely by changing the circuit connections. For motor operation at 115 V, the two main windings are connected in parallel, thereby providing 115 volts to each winding. For 230 volt operation, the two main windings are connected in series.

Example 4.6. A $\frac{1}{4}$ hp, split-phase motor draws its starting winding current of 4 A lagging the supply voltage by 15º elec. and its running winding current is 6 A lagging by 40º elec. Calculate

(*i*) the total current and power factor (at steady state),

(*ii*) the component of starting winding current in phase with supply voltage

(*iii*) the phase angle between the starting and running winding currents

(*iv*) the component of running winding current that lags the supply voltage by 90º.

Solution : Starting winding current, $\mathbf{I}_s = 4 \angle -15^\circ$ A = (3.8637 – *j* 1.0353) A

Running winding current, $\mathbf{I}_m = 6 \angle -40^\circ$ A = (4.5963 – *j* 3.8567) A

(*i*) Total current $\mathbf{I}_L = \mathbf{I}_s + \mathbf{I}_m$ = (3.8637 – *j* 1.0353) + (4.5963 – *j* 3.8567)

= (8.46 – *j* 4.892) A or $9.77 \angle -30^\circ$ A

Total current in magnitude, I_L = 9.77 A **Ans.**

Power factor, $\cos \phi = \text{Cos}(-30^\circ) = 0.866$ (lag) **Ans.**

(*ii*) The component of starting winding current in phase with supply voltage

= Active component of starting current = 3.8637 A **Ans.**

(*iii*) The phase angle between the starting and running winding currents,

$\theta = \phi_m - \phi_s = 40^\circ - 15^\circ = 25^\circ$ **Ans.**

(*iv*) The component of running winding current that lags behind the supply voltage by 90°

= Reactive or *j*-component of running winding current

= 3.8567 A **Ans.**

4.10. CAPACITOR-START SINGLE PHASE INDUCTION MOTOR

The schematic diagram of a capacitor-start single phase induction motor is given in Fig. 4.14 (*a*). It is also known as *capacitor-start, induction-run motor.* This is an improved form of a resistance-start split-

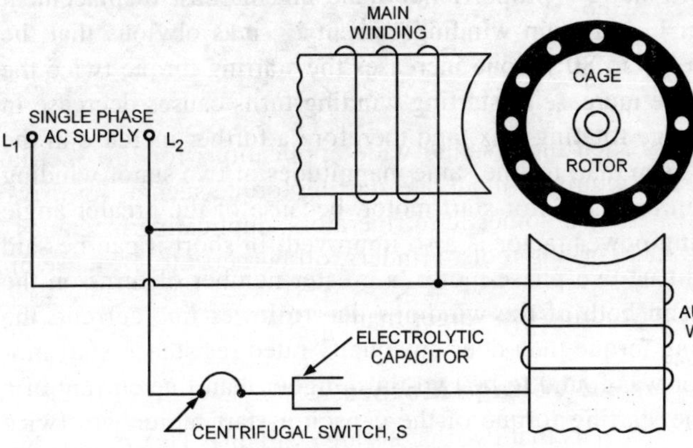

(*a*) *Schematic Diagram*

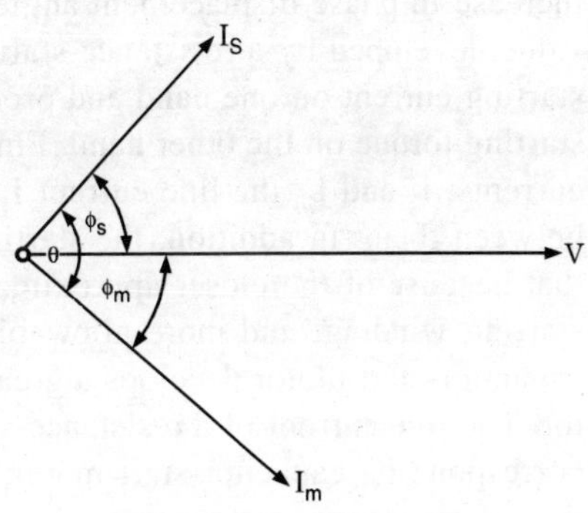

(*b*) *Phasor Diagram*

phase motor. Fundamentally this motor is very similar to the resistance-start split -phase motor, except that the starting winding has a few more turns and consists of a heavier wire than the starting winding of a resistance-start motor; there is also a large electrolytic capacitor in place of resistor, connected is series with the starting winding. By using a capacitor of proper value, the starting winding current I_s at standstill, can be made to lead the main winding current I_m by 90° (*i.e.* θ = 90°) . The phasor diagram of a capacitor-start motor is given in Fig. 4.14 (*b*).

The capacitor is usually mounted in a metal casing on the top of the motor although it may be mounted in any convenient external position on the motor frame. In some cases, the capacitor is mounted inside the motor housing. The capacitor used in the capacitor-start motor is of the electrolytic type and is designed for extremely short-duty service and is guaranteed for not

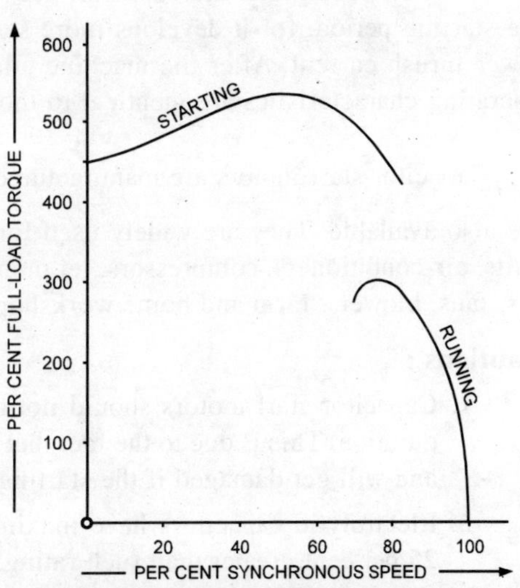

(*c*) *Torque-Speed Characteristic*

Fig. 4.14. ***Capacitor-Start Single Phase Induction Motor***

more than 20 periods of operation per hour, each period not to exceed 3 seconds. The size of the capacitor depends upon the rating of the motor and has a value around 50 μF to a few hundred microfarads. It is primarily due to the development of the inexpensive yet reliable ac electrolytic* capacitor that the capacitor-start motor has largely replaced all other single-phase motors where high starting torque is the essential requirement.

The capacitor-start motor, like resistance–start motor, has the starting winding disconnected by means of a centrifugal switch as the motor picks up speed. Because of short-duty service rating of electrolytic capacitor used in capacitor-start motor, proper operation of the starting (or centrifugal) switch is doubly important. A faulty switch may keep the capacitor in the circuit for too long time during each starting period, and thereby appreciably shorten its life. Of course, the starting winding itself is affected by the switch in the same manner as in the resistance-start motor.

The direction of rotation of a capacitor-start induction motor may be reversed by interchanging the connections to the supply of either the main or starting winding, but it is subject to the same limitations, as given in Art. 4.8.

Since the torque developed by a phase- split motor is proportional to the sine of phase displacement angle θ between the starting winding current I_s and main winding current I_m, it is obvious that the increase in phase displacement angle θ (from 30° to 80°) alone increases the starting torque twice the value developed by a resistance-start motor. The increase in starting winding turns causes decrease in starting current on one hand and produces a large rotating flux, and therefore a further increase in the starting torque on the other hand. Finally it is seen that, for the same magnitudes of two stator winding currents, I_s and I_m the line current I_L is less in the capacitor-start motor, because of the greater angle between them. In addition, the starting winding power factor is also improved. In short it can be said that because of the closer approximation to a true two-phase motor, a greater number of turns on the starting winding, and more allowable current in both of the windings due to lower line current, the capacitor-start motor develops a greater starting torque than does an equally rated resistance-start motor. The line current of a resistance-start motor was found to be two-third higher than line current of a corresponding capacitor-start motor, while the starting torque of the capacitor-start motor was twice that of resistance-start motor.

It is noteworthy that the capacitor-start motor differs from the resistance-start motor only during the starting period, for it develops more starting torque (about 3.5-4.5 times full-load torque) with a lower inrush current. After the machine attains normal speed and the starting winding is cut out, its operating characteristics are identical to those of a corresponding resistance-start motor.

Capacitor-start motors are manufactured in rating ranging from $\frac{1}{10}$ kW to $\frac{3}{4}$ kW, but larger sizes are also available. They are widely used for heavy duty general purpose applications like refrigerator units, air-conditioners, compressors, jet pumps, sump pumps, centrifugal pumps, compressors, conveyors, fans, blowers, farm and home workshop tools, oil burners, etc.

Cautions :

1. Capacitor-start motors should not be used for applications having starting periods of longer duration. This is due to the fact that electrolytic capacitors are rated for short-duty service only and will get damaged if the starting load does not allow the motor to pickup speed quickly.
2. Electrolytic capacitors have maximum voltage rating and if subjected to voltage exceeding 25 per cent greater than such rating, they will get damaged. This voltage limitation means that

* AC electrolytic capacitor is not the same as the dc electrolytic capacitor often employed in a filter in radios, amplifiers, and TV sets etc.

a centrifugal switch must positively open at its preset value of speed; if the spring is not properly adjusted or if the load keeps the speed near the point at which the switch flutters, *i.e.*, opens and closes rapidly, it is possible for the capacitor voltage to be twice rated value for an instant, in which even the capacitor will fail. This double voltage may be caused if the capacitor is fully charged at the instant the switch opens and then closes on the ac voltage wave when it is a maximum in the opposite direction.

3. The capacitor-start motor should not be started more than 20 times an hour. If the motor is started too frequently in a short time interval, the dielectric of the capacitor will get damaged by the frequent current surges. As a result, the dielectric will rupture or break down and short the capacitor plates and cause the fuse protection for the branch motor circuit to blow. If the fuse ratings are quite high and do not interrupt the power supply to the motor, the starting winding may burn out.

Example 4.7. A 250 watt, 230-V, 50-Hz single- phase capacitor -start induction motor has the following constants for the main and auxiliary windings. Main winding Z_m = (4.5 + *j* 3.7) Ω, auxiliary winding Z_a = (9.5 + *j* 3.5) ohm. Determine the value of the starting capacitor that will place the main and auxiliary winding currents in quadrature at starting.

Solution : Main winding impedance, $\mathbf{Z}_m = (4.5 + j\,3.7)\ \Omega$ or $5.826\ \angle 39.43°\ \Omega$

Obviously, main winding current, I_m lags behind the applied voltage V by 39.43°

Auxiliary winding impedance, $\mathbf{Z}_a = (9.5 + j\,3.5)\ \Omega$

Since, time phase angle between auxiliary winding current I_a and main winding current I_m is 90°, so auxiliary winding current I_a must lead the applied voltage by (90° – 39.43°) or 50.57°.

If X_c is the capacitive reactance of the capacitor connected in series with the auxiliary winding, then impedance of the auxiliary winding will be given as

$$\mathbf{Z}_a = (9.5 + j\,3.5 - j\,X_c) \text{ or } [9.5 + j\,(3.5 - X_c)] \text{ ohm}$$

For auxiliary winding,

$$\tan \phi_a = \frac{3.5 - X_c}{9.5}$$

$$\text{or } X_c = 3.5 - 9.5 \tan \phi_a = 3.5 - 9.5 \tan(-50.57)°$$

$$= 3.5 + 11.553 = 15.053\ \Omega$$

$$\text{Capacitance of capacitor, } C = \frac{1}{2\pi f X_c} = \frac{1}{2\pi \times 50 \times 15.053} = 211.5\ \mu\text{F}\ \textbf{Ans.}$$

Example 4.8. The following data shown result on a 230 V, 50 Hz capacitor start single-phase induction motor at standstill :

Main winding : 100 V, 2 A, 40 W

Auxiliary winding : 80 V, 1.0 A, 50 W

Determine the value of capacitance for obtaining the maximum starting torque.

Solution : Equivalent impedance with main winding, $Z_{01m} = \frac{V_m}{I_m} = \frac{100}{2} = 50\ \Omega$

$$\text{Equivalent resistance with main winding, } R_{01m} = \frac{P_m}{I_m^2} = \frac{40}{2^2} = 10\ \Omega$$

Assuming rotor resistance and reactance referred to stator, R_2' and X_2' equal to stator resistance R_1 and reactance X_1 respectively

$$R_{1m} = R_2' = \frac{R_{01m}}{2} = \frac{10}{2} = 5\ \Omega$$

$$\text{and } X_{1m} = X_2' = \frac{X_{01m}}{2} = \frac{\sqrt{50^2 - 10^2}}{2} = 24.5\ \Omega$$

Angle of lag of main winding current, $\phi_m = \text{Tan}^{-1} \dfrac{X_{1m}}{R_{1m}} = \tan^{-1} \dfrac{24.5}{5} = 78.46°$

Equivalent impedance with auxiliary winding,

$$Z_{01a} = \frac{V_a}{I_a} = \frac{80}{1} = 80\ \Omega$$

Equivalent resistance with auxiliary winding,

$$R_{01a} = \frac{P_a}{I_a^2} = \frac{50}{1^2} = 50\ \Omega$$

Equivalent reactance with auxiliary winding,

$$X_{01a} = \sqrt{80^2 - 50^2} = 62.45\ \Omega$$

Auxiliary winding resistance $R_a = R_{01a} - R_2' = 50 - 5 = 45\ \Omega$

Auxiliary winding reactance, $X_a = X_{01a} - X_2' = 62.45 - 24.5 = 37.95\ \Omega$

Torque developed at start will be maximum, if the phase angle between main winding current I_m and auxiliary winding current I_a is 90° or the auxiliary winding current leads the applied voltage by (90° – 78.46°) or 11.54°.

If X_c is the capacitive reactance of the capacitor connected in series with the auxiliary winding, then impedance of the auxiliary winding will be given as

$$\mathbf{Z}_a = R_a + X_a - X_c = 45 + j\,37.95 - j\,X_c = [45 + j\,(37.95 - X_c)] \text{ ohms}$$

For auxiliary winding

$$\text{Tan}\ \phi_a = \frac{37.95 - X_c}{45}$$

$$\begin{aligned}\text{or } X_c &= 37.95 - 45 \tan \phi_a \\ &= 37.95 - 45 \tan(-11.54°) \\ &= 37.95 + 9.19 = 47.138\ \Omega\end{aligned}$$

Capacitance of capacitor, $C = \dfrac{1}{2\pi f X_c} = \dfrac{1}{2\pi \times 50 \times 47.138} = 67.5\ \mu\text{F}$ **Ans.**

4.11. CAPACITOR-START, CAPACITOR-RUN SINGLE PHASE INDUCTION MOTOR

In order to have the advantages that result from the use of a capacitor during the normal running operation as well as starting period, the capacitor-start, capacitor-run motors have been developed. The capacitor-start, capacitor-run motor is similar to capacitor-start, induction-run motor, except that the auxiliary winding and a capacitor remain connected in the circuit at all times. Since the value of the capacitance which will result in optimum running performance is not the same as the value which will given the best starting torque conditions and also the capacitor employed for normal running should be of continuous-duty rating and that employed for starting should be of short-duty rating, therefore, in such a motor two different capacitors—one small value oil impregnated paper continuous rating capacitor C_R and another much larger value electrolytic short-duty capacitor C_S are employed. It is also necessary to provide a centrifugal switch to take out the electrolytic capacitor out of auxiliary winding circuit when the motor has attained a predetermined speed. The schematic diagram of a capacitor-start, capacitor-run motor is given in Fig. 4.15 (*a*). When the motor attains 75 per cent of synchronous speed, the starting capacitor C_S is taken out of the circuit by the operation of centrifugal or transfer switch, which is normally closed.

It is to be noted that when the centrifugal switch is closed during the starting period, the two capacitors C_S and C_R are in parallel, so that total capacitance in the auxiliary winding circuit is sum of their values; after the switch opens, only the continuous oil type capacitor C_R remains in the auxiliary winding circuit. In practice starting capacitor C_S is about 10-15 times as large as the running capacitor C_R.

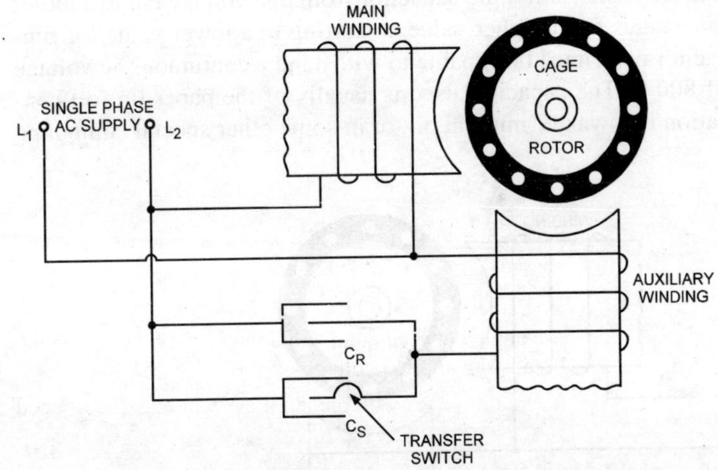

(a) Schematic Diagram

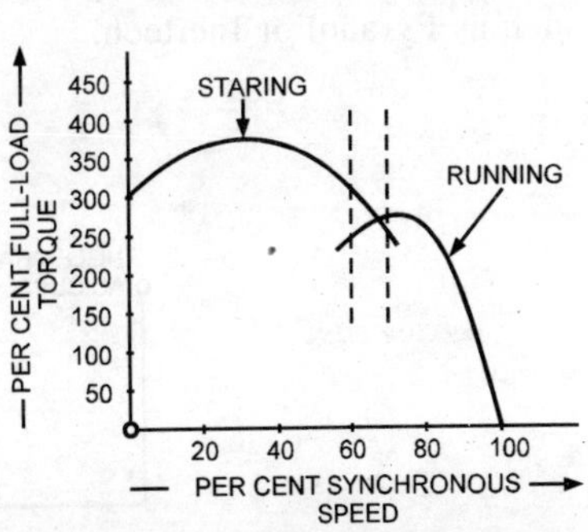

(b) Torque-Speed Characteristic

Fig. 4.15. ***Capacitor-Start, Capacitor-Run Single Phase Induction Motor***

Such a motor operates as a two-phase motor from single phase supply, thereby, producing a constant torque and not a pulsating torque, as in other single phase motors. Besides their ability to start heavy loads they are extremely quiet in operation, have better efficiency (55-65%) and power factor (0.8-0.95) when loaded and develop up to 25% greater overload capacities. They are indeed splendid machines where the load requirements are severe as in case of compressors, refrigerators and fire strokers, pumps, conveyors etc. The disadvantage of such machines is only high cost. These motors are often employed requiring a quiet operating motor such as in hospitals, studios etc.

In recent years, home air-conditioners have become extremely popular. In many cases, the electrical system in the home may not have the capacity to handle the electrical load of the air-conditioner(s). For this reason, capacitor-start capacitor-run motor have found more wide spread use in the compressors. Both the starting and running currents are low—full-load current may be $\frac{3}{4}$ th of that for an equivalent capacitor-start motor.

Such motors have become increasingly popular and are now manufactured in the larger sized single phase motors where previously only repulsion-start motors were available. A typical torque-speed characteristic is shown in Fig. 4.15 (*b*). The direction of rotation of motor may be reversed by interchanging the connections to the supply of either the main or auxiliary winding.

4.11.1. Capacitor-Start, Capacitor-Run Motor With an Auto-Transformer

Capacitor-start, capacitor-run motors are also manufactured in which an auto-transformer and a single capacitor unit are used. The auto-transformer and a capacitor are sealed in a rectangular iron box and mounted on the top of the motor. The idea is based on the principle that a capacitor of value C, connected across the secondary of a step-up transformer, appears to the primary as though it had a value of K^2C where K is a voltage transformation ratio. For example, if the actual capacitor connected across the high voltage secondary is of 6 μF and the voltage transformation ratio is 4, the low-voltage primary acts as though it had a ($4^2 \times 6$) or 96 μF capacitor connected to its terminals. Thus the effectiveness of a low value capacitor is increased largely by the use of a step-up transformer. Figure 4.16 illustrates how this is done in one design in which the stepping-up is accomplished by means of an auto-transformer. The transfer switch is now a single pole double-throw switch, rather than a simple on-off switch. As the motor picks up the speed, the transfer switch shifts the capacitor from one voltage tap to another so that the voltage transformation ratio changes from higher value at starting to a lower value for running. It is noteworthy here that the capacitor used must be capable to withstand a continuous ac voltage of 330 V and an intermittent voltage of 800 V. The capacitor used is usually of the paper-tin foil construction immersed in high grade insulation like wax or mineral oil, or in some other special compound such as Pyranol or Inerteen.

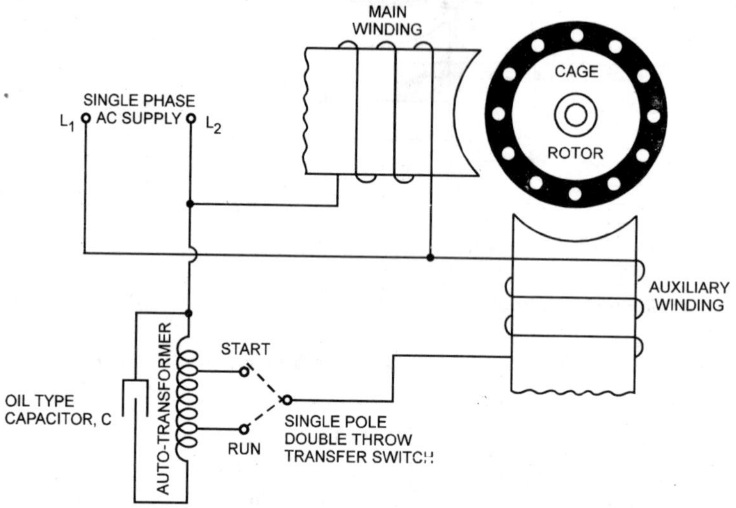

Fig. 4.16. *Capacitor-Start, Capacitor-Run Motor With An Auto-Transformer*

Since an auxiliary winding is permanently connected in the motor circuit, a faulty starting switch will not harm it. If, however, the switch fails to open at proper time, the electrolytic capacitor connected in auxiliary winding circuit may get destroyed. In the case of the transformer capacitor type motor, the voltage across the capacitor may become dangerously high as the motor picks up speed and the dielectric may breakdown. If the switch remains in the open position, the motor using two capacitors will have only low-value oil-filled capacitor in the circuit at starting and thus low starting torque will be developed. In case of a transformer capacitor motor, if the transfer switch remains in the running position, the motor will give low starting torque.

The capacitor-start, capacitor-run motor with an auto-transformer has the advantage of greater design flexibility because any desired capacitance values can be had by a proper choice of transformer turn-ratio. The drawbacks of such a motor are the size and the cost of the auto-transformer and the increased cost of the transfer switch.

Both types of the motors, described above, operate with both the windings while running, and so the direction of rotation can be reversed while running. This, however, causes the transfer switch to return to the starting position, and then again to the running position, when the motor rotates in the opposite direction. This is not desirable.

4.12. PERMANENT CAPACITOR SINGLE PHASE INDUCTION MOTOR

Permanent capacitor single phase induction motor, also sometimes called the *single-value capacitor–run motor*, has two stator windings placed mutually 90 electrical degrees apart. The main or running winding is connected directly across the single phase ac supply. A capacitor in series with the auxiliary winding is also connected across the supply lines. There is no centrifugal switch, since the auxiliary winding is energized at all times when the motor is in operation. The auxiliary winding in such motors usually has almost the same size of wire and almost as many turns as the main winding; in fact, in some motors the two windings are identical. The schematic diagram of a permanent capacitor motor is given is Fig. 4.17 (*a*).

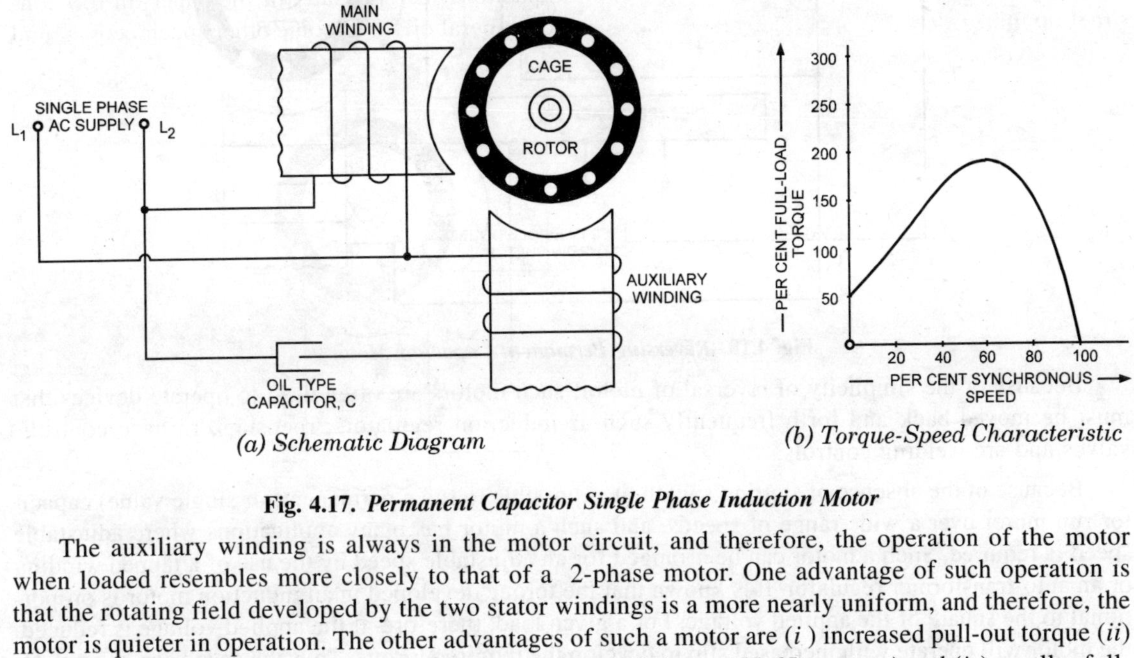

(*a*) *Schematic Diagram* (*b*) *Torque-Speed Characteristic*

Fig. 4.17. ***Permanent Capacitor Single Phase Induction Motor***

The auxiliary winding is always in the motor circuit, and therefore, the operation of the motor when loaded resembles more closely to that of a 2-phase motor. One advantage of such operation is that the rotating field developed by the two stator windings is a more nearly uniform, and therefore, the motor is quieter in operation. The other advantages of such a motor are (*i*) increased pull-out torque (*ii*) a higher power factor (0.8-0.95), (*iii*) higher operation efficiency (55-65 per cent) and (*iv*) smaller full-load line current.

Such motors have some drawbacks also. Since in such motors the same capacitor is used for starting and running, therefore, neither optimum starting nor running performance can be obtained. A typical performance characteristic for such a motor is given in Fig. 4.17 (*b*). Generally capacitors of 2-20 μF capacity are employed. The electrolytic capacitor cannot be employed owing to the requirement of continuous operation, and therefore, more expensive oil or Pyranol insulated foil-paper capacitors are to be used. Under starting conditions, the capacitor is too small and so the starting torque tends to be low (50 to 80 per cent of full-load torque); but the manufacturers can partly compensate for this by using a high resistance squirrel cage. The permanent capacitor induction motor is more expensive than the equivalent resistance-start or capacitor -start induction motor mainly owing to the increased cost of auxiliary winding which is now wound with thicker wire and consists of as many turns as the main winding.

A unique feature of such a motor is that it can be very easily reversed by an external switch, provided the two windings are identical. Because of absence of starting switch it is possible to reverse the leads of either winding while the motor is running. The torque is thus reversed, the motor comes to a standstill, and then starts in the opposite direction. If the two stator windings are identical, it is possible to place the capacitor in series with either of the two windings, making that particular winding the auxiliary or leading one. Circuit for such an operation is given in Fig. 4.18. When the SPDT switch is in position 1 winding A serves as a main winding and winding B as an auxiliary winding. When the SPDT switch is moved in position 2, winding A becomes the auxiliary winding and winding B the main one.

Such motors are widely used in table and ceiling fans, blowers, oil burners, room coolers, portable tools, other domestic and commercial electrical appliances, where low starting torque is required.

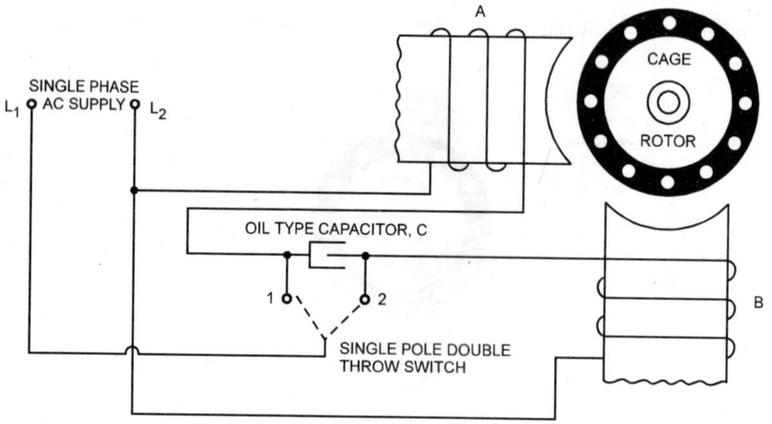

Fig. 4.18. ***Reversing Permanent Capacitor Motor***

Because of the simplicity of reversal of motor, such motors are often used to operate devices that must be moved back and forth frequently such as induction regulators, rheostats, furnace controls, valves and arc-welding controls.

Because of the absence of starting switch, it is possible to run a permanent (or single value) capacitor run motor over a wide range of speeds, and such a motor has many applications where adjustable speed is required. Such a motor can be arranged for an adjustable speed by the use of a tapped winding or an auto-transformer regulator. It is known that the torque developed in an induction motor is proportional to the square of the applied voltage. For a given load, therefore, if the applied voltage is reduced, the motor will operate with increased slip to develop the required torque. Thus the speed can be controlled by varying the voltage applied to the motor.

4.13. SHADED-POLE MOTOR

Shaded-pole motor is a split-phase type single phase induction motor. The stator has salient poles, each provided with its own exciting coil. A part $\left(\text{about } \frac{1}{4} \text{ to } \frac{1}{3}\right)$ of each pole is wrapped by a copper strap forming a closed loop, known as a *shading coil,* as shown in Fig. 4.19 (*a*). Stator usually has 2 or 4 poles giving synchronous speed of 3,000 or 1,500 rpm for 50 Hz power supply. The rotor is a simple squirrel cage type with a 60º skew for optimum starting torque and for limiting the torque dip during run-up. A low cage resistance impairs both the starting torque and the dip, and usually the resistance is greater than for any other type of single phase cage motor, giving a starting torque in the range of 40-80 per cent of full-load torque and pull-out torque of about 110-140 per cent of full-load torque. The gap length is of the order of 0.25-0.5 mm. Too short air gap may result in starting-torque variations due to rotor slotting. It has no commutator, brushes, collector rings, contactors, capacitors or moving switch parts, so it is relatively cheaper, simpler and extremely rugged in construction and reliable (absence of centrifugal switch or starting mechanical auxiliary eliminates the possibility of motor failure due to faulty centrifugal switch mechanisms).

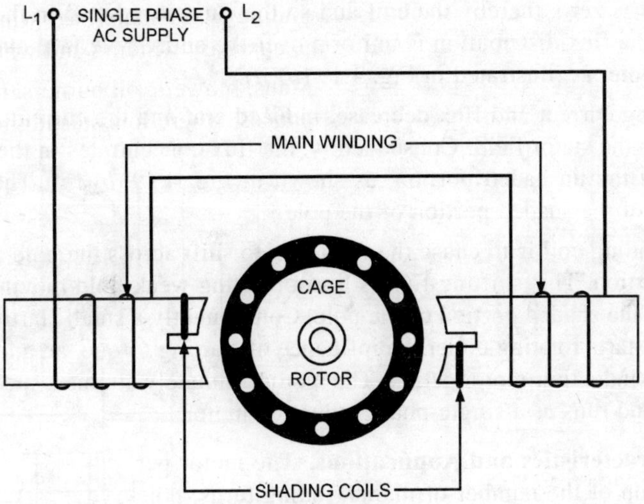

(a) Schematic Diagram

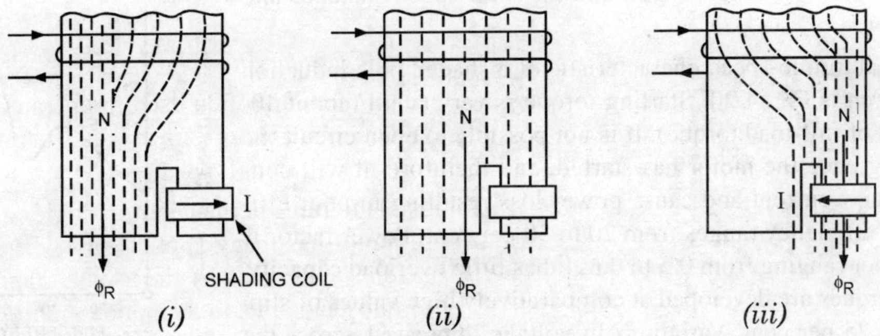

(b) Shifting of Air-Gap Flux

Fig. 4.19. ***Shaded-Pole Single Phase Induction Motor***

Action. When the exciting winding is connected to a single phase ac supply source an alternating current flows through it and sinusoidally varying flux is set up in the core. This change of flux induces short-circuit currents in the shading coil in such a direction as to oppose the core flux. The flux in the shaded part of the pole thus lags behind the flux in the main part of the pole. At the same time, the main flux and shaded-pole flux are obviously displaced in space, although by less than 90º. Since two fields have both space and time displacement, the conditions for setting up a rotating field are present, no matter how imperfectly. Torque is thus developed in the squirrel cage rotor.

The movement of the flux around the stator may be more clearly illustrated in the following discussion. In the following discussion the exciting current will be assumed to be varying sinusoidally. The discussion will be made for three parts of current (*i*) from zero to a nearly positive maximum (*ii*) in the region of maximum current and (*iii*) from a nearly positive maximum to zero.

As the current in the exciting winding increases, a voltage is induced in the shading coil by transformer action. The current in the low resistance shading coil is comparatively high and is in such a direction as to oppose the rapidly changing flux, created by it, in the shaded portion. Thus, when the exciting current increases, the greater portion of flux passes through the unshaded part of each pole and the resultant lies near the centre of the unshaded part of the pole, as illustrated in Fig. 4.19 (*b*)(*i*).

When the exciting (or stator) current and flux attain their maximum values, the rate of change in either current or flux is zero, thereby the emf and so the current induced in the shading coil becomes zero. Consequently the flux distribution is uniform over the entire pole face and the resultant flux lies at the centre of the pole, as illustrated in Fig. 4.19 (*b*)(*ii*).

When the exciting current and flux decrease, induced emf and the current in the shading coil sets up a mmf which aids the stator field. Consequently, the flux concentrates in the shaded portion of the pole and weakens in the unshaded portion, as shown in Fig. 4.19 (*b*)(*iii*). The resultant flux is thus shifted to the centre of the shaded portion of the pole.

The effect of shading coil is to cause the field flux to shift across the pole face from the unshaded portion to shaded portion. This shifting flux is like a rotating weak field moving in the direction from unshaded portion to the shaded portion of the pole. Consequently a small starting torque is developed. As soon as the rotor starts rotating under the influence of starting torque, an additional torque is developed by single phase induction-motor action. The motor ultimately attains a speed slightly less than the synchronous speed and runs as a single-phase induction motor.

Operating Characteristics and Applications. The motor performance is a function of the number of main-winding turns, pole-arc / pole-pitch ratio, the proportion of the pole-arc in the shaded region, the shading coil resistance, and the rotor cage resistance and the slot skew.

A typical torque-speed characteristic of a shaded pole induction motor is given is Fig. 4.20. Starting torque is very small (about 40-80 per cent of full-load torque). It is not possible to open circuit the shading ring once the motor has started, and therefore, it will continuously carry current and cause power loss resulting in poor efficiency. The efficiency ranges from 20 to 50 per cent. Power factor is also very poor ranging from 0.5 to 0.6. It has little overload capacity. Full-load torques are developed at comparatively high values of slip, about 10 to 25 per cent. Variations in voltage impressed across the stator winding affects the motor operating speed under load *i.e.*, speed falls with the decrease in applied voltage.

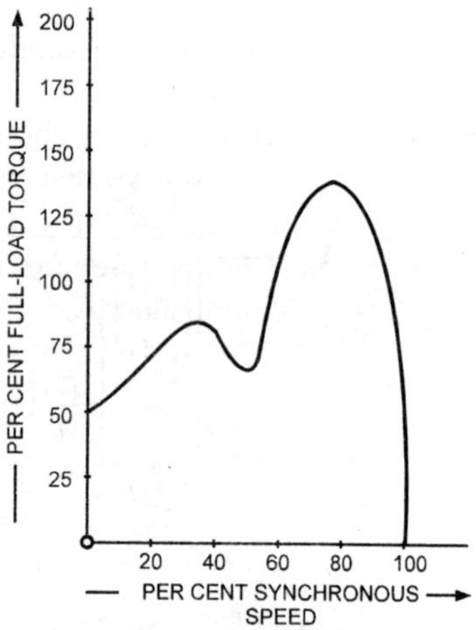

Fig. 4.20. ***Torque-Speed Characteristic of Shaded Pole Motor***

The direction of rotation of motor depends upon the position of the shading coil on the pole i.e which portion of the pole is encircled by the shading coil. Unless the machine is constructed so that the shading coil can be shifted to the other side of the pole, the direction of rotation of such a machine cannot be reversed.

The speed of a shaded-pole motor can be varied by varying the voltage applied across its stator winding. The voltage applied across the stator winding can be varied in three ways viz (*i*) by using an auto-transformer tapped for several voltages (*ii*) by using a tapped reactance coil (sometimes a resistance) that incurs a voltage drop in the line and (*iii*) by using tapped exciting winding, so that the constant voltage source may be impressed across the entire winding or some part of it.

Because of their poor starting torque, low power factor and poor efficiency, such motors are only suitable for low power applications such as for toys, small fans, electric clocks, hair dryers, time phonographs, slide projectors, humidifiers, small business machines (photocopying machines, vending machines, advertising displays etc.) and other similar applications. They are manufactured in small sizes ranging from below 5 W to about 100-150 W. They are available with built-in gear reducers to give almost any speed (as low as below 1 rpm.)

Example 4.9. The name plate of a 1-phase, 4-pole induction motor gives the following data:

Output 373 W; 230 V; Frequency 50 Hz; Input current 2.9 A; Power factor 0.71; Speed 1,410 rpm. Calculate (*a*) the efficiency and (*b*) the slip of the motor when delivering the rated output.

Solution : Power output, $P_0 = 373$ watts

Power input, $P_{in} = VI \cos \phi = 230 \times 2.9 \times 0.71 = 473.57$ W

(*a*) Motor efficiency, $\eta = \dfrac{P_0}{P_{in}} \times 100 = \dfrac{373}{473.57} \times 100 = 78.76\%$ **Ans.**

Synchronous speed, $N_s = \dfrac{120\,f}{P} = \dfrac{120 \times 50}{4} = 1{,}500$ rpm

(*b*) Slip, $s = \dfrac{N_s - N}{N_s} \times 100 = \dfrac{1{,}500 - 1{,}410}{1{,}500} \times 100 = 6\%$ **Ans**

4.14. SPEED CONTROL OF SINGLE PHASE INDUCTION MOTORS

The stator winding of a single phase induction motor can easily be arranged to give two synchronous speeds, one double the other. In Fig. 4.21 the connections for two pole motors are shown in which two coils A and B have been connected in series with their mmfs helping each other. If the connections to coil B are reversed so that both the coils are again in series but their mmfs opposing each other, as shown in Fig. 4.22, 4 poles are produced and the synchronous speed is reduced to half. Similarly a four -pole motor can be re-connected to give 8 poles. The change from one speed to the other is easily and quickly accomplished by means of a two-pole double-throw switch.

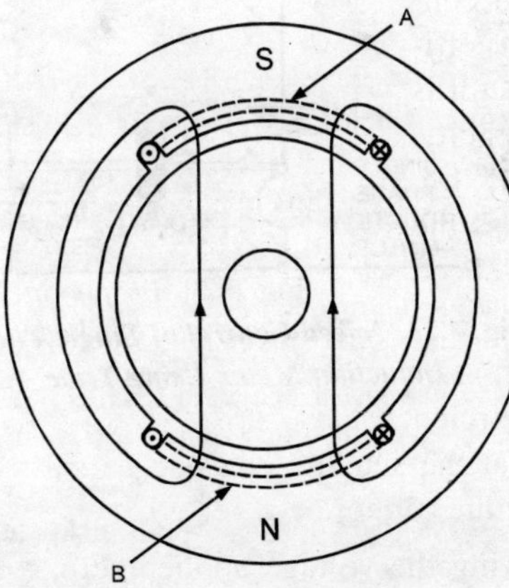

Fig. 4.21. ***2-Pole Connection***

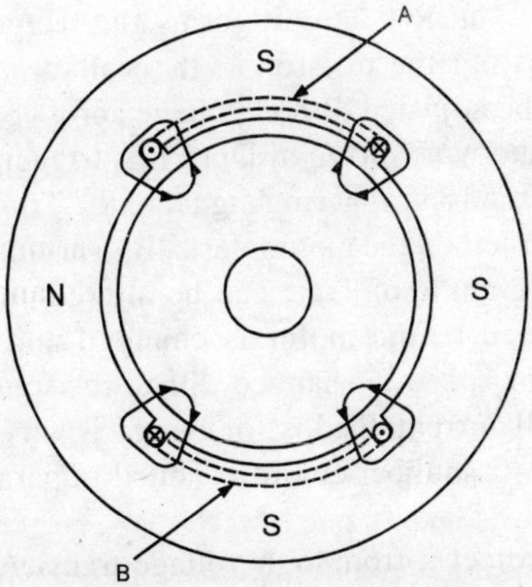

Fig. 4.22. ***4-Pole Connection***

Furthermore, because of the comparatively small power handled by single phase induction motors, excessive slip to give a slight reduction in speed is not serious, from the stand point of waste of power, as is case of a 3-phase induction motors. For that reason in various cases, such as for fan and blower applications where speeds of 1,100 or 1,300 rpm might be required, it is not unusual to employ 4-pole motor with high slip. This can be accomplished (*i*) by using high rotor resistance, in which case no variation in speed is introduced, but the motor displays a drooping speed curve; or (*ii*) by reducing the voltage applied to the stator winding. For this purpose variable resistance or tapped reactance coil in series with the motor can be employed for speed regulation. This method is inefficient, gives large variations in speed with the variations in load, but it is much used inspite of these drawbacks. Alternatively, the voltage applied to the motor stator winding can be varied either by means of a transformer with taps on its secondary or by means of a variac, which provides the maximum possible number of taps and thus gives finest speed adjustment. This is one of the simplest methods of speed control of fractional kilowatt motors. The speed of the single phase-induction motor can also be controlled by employing tapped exciting winding, so that the constant voltage source may be impressed across the entire winding or some part of it. By this means, a normal speed and a single reduced speed are provided.

Speed control because of its more general use, has brought about the standardization of fan speeds, allowing the use of high slip motors. A squirrel cage motor with 8-10 per cent slip and with low pull-out torque operates satisfactorily not only for constant speed drives, but also for adjustable speed drives through the use of voltage control on the stator winding. With this method of speed control, the torque-speed characteristics resembles with those of a wound-rotor motor with different number of external resistances in the motor circuit.

4.15. SPEED CONTROL OF SINGLE PHASE INDUCTION MOTOR USING TRIAC (FAN REGULATOR)

The conventional speed regulator for a fan uses a resistance regulator. The regulator resistance is in series with the fan motor. The speed of the fan is reduced, whenever desired, by increasing the regulator resistance. Thus the voltage drop across the regulator resistance increases and, therefore, voltage applied to the motor is reduced and so the speed. The resistance regulator causes loss of energy. This loss of energy becomes significant at low speeds.

Figure 4.23 shows a speed control circuit using a Diac-Triac pair. The Diac is meant to trigger the Triac into conduction. It is preferable to use a matched Diac-Triac pair. The R-C circuit forms the triggering network. By adjusting the resistor *R*, the voltage across capacitor *C* can be adjusted. When voltage across capacitor *C* exceeds breakdown voltage of Diac, it is triggered into conduction and sends a triggering signal to the Triac gate. When Triac is turned on the motor starts. By varying resistance R, The firing angle of Triac can be altered and thus the voltage applied to the motor is changed and consequently the motor speed is changed. Since resistance R, carries very small current, the loss of energy is very small. Sometimes an R-C snubber circuit is added in parallel with the Triac to protect it from high voltage transients and high $\frac{dv}{dt}$.

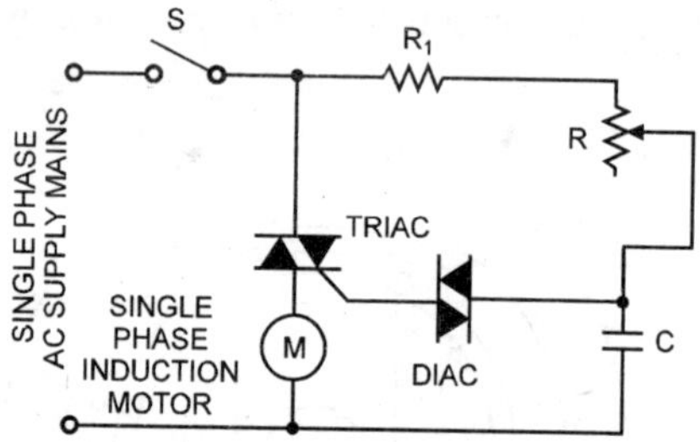

Fig. 4.23. ***Speed Control of Single Phase Induction Motor Using Triac***

4.16. COMPARISON OF METHODS OF STARTING OF SINGLE PHASE INDUCTION MOTORS

The various types of single phase induction motors, classified on the basis of methods of their starting, are arranged in ascending order of capital cost as follows.

1. Shaded pole 2. Resistance-start 3. Permanent capacitor 4. Capacitor-start 5. Capacitor-start, capacitor-run 6. Repulsion-start.

If the motors placed at no. 2 and 3 are interchanged in the above list, the motors are then in the ascending order of starting torque, except that motors placed at no. 4 and 5 should be equal.

The repulsion-start induction motor was at one time the most widely used single phase induction motor, but it has now been almost completely replaced in the fractional kilowatt ratings by the capacitor motors. However, with the increase in output rating, the increase in cost of repulsion start motors is not so rapid as in case of capacitor motors and above the ratings of 1 kW the repulsion-start motors are cheaper than capacitor–start and capacitor-start, capacitor-run motors. The repulsion-start motors are, therefore, still used in ratings above 1 kW, although they have the drawback of being somewhat noisier than the capacitor motors.

HIGHLIGHTS

1. A single phase split-induction motor has a squirrel cage rotor and a stator which has two windings. One is running winding or main winding and the other is the starting or auxiliary winding which is disconnected from the supply by a centrifugal switch when rotor attains 70 or 80 per cent of its rated speed.
2. *Operating Principle.* When a single phase induction motor is connected to a single phase supply, two fields of magnitude $\phi_{max}/2$ are developed which rotate in opposite directions. An equal and opposite torque is set up by the two fields and the resultant torque is zero at start. Thus single phase induction motor is inherently not a self starting motor. However, if the rotor is rotated in either direction by some external means, torque will be developed and the rotor will pick up the speed in that direction.

   To obtain starting torque in a single phase induction motor, the single winding is splitted into two windings displaced in space by 90º and having different resistance and inductance. They carry currents of different phase angle which develop a resultant field revolving in space at synchronous speed.
3. Single phase split-phase induction motors are of two types viz resistance-start induction motor and capacitor-start induction motor.

   The *resistance-start induction motor* has a relatively poor starting torque because the phase angle between the two winding currents is only about 30-50 electrical degrees.

   The capacitor-start induction motor has a comparatively good starting torque because phase angle between the currents flowing in the two windings (running and starting winding) can be as large as 90º.
4. *Capacitor motors* may be (*i*) capacitor-start (*ii*) capacitor-start, capacitor-run and (*iii*) permanent capacitor motors.

   The capacitor-start capacitor-run motor is similar to capacitor-start motor except that the auxiliary winding and a capacitor remain connected in the circuit at all times.

   The permanent capacitor motor has no centrifugal switch and the auxiliary winding remains energized at all times when the motor is in operation.
5. *Shaded pole motors* have projected poles with one-third pole part wrapped with a copper strip. This portion of the pole is called shaded part of the pole and because of this a revolving field is produced in the stator and starting torque is developed.

   The shaded pole motors have poor starting torque, poor efficiency (20-50%), very poor power factor (0.5–0.6 lagging) and little overload capacity.

EXERCISES

1. Why in its elementary form, the single phase induction motor is not self starting ? Describe various methods of starting of single phase induction motors.
2. Give applications of 3-phase induction motor and single phase induction motor. Why is 3-phase induction motor used for medium and large load and single phase induction motor for small load applications ?
3. Show that the starting torque of a single phase induction motor is zero.
4. State the reasons for the inferior performance of single phase induction motors compared to three phase induction motors.
5. Using double revolving field theory, explain why a single phase induction motor is not a self starting motor and describe briefly the various methods of producing starting torque in a single phase induction motor. **[M.D. Univ. Electrical Machines-II, May 2009]**
6. Using double revolving field theory explain the working of a single phase induction motor.
7. Why is it advantageous to use double revolving field theory for determining the running performance of a single-phase induction motor?

   Draw torque-speed characteristics of a single-phase induction motor based on double revolving field theory and discuss about the magnitude of torque at zero speed and synchronous speed.

   Sketch waveform of rotor current under normal running conditions of the single-phase induction motor.
8. Explain the cross-field theory as applied to a single phase induction motor.
9. Write short note on "cross-field theory of 1-phase induction motor."

   [M.D. Univ. Electrical Machines-II, December 2010]
10. Draw equivalent circuit of single phase induction motor, describing all the parameters.
11. Discuss the different methods of starting of 1-ϕ induction motor. Derive the equivalent circuit of a single phase induction motor. How can the performance of the motor be analysed ?
12. Describe no-load test and blocked rotor test on a 1-ϕ induction motor and explain how you obtain data for parameters of equivalent circuit of 1-ϕ induction motor. Draw the equivalent circuit of 1-ϕ induction motor both at no-load condition and full-load condition.
13. Explain the construction and working principle of a single-phase induction motor. Why this type of motor has no self-starting torque ?
14. Discuss why single phase induction motors do not have starting torque. Explain its principle of operation and list various methods of starting.
15. Explain why do pure, 1-phase induction motors have no starting torque ? What modifications are being done to have starting torque in normal 1-phase fan motors ? Give brief explanations.
16. Explain the operation of single phase capacitor-start motor.
17. Explain the principle of operation of capacitor-start and capacitor run single phase induction motors. Discuss about the torque-speed characteristic of each.
18. Explain construction and working of any one type of capacitor split-phase induction motor. What type of capacitors are used?
19. Explain with neat diagrams the following types of single phase induction motors.

    (*a*) Split- phase induction motor

    (*b*) Capacitor- start induction-run motor.

    Also draw their torque-speed characteristics.
20. Discuss the performance of (*a*) shaded pole single phase induction motor (*b*) single phase capacitor motor. Mention their relative advantages and disadvantages.

    Mention a few applications of a single phase induction motor.

21. How is a single phase induction motor made self starting ? Discuss the working principle of shaded pole motor with the help of neat sketch.

22. Describe shaded-pole method of starting a single phase induction motor.

23. Indicate the slip-torque characteristics of different types of single phase induction motors in one diagram and compare. State the reasons for their deviation.

24. Explain the speed control of single phase induction motor.

SHORT ANSWER TYPE QUESTIONS WITH ANSWERS

Q. 1. How will you resolve a single pulsating field ?

Ans. A single pulsating field can be resolved into two components of half its amplitude and rotating in opposite directions with equal speed.

Q. 2. State the principle that the double revolving field theory makes use of.

Ans. The double revolving field theory makes use of the principle that pulsating field produced in a single phase motor can be resolved into two components of half its amplitude and rotating in opposite directions with synchronous speed.

Q. 3. Name the two theories regarding single phase induction motor.

Ans. The two theories regarding single phase induction motor are (*i*) double revolving field theory and (*ii*) cross-field theory.

Q. 4. Is the starting torque of a single phase induction motor zero ?

Ans. Yes, the starting torque of a single phase induction motor is zero.

Q. 5. Is the torque developed by a single phase induction motor at synchronous speed zero ?

Ans. No, it is negative.

Q. 6. What is the effect of increasing rotor resistance in a single phase induction motor ?

Ans. Increase in rotor resistance of a single phase induction motor reduces its breakdown torque, lowers the efficiency and increases the slip corresponding to the maximum torque.

Q. 7. Why single phase induction motors are usually set on rubber spring mounts ?

Ans. In a single phase induction motor, pulsating torque is developed as a direct consequence of the pulsating power in a single phase circuit.

Q. 8. What is a split-phase motor ?

Ans. This is a single phase induction motor equipped with an auxiliary winding, also called the starting winding, connected in parallel with the main winding on stator and differing from it in both phase and space. The starting winding is usually opened by a centrifugally operated device when the motor attains a predetermined speed.

Q. 9. A single phase induction motor is provided with a main winding and an auxiliary winding. Which of the two windings should be more resistive ?

Ans. Auxiliary winding should be more resistive.

Q. 10. In the design of resistance-start single phase induction motor the starting winding is disconnected once the slip reaches 0.25 or less. Explain.

Ans. The starting winding of a resistance-start motor is not designed for continuous operation and so it is necessary to disconnect it from the motor circuit when the motor attains a definite speed. The starting winding is usually taken out of circuit at a speed of about 75% of synchronous speed because at this speed the torque developed by the main winding alone is greater than that developed by both windings.

Q. 11. How is the direction of rotation of a single phase induction motor reversed ?

Ans. The direction of rotation of a single phase induction motor (split-phase type) is reversed by reversing the leads to the main or starting winding, but not both.

Q. 12. How does a capacitor-start motor differ from a resistance-start motor ?

Ans. Fundamentally the capacitor-start motor is quite similar to a resistance-start motor, except that its starting winding has a few more turns and is wound with comparatively thick wire than the starting winding of a resistance-start motor and also the external resistance, connected in series with the starting winding, is replaced by a large electrolytic capacitor.

Q. 13. How the direction of rotation of a capacitor-start induction motor is changed ?

Ans. The direction of rotation of a capacitor-start, induction run motor is reversed by reversing either the starting or running winding leads, not both, to the line.

Q. 14. What happens to a capacitor-start induction motor if the auxiliary winding is left in the circuit after it has been used to start the motor ?

Ans. The auxiliary winding of a capacitor-start induction motor is not designed for continuous operation and if left in the circuit after it has been used to start the motor, it will burn out. In addition to it, the short-duty rated electrolytic capacitor connected in series with the starting winding will also get damaged.

Q. 15. Are capacitor-start induction-run motors usually dual voltage or single voltage motors ?

Ans. Capacitor-start, induction-run motors are usually dual voltage motors.

Q. 16. If a capacitor-start motor is rated as 115/230-V motor (the capacitor is usually rated at 115 V), how is it connected, when the motor is used on 230 V ?

Ans. The main or running winding of a dual voltage motor is in two sections so that they may be connected in parallel for 115-V operation and in series for 230-V operation. The capacitor is connected in series with the starting winding; therefore, when operating at 115 V, the winding leads are connected across the line at start, but when operating on 230 V, one lead is connected to the one side of the line and the other lead to the mid point of the running windings.

Q. 17. What is the advantage of using a capacitor-start motor over a resistance-start split-phase motor ?

Ans. The capacitor-start motor develops more starting torque than does an equally rated resistance-start split-phase motor with a lower inrush current.

Q. 18. What is a capacitor-start, capacitor-run motor ?

Ans. This is the same motor as the capacitor-start motor, except that it has an extra oil type capacitor connected in series with the starting winding, which always remains in the circuit.

Q. 19. In a 2-phase capacitor motor which capacitor is of fairly high value and to what order ?

Ans. In a 2-phase capacitor motor the starting capacitor is about 10 – 15 times as large as the running capacitor.

Q. 20. State the advantage of capacitor-run over capacitor-start motor.

Ans. Capacitor-start, capacitor-run motors are quieter in operation, develop more pull-out torque, operate with higher power factor and efficiency, and consequently draw less current than the capacitor-start motors.

Q. 21. What happens when the auxiliary winding of a capacitor motor is disconnected during running condition ?

Ans. The motor will continue running but will develop small torque. Once it stops, it will not start again.

Q. 22. Name the motor being used in a ceiling fan ?

Ans. Split-phase capacitor type single phase induction motor.

Q. 23. What is a shaded-pole motor ?

Ans. Shaded-pole motor is a salient-pole, split-phase type, squirrel cage rotor single phase induction motor provided with one or more auxiliary short-circuited stator windings displaced in magnetic position from the main winding.

Q. 24. Does a shaded-pole motor have much starting torque ?

Ans. No, it has very little starting torque.

Q. 25. In what direction do shaded pole motors turn ?

Ans. Shaded pole motor rotates in a direction from the unshaded part of the pole to the shaded part.

Q. 26. What role do 'magnetic bridges' play in the operation of a shaded pole motor ?

Ans. The magnetic bridges cause the flux in the shaded portion of the pole to lag behind the flux in the unshaded portion. This gives, in effect a motion of flux across the pole face and under the influence of this moving flux a starting torque is developed.

Q. 27. State the applications of shaded-pole single phase induction motors.

Ans. Shaded pole motor develops very little starting torque and, therefore, it is used only for small fans, electric clocks, hair dryers, time phonographs, advertising displays, toys, instruments and such other applications.

Q. 28. How can a shaded-pole motor be reversed in direction of rotation ?

Ans. Unless the machine is constructed so that the shading coil can be shifted to the other side of the pole, the direction of rotation of such a machine cannot be reversed.

Q. 29. Which is preferable, a small or a large number of rotor slots, in case of single phase motors ?

Ans. Large number of slots are preferable in the stator of a single phase motor.

Q. 30. Why is the normal full-load slip of a single phase induction motor higher than that of a three-phase motor ?

Ans. The normal full-load slip of a single phase induction motor is higher than that of a 3-ϕ motor, owing to development of backward rotating field. The power is to be delivered to the backward field from the power converted into mechanical power by the forward field.

PROBLEMS

1. A 150 W, 4-pole, 110 V, 50 Hz, 1-ϕ induction motor delivers rated output at a speed of 1,425 rpm. The total copper loss at full load is 30 W. Determine the full-load slip and full-load efficiency. The rotational losses equal 25 watts and neglect stator copper loss. Also determine the copper losses caused by the two fields.

[**Ans.** 5 %, 73.2 %, 9.73 W, 20.27 W]

2. Discuss the double revolving field theory of single phase induction motors. Find the mechanical power output at a slip of 0.05 of the 185 W, 4-pole, 110 V, 60 Hz single-phase induction motor, whose constants are given below:

Resistance of stator main winding, $R_1 = 1.86\ \Omega$

Reactance of stator main winding, $X_1 = 2.56\ \Omega$

Magnetizing reactance of the stator main winding, $X_m = 53.5\ \Omega$

Rotor resistance at standstill, $R_2 = 3.56\ \Omega$

Rotor reactance at standstill, $X_2 = 2.56\ \Omega$ [**Ans.** 201.5 watts]

3. Find the mechanical power output of 185 W, 4-pole, 110 V, 60 Hz single phase induction motor, whose constants are given below at a slip of 0.05

$R_1 = 1.86\ \Omega$, $X_1 = 2.56\ \Omega$, $X_\phi = 53.5\ \Omega$, $R_2 = 3.56\ \Omega$, $X_2 = 2.56\ \Omega$, core loss = 35.0 W, Friction and windage loss = 13.5 W [**Ans.** 183 watts]

4. Estimate the parameters of the equivalent circuit of a single phase induction motor from the following data:

Light running test : 220 V, 1.3 A, 60 W

Blocked rotor test : 220 V, 5.5 A, 840 W

Stator resistance : 11 Ω (dc)

Rotor resistance (at standstill/at running) : 1.2.

[**Ans.** $R_1 = 11\ \Omega$; $X_1 = 24.347\ \Omega$; $X_m = 263.5\ \Omega$; $R_2 = 13.97\ \Omega$; $X_2 = 24.347\ \Omega$]

5. A 230 V, 380 W, 50-Hz, 4-pole, 1-phase induction motor gave the following test results:

No-load test:	230 V,	84 W,	2.8 A
Blocked rotor test :	110 V,	460 W,	6.2 A

The stator winding resistance is 4.6 Ω and during the blocked rotor test, the auxiliary winding is open. Determine the equivalent circuit parameters. [**Ans.** $R_2' = 7.067\ \Omega$; $X_2' = X_1 = 6.68\ \Omega$, $X_m = 143.75\ \Omega$]

6. A 1-ϕ induction motor has stator windings in space quadrature and is supplied with a single phase voltage of 200 V at 50 Hz. The standstill impedance of the main winding is $(5.2 + j\,10.1)\ \Omega$ and that of the auxiliary winding is $(19.7 + j\,14.2)\ \Omega$. Find the value of capacitance to be inserted in the auxiliary winding for maximum starting torque.

[Ans. 131 μF]

7. A 230-V, 4-pole, 50-Hz single phase induction motor has the following data at standstill.

Main winding $(1.5 + j\,4.0)\ \Omega$; starting winding $(2.2 + j\,5.5)\ \Omega$. For making the motor develop maximum starting torque, find the value of (*i*) resistor (*ii*) capacitor in series with the starting winding. **[Ans.** (*i*) 5.74 Ω (*ii*) 503 μF]

[**Hint :** starting torque in case of a resistance-start motor will be maximum when $\phi_s = \dfrac{\phi_m}{2}$]

❑❑❑

SECTION C

SYNCHRONOUS GENERATOR

CONTENTS OF SECTION-C

SYNCHRONOUS GENERATOR : Principle; Construction of Cylindrical Rotor and Salient Pole Machines; Winding; EMF Equation; Armature Reaction; Testing; Model of the Machine; Regulation; Synchronous Reactance Method; Rothert's MMF Method; Potier Triangle Method; Output Power Equation; Power Angle Curve; Two Reactance Theory; Slip Test; Transient and Subtransient Reactances; Synchronisation; Parallel Operation.

5

CHAPTER

Synchronous Generator

5.1. INTRODUCTION

A synchronous machine is an ac machine in which the rotor moves at a speed which bears a constant relationship to the frequency of currents, in the armature winding. A synchronous machine is one of the important types of electric machines. Large ac networks operating at constant frequency of 50 Hz (or 60 Hz) rely almost exclusively on *synchronous generators,* also called the *alternators*, for the supply of electrical energy, and may have *synchronous compensators* at key points for control of reactive power. Private, stand-by and peak load plants with diesel or gas-turbine prime movers also have synchronous generators. Non-land-based synchronous plants (with a generating capacity of 50 MVA or more) are found on oil rigs, on large aircrafts (with hydraulically driven generators working at 400 Hz), and on ships for variable frequency supply to synchronous propeller motors. Synchronous motors provide constant speed industrial drives with the possibility of power factor correction.

Synchronous machines are generally constructed in larger sizes. Small size alternators are not economical. The modern trend is to build alternators of very large sizes capable of generating 500 MVA or even more. The synchronous motor is rarely built in small sizes owing to superior performance characteristics and economical construction of induction motors.

INSIDE THIS CHAPTER

5.1.1. Operating Principle

The operating principle of a synchronous machine is fundamentally the same as that of a dc machine, but, unlike the latter, in the synchronous machine there is no need to rectify the time varying emf which is induced in the armature winding. Consequently a synchronous machine does not require a commutator. It is, in fact, quite possible to use a dc generator as an alternator by placing a set of collector rings on the shaft and connecting these rings to the proper points on the armature winding; brushes riding on the rings can then be connected to the load. But unlike dc generator, they are to be driven at a very definite constant speed as the frequency of generated emf is determined by that speed. The latter is usually referred to as the *synchronous speed,* for which reason these machines are frequently called the synchronous generators.

Synchronous generators, because of absence of commutator, are comparatively simple and possess several important advantages over the dc generators.

5.1.2. Classification of Synchronous Machines

Synchronous machines, according to their applications, may be *synchronous generators, synchronous motors* or *synchronous compensators.* A synchronous generator is a synchronous machine which receives mechanical energy from a prime mover (steam turbine, hydraulic turbine or diesel engine) to which it is mechanically coupled and delivers electrical energy. A synchronous motor receives electrical energy from ac supply main and drives mechanical load. Synchronous compensator (or phase modifier) is a synchronous machine designed to operate on no load with its shaft connected neither to a prime mover nor to a mechanical load and is used to control reactive power in power supply networks. Synchronous compensators are designed for ratings up to 100 MVAR and speeds up to 3,000 rpm.

The synchronous machines may be *single, two* or 3-*phase* types. For the single phase type, the armature has all the coils connected for addition of individual voltages; this provides a pair of output terminals and a single ac output. For the two phase type, the armature has two sets of windings placed so the outputs from three terminals (one common) are 90° out of phase. For the three phase type, the armature has three sets of windings placed so that three outputs with a mutual phase difference of 120 ° are available. Three windings may be connected either in star or delta. Large ac machines are invariably 3-phase type.

Based on the construction of the machines, the synchronous machines may be classified as (*i*) rotating armature type and (*ii*) rotating field type.

Since it is immaterial for generation of an induced emf whether a conductor moves across a magnetic field or vice versa, synchronous generators may be constructed with either the armature or the field structure as the revolving member.

Rotating armature type alternator looks very much like a dc generator except that there are 3 slip rings in place of commutator (or 4 slip rings if it is desired to provide a connection to the generator neutral). In such generators the required magnetic field is produced by dc electromagnets placed on the stationary member called the *stator*, and the current generated is collected by means of brushes and slip rings on the revolving member, called the *rotor*. Such an arrangement is economical for the small low-voltage generators. Rotating armature type alternators are built only in small ratings up to about 200 or 250 kVA because the voltage generated is comparatively low and the current to be collected by the brushes small, no difficulty being experienced in collecting such a current. Such machines are suitable for small power plants, isolated lighting plants, where medium or small size machines are required.

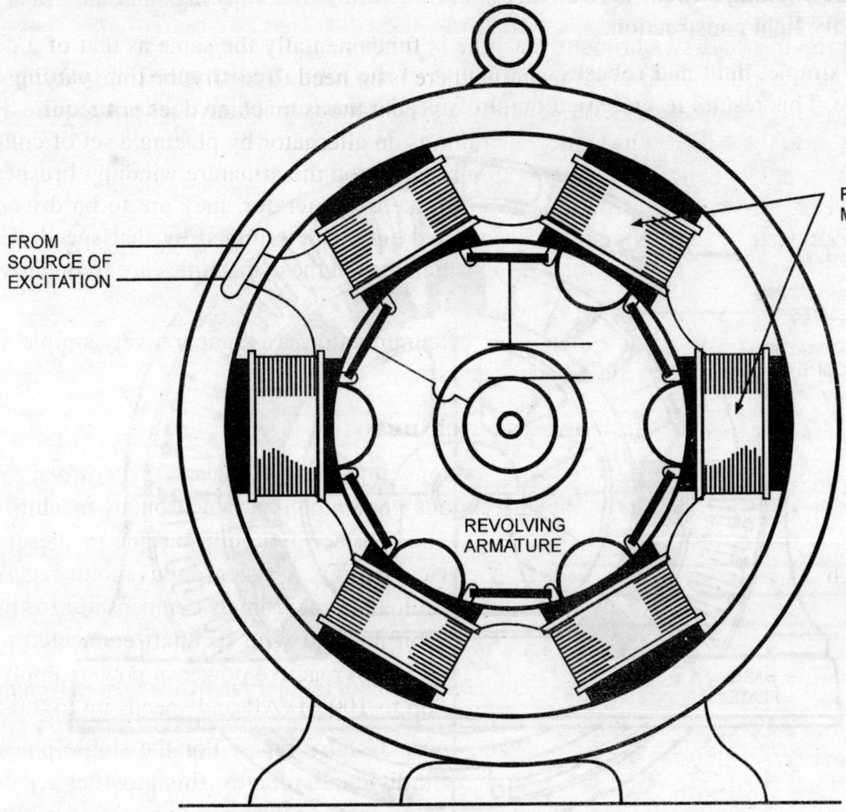

Fig. 5.1. ***Revolving Armature Alternator***

Practically all medium and large machines are always constructed with revolving field. The advantages of stationary armature and revolving field system are given below:

1. It is easier to insulate stationary armature winding for very high voltage e.g. as high as 33,000 volts because insulation of stationary armature is not subjected to mechanical stresses due to centrifugal action and more space is available on the stator for providing more insulation as the stator is outside the rotor.

2. The load circuit can be connected directly with the fixed terminals of the stator (or armature windings) without passing through slip rings and brushes. In the case of rotating armature the load circuit is required to be connected with the armature winding by means of slip rings and brushes in which case a difficulty is experienced in collection of heavy currents at very high voltages.

3. The armature winding can be more easily braced in a rigid frame to prevent any deformation which could be developed by the mechanical stresses set up due to short-circuit currents and the high centrifugal forces brought into play.

4. The armature winding is cooled more readily because the stator core can be made large enough with many air passages or cooling ducts for forced air circulation.

5. Only two slip rings are required for the supply of direct current to the rotor and since the exciting current is to the supplied at low voltage of 125 or 250 V, there is no difficulty in insulating them.

6. Since the exciting current is relatively small, therefore, the slip rings and the brush gear need be of only light construction.
7. Due to simple, light and robust construction of the rotor, higher speed of rotating dc field is possible. This results in increased output from the machine of given dimensions.

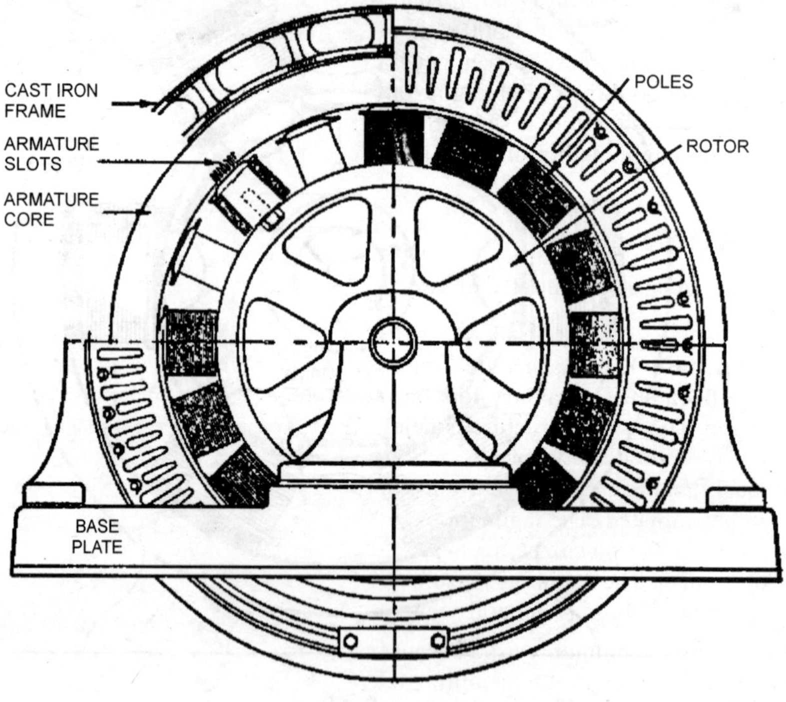

Fig. 5.2. ***Revolving Field Salient Pole Type Alternator***

Since in dc machines, the commutator makes necessary that either the armature should be rotating one, or brushes should revolve with the field, so it is convenient to have armature as a rotating member in dc machines.

Revolving field alternator with essential parts is shown in Fig. 5.2.

The synchronous machines may be classified as (*i*) salient pole machines and (*ii*) cylindrical rotor machines depending upon the type of construction used for the rotor. The salient pole construction is used for generators and motors of all ranges of output and up to all but the higher speeds. Medium-and large-sized generators for the highest speeds are of the cylindrical rotor type.

The synchronous generators, based on the type of prime movers to which they are mechanically coupled, may be classified as (*i*) turbo-generators, (*ii*) hydro-generators and (*iii*) engine driven generators.

Turbo-generators are driven by steam turbines. The efficiency of steam turbines is high at large speeds, and therefore, synchronous machines driven by steam turbines (*i.e.* turbo-generators) are high speed machines. The maximum speed of turbo-generators is 3,000 rpm corresponding to 2 poles and 50 Hz. These have small diameters (limited to about 5.2 m) and horizontal configurations. Cylindrical rotor construction is used as the salient pole construction is impractical owing to large mechanical forces. Two-pole construction with a speed of 3,000 rpm for 50 Hz is commonly used for turbo-generators. The 4-pole construction, with a synchronous speed of 1,500 rpm is now obsolete. Turbo-alternators have ratings up to 1,000 MVA

The synchronous generators driven by water turbines are called the *water-wheel generators* or *hydro-generators*. Such generators have to match the specific speed requirements of the turbines driving them. These generators are low speed machines with speeds between 1,500 and 100 rpm (*i.e.* number of poles ranging from 4 to 60) and of rating up to 750 MVA. Hydroelectric power plants usually have vertical shaft.

Diesel-engine driven synchronous generators are of small ratings from 100 kVA or less to 1,000 kVA. The speed is generally between 375 to 1,500 rpm, *i.e.* the number of poles is 16 to 4. Generally 4 to 6 pole alternators are common for diesel driven generators. The diesel engines are manufactured as horizontal type, and therefore, both the diesel engine and the synchronous generator are mounted horizontally and are connected by a horizontal shaft. Such generators employ salient pole type construction. The torque of the diesel engine is not uniform and this makes the synchronous generator sensitive to torque variations.

5.2. CONSTRUCTION

Synchronous machine consists essentially of two parts namely the armature (or stator) and the field magnet system.

1. Armature (or Stator). The armature is an iron ring, formed of laminations of special magnetic iron or steel alloy (silicon steel*) having slots on its periphery to accommodate armature conductors and is known as *stator*. The whole structure is held in a frame** which may be of cast iron or welded steel plate. Since the field rotates in between the stator, so that flux of the rotating field cuts the core of the stator continuously and causes eddy current loss in the stator core. To minimize the eddy current loss, the stator core is laminated. The laminations (usually of thickness 0.5 mm or less) are stamped out in complete rings (for smaller machines) or in segments (for larger machines) and insulated from each other with paper or varnish. The stampings also have openings which make axial and radial ventilating ducts to provide efficient cooling. A general view of stator and frame is shown in Fig. 5.3.

The open slots are more commonly used because the coils can be form wound and insulated prior to being placed in the slots giving least expenditure and more satisfactory winding method. This type of slots also facilitate in removal and replacement of defective coils. But this type of slots have disadvantage of distributing the air gap flux into branches or tufts which tends to produce ripples in the emf wave. The semi-closed type slots are better in this respect but do not permit the use of form-wound coils. Totally closed slots are rarely used.

Number of slots per pole per phase should be large to give sinusoidal generated emf. In general, the number of slots per pole per phase is 3 to 4 for small machines and more than 5 for large machines. Small machines may have integral slot winding with enough chording. For large machines fractional slot winding is used and so the number of slots per pole per phase may not be an integer.

The minimum tooth width should be such that the maximum flux density in it does not exceed the limit of flux density in teeth permissible to avoid saturation. The depth of the slot and shape is such as to accommodate the required number of conductors per slot for the rated voltage to be produced at the rated speed. Generally depth is 4 to 6 times the width. If the depth is increased, synchronous reactance in slot increases, core losses in teeth are larger and more excitation will be required for the machine.

2. Field Magnet System or Rotor. The magnetic field required for the generation of alternating voltage is provided by field magnets similar to those of a dc machine and these need to be supplied with

* Although grain oriented steels give lower core loss, the orientation is not everywhere favourable, and cold-rolled non-oriented steels containing 3 per cent silicon are usually used.

** In a dc machine the outer frame (or yoke) serves to carry the magnetic flux but in an alternator, it is merely used for holding the armature stampings and coils in position.

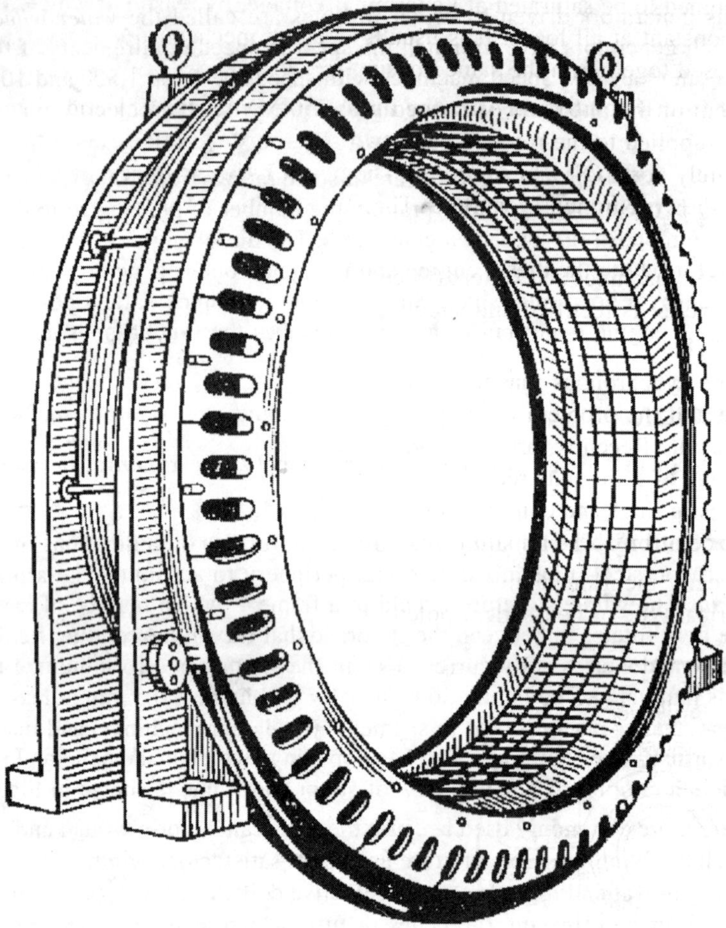

Fig. 5.3. *Stator of a Synchronous Generator*

direct current. This may be had from some external source, but so as to make the alternator self-contained and independent of other source of supply, it is more usual to install a small dc generator (usually a flat compound wound type) for this purpose. The generator is called the '*exciter*' and is coupled to the engine or turbine driving the alternator. The exciter armature is connected directly to the alternator field winding, usually without any kind of controlling resistance, so that except for variations due to temperature changes, the resistance of the exciter load is constant. For the successful operation of alternator, it is necessary to be able to vary the exciting current over a wide range from practically zero to the full output of the exciter, and this is achieved by varying the resistance in exciter field circuit. For making the field current entirely independent of the armature terminal voltage, a small

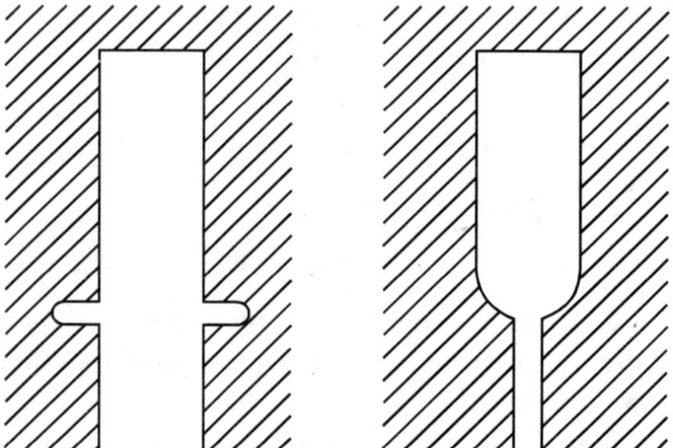

(*a*) *Open Slot* (*b*) *Semi-Closed Slot*

Fig. 5.4

pilot exciter is designed to be saturated at its terminal voltage, by reason of which its terminal voltage remains sensibly constant at all loads. It is usually coupled mechanically to the main exciter and the two are often enclosed together in the same housing.

The field system of the alternator is rotated within the armature ring and is known as *rotor*. The exciting current is supplied to the rotor through two slip rings and brushes. The polarities of the field produced is alternately north (N) and south (S). The power rating of the exciter is ordinarily 0.5 to 1% of power rating of the synchronous generator. The voltage rating of the exciter is usually between 125 and 250 V.

For medium sized machines ac exciters (three-phase ac generators) are used in place of dc exciters. The output of the exciter is rectified and supplied to the rotor winding of the main alternator through brushes and slip rings.

In case of large synchronous machines, of the rating of few hundred MW, the excitation needs become very large and the problem of supplying such amounts of power through high-speed sliding contacts becomes acute. Recently, brushless excitation system has been developed in which a 3-phase ac exciter and a group of solid-state rectifiers supply the excitation directly to the alternator field winding. Thus, brushes, slip rings and commutator are eliminated. A brushless excitation needs less maintenance due to absence of brushes and slip rings.

Rotors, as already mentioned in Art. 5.1.2, are of two types namely (*i*) salient pole type and (*ii*) smooth cylindrical type or non-salient pole type. Both types are shown in Fig. 5.5.

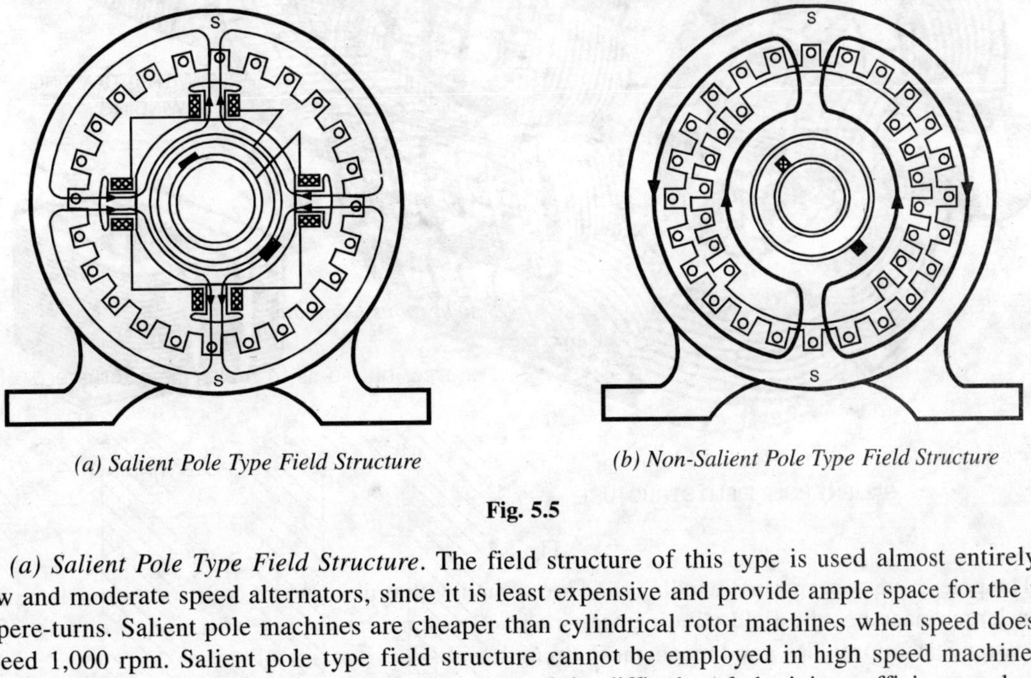

(a) Salient Pole Type Field Structure *(b) Non-Salient Pole Type Field Structure*

Fig. 5.5

(a) Salient Pole Type Field Structure. The field structure of this type is used almost entirely for slow and moderate speed alternators, since it is least expensive and provide ample space for the field ampere-turns. Salient pole machines are cheaper than cylindrical rotor machines when speed does not exceed 1,000 rpm. Salient pole type field structure cannot be employed in high speed machines on account of high peripheral speed (100 to 170 m/s) and the difficulty of obtaining sufficient mechanical strength. The salient field poles would also cause an excessive windage loss and produce noise if driven at high speed.

The rotor of a low-speed machine resembles a flywheel with electromagnets, similar to the field magnets of a dc machine, bolted on all round the periphery. It is usually made of cast iron or cast steel and has a heavy central hub, a number of radial arms, and a heavy rim which forms the yoke of the magnet system. The flywheel type of rotor is constructed to give adequate mechanical strength to withstand centrifugal and driving forces and at the same time to provide the necessary paths for the magnetic fluxes. The poles may be either solid or laminated but in any case, the pole shoes are laminated because the armature teeth between the slots cause concentrations of flux which generate emfs in the pole shoes. If they were not laminated the resulting eddy currents would cause excessive loss and heating. The pole faces are so shaped that the radial air-gap length increases from the pole centre to the pole-tips, as shown in Fig. 5.6, so that the flux distribution over the armature surface is sinusoidal and waveform of generated emf is sinusoidal.

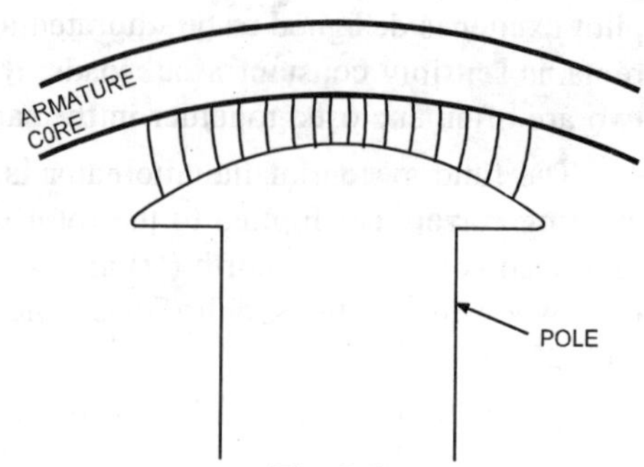

Fig. 5.6

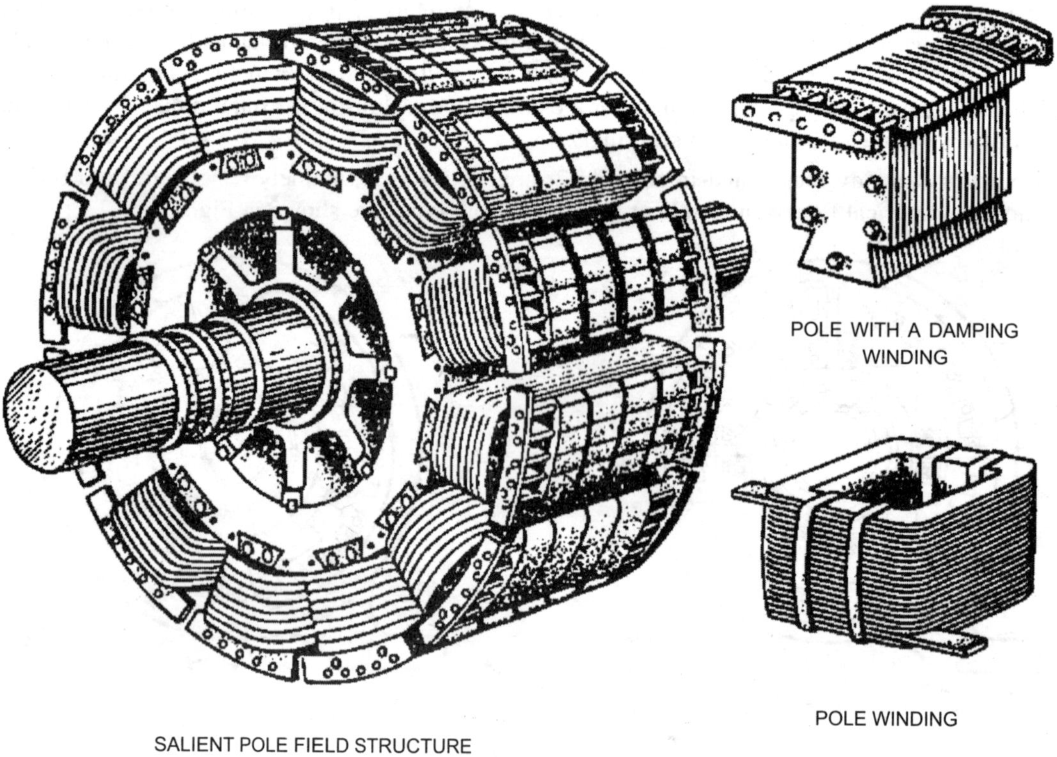

Fig. 5.7

The poles have concentrated windings and are wound with ordinary wire coils for the smaller sizes, but in larger machines the field windings are made of rectangular copper strip wound on edge. The latter arrangement provides good mechanical strength and saves wastage of space.

Salient pole machines are frequently provided with a damper winding on the rotor to damp rotor oscillations during transient conditions and to facilitate operation under unbalanced load conditions. The damper winding consists of copper bars laid in slots on the faces of the pole pieces and connected to each other at the ends by means of short-circuiting segments. The segments on individual poles are generally joined together to form common rings resulting-in a short-circuited squirrel cage winding similar to that used in induction machines with squirrel cage rotors.

It should be clearly understood that under normal operating conditions, damper winding does not carry any current because rotor runs at synchronous speed. The damper winding also tends to maintain balanced 3-phase voltage under unbalanced load conditions.

Since the speed of salient pole machines is quite low, so to give required frequency, such machines have large number of poles (4 to 60). To accommodate such a large number of poles, these machines have large diameter and small length, because the output of a machine is roughly proportional to its volume.

The salient pole field structure has the following special features :

(*i*) They have large diameter and short axial length.

(*ii*) The pole shoes cover about 2/3 of pole pitch.

(*iii*) Poles are laminated in order to reduce eddy current losses.

(*iv*) These are employed with hydraulic turbines or diesel engines. The speed is 100 to 375 rpm. A general view of salient pole field structure is shown in Fig. 5.7.

(*b*) *Smooth Cylindrical or Non-salient Pole Type Field Structure.* The field structure of this type is used in very high speed alternators (turbo-alternators). To reduce the peripheral velocity the diameter of the rotor is reduced and axial length is increased. The ratio of diameter to axial length ranges from 1/3 to 1/2. A small number of poles is required to give the required frequency.

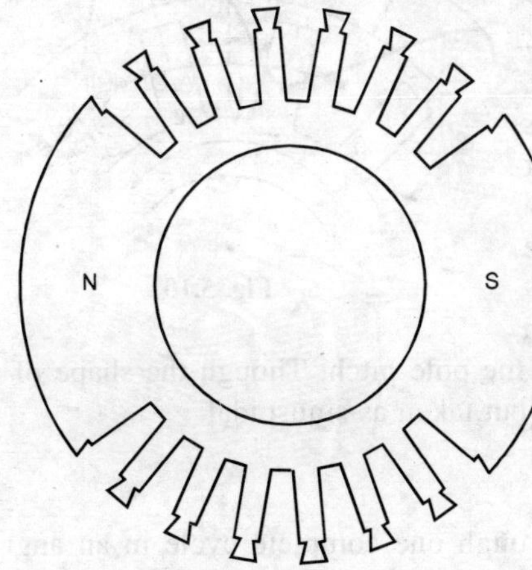

Fig. 5.8

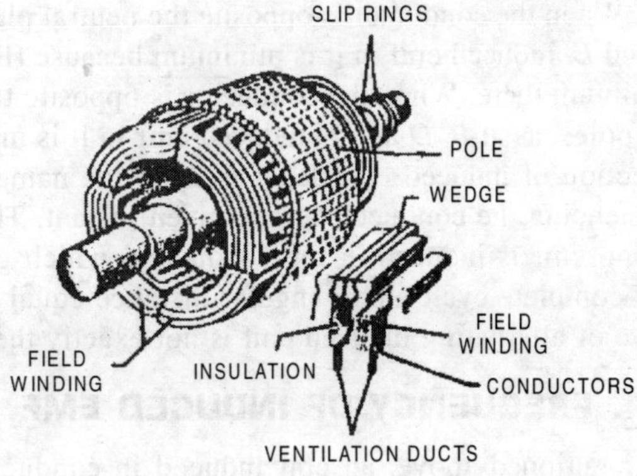

Fig. 5.9

Due to high peripheral speed, the rotating part of the turbo-generator is subjected to high mechanical stresses. As a result the rotor of a large turbo-generator is normally built from solid steel forging. Chromium-nickel-steel or special chrome-nickel-molybdenum steel is used for the rotors of turbo-alternators. The forging has radial slots in which the field copper, usually in strip form, is placed. The coils are held in place by steel or bronze wedges and the coil ends are fastened by metal rings. Normally two-thirds of the rotor is slotted for the field winding and one-third is left without slots so as to form the pole faces, as shown in Fig. 5.8. Rectangular slots with tapered teeth are milled out in the rotor so that rectangular conductors can be used for the field winding. Each slot is provided with a ventilation hole at the bottom. To reduce harmful tooth ripples, either stator slots or poles are skewed. General view of an assembled smooth cylindrical rotor is shown in Fig. 5.9.

The non-salient field structure has the following special features.

(*i*) They are of smaller diameter and of very long axial length.

(*ii*) Robust construction and noiseless operation.

(*iii*) Less windage (air-resistance) loss.

(*iv*) Better in dynamic balancing.

(*v*) High operating speed (3,000 rpm).

(*vi*) Nearly sinusoidal flux distribution around the periphery, and therefore, gives a better emf wave-form than that obtainable with salient pole field structure.

(*vii*) There is no need of providing damper windings (except in special cases to assist in synchronising) because the solid field poles themselves act as efficient dampers.

Air gap. A very small air gap increases the stray-load loss; also the eccentricity in the air gap can result from mechanical difficulties. This also increases the synchronous reactance X_d. A large air gap needs larger excitation current. A compromise has to be made. Generally the ratio of air gap to the pole pitch is between 0.008 to 0.02.

5.3. PRODUCTION OF SINUSOIDAL ALTERNATING EMF

When the rotor is rotated by means of some prime mover, the armature conductors cut the magnetic flux, therefore, an emf is induced in the armature conductors, due to electromagnetic induction effect.

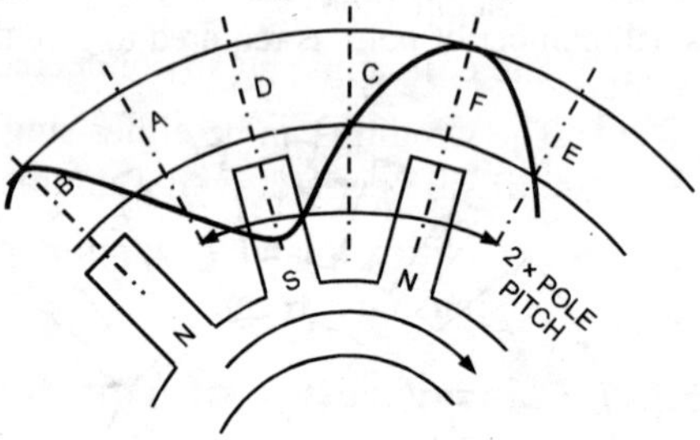

Fig. 5.10

When the conductor is opposite the neutral planes, as at *A*, *C* and *E*, induced emf in it is minimum because flux density is minimum there. When the conductor is opposite the middle of the poles, as at *B*, *D* and *F*, induced emf in it is maximum, the direction of induced emf depends upon the name of the pole influencing the conductor at any given instant. Thus an alternating emf is induced in the conductors which goes through one complete cycle in an angular distance equal to twice of the pole pitch. Though the shape of the wave of alternating induced emf is not exactly the sinusoidal but taken as sinusoidal.

5.4. FREQUENCY OF INDUCED EMF

As mentioned above, an emf induced in conductor goes through one complete cycle in an angular distance equal to twice the pole pitch.

If the number of poles on rotor of an alternator is P, then P/2 cycles of emf are completed in one revolution. The number of cycles per second, *known as frequency f*, will be equal to the product of number of cycles of emf per revolution and number of revolutions made per second by the rotor.

$$\text{i.e. Frequency, } f = \frac{P}{2} \times n$$

where n is the number of revolutions made per second by rotor

$$\text{or } f = \frac{P}{2} \times \frac{N}{60} = \frac{PN}{120} \qquad ...(5.1)$$

where N is the number of revolutions made per minute by the rotor.

Hence frequency of induced emf or current induced in stator conductors depends upon the number of poles and speed of rotor.

Example 5.1. A 6-pole ac generator is running and producing the frequency of 60 Hz. Calculate the revolutions per minute of the generator. If the frequency is decreased to 20 Hz, how many number of poles will be required if the generator is to be run at the same speed.

Solution : Frequency generated, $f = 60$ Hz

Number of poles, P = 6

$$\text{Speed, N} = \frac{120f}{P} = \frac{120 \times 60}{6} = 1{,}200 \textbf{ Ans.}$$

When frequency, $f' = 20$ Hz

$$\text{Number of poles required, P}' = \frac{120f'}{N} = \frac{120 \times 20}{1{,}200} = 2 \textbf{ Ans.}$$

5.5. ARMATURE WINDINGS

The armature windings of dc machines are usually closed circuit windings but alternators windings may be either closed giving *delta connections* or open giving *star connections*.

The general principles governing windings for dc machines also hold good for alternator windings e.g.

(*i*) The span of each coil must be equal to pole pitch *i.e.* the two sides of any coil must be under adjacent poles.

(*ii*) The coils must be so connected that emfs induced in them help each other.

(*iii*) The winding can be either single layer or double layer. The single layer winding will have one coil side per slot, while double layer winding will have two coil sides/slot.

(*iv*) In addition to above, winding must be so designed that it may approximately give induced emf of sinusoidal form.

5.5.1. Classification of Armature Windings

The classification may be made according to several points of view e.g.

(*a*) Single phase windings and polyphase windings.

(*b*) Concentrated windings and distributed windings.

(*c*) Half coiled (or hemitropic) windings and whole coiled windings.

(*d*) Single layer windings and double layer windings.

(*e*) Lap windings, wave windings and concentric (or spiral) windings.

(*f*) Full pitched coil windings and fractional pitched coil windings.

(*g*) Integral slot winding and fractional slot winding.

1. ***Single Phase Armature Windings.*** There are two types of single phase armature windings, namely

(*a*) concentrated and (*b*) distributed windings.

(*a*) *Concentrated Winding* : If one slot per pole or slots equal to number of poles are employed, then *concentrated winding* is obtained. Such windings give maximum induced emfs for a given number of conductors but the waveform of induced emf is not exactly of sinusoidal form.

Simplest possible winding known as "*skelton wave winding*" is shown in Fig. 5.11 (*a*). In this winding the number of conductors or coil sides is equal to the number of poles. The direction of induced emf in the conductor under north pole is shown upward while in the conductor under south pole downward. The conductor no. 1 under north pole is connected to conductor no. 2 under south pole, at the back and conductor no.2 is connected to conductor no. 3 at the front and so on. It is single layer, full pitched coil winding.

Single layer winding requires considerable space for the end connections of the coils and thus it is rarely used in practice.

If single turn coils of skelton wave winding are replaced by multi-turns coils in order to obtain higher emf, the multi-turn half coiled (*hemitropic*) winding is obtained. Since coils only cover one half of the armature periphery, therefore, this winding is known as *half coiled or hemitropic* winding.

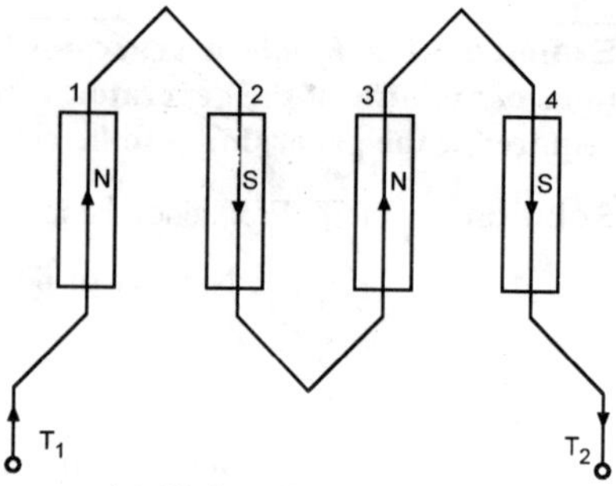

(a) Skelton Wave Winding

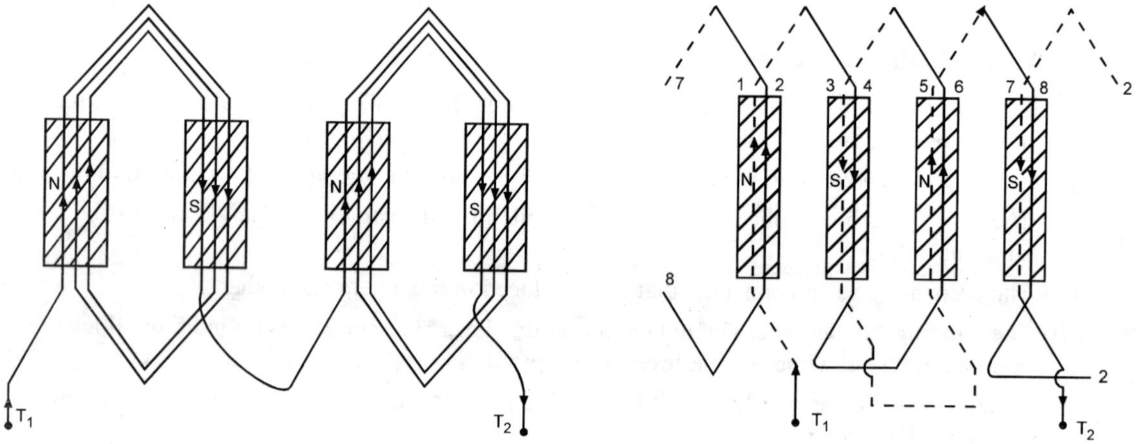

(b) Half Coiled (or Hemitropic) Winding

(c) Whole Coiled Winding

Fig. 5.11

If the coils are distributed over the whole of the periphery, the winding so obtained is called *whole coiled winding*.

In this winding the number of coils is equal to the number of poles. One side of each coil lying on the top is shown by continuous (or solid) line and the other side lying in the bottom of the slot is shown by dotted line in Fig. 5.11 (*c*). It is double layer winding.

(*b*) *Distributed Winding*. If the conductors are placed in several slots under one pole, the winding is known as *distributed winding*. The winding may be partially or completely distributed accordingly as the slots are spread over only a portion of it or over the entire armature surface. Though it reduces the induced emf it yet is most commonly employed due to its manifold advantages.

Advantages. (*i*) The harmonic emfs are reduced and so the waveform is improved.

(*ii*) Certain distorting harmonics can be altogether eliminated.

(*iii*) The distributed winding diminishes armature reaction and armature reactance.

(*iv*) Higher current density in the copper can be adopted due to even distribution of copper over the surface of armature resulting in even distribution of copper loss and, therefore, in efficient cooling.

(*v*) The core is better utilised as a number of small slots evenly spaced are employed.

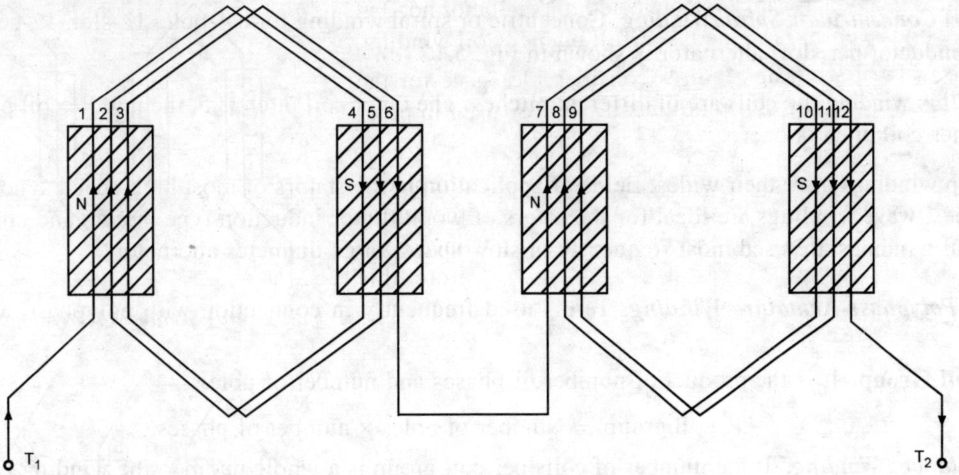

Fig. 5.12 ***(a) Single Layer Lap Winding***

(*i*) *Lap Winding*. Full pitched lap winding for a 4-pole, 12-slot, 12–conductor (one conductor per slot) alternator is shown in Fig. 5.12 (*a*). The back pitch of the winding is equal to the number of conductors per pole *i.e.* = 3 and front pitch is equal to back pitch minus one. The winding is completed per pair of pole and then connected in series, as shown in the figure.

(*ii*) *Wave Winding*. Wave winding for a 4-pole, 12-slot, 12-conductor (one conductor per slot), alternator is shown in Fig. 5.12 (*b*). The back pitch and front pitch, each is equal to number of conductors per pole.

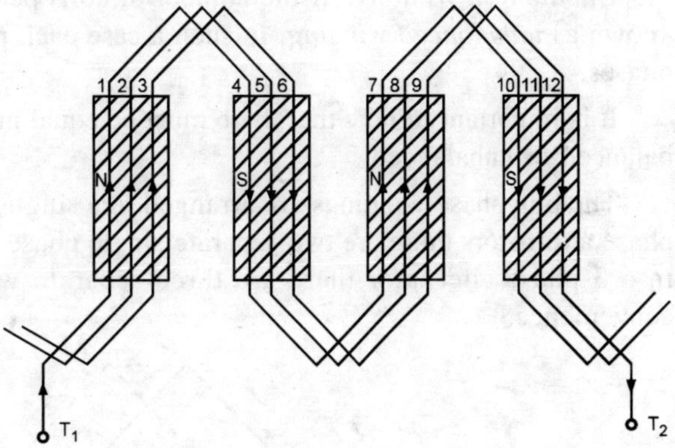

Fig. 5.12 ***(b) Single Layer Wave Winding***

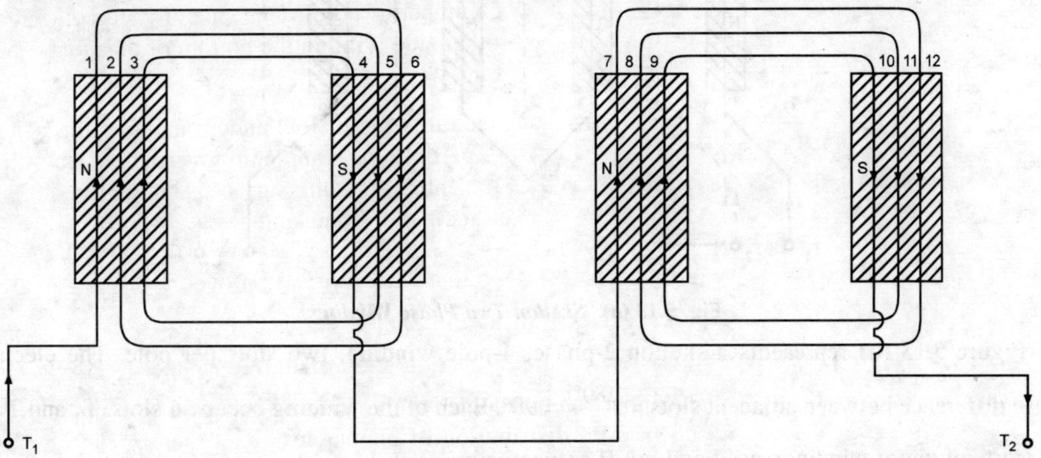

Fig. 5.12 ***(c) Single Layer Spiral Winding***

(*iii*) *Concentric or Spiral Winding*. Concentric or spiral winding for a 4-pole, 12-slot, 12-conductor (one conductor per slot) alternator is shown in Fig. 5.12 (*c*).

In this winding the coils are of different pitches. The outer coil pitch is 5, the middle coil pitch is 3 and inner coil pitch is one.

Lap windings have their widest fields of application in the stators of most high-speed synchronous machines, wave windings are ideal for the rotors of wound rotor induction type motors and concentric or spiral windings are used most frequently in slow speed, large diameter alternators :

2. *Polyphase Armature Winding.* Terms used frequently in connection with polyphase windings are:

Coil Group. It is the product of number of phases and number of poles

i.e. coil group = Number of poles × number of phases.

Balanced Winding. If the number of coils per coil group is a whole number, the winding is known as *balanced winding.* Each pole contains equal number of coils of different phases in such a case.

Unbalanced Winding. If the number of coils per coil group is not a whole number, the winding is known as *unbalanced winding*. In such a case each pole contains unequal number of coils of different phases.

It is important to note that there must be equal number of coils in each phase, either the winding is balanced or unbalanced.

The polyphase windings are arranged like single phase windings, the only difference is that in two phase alternators there are two separate single phase windings placed 90º (electrical degrees) apart and in a 3-phase alternator there are three separate windings placed 60° electrical degrees apart for convenience.

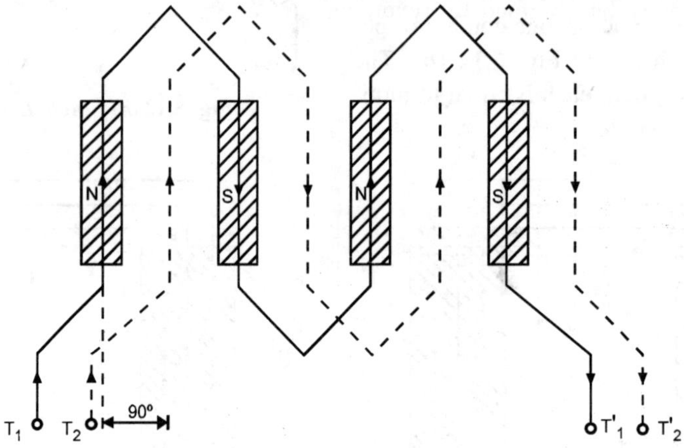

Fig. 5.13 (*a*). *Skelton Two Phase Winding*

Figure 5.13 (*a*) represents a skelton 2-phase, 4-pole winding, two slots per pole. The electrical phase difference between adjacent slots = $\frac{180^\circ}{2}$ = 90º. Each of the winding occupy 4 slots. T_1 and T_2 are the starts of phase windings no. I and no. II respectively.

T'_1 and T'_2 are the finish of phase windings no. I and no. II respectively.

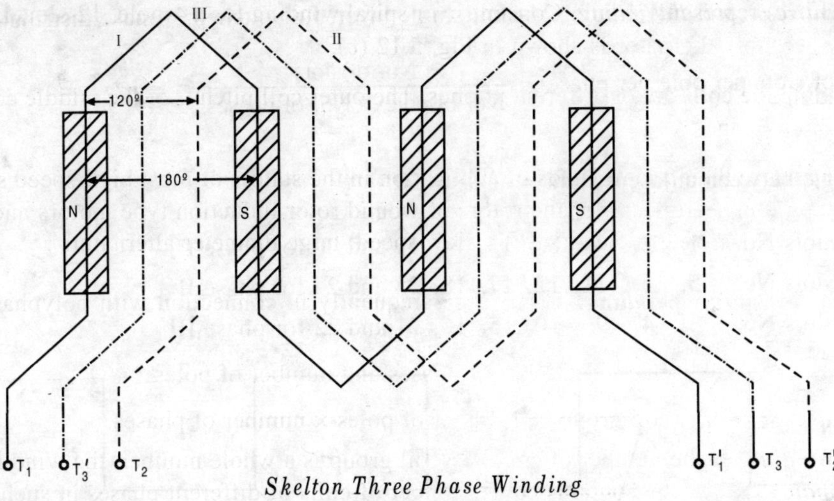

Skelton Three Phase Winding

Fig. 5.13 (*b*). *Skelton Three Phase Winding*

Figure 5.13 (*b*) represents a skelton 3-phase, 4-pole winding, 3 slots per pole. The electrical phase difference between adjacent slots = $\frac{180^\circ}{3} = 60^\circ$. Each of the winding occupies 4 slots, T_1, T_2, and T_3 are the starting terminals and T_1', T_2' and T_3' are the finishing terminals of the three phases.

First winding starts at slot no. 1 and ends over slot no. 10. Second winding starts at slot no. 2 and ends over slot no. 11. Third winding starts at slot no. 3 and ends over slot no. 12. The phase difference of the emfs induced in windings first and second, second and third, and third and first is 60º, 60º and 240º respectively. Since in three phase system the phase difference between the adjacent windings must be 120º; this can easily be obtained by reversing the connections of second winding, as shown in Fig. 5.13 (*b*).

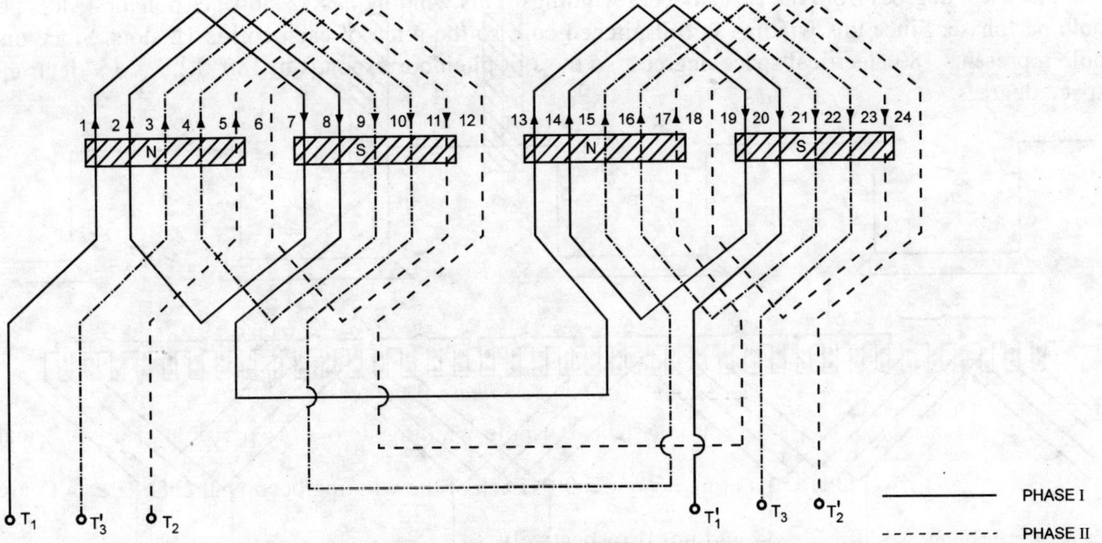

Fig. 5.13 (*c*). *3-Phase Distributed Winding*

Figure 5.13 (*c*) represents 4-pole, 24-slot, single layer, full-pitched 3-phase distributed winding.

$$\text{No. of slots per pole per phase} = \frac{\text{No. of slots}}{\text{No. of poles} \times \text{No. of phases}} = \frac{24}{4 \times 3} = 2$$

$$\text{Phase angle between adjacent slots} = \frac{180^\circ}{6} = 30^\circ$$

Hence slots No. 1, 2, 7, 8, 13, 14, 19 and 20 for phase I
slots No. 5, 6, 11, 12, 17, 18, 23 and 24 for phase II
and slots No. 3, 4, 9, 10, 15, 16, 21 and 22 for phase III

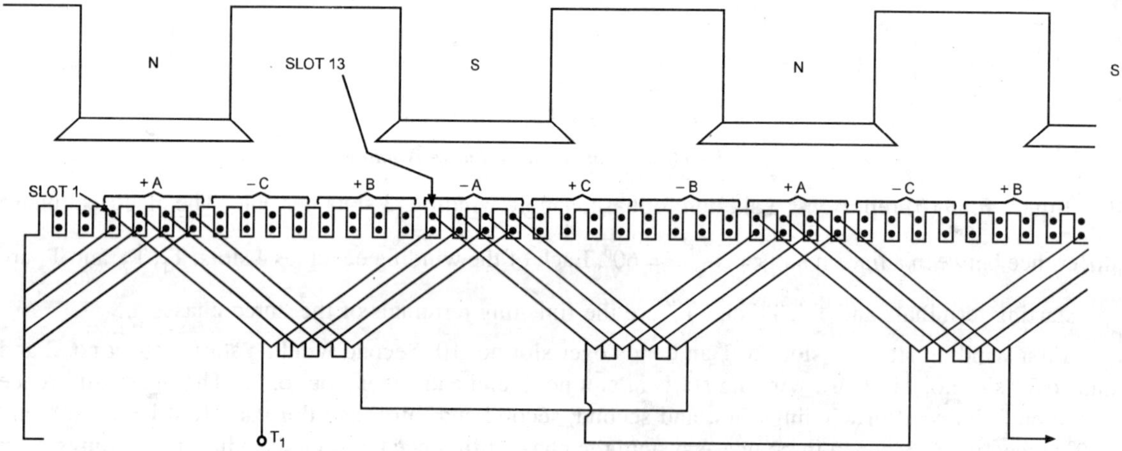

Fig. 5.13 (*d*). *Three-Phase Full-pitch Double Layer Lap Winding*

Figure 5.13 (*d*) shows three-phase full-pitched, double layer lap winding. Each winding is spaced 120 electrical degrees from the two adjacent windings. This winding has 12 slots per pole or 4 slots per pole per phase. Since this winding is full-pitched coil, so the pitch of each coil is 12 slots. Since one pole represents 180 electrical space degrees, so the slot pitch corresponds to 180°/12, *i.e.* 15 electrical space degrees.

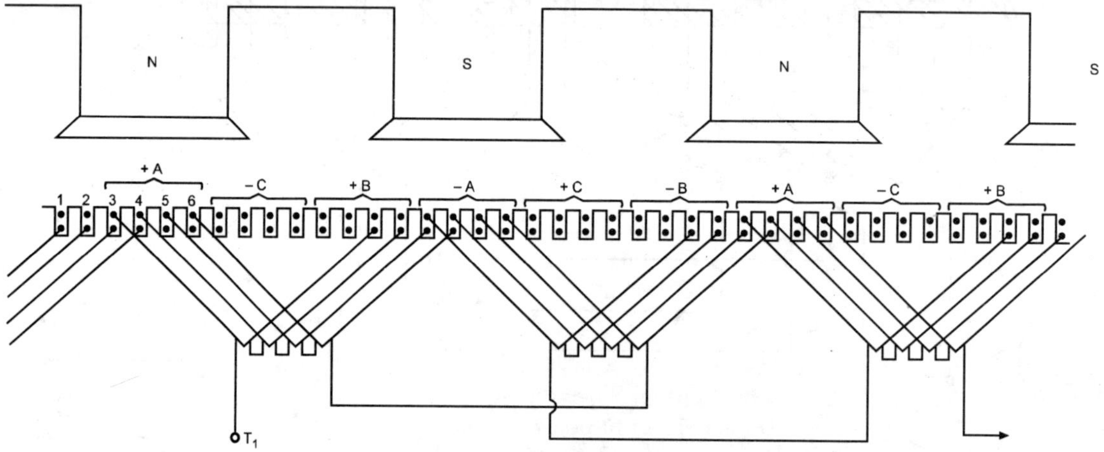

Fig. 5.13 (*e*). *Three-Phase Five-sixths-pitch Double Layer Lap Winding*

Figure 5.13 (*e*) shows three-phase five-sixths pitch, double layer lap winding having 4 slots per pole per phase.

In a fractional pitch winding, the coil span is made less than 180° electrical space degrees. In Fig. 5.13 (*e*) a coil instead of having a pitch of 12 slots, now has a pitch of 10 slots, so that its spread is no longer equal to pole pitch. Except pitch, the winding is similar to the winding shown in Fig. 5.13 (*d*) in every respect.

This type of winding improves the waveform of the induced emf, saves copper in the connections of coil ends and reduces the inductance of the windings. Such winding induces slightly less emf than full pitch winding under the same conditions.

Full Pitch and Short Pitch Windings. When the two coil sides forming a complete coil of a winding are 180 electrical space degrees apart, the winding is known as *full pitch winding*, as shown in Fig. 5.13 (*d*). In this type of winding when one side of the coil is under N pole, the other side is in corresponding position under S pole. The induced emfs differ by 180° in phase, but the coil is connected in such a way that emfs add giving resultant emf E, as shown in Fig. 5.14.

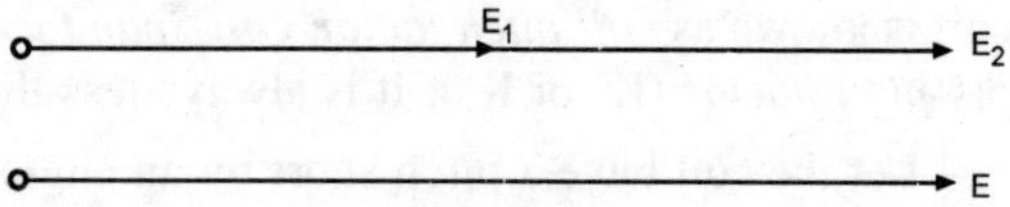

Fig. 5.14

When the coil span of the winding is less than 180 electrical space degrees, *i.e.* the two coil sides forming a complete coil of the winding are less than 180 electrical space degrees apart, the winding is known as *fractional pitch winding*. In this type of winding, the induced emf in each coil side is not in phase, so the resultant emf is the phasor sum of induced emfs in the coil sides, which is slightly less than their arithmetic sum, hence induced emf in short pitch winding is less than that induced in full pitch winding under the same conditions.

The advantages of short-pitch windings are :

(*i*) Waveform of the induced emf is improved and the distorting harmonic can be reduced or totally eliminated.

(*ii*) Copper is saved in the coil ends due to less span.

(*iii*) The inductance of the winding is reduced due to lesser length of coil ends as well as lesser mutual inductance between those conductors which lie in slots containing also conductors of either of the other two phases.

(*iv*) Fractional number of slots per pole can be used which in turn reduces the tooth ripples.

(*v*) Mechanical strength of the coil is increased.

Integral Slot and Fractional Slot Windings. When the number of slots per pole per phase, m is an integer, the winding is called the *integer slot* winding but when the number of slots per pole per phase, m is a fractional number the winding is referred to as *fractional slot winding*. Of course in both cases the number of slots per phase must be whole number so that each phase has the same number of coils.

Fractional slot winding is practicable only with double layer arrangement. It limits the number of parallel circuits available, because phase group under several poles must be connected in series before a unit is formed and the winding repeats the pattern to give a second unit that can be put in parallel with the first.

Though the fractional slot winding may appear to be little complicated, it has certain technical advantages as well as ease in manufacturing. Since the number of armature slots chosen need not necessarily be an integral multiple of the number of poles, one may choose a particular number of slots for which the lamination notching gear is available. This results in saving in cost of drawings and equipment. The flexibility can be effectively employed where the number of poles varies over a wide range as in the case of synchronous machines. Further mmf harmonics, leakage reactance of windings and cost of copper is reduced owing to chording, which is must in case of fractional slot winding and shorter end connections.

5.6. WINDING FACTOR

The product of pitch factor K_p and distribution factor K_d is referred to the *winding factor*

$$i.e.\ K_w = K_p\, K_d. \qquad \text{...(5.2)}$$

The pitch factor and distribution factor are explained below.

5.6.1. Pitch Factor

In a short-pitch winding the induced emfs in the two sides of the coil are not in phase, as discussed in Art. 5.5, hence their resultant given by the phasor sum is always less than their arithmetic sum. The ratio of phasor sum of induced emfs per coil to the arithmetic sum of induced emfs per coil is known as *coil pitch factor, coil-span factor* or simply *pitch factor* (K_c or K_p). It is always less than unity.

Let the coil have a pitch short by an angle α electrical space degrees from full pitch and induced emf in each coil side be E.

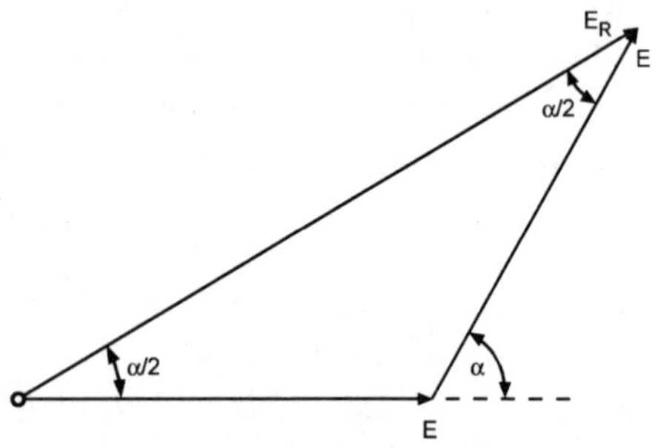

Fig. 5.15

Now if the coil would have been full-pitched, then total induced emf in the coil would have been 2E.

If the coil is short-pitched by α electrical space degrees the resultant induced emf, E_R in the coil is phasor sum of two voltages α° (electrical) apart

$$i.e.\ E_R = 2\,E \cos \frac{\alpha}{2}$$

$$\text{Pitch factor, } K_p = \frac{\text{Phasor sum of coil side emfs}}{\text{Arithmetic sum of coil side emfs}} = \frac{2\,E \cos \frac{\alpha}{2}}{2\,E} = \text{Cos}\ \frac{\alpha}{2} \qquad \text{...(5.3)}$$

If the coil span is reduced by one slot, then the phase angle α between the induced emfs in the two sides of the coil is given as

$$\alpha = \frac{180^\circ}{n} \quad \text{where } n \text{ is the number of slots per pole.}$$

The pitch factor given by Eq. (5.3) is for the fundamental component of emf. The flux wave may consist of space-field harmonics also, which give rise to the corresponding time harmonics in the generated emf wave. A 3rd harmonic component of the flux wave, may be imagined as produced by 3 poles as compared to one pole for the fundamental component. In view of this, the chording angle for the *rth* harmonic becomes *r* times the chording angle for the fundamental component and the pitch factor for the *rth* harmonic is given as

$$\text{Pitch factor, } K_{pr} = \text{Cos}\ \frac{r\alpha}{2} \qquad \text{...(5.4)}$$

The *rth* harmonic emf will be reduced to zero if chording angle α is made such that

$$\text{Cos}\ \frac{r\alpha}{2} = 0 \ \text{ or } \ \frac{r\alpha}{2} = 90^\circ$$

In case of three phase alternators, the third harmonic emf can be suppressed by star or delta connections, as in the case of 3-phase transformers, Hence while designing a winding, the attention is mainly given for the attenuation of 5th and 7th order harmonics, by adopting a suitable chording angle. A chording angle of $\alpha = 30°$ (coil pitch 150° electrical) is most useful, as it gives the pitch factors of 0.966 for the fundamental, 0.707 for 3rd and 9th harmonics and 0.259 for 5th and 7th harmonics.

Three phase alternators are invariably star-connected to suppress 3rd and 9th (*i.e.* triplen) harmonics from the induced emfs; though 5th and 7th harmonic emfs of reduced magnitude are present in the lines.

5.6.2. Distribution Factor

As already stated if all the coil sides of any one phase under one pole are bunched in one slot, the winding obtained is known as *concentrated winding* and the total emf induced is equal to arithmetic sum of the emfs induced in all the coils of one phase under one pole.

But in order to obtain a better wave shape the coils are not concentrated or bunched in one slot, but are distributed in a number of slots to form *polar groups* under each pole. The winding so obtained is known as *distributed winding* and in such winding the coils/phase are displaced from each other by a certain angle. The result is that the emfs induced in the coil sides of one phase in the adjacent slots under one pole are not in phase with each other but differ by an angle equal to the angular displacement of the slots. So the resultant emf of all the coils of one phase under one pole would be the phasor sum of the individual coil emfs and will be less than that if all the coil sides of one phase under one pole were bunched in one slot.

The ratio of the phasor sum of the emfs induced in all the coils distributed in a number of slots under one pole to the arithmetic sum of the emfs induced (or to the resultant of emfs induced in all coils concentrated in one slot under one pole) is known as breadth factor (K_b) or distribution factor (K_d)

$$\text{or distribution factor, } K_d = \frac{\text{EMF induced in distributed winding}}{\text{EMF induced if the winding would have been concentrated}}$$

$$= \frac{\text{Phasor sum of component emfs}}{\text{Arithmetic sum of component emfs}}$$

The distribution (or breadth) factor is always less than unity.

Let number of slots per pole $= n$

and number of slots per pole per phase $= m$

Induced emf in each coil side $= E_c$

Angular displacement between the slots,

$$\beta = \frac{180°}{n}$$

The emfs induced in different coils of one phase under one pole are represented by side AC, CD, DE, EF, ...which are equal in magnitude (say each equal to E) and differ in phase (say by β°) from each other.

If bisectors are drawn on AC, CD, DE, EF... they would meet at common point (say at O). The point O would be the circum centre of the circle having AC, CD, DE, EF...as the chords and representing the emfs induced in the coils in different slots.

EMF induced in each coil side, $E = AC = 2\,OA \sin\frac{\beta}{2}$

$$\text{Arithmetic sum} = m \times 2 \times OA \sin\frac{\beta}{2}$$

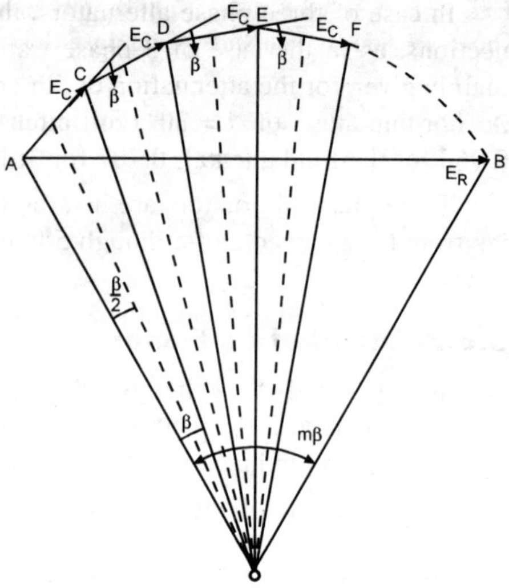

Fig. 5.16

The resultant emf induced in one polar group of one phase would be the phasor sum as represented by the phasor AB shown in Fig. 5.16.

∴ The resultant emf,

$$E_R = AB$$

$$= 2 \times OA \sin\frac{AOB}{2}$$

$$= 2 \times OA \sin\frac{m\beta}{2}$$

and distribution factor,

$$K_d = \frac{\text{Phasor sum of components emfs}}{\text{Arithmetic sum of components emfs}}$$

$$= \frac{2 \times OA \sin\frac{m\beta}{2}}{m \times 2 \times OA \sin\frac{\beta}{2}} = \frac{\sin\frac{m\beta}{2}}{m \sin\frac{\beta}{2}} \qquad \text{...(5.5)}$$

$m\beta$ is also known as the *phase spread* and is spread in electrical degrees or radians.

The distribution factor K_d given by Eq. (5.5) is for the fundamental component of emf. If the flux distribution contains space-harmonics, the slot angular pitch β on the fundamental scale, would become $r\beta$ for the *r th* harmonic component and thus the distribution factor for the *r th* harmonic would be

$$K_{dr} = \frac{\sin\frac{rm\beta}{2}}{m \sin\frac{r\beta}{2}} \qquad \text{...(5.6)}$$

Example 5.2. Calculate the distribution factor for a 36-slot, 4-pole, single layer 3-phase winding.

Solution : Number of slots per pole, $n = \frac{36}{4} = 9$

Number of slots per pole per phase, $m = \frac{n}{\text{Number of phases}} = \frac{9}{3} = 3$

Angular displacement between the slots, $\beta = \frac{180°}{n} = \frac{180°}{9} = 20°$ (electrical)

Distribution factor, $K_d = \frac{\sin\frac{m\beta}{2}}{m \sin\frac{\beta}{2}} = \frac{\sin\frac{3 \times 20°}{2}}{m \sin\frac{20°}{2}} = \frac{1}{3}\frac{\sin 30°}{\sin 10°} = 0.96$ **Ans.**

Example 5.3. An 8-pole, 3-phase, 60° spread, double layer winding has 72 coils in 72 slots. The coils are short-pitched by two slots. Calculate the winding factor for the fundamental and third harmonic.

Solution :

$$\text{Number of slots per pole, } n = \frac{72}{8} = 9$$

$$\text{Number of slots per pole per phase, } m = n \times \frac{\text{phase spread}}{180} = 9 \times \frac{60}{180} = 3$$

$$\text{Angular displacement between the slots, } \beta = \frac{180^\circ}{n} = \frac{180^\circ}{9} = 20^\circ \text{ (electrical)}$$

$$\text{Coil span} = \frac{180^\circ \times \text{coil span in terms of slots}}{\text{Number of slots per pole}}$$

$$= \frac{180\,(9-2)}{9} = 140^\circ$$

$$\text{Chording angle, } \alpha = 180^\circ - \text{coil span} = 180^\circ - 140^\circ = 40^\circ$$

For the fundamental component

$$\text{Distribution factor, } K_d = \frac{\text{Sin}\,\frac{m\beta}{2}}{m\sin\frac{\beta}{2}} = \frac{\text{Sin}\,\frac{3\times 20^\circ}{2}}{3\sin\frac{20^\circ}{2}} = 0.96$$

$$\text{Pitch factor, } K_p = \text{Cos}\,\frac{\alpha}{2} = \text{Cos}\,\frac{40^\circ}{2} = 0.94$$

$$\text{Winding factor, } K_w = K_d \times K_p = 0.96 \times 0.94 = 0.9 \textbf{ Ans.}$$

For the third harmonic component (*i.e.* $r = 3$)

$$\text{Distribution factor, } K_{d_3} = \frac{\text{Sin}\,\frac{rm\beta}{2}}{m\sin\frac{r\beta}{2}} = \frac{\text{Sin}\,\frac{3\times 3\times 20}{2}}{3\sin\frac{3\times 20^\circ}{2}} = 0.666$$

$$\text{Pitch factor, } K_{p_3} = \text{Cos}\,\frac{r\alpha}{2} = \text{Cos}\,\frac{3\times 40^\circ}{2} = 0.5$$

$$\text{Winding factor, } K_{w_3} = K_{d_3} \times K_{p_3} = 0.666 \times 0.5 = 0333 \textbf{ Ans.}$$

5.7. EMF EQUATION

Let Z_p = Number of conductors in series per phase

P = Number of poles

ϕ = Useful flux per pole in webers

N = Rotational speed in rpm

and f = Frequency in hertz and is equal to $\frac{NP}{120}$

The flux cut by any conductor while passing from the centre of one inter polar gap to the centre of the next is ϕ webers and since during the movement, the emf wave completes half cycle *i.e.* the time taken is $\frac{1}{2f}$ seconds, therefore, the average rate of cutting the flux,

$$\frac{d\phi}{dt} = \frac{\phi}{1/2f} = 2f\phi \text{ Wb/s}$$

Hence average emf induced in each conductor = $2f\phi$ volts

Average emf per phase, E_{av}/phase = Number of conductors in series/phase arranged in one slot/pole × average emf induced per conductor

$= Z_p \times 2f\phi = 2T \times 2f\phi = 4\phi f T$ volts $\quad \because Z_p = 2T$ (turns)

For distributed winding the average value of emf per phase will be K_d times the above value

i.e. E_{av} / phase $= 4 K_d \phi f T$ volts

For short pitch winding the true average value of emf per phase will be K_p times the above value

E_{av}/phase $= 4 K_d K_p \phi f T$ volts

and E_{rms}/phase = Form factor, $K_f \times E_{av}$/phase $= 4 K_f K_d K_p \phi f T$ volts

For sinusoidal wave of emf, $K_f = 5.11$

$\therefore E_{rms}$ / phase $= 4.44 K_d K_p \phi f T$ volts ...(5.7)

If the alternator is star-connected, as is usually the case, the line voltage is $\sqrt{3}$ times the phase value.

So line induced emf, $E_L = \sqrt{3} \times 4.44 K_d K_p \phi f T$ volts ...(5.8)

For full-pitched and concentrated windings, $K_p = 1$ and $K_d = 1$

Example 5.4. The armature of a 1-phase alternator is completely wound with 100 single-turn coils distributed uniformly. The induced emf in each turn is 2 V. What is the emf induced in the whole winding with all 100 coils connected in series?

Solution : In case of single phase winding, the number of slots per pole per phase,

m = Number of slots per pole, n

and distribution factor is given as

$$K_d = \frac{\text{Sin}\frac{m\beta}{2}}{m\sin\frac{\beta}{2}} = \frac{\text{Sin}\, n\times\frac{180°/n}{2}}{n\times\sin\frac{180°/n}{2}} \qquad \because m = n \text{ and } \beta = \frac{180°}{n}$$

$$= \frac{\text{Sin } 90°}{n\times\frac{\pi}{2n}} \qquad \because 180° = \pi \text{ radians and for very small angle } \sin\theta = \theta \text{ in radians}$$

$$= \frac{1}{\pi/2} = \frac{2}{\pi}$$

EMF induced in the whole winding = Number of turns × induced emf in each turn × K_d

$= 100 \times 2 \times \frac{2}{\pi} = \frac{400}{\pi}$ or 127.32 V **Ans.**

Example 5.5. A 3-ϕ, 6-pole, star-connected synchronous generator revolves at 1,000 rpm. The stator has 90 slots and 8 conductors per slot. The flux per pole is 0.05 Wb. Calculate the voltage generated, if the winding factor is 0.96.

Solution : Flux per pole, $\phi = 0.05$ Wb

$$\text{Frequency, } f = \frac{PN}{120} = \frac{6 \times 1{,}000}{120} = 50 \text{ Hz}$$

Number of conductors connected in series per phase,

$$Z_p = \frac{\text{Number of conductors per slot} \times \text{number of slots}}{\text{Number of phases}}$$

$$= \frac{90 \times 8}{3} = 240$$

$$\text{Number of turns per phase, T} = \frac{Z_p}{2} = \frac{240}{2} = 120$$

$$\text{Winding factor, } K_w = K_d K_p = 0.96$$

$$\text{Voltage generated between lines, } E_L = \sqrt{3} \times 4.44 \times K_w \times \phi \times f \times T \text{ volts}$$

$$= \sqrt{3} \times 4.44 \times 0.96 \times 0.05 \times 50 \times 120 = 2{,}215 \text{ V Ans.}$$

Example 5.6. A 3-phase, 16-pole, star-connected alternator has 144 slots and 6 conductors per slot. The flux per pole is 0.03 Wb sinusoidally distributed and the speed is 375 rpm. If the coil span is 160°, determine:
(*i*) Frequency (*ii*) Pitch factor (*iii*) Distribution factor (*iv*) Phase and line emfs.

Solution :

$$\text{Number of slots per pole, } n = \frac{144}{16} = 9$$

$$\text{Number of slots per pole per phase, } m = \frac{9}{3} = 3$$

Number of conductors connected in series per phase,

$$Z_p = \frac{144 \times 6}{3} = 288$$

$$\text{Number of series turns per phase, T} = \frac{Z_p}{2} = \frac{288}{2} = 144$$

$$\text{Angular displacement between the slots, } \beta = \frac{180^\circ}{9} = 20^\circ$$

$$(i) \text{ Frequency, } f = \frac{PN}{120} = \frac{16 \times 375}{120} = 50 \text{ Hz } \textbf{Ans.}$$

$$(ii) \text{ Pitch factor, } K_p = \text{Cos } \alpha = \text{Cos } \frac{180^\circ - 160^\circ}{2} = 0.9848$$

$$(iii) \text{ Distribution factor, } K_d = \frac{\text{Sin } m \times \frac{\beta}{2}}{m \sin \frac{\beta}{2}} = \frac{\text{Sin } 3 \times \frac{20^\circ}{2}}{3 \sin \frac{20^\circ}{2}} = 0.9598 \textbf{ Ans.}$$

(*iv*) Phase emf, $E_p = 4.44 \times K_d K_p \times \phi \times f \times T$ volts

$= 4.44 \times 0.9598 \times 0.9848 \times 0.03 \times 50 \times 144$

$= 906.5$ V **Ans.**

Line emf, $E_L = \sqrt{3} E_p = \sqrt{3} \times 906.5 = 1{,}570$ V **Ans.**

Example 5.7. Calculate the rms value of the induced emf per phase of a 10-pole, 3-ϕ, 50 Hz alternator with 2 slots per pole per phase and 4 conductors per slot in two layers. The coil span is 150°. The flux per pole has a fundamental component of 0.12 Wb and a 20 % third harmonic component.

Solution : Number of slots / pole / phase, $m = 2$

Number of slots / pole, $n = 2 \times 3 = 6$

Number of slots / phase $= 2 \times 10 = 20$

Number of conductors connected in series, $Z_p = 20 \times 4 = 80$

Number of series turns/phase, $T = \frac{Z_p}{2} = \frac{80}{2} = 40$

Angular displacement between adjacent slots, $\beta = \frac{180°}{n} = \frac{180°}{6} = 30°$

$$\text{Distribution factor, } K_d = \frac{\text{Sin } m\frac{\beta}{2}}{m \sin\frac{\beta}{2}} = \frac{\text{Sin}\frac{2\times 30°}{2}}{\text{Sin}\frac{30°}{2}} = 0.966$$

$$\text{Coil span factor, } K_p = \text{Cos}\frac{\alpha}{2} = \text{Cos}\frac{180° - 150°}{2} = \text{Cos } 15° = 0.966$$

Induced emf per phase (fundamental component),

$$E_1 = 4.44\, K_d\, K_p\, \phi f T$$

$$= 4.44 \times 0.966 \times 0.966 \times 0.12 \times 50 \times 40 = 994.4 \text{ V.}$$

For third harmonic component of flux

$$\text{Distribution factor, } K_{d_3} = \frac{\text{Sin}\frac{mr\beta}{2}}{m \sin\frac{r\beta}{2}} = \frac{\text{Sin}\frac{2\times 3\times 30°}{2}}{2 \sin\frac{3\times 30°}{2}} = 0.707$$

$$\text{Coil span factor, } K_{p_3} = \text{Cos } 3 \times \frac{180° - 150°}{2} = \text{Cos } 45° = 0.707$$

Frequency, $f_3 = 3 \times f = 3 \times 50 = 150$

Flux per pole, $\phi_3 = \frac{1}{3} \times 0.12 \times \frac{20}{100} = 0.008$ Wb

Induced emf per phase (third harmonic component)

$$E_3 = 4.44\, K_d\, K_p\, \phi_3 f_3 T$$

$$= 4.44 \times 0.707 \times 0.707 \times 0.008 \times 150 \times 40 = 106.56 \text{ V}$$

Induced emf per phase, $E_p = \sqrt{E_1^2 + E_3^2} = \sqrt{(994.4)^2 + (106.56)^2} = 1{,}000$ V **Ans.**

Example 5.8. The flux distribution curve of a smooth-core, 50 Hz generator is B = 10 sin θ + 2 sin 3θ + 2 sin 5θ + 2 sin 7 θ kilo-gauss when θ is measured from the neutral axis. The pole pitch is 35 cm, the core length 32 cm, and the stator coil span four-fifths of the pole pitch. Determine the equation for the emf induced in one turn, and its rms value.

Solution : Area of pole pitch $= 0.35 \times 0.32 = 0.112\ m^2$

Fundamental component of flux per pole, ϕ_1 = Average flux density × area

$$= \frac{2}{\pi} \times 10 \times 1{,}000 \times 10^{-4} \times 0.112 = 0.0713 \text{ Wb}$$

$\because$ Average value of flux density $= \frac{2}{\pi}$ peak value of flux density

and 1 kilo-gauss $= 10^3 \times 10^{-4}$ Wb/m^2

RMS value of fundamental voltage per conductor $= 1.11 \times 2f \times \phi_1 = 1.11 \times 2 \times 50 \times 0.0713 = 7.914$ V

Peak value of fundamental voltage per conductor $= \sqrt{2} \times 7.914 = 11.193$ V

Since harmonic conductor voltages are in proportion to their flux densities, so

Third harmonic voltage (maximum value) $= 11.193 \times \frac{2}{10} = 2.24$ V

Fifth harmonic voltage (maximum value) $= 11.193 \times \frac{2}{10} = 2.24$ V

Seventh harmonic voltage (maximum value) $= 11.193 \times \frac{2}{10} = 2.24$ V

Coil span $= \frac{4}{5} \times 180° = 144°$

So chording angle, $\alpha = 180° - 144° = 36°$

Pitch factor for fundamental component, $K_{p_1} = \text{Cos}\ \frac{\alpha}{2} = \text{Cos}\ \frac{36°}{2} = 0.951$

Pitch factor for third harmonic component, $K_{p_3} = \text{Cos}\ \frac{3\alpha}{2} = \text{Cos}\ \frac{3 \times 36°}{2} = 0.588$

Pitch factor for fifth harmonic component, $K_{p_5} = \frac{\text{Cos}\ 5\alpha}{2} = \frac{\text{Cos}\ 5 \times 36°}{2} = 0$

Pitch factor for seventh harmonic component, $K_{p_7} = \frac{\text{Cos}\ 7\alpha}{2} = \frac{\text{Cos} \times 7 \times 36°}{2} = -0.588$

Since peak values of induced emfs in one turn $= 2 \times K_{p_r} \times$ peak value of induced emf per conductor

So peak values of induced emfs in one turn

(*i*) for fundamental component, $E_{1max} = 2 \times 0.951 \times 11.193 = 21.3$ V

(*ii*) for third harmonic component, $E_{3\ max} = 2 \times 0.588 \times 2.24 = 2.64$ V

(*iii*) for 5th harmonic component, $E_{5\ max} = 2 \times 0 \times 2.24 =$ zero

(*iv*) for 7th harmonic component, $E_{7\ max} = 2 \times (-0.588) \times 2.24 = -2.64$ V

Hence the equation for the emf induced in one turn is given as

$$e = 21.3 \sin \theta + 2.64 \sin 3\theta - 2.64 \sin 7\theta \text{ volts } \textbf{Ans.}$$

$$\text{RMS value of emf induced in one turn, } E_{rms} = \sqrt{\left(\frac{21.3}{\sqrt{2}}\right)^2 + \left(\frac{2.64}{\sqrt{2}}\right)^2 + \left(\frac{2.64}{\sqrt{2}}\right)^2} = 15.29 \text{ V } \textbf{Ans.}$$

5.8. WAVE SHAPE

The flux distribution curve in the air gap of an alternator (Fig. 5.17) is not usually so rectangular as in a dc machine, but it is never perfectly sinusoidal unless the machine has a salient pole rotor with specially shaped pole shoes graduating the air gap (Fig. 5.6), or a cylindrical rotor with a sinusoidally distributed field winding. The induced emf wave per conductor is similar to flux wave since induced emf per conductor is directly proportional to flux density at constant speed. (e being equal to Blv volts, where l, the length of conductor is constant and at constant speed v is also constant).

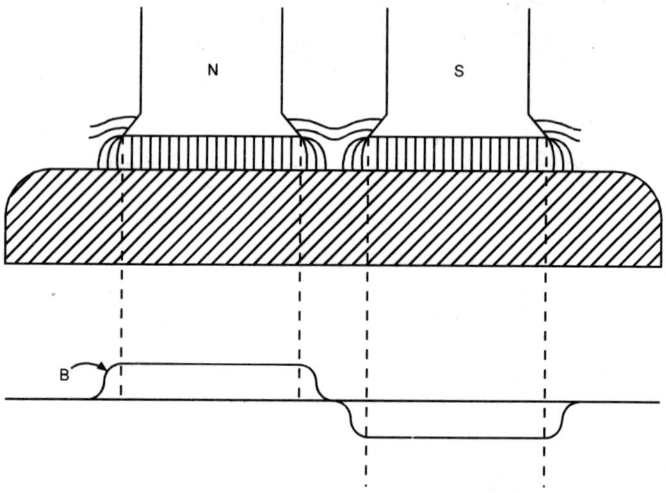

Fig. 5.17

If the winding is full-pitched one, the emfs, induced in two coil sides of each coil will be in phase and of the same magnitude because at any instant both the sides of the coil lie under corresponding positions of opposite poles. Hence for full-pitched winding, the induced emf wave in each coil will have the same shape as the emf induced in each coil side. If concentrated winding is employed, the resulting emf wave will be of the same shape as the flux density curve, shown in Fig. 5.17, which is flat topped.

In Fig. 5.18, a phase belt consisting of 3 coils of a 3 phase alternator having 3 slots per pole per phase is shown. The shape of emf wave for each of the 3 full-pitch coils giving one phase of the winding will be of the shape, as shown in Fig. 5.19 by curve I, II and III respectively. The displacement angle, β between the adjacent slots is $\frac{180^\circ}{9}$ = 20 electrical space degrees. Hence three emfs in the three successive coils are 20 electrical space degrees apart, as shown in Fig. 5.19. The resultant emf per phase would be the phasor sum of the induced emfs in the three coils, which may be obtained by adding the ordinates of the three emfs waves. The resultant emf per phase E_p wave is not flat topped but is almost sine wave. Hence the distributed winding is employed to give emf wave nearly sinusoidal. By using distributed winding, breadth factor for harmonics is very much less than for the fundamental as explained in Art. 5.6.2, hence harmonics are reduced in the resulant emf wave.

For a fractional pitch winding, the emf in each coil may be obtained first by adding the emfs of each coil side graphically and then resultant emf per phase may be obtained by adding the coil emfs, as shown in Fig. 5.19. The resultant emf wave for a fractional pitch winding will be more nearly sinusoidal than for full pitch winding. Pitch factor or coil span factor for harmonics is very much less than for the fundamental component, as already explained in Art. 5.6.1.

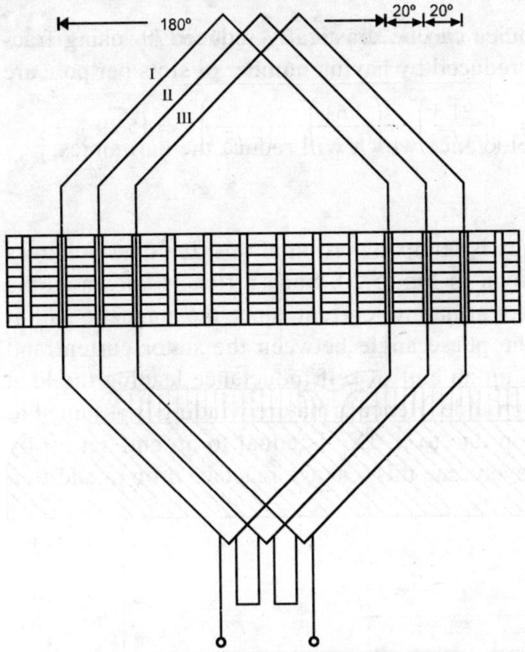

Fig. 5.18

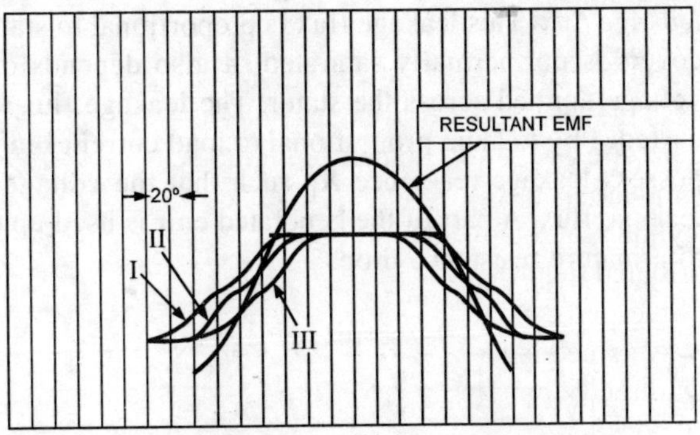

Fig. 5.19

5.9. HARMONICS IN VOLTAGE WAVEFORM

The sources of harmonics in the output voltage waveform are the (*i*) non-sinusoidal waveform of the field flux, as already explained in the foregoing article, and (*ii*) variation in reluctance of the air gap due to the slotting of the stator core.

5.9.1. Suppression of Harmonics

In case of salient pole machines if the air gap is made to vary sinusoidally around the machine, the field produced will be sinusoidal. However, it suffices to increase the air gap gradually from centre toward the pole tips, as shown in Fig. 5.6. An approximate of sinusoidal field form can be obtained by skewing the pole faces.

In cylindrical rotor machines, the length of air gap is uniform throughout, therefore, to have field of sinusoidal form, the mmf of the field winding is made to vary as nearly sinusoidal as possible. This is achieved by distributing the winding in slots.

It may be mentioned here that on account of saturation in iron parts, the sinusoidal field form cannot be obtained in salient pole machines even if the air-gap length is varied sinusoidally or in cylindrical rotor machines even if the mmf distribution in space is made sinusoidal. An ideal sinusoidal field form is very difficult to achieve, and therefore, harmonic emfs will be induced in the winding. The redeeming feature is that the harmonics can be easily eliminated from the output voltage waveform by properly designing the windings. The various methods for elimination of harmonics from the output voltage are:

(*i*) *Distributing* the armature winding as already explained in the foregoing Arts.

(*ii*) *Short-chording* the armature winding by making the coil span less than a full pole-pitch, as already explained in the foregoing Arts.

(*iii*) *Skewing the poles* usually through one slot-pitch.

(*iv*) *Fractional slot winding* : Higher order harmonics can be drastically reduced by using fractional slot winding. The flux pulsations can be reduced by having number of slots per pole arc as an integer plus 1/2.

(*v*) *Larger air-gap length* causes the increase in reluctance which will reduce the harmonics.

5.10. LEAKAGE REACTANCE

When current flows through the stator conductors, the flux is set up, a portion of this flux does not cross the air gap, but completes its path in the stator, as shown in Fig. 5.20. Such a flux is known as the *leakage flux.* This leakage flux is proportional to stator (or armature) current, since the magnetic path it covers is not normally saturated. It also depends on the phase angle between the stator current and voltage applied across the stator. The leakage flux sets up an emf of self inductance leading the load current I by $\pi/2$ and proportional to load current I in magnitude. Hence armature winding is assumed to possess leakage reactance X_L such that the voltage drop due to it, IX_L is equal to an emf set up by leakage flux. A part of the generated emf is used up to overcome this leakage reactace drop in addition to armature resistance drop.

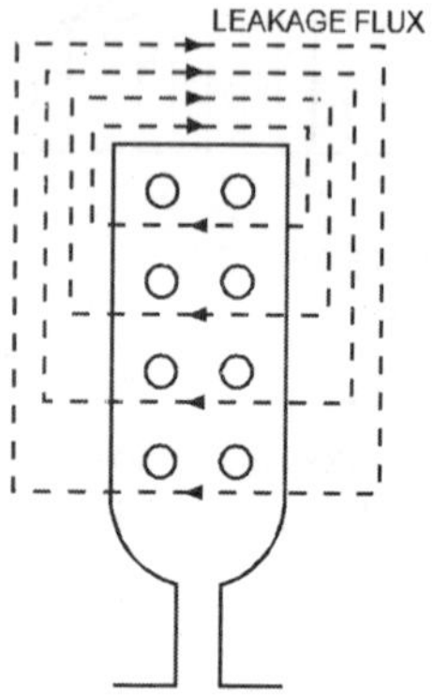

Fig. 5.20

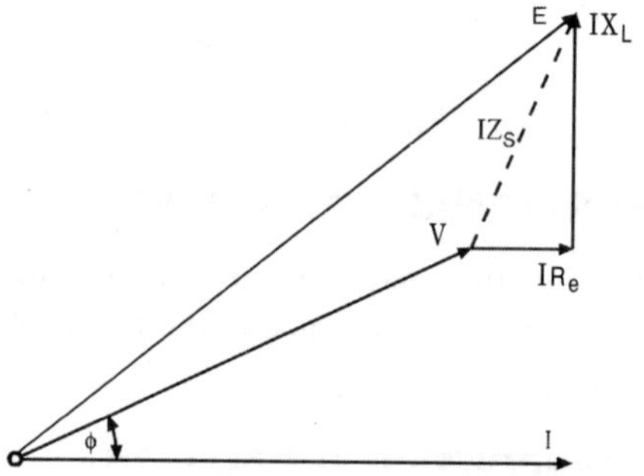

Fig. 5.21

i.e. Generated emf = Phasor sum of terminal voltage, armature resistance drop and leakage reactance drop.

$$i.e.\ \mathbf{E} = \mathbf{V} + \mathbf{I}R_e + j\,\mathbf{I}\,X_L = \mathbf{V} + \mathbf{I}\,(R_e + j\,X_L) \quad ...(5.9)$$

Phasor diagram illustrating this fact is shown in Fig. 5.21.

5.11. ARMATURE REACTION

In a dc machine, the armature mmf (ampere-turns) act on the magnetic circuit of the machine in such a way as to distort the air-gap flux and to change its magnitude. For a given armature current, the direction and magnitude of this armature reaction depend on the position of brushes, as already discussed in chapter 4 of Electrical Machines-I. In an alternator, somewhat similar conditions exist. For a given armature current, the magnitude and direction of the armature reaction cannot depend on the brush position, as in case of a dc machine, but depends on the phase displacement of the stator current I with respect to the emf E induced in the stator winding by the field winding flux.

In an alternator the phase displacement or phase angle ϕ between current I and the emf E can be within the limits of $\frac{-\pi}{2} \le \phi \le \frac{+\pi}{2}$.

We will consider three cases (*i*) when $\phi = 0$ *i.e.* when the power factor of load is unity (*ii*) when $\phi = \frac{+\pi}{2}$ *i.e.* when the power factor is zero and lagging and (*iii*) when $\phi = \frac{-\pi}{2}$ *i.e.* when the power factor of the load is zero and leading.

(*a*) At Unity Power Factor. The distribution of the stator currents and fluxes of an alternator with zero phase displacement between current I and emf E (*i.e.* when pf is unity), assuming the rotation of the poles to be clockwise, is shown in Fig. 5.22 (*a*). The crosses (+) and dots (•) appearing in the conductor cross sections indicate the instantaneous direction of the emfs induced and currents flowing—cross indicating the inward direction and dot indicating the outward direction. The maximum of the fundamental wave of the field will be opposite to the pole centres, and at the same points the conductors have their maximum induced emf; with $\phi = 0$ the conductors carrying the maximum current will also be at the same points, as shown in the figure. The armature reaction mmf is directed perpendicular to the main field mmf, as in case of a dc machine with brushes on the neutral axis. This causes the distortion of the flux due to main field and asymmetrical distribution of the flux density under the pole shoe. The flux density under the trailing pole tips increases somewhat, while under the leading pole tips it decreases. The axis of the resultant field is displaced under the action of armature reaction mmf in a direction opposite to that of rotation of the rotor.

Hence armature reaction at unity power factor has got distorting effect.

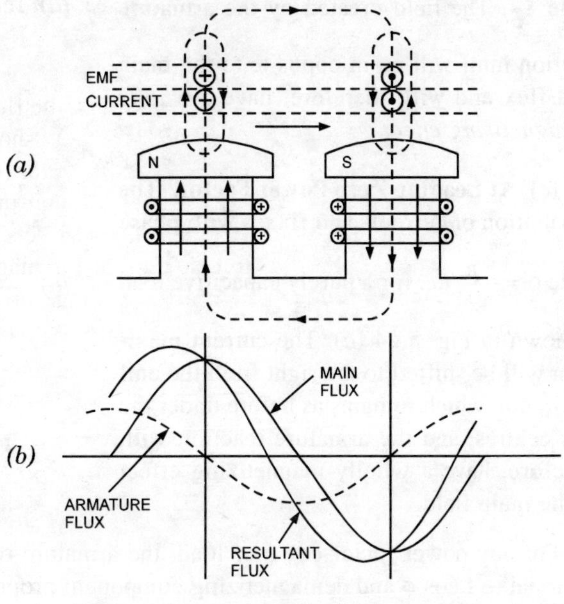

Fig. 5.22. ***Armature Reaction At Unity Power Factor***

(*b*) At Lagging Zero Power Factor. The distribution of currents and fluxes with phase angle $\phi = +\frac{\pi}{2}$ radians *i.e.*, for a purely inductive load is shown in Fig. 5.23 (*a*). The current maximum will be shifted in space by an angle $\frac{\pi}{2}$ from the emf maximum, which coincides with the centres of the poles. This shift will be opposite to the direction of rotation, because the fun-

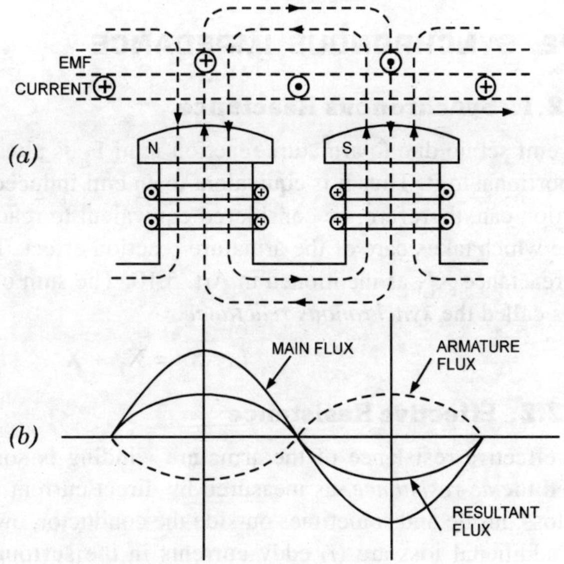

Fig. 5.23. ***Armature Reaction At Zero Power Factor (Lagging)***

damental armature reaction wave rotates in step with the field poles, while when $\phi = +\frac{\pi}{2}$, current wave lags behind the emf wave by an angle $\frac{\pi}{2}$. The field created by the armature reaction mmf will be in opposite to the main field flux and will, therefore, have a *wholly demagnetising effect.*

(c) At Leading Zero Power Factor. The distribution of currents and fluxes with phase angle $\phi = -\frac{\pi}{2}$ *i.e.* for a purely capacitive load is shown in Fig. 5.24 (*a*). The current maximum will be shifted to the right from the emf maximum, which remains as before under the pole centres, and the armature reaction will, therefore, have a wholly magnetizing effect on the main field.

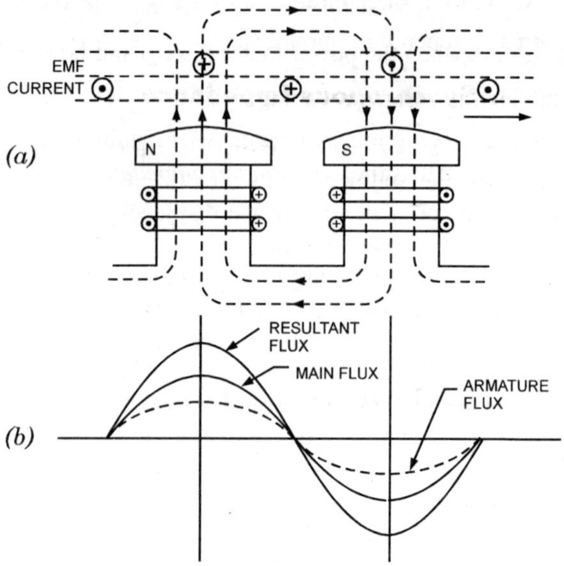

Fig. 5.24. *Armature Reaction At Zero Power Factor (Leading)*

For any power factor cos ϕ of load, the armature reaction has cross-magnetizing component proportional to I cos ϕ and demagnetizing component proportional to I sin ϕ, and is taken + ve for lagging power and – ve for leading power factor.

The phasor sum of the fundamental armature winding and field winding mmf waves constitutes, in a synchronous machine, the mmf creating the resultant magnetic flux.

5.12. SYNCHRONOUS IMPEDANCE

5.12.1. Synchronous Reactance

The emf set up due to armature reaction mmf F_a is always in quadrature with the load current I and is proportional to it. Thus it is equivalent to an emf induced in an inductive coil and the effect of armature reaction can, therefore, be considered equivalent to reactance drop IX_a where X_a is the fictitious reactance which takes care of the armature reaction effect. The armature winding possesses a certain leakage reactance X_L, as mentioned in Art. 5.10. The sum of leakage reactance X_L and fictitious reactance X_a is called the *synchronous reactance*.

$$i.e.\ X_s = X_L + X_a \quad ...(5.10)$$

5.12.2. Effective Resistance

The effective resistance of the armature winding is somewhat greater than the conductor resistance, called the *dc resistance,* as measured by direct current. This is due to additional loss, over the purely I^2R loss, inside and sometimes outside the conductor, owing to alternating current. The main sources of this additional loss are (*i*) eddy currents in the surrounding material; (*ii*) magnetic hysteresis in the surrounding material; and (*iii*) eddy currents or unequal current distribution in the conductor itself.

In many cases it is sufficiently accurate to measure the armature resistance by direct current and increase it to a fictitious value, called the *effective resistance*, R_e which is large enough to take care of these additional losses. Effective resistance, R_e can vary widely from 1.25 to 1.75 or more times the dc resistance, depending upon design.

5.12.3. Synchronous Impedance

When the synchronous reactance X_s is combined with the armature effective resistance R_e, the quantity obtained is called the *synchronous impedance.*

$$i.e.\ \mathbf{Z}_s = R_e + jX_s \quad ...(5.11)$$

Armature winding effective resistance R_e in alternators is usually very small in comparison to synchronous reactance X_s, and therefore, synchronous impedance Z_s may be assumed equal to the synchronous reactance for many purposes.

5.13. ALTERNATOR ON LOAD

If the excitation of an alternator is adjusted to give normal voltage at no load and then a load is applied, the terminal voltage of the alternator changes, even though the speed is kept constant. The voltage usually falls, but in certain circumstances it may rise even.

The change in terminal voltage of an alternator with the change in load supplied by it is due to the following three reasons :

(*i*) Voltage drop on account of armature effective resistance.

(*ii*) Voltage drop on account of armature leakage reactance.

(*iii*) Voltage drop on account of armature reaction.

5.13.1. Phasor Diagram

Let the effective armature resistance of an alternator be R_e, leakage reactance X_L and voltage drop due to armature reaction be equal to voltage drop on account of fictitious reactance X_a. All the values are referred to phase values.

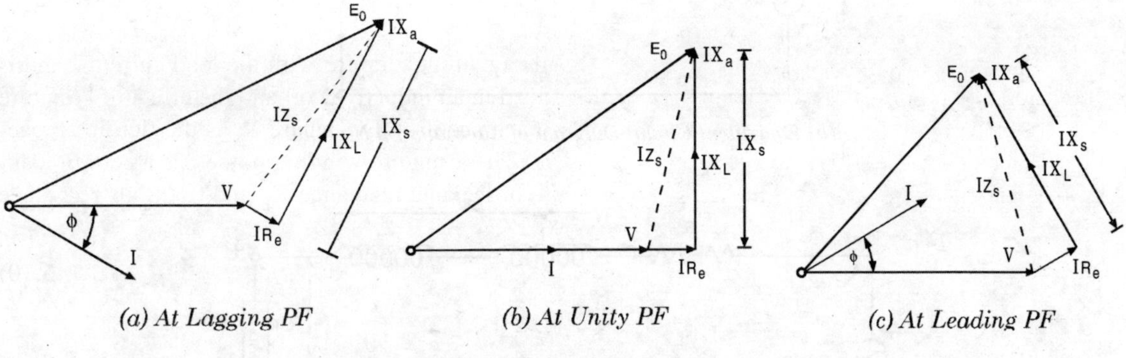

Fig. 5.25. ***Phasor Diagrams For a Loaded Alternator***

The phasor diagrams for inductive, purely resistive and capacitive loads are shown in Fig. 5.25 (*a*), 5.25 (*b*) and 5.25 (*c*) respectively. All these phasor diagrams apply to one phase of a 3-phase machine.

Let the terminal voltage per phase be V volts and load current per phase be I amperes and power factor cos ϕ. The voltage drop due to armature effective resistance, I R_e will be in phase with current I, the voltage drop due to leakage reactance, I X_L and voltage drop due to armature reaction I X_a will be in quadrature with the load current I. The terminal voltage V differs from the terminal voltage on no load, also called the *excitation voltage,* E_0 by the phasor sum of above voltage drops *i.e.* $\mathbf{IZ}_s$, where $\mathbf{Z}_s$ is the synchronous impedance.

$$\text{Total voltage drop on load} = \mathbf{I}\,R_e + j\,(\mathbf{I}\,X_L + \mathbf{I}\,X_a) = \mathbf{I}\,[R_e + j\,(X_L + X_a)] = \mathbf{I}\,\mathbf{Z}_s \quad ...(5.12)$$

$$\text{and voltage on no load } \mathbf{E}_0 = \mathbf{V}_0 + \mathbf{I}\,\mathbf{Z}_s \quad ...(5.13)$$

5.13.2. Equivalent Circuit (or Circuit Model)

When the alternator is operating on no load (*i.e.* the rotor is rotating and energized and stator is open circuited) its circuit diagram is shown in Fig. 5.26 (*a*). Each phase generates an emf E_0 per phase. The equivalent circuit of an alternator for one phase is shown in Fig. 5.26 (*b*). At no load, the terminal voltage V per phase is equal to excitation or generated emf E_0 per phase *i.e.* $V = E_0$.

The per phase equivalent circuit of an alternator on load may be represented as given in Fig. 5.26.

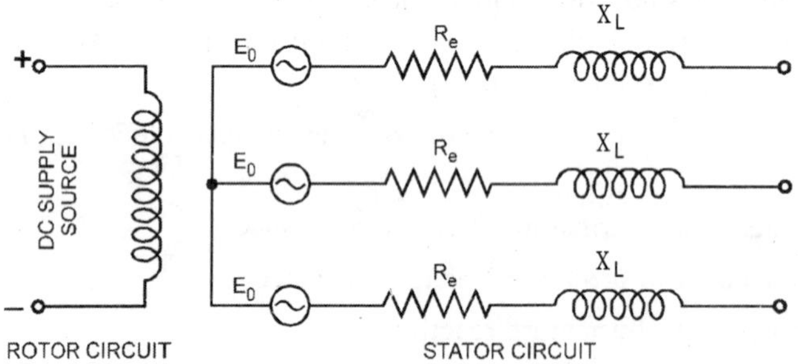

(a) Circuit Diagram

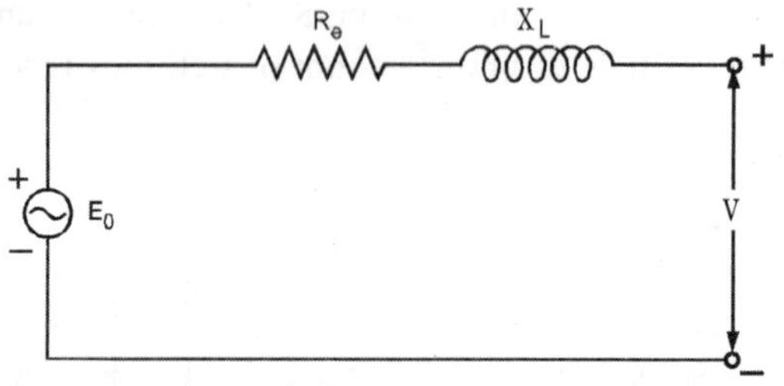

(b) Equivalent Circuit Diagram of Alternator on No Load

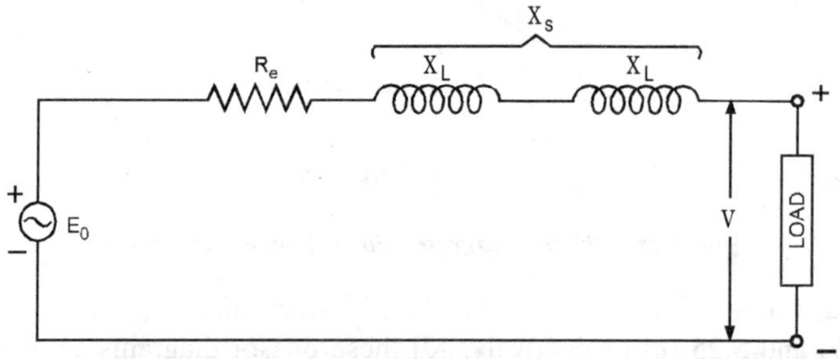

(c) Equivalent Circuit Diagram of Alternator on Load

Fig. 5.26

5.13.3. Load Characteristic

While the exciting current and the speed remain constant, the terminal voltage of an alternator changes with the change in the load or armature current, as already mentioned above. The relationship between the terminal voltage, V and load current, I of an alternator is known as its *load characteristic*. The curves showing variation of terminal voltage V with load current I for constant excitation (or no-load voltage E_0) for three different power factor loads are given in Fig. 5.27. The curves are plotted in terms of percentage values because general operating conditions can be illustrated better in this way.

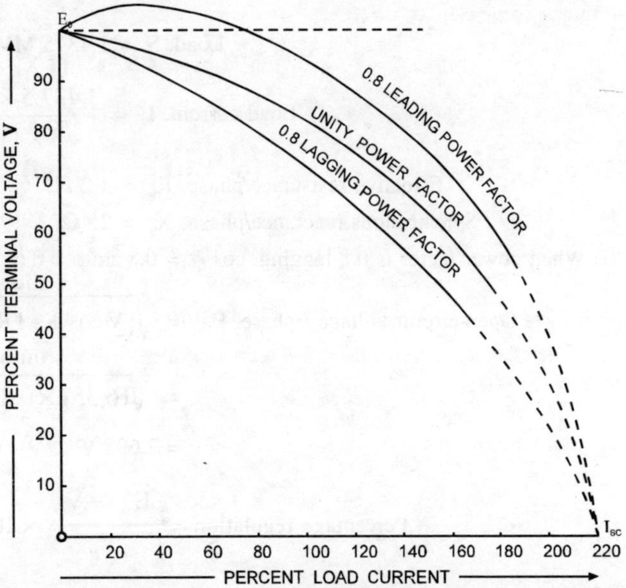

Fig. 5.27

Normally the terminal voltage falls with the increase in load current, but when the power factor is leading one, the load characteristic curve may rise at first. Each curve is nearly straight at the beginning but tends to droop because, with the increase in load current, the angle of lag between current and emf, owing to the original field, increases.

The highest current is obtained when the alternator terminals are short circuited, the value being given as $I_{sc} = \dfrac{E_0}{Z_s}$ amperes where E_0 is no-load terminal voltage and Z_s is the synchronous reactance. All curves meet at the short-circuit point.

In modern alternators, the steady short-circuit current is not much greater than full-load rated current. This is purposely arranged to prevent excessive current in the event of a short circuit.

The synchronous impedance can be determined from open-and short-circuit tests, as will be discussed in the following article (Art. 5.14).

5.13.4. Effect of Variation of Power Factor On Terminal Voltage (Load Current, Excitation and Speed Remaining Unchanged)

On changing the power factor from leading to lagging one, if the load current, excitation and speed are kept constant, the terminal voltage falls. Since when the power factor is leading (highly capacitive load) the effect of armature flux is to help the main flux, hence to generate more emf but when power factor is lagging (highly inductive load) the effect of armature flux is to oppose the main flux, hence to generate less emf. Thus at lagging power factor the terminal voltage falls from that on leading power factor mainly owing to decrease in generated emf.

Example 5.9. A 3-phase star-connected synchronous generator is rated at 1.5 MVA, 11 KV. The armature effective resistance and synchronous reactance are 1.2 Ω and 25 Ω respectively per phase. Calculate the percentage voltage regulation for a load of 1.4375 MVA at (*i*) 0.8 pf lagging and (*ii*) 0.8 pf leading. Also find out the pf at which the regulation becomes zero.

Solution : Phase voltage, $V_p = \dfrac{11{,}000}{\sqrt{3}} = 6{,}351$ V

Load, S = 1.4375 MVA = 1,437.5 kVA

Load current, $I = \dfrac{1{,}437.5 \times 1{,}000}{\sqrt{3} \times 11{,}000} = 75.45$ A

Effective resistance/phase, $R_e = 1.2\ \Omega$

Synchronous reactance/phase, $X_s = 25\ \Omega$

(*i*) When power factor is 0.8 lagging, $\cos\phi = 0.8 \sin\phi = 0.6$

Open-circuit voltage / phase, $E_{0p} = \sqrt{(V\cos\phi + IR_e)^2 + (V\sin\phi + IX_s)^2}$

$$= \sqrt{(6{,}351 \times 0.8 + 75.45 \times 1.2)^2 + (6{,}351 \times 0.6 + 75.45 \times 25)^2}$$

= 7,695 V

Percentage regulation $= \dfrac{E_{op} - V_p}{V_p} \times 100 = \dfrac{7{,}695 - 6{,}351}{6{,}351} \times 100 = 21.6\%$ **Ans.**

(*ii*) When power factor is 0.8 leading, $\cos\phi = 0.8 \sin\phi = -0.6$

Open-circuit voltage per phase, $E_{op} = \sqrt{(V\cos\phi + IR_e)^2 + (V\sin\phi + IX_s)^2}$

$$= \sqrt{6{,}351 \times 0.8 + 75.45 \times 1.2)^2 + [6{,}351 \times (-0.6) + 75.4 \times 25]^2}$$

= 5,518.5 V

Percentage regulation $= \dfrac{5{,}518.5 - 6{,}351}{6{,}351} \times 100 = -13.1\ \%$ **Ans.**

Open-circuit voltage / phase, $E_0 = V_p + I R_e \cos\phi + I X_s \sin\phi$

In phasor diagram shown in Fig. 5.28

$\phi = \text{Cos}^{-1}\ 0.8 = 36.87°$

$\alpha = \text{Tan}^{-1} \dfrac{IR_e}{IX_s} = \text{Tan}^{-1} \dfrac{1.2}{25} = 2.75°$

OC = E_0

OA = V

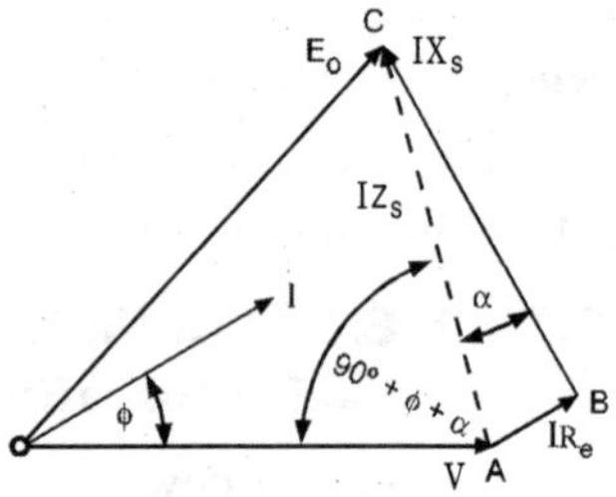

Fig. 5.28. ***Phasor Diagram***

$AC = \sqrt{IR_e^2 + (IX_s)^2}$

$$= \sqrt{(75.45 \times 1.2)^2 + (75.45 \times 25)^2} = 1{,}886.27 \text{ V}$$

Now $OC^2 = OA^2 + AC^2 + 2OA \times AC \times \sin(\phi + \alpha)$

or $\sin(\phi + \alpha) = \dfrac{-AC^2}{2 \times OA \times AC}$ ∵ For zero regulation E_0 = V or OC = OA

$$= \frac{-AC}{2 \times OA} = -\frac{1{,}886.27}{2 \times 6{,}351} = -0.1485$$

$$\text{or } \phi + \alpha = \text{Sin}^{-1}(-0.1485) = -8.54°$$

$$\text{or } \phi = -8.54 - \alpha = -8.54° - 2.75° = -11.29°$$

$$\text{Power factor} = \text{Cos } \phi = \text{Cos } (-11.29°) = 0.981 \text{ leading } \textbf{Ans.}$$

5.14. LABORATORY METHOD OF DETERMINATION OF SYNCHRONOUS REACTANCE (TESTING OF SYNCHRONOUS GENERATOR)

As already defined, the sum of leakage reactance X_L and fictitious reactance X_a, to account for the voltage drop due to armature reaction is known as synchronous reactance X_s and it is given as

$$X_s = \sqrt{Z_s^2 - R_e^2}$$

where Z_s is the synchronous impedance and R_e is the effective resistance.

Determination of Effective Resistance of Armature. Armature resistance per phase can be measured directly by voltmeter-ammeter method or by using Wheatstone bridge. The connection diagram for measuring armature resistance by ammeter-voltmeter method is shown in Fig. 5.29.

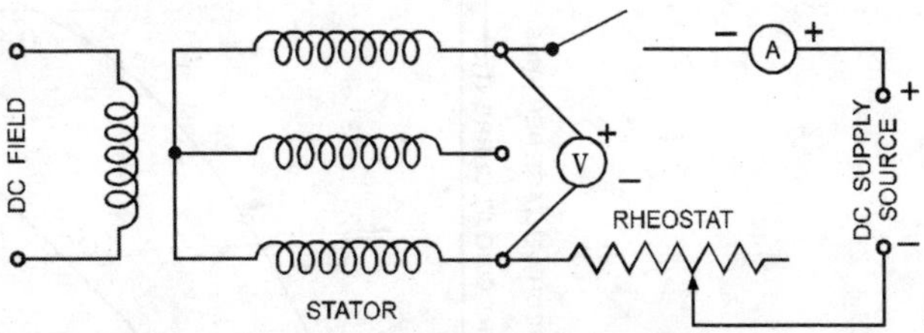

Fig. 5.29

DC resistance between pair of terminals, with the dc field winding open, is measured. The voltmeter reading divided by ammeter gives the value of resistance of two phases connected in series. So resistance of each phase, R_a will be equal to half the resistance measured. Since the resistance under measurement is very low so it will be necessary to use a high current rheostat to adjust the current to about rated value, a dc ammeter of proper range, and a low-range voltmeter. The alternator should be at rest. For operating conditions this measured value of armature resistance R_a is increased by about 60% or so to allow the skin effect and give effective value of armature resistance, R_e.

Synchronous impedance is determined from the open-circuit and short-circuit tests, as explained below :

1. Open-Circuit Test. The connections for the open-circuit test are shown in Fig. 5.30. With the armature- winding circuit open, the machine is run at rated speed. The field is connected to a dc source in series with a field rheostat and an ammeter so that the field current be adjusted and noted. The field current is raised in suitable steps until the voltage between any pair of terminals of the armature winding is somewhat above rated emf and corresponding values of voltmeter V indicating no-load line voltage are noted. The open-circuit voltages per phase, E_0 are obtained by dividing the voltmeter readings by $\sqrt{3}$. Now the curve is drawn between E_0 and field current I_f, as shown in Fig. 5.31. The curve so drawn is known as *open-circuit characteristic.* The initial straight part of the curve yields the air-gap characteristic since at that low excitation the iron is unsaturated and its reluctance is negligible.

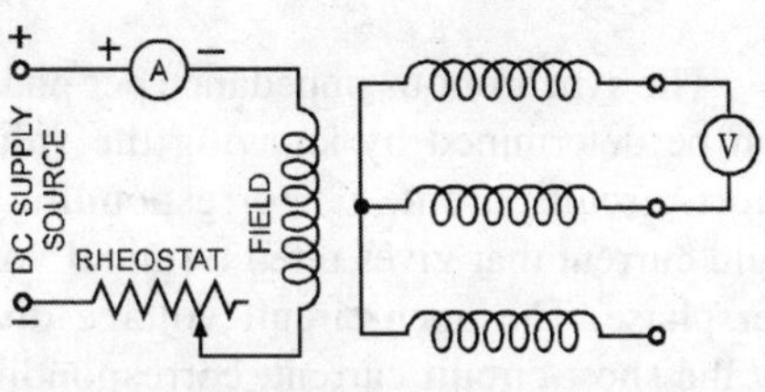

Fig. 5.30. ***Open-Circuit Test On a 3-Phase Alternator***

2. Short-Circuit Test. The connections for the short-circuit test are shown in Fig. 5.32. All the three phases are shorted, as shown in Fig. 5.32. Since the three currents will be equal, only one of them need be measured and so one ammeter is connected in any one of the three phases. Rheostat of sufficiently high ohmic value is inserted in the dc field circuit to keep the current in the circuit very low. The machine is run at synchronous speed, the field excitation is raised from zero to an amount sufficient to circulate 150 to 200 % of full-load current. The readings of the ammeters connected in the field circuit and armature circuit A_1 and A_2 are noted. The short-circuit characteristic is drawn by plotting a curve between short-circuit current I_{sc} and field current I_f, as shown in Fig. 5.31. The short-circuit characteristic is normally a straight line through the origin because the net excitation is so small that there is no saturation in the magnetic circuit. Further, the short-circuit characteristic is not affected due to slight deviation in speed from the rated.

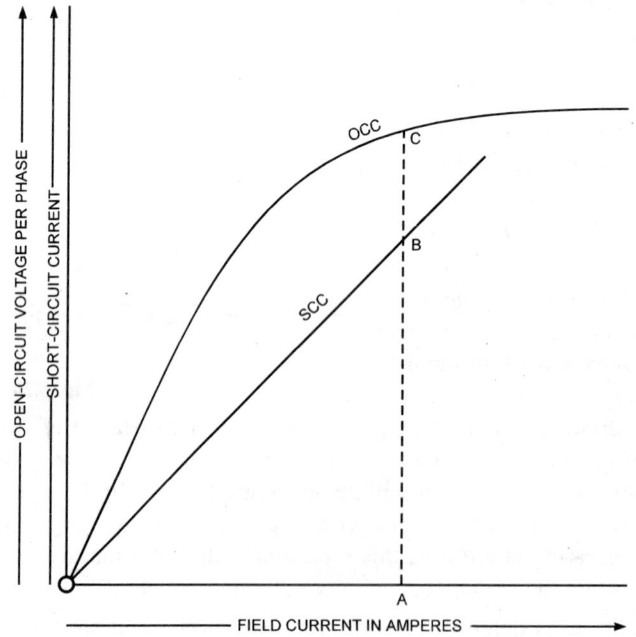

Fig. 5.31

The synchronous impedance per phase Z_s can be determined by knowing the value of short-circuit curent, I_{sc} corresponding to a field current that gives rated terminal voltage per phase. The open-circuit voltage divided by the short-circuit current, corresponding to field current which gives rated emf per phase, gives the value of synchronous impedance per phase Z_s.

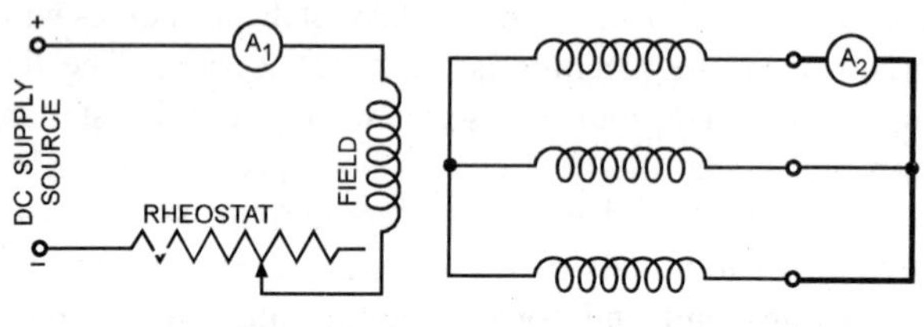

Fig. 5.32. ***Short-Circuit Test On a 3 Phase Alternator***

If OA is the field current that gives the rated emf per phase represented by AC and AB gives the short-circuit current I_{sc} corresponding to field current OA. Since on short circuit the terminal voltage is zero and whole of the induced emf is utilized to create a short circuit current I_{sc} through the synchronous impedance Z_s . That is

$$E_0 = I_{sc} Z_s$$

or synchronous impedance per phase, $Z_s = \dfrac{E_0}{I_{sc}} = \dfrac{\text{AC (in volts)}}{\text{AB (in amperes)}}$

and synchronous reactance, $X_s = \sqrt{Z_s^2 - R_e^2}$

Thus synchronous reactance can be determined in laboratory.

In general R_e is so much smaller than X_s that synchronous reactance X_s is practically numerically equal to synchronous impedance Z_s.

It may be seen from Fig. 5.31 that the synchronous reactance decreases with the increase in excitation. This is due to magnetic saturation effect. With the increase in field current, the armature current increases, and the components of flux created by these two currents both increase. As long as the resultant flux density is below the knee of the saturation curve, the flux produced per ampere of armature current is approximately constant, and therefore, the synchronous reactance X_s is constant. For higher flux densities, the flux produced per ampere decreases, and consequently the synchronous reactance decreases.

The fact that the synchronous reactance is not actually a constant but varies considerably with both saturation and power factor makes it difficult to obtain accurate results in the calculations, because there is no way of knowing what value of X_s is to be taken in calculations. For rough calculations, one may use the value of X_s corresponding to rated no-load voltage.

Example 5.10. The effective resistance of a 2200 V, 50 Hz, 440 kVA, 1-phase alternator is 0.5 Ω, on short circuit a field current of 40 A gives the full load current of 200 A. The emf on open circuit with the same excitation is 1,160 V. Calculate the synchronous impedance and reactance.

Solution : For a field current of 40 A

Short-circuit current, $I_{sc} = 200$ A

Open-circuit voltage, $E_0 = 1{,}160$ V

$$\text{Synchronous impedance, } Z_s = \frac{\text{OC voltage}}{\text{SC current}} = \frac{1{,}160}{200} = 5.8\ \Omega \textbf{ Ans.}$$

$$\text{Synchronous reactance, } X_s = \sqrt{Z_s^2 - R_e^2} = \sqrt{(5.8)^2 - (0.5)^2} = 5.778\ \Omega \textbf{ Ans.}$$

5.15. SHORT-CIRCUIT RATIO (SCR)

The short-circuit ratio (SCR) of a synchronous machine is defined as the ratio of field current to produce rated voltage on open circuit to field current required to circulate rated current on short circuit while the machine is mechanically driven at synchronous speed.

From OCC and SCC shown in Fig. 5.33, according to definition

$$\text{Short-circuit ratio, SCR} = \frac{I_{f1}}{I_{f2}} = \frac{OA}{OD} = \frac{AB}{DE} = \frac{AB}{AC} = \frac{1}{AC/AB}$$

$$\because \frac{AC}{AB} = \frac{\text{PU voltage on open-circuit}}{\text{Corresponding PU current on short-circuit}} = X_s$$

$$\text{So SCR} = \frac{1}{X_s} \quad \text{...(5.14)}$$

Thus SCR is reciprocal of per unit synchronous reactance X_s. The value of synchronous reactance depends upon saturated conditions of the machine but SCR is specific and defined at rated voltage.

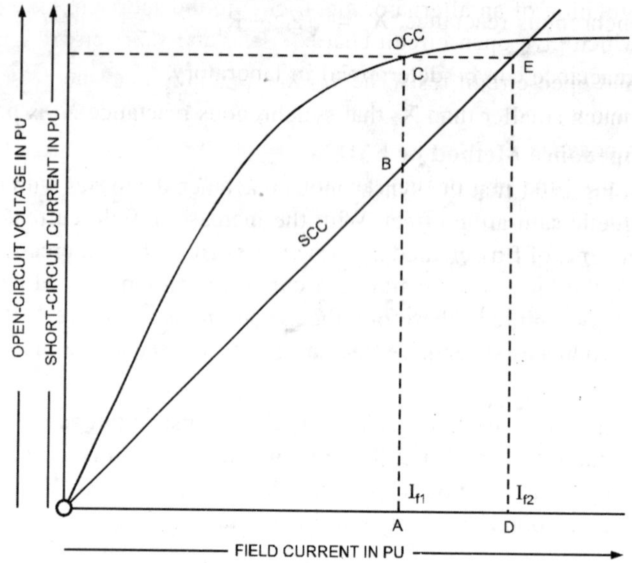

Fig. 5.33

A small value of SCR indicates a smaller value of current under short-circuit conditions owing to large value of synchronous reactance. The SCR of a high speed turbo-alternator usually lies between 0.5 and 0.75 while that of low speed salient pole generators lies between 1.0 and 1.5.

With high value of SCR, stability limit increased, voltage regulation improved and machines with a low value of SCR have difficulty during parallel operation owing to small value of synchronising power.

5.16. VOLTAGE REGULATION

From the previous discussions it is obvious that the terminal voltage of an alternator changes from no load to full load. The magnitude of this change in terminal voltage depends not only on load current but also on the power factor of the load.

Voltage regulation of an alternator is defined as the increase in terminal voltage, expressed as percentage (or per unit) of the rated terminal voltage, when the load at a given power factor is thrown off, with speed and field current remaining the same.

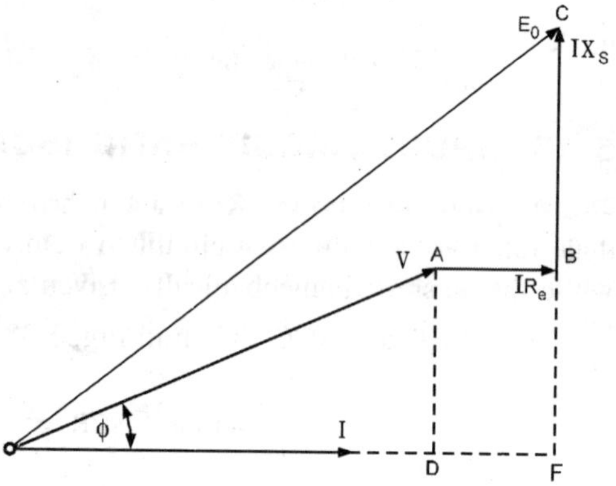

Fig. 5.34

i.e. voltage regulation $= \dfrac{E_0 - V}{V}$ per unit ...(5.15)

$= \dfrac{E_0 - V}{V} \times 100$ in percentage ...(5.16)

where E_0 is no-load terminal voltage and V is the full-load terminal voltage.

Note : In case of leading power factor E_0 is less than V, so voltage regulation will be negative.

Voltage rise in terminal voltage when the full load is thrown off is not the same as the fall in terminal voltage when the full load is applied.

To determine the regulation of an alternator, open-circuit and short-circuit tests are performed, as discussed in Art. 5.14, which give open-circuit characteristic and short-circuit characteristic.

From short-circuit and open-circuit tests, the voltage regulation can be determined by any one of the two methods given below.

1. Synchronous Impedance Method or EMF Method. After determining R_e and X_s from open-circuit and short-circuit tests, the phasor diagram for any load of any power factor can be drawn, as shown in Fig. 5.34.

Let load current be I lagging behind the terminal phase voltage V by ϕ

No-load terminal voltage per phase, $E_0 = OC$

Full-load terminal voltage, $V = OA$

Armature resistance drop, $I R_e = AB$

Synchronous reactance drop $I X_s = BC$

$$\text{Now } OC^2 = OF^2 + FC^2 = (OD + DF)^2 + (FB + BC)^2$$

$$\text{or } E_0^2 = (V \cos \phi + I R_e)^2 + (V \sin \phi + IX_s)^2$$

[Since $BF = AD = V \sin \phi$]

$$\text{or } E_0 = \sqrt{(V \cos \phi + I R_e)^2 + (V \sin \phi + I X_s)^2}$$

$$\text{And \% regulation} = \frac{E_0 - V}{V} \times 100$$

Notes. 1. For leading pf taking ϕ as negative and for unity pf taking $\phi = 0$ voltage regulation can be determined at leading pf and unity pf respectively.

2. Of various methods for determination of regulation, this one yields results most likely to be inexact. The regulation obtained by this method is always higher than actual value and, therefore, this is called the *pessimistic* method. The results are, however, more likely to be 'on safe side'. However, this method is theoretically accurate for non-salient pole machines with distributed field windings when saturation is not considered. The reason of error is that in this method synchronous impedance or reactance is assumed to remain constant while actually it is not. The value of synchronous impedance (or reactance) varies with the saturation. At low saturation its value is larger because the effect of armature reaction is greater than that at high saturation. Now under short-circuit conditions, saturation is very low and, therefore, the value of synchronous impedance measured is higher than that in actual working conditions.

2. *Ampere-Turn (or MMF) Method.* In this method, the datas determined from open-circuit tests are utilized.

From the above two tests the open-circuit and short-circuit characteristics are drawn, as shown in Fig. 5.31.

From the above characteristic field current I_{f1} is determined to give rated voltage V on no load, neglecting armature resistance drop, and field current I_{f2} is determined to cause short-circuit current, equal to full-load current, on short circuit.

On short circuit, the field excitation I_{f2} balances the impedance drop in addition to armature reaction on full load. But since R_e is usually very small and X_L is also small for low voltage on short circuit, so impedance drop can be neglected. Hence pf on short circuit is almost zero lagging and the field amp-turns are used entirely to overcome the armature reaction. Therefore, I_{f2} gives demagnetising amp-turns at full load.

Now let the alternator supply full load current at a pf of cos ϕ. Draw OA representing I_{f1} to give full load rated voltage, V [or more exactly $V + I R_e \cos\phi$] then draw AB at an angle (90° ± ϕ) representing I_{f2} to give full load current on short circuit; + ve sign for lagging pf and –ve sign for leading pf. Now find field current I_f, measuring OB, which will give open-circuit EMF E_0, which can be determined from OCC.

Then percentage regulation can be determined from the relation,

$$\% \text{ Regulation} = \frac{E_0 - V}{V} \times 100.$$

This method of determination of synchronous impedance is known as *optimistic method* since it gives values lower than actual values. The reason of it is that the excitation to overcome armature reaction is determined on unsaturated part of the saturation curve.

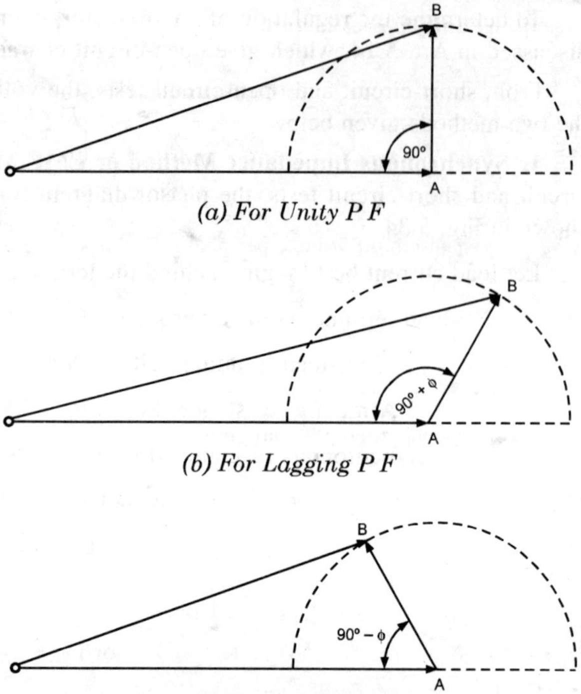

(a) *For Unity P F*

(b) *For Lagging P F*

(c) *For Leading P F*

Fig. 5.35

Example 5.11. A 3-phase star-connected alternator is rated at 1,600 kVA, 13.5 kV. The per phase armature effective resistance and synchronous reactance are 1.5 Ω and 30 Ω respectively. Calculate voltage regulation for a load of 1.280 MW at power factors of (*i*) 0.8 leading (*ii*) unity and (*iii*) 0.8 lagging.

Solution :

$$\text{Load, } P = 1.28 \text{ MW}$$

$$\text{Phase voltage, } V_P = \frac{V_L}{\sqrt{3}} = \frac{13.5 \times 1{,}000}{\sqrt{3}} = 7{,}794 \text{ V}$$

$$\text{Effective resistance per phase, } R_e = 1.5\ \Omega$$

$$\text{Synchronous reactance per phase, } X_s = 30\ \Omega$$

(*i*) At power factor 0.8 leading.

$$\text{Load current, } I = \frac{\text{Load in MW} \times 10^6}{\sqrt{3}\, V_L \cos\phi} = \frac{1.28 \times 10^6}{\sqrt{3} \times 13.5 \times 10^3 \times 0.8} = 68.4 \text{ A}$$

$$\text{Power factor, } \cos\phi = 0.8$$

$$\text{Sin } \phi = \sqrt{1 - 0.8^2} = -0.6 \qquad \text{minus sign for leading pf}$$

$$\text{Open-circuit voltage per phase, } E_{op} = \sqrt{(V_P \cos\phi + IR_e)^2 + (V_P \sin\phi + IX_s)^2}$$

$$= \sqrt{(7{,}794 \times 0.8 + 68.4 \times 1.5)^2 + [7{,}794 \times (-0.6) + 68.4 \times 30]^2}$$

$$= 6{,}860 \text{ V}$$

$$\text{Percentage regulation} = \frac{E_{op} - V_P}{V_P} \times 100 = \frac{6{,}860 - 7{,}794}{7{,}794} \times 100 = -11.98\ \%$$ **Ans.**

(*ii*) At unity power factor

$$\text{Load current, I} = \frac{1.28 \times 10^6}{\sqrt{3} \times 13.5 \times 10^3 \times 1.0} = 54.74 \text{ A}$$

$$\text{Cos } \phi = 1.0$$

$$\text{Sin } \phi = 0$$

$$\text{Open-circuit voltage per phase, } E_{op} = \sqrt{(7{,}794 \times 1.0 + 54.74 \times 1.5)^2 + (7{,}794 \times 0 + 54.74 \times 30)^2}$$

$$= \sqrt{(7{,}794 + 82.11)^2 + (0 + 1{,}642.2)^2} = 8{,}045.5 \text{ V}$$

$$\text{Percentage regulation} = \frac{8{,}045.5 - 7{,}794}{7{,}794} \times 100 = 3.227 \text{ \% } \textbf{Ans.}$$

(*iii*) At power factor 0.8 lagging

Load current, I = 68.4 A, same as in case (*i*)

$$\text{Cos } \phi = 0.8$$

$$\text{Sin } \phi = 0.6$$

$$\text{Open-circuit voltage per phase, } E_{op} = \sqrt{(7{,}794 \times 0.8 + 68.4 \times 1.5)^2 + (7{,}794 \times 0.6 + 68.4 \times 30)^2}$$

$$= 9{,}243 \text{ V}$$

$$\text{Percentage regulation} = \frac{9{,}243 - 7{,}794}{7{,}794} \times 100 = 18.6 \text{ \% } \textbf{Ans.}$$

Example 5.12. In 50 kVA, star-connected, 440 V, 3-phase, 50 Hz alternator, the effective armature resistance is 0.25 Ω per phase. The synchronous reactance is 3·2 Ω /phase and the leakage reactance is 0.5 Ω / phase. Determine at rated load and unity power factor

(*a*) internal emf E_a (*b*) no-load emf E_0 (*c*) percentage regulation on full load (*d*) value of synchronous reactance which replaces armature reaction.

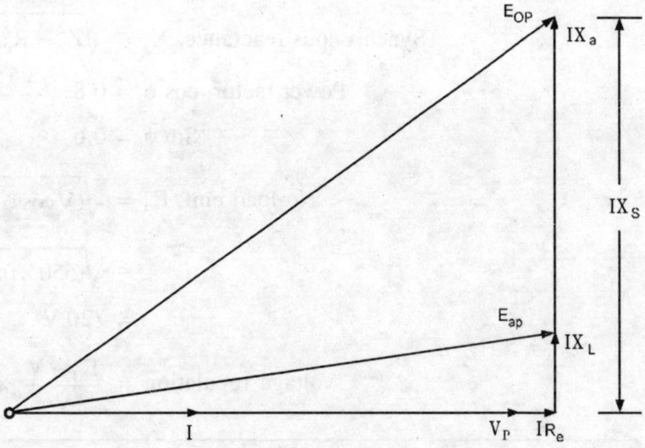

Fig. 5.36

Solution : Rated load = 50 kVA

= 50,000 volt-amperes

Line voltage, V_L = 440 V

$$\text{Phase voltage } V_p = \frac{440}{\sqrt{3}} = 254 \text{ V}$$

$$\text{Full load current, I} = \frac{50{,}000}{\sqrt{3} \times 440} = 65.6 \text{ A}$$

Effective armature resistance per phase, R_e = 0.25 Ω

Leakage reactance per phase, X_L = 0.5 Ω

Synchronous reactance per phase, X_s = 3.2 Ω

Power factor, cos ϕ = 1

From phasor diagram (Fig. 5.36)

$$\text{Internal emf per phase, } E_{ap} = \sqrt{(V_p + IR_e)^2 + (IX_L)^2}$$

$$= \sqrt{(254 + 65.6 \times 0.25)^2 + (65.6 \times 0.5)^2} = 272 \text{ V}$$

(*a*) Line value of internal emf, $E_a = \sqrt{3} \times 272 = 472$ V **Ans.**

$$\text{No-load emf per phase, } E_{op} = \sqrt{(V_p + IR_e)^2 + (IX_s)^2}$$

$$= \sqrt{(254 + 65.6 \times 0.25)^2 + (65.6 \times 3.2)^2} = 342 \text{ V}$$

(*b*) Line value of no-load emf, $E_{0L} = \sqrt{3} \times 342 = 593$ volts **Ans.**

(*c*) Percentage regulation $= \dfrac{593 - 440}{440} \times 100 = 34.6$ % **Ans.**

(*d*) Fictitious reactance/phase, $= X_a = X_s - X_L = 3.2 - 0.5 = 2.7\ \Omega$ **Ans.**

Example 5.13. A 550 V, 55 kVA, single-phase synchronous generator has an effective resistance of 0.2 Ω. A field current of 10 amps produced an armature current of 200 A on short circuit and an emf of 450 V on open circuit. Calculate synchronous impedance and voltage regulation at full load with pf 0.8 (lagging).

Solution : Effective armature resistance, $R_e = 0.2\ \Omega$

$$\text{Synchronous impedance, } Z_s = \frac{\text{OC voltage}}{\text{SC current}} = \frac{450}{200} = 2.25\ \Omega \textbf{ Ans.}$$

Terminal voltage, V = 550 V

$$\text{Full-load load current, } I = \frac{\text{Rated kVA} \times 1{,}000}{V} = \frac{55 \times 1{,}000}{550} = 100 \text{ A}$$

$$\text{Synchronous reactance, } X_s = \sqrt{Z_s^2 - R_e^2} = \sqrt{(2.25)^2 - (0.2)^2} = 2.24\ \Omega$$

Power factor, $\cos\phi = 0.8$

$\sin\phi = 0.6$

$$\text{No-load emf, } E_0 = \sqrt{(V\cos\phi + IR_e)^2 + (V\sin\phi + IX)^2}$$

$$= \sqrt{(550 \times 0.8 + 100 \times 0.2)^2 + (550 \times 0.6 + 100 \times 2.24)^2}$$

$$= 720 \text{ V}$$

$$\text{Voltage regulation} = \frac{E_0 - V}{V} \times 100 = \frac{720 - 550}{550} \times 100 = 30.91\ \% \textbf{ Ans.}$$

Example 5.14. The data obtained on 100 kVA, 1100 V, 3-phase alternator is :

DC resistance test, E between line = 6 V dc, I in lines = 10 A dc

Open-circuit test, Field current = 12.5 A dc, Line voltage= 420 V ac

Short-circuit test, Field current = 12.5 A, Line current = rated value. Calculate the voltage regulation of alternator at 0.8 pf lagging.

Solution : Assume alternator to be star-connected, as usually.

$$\text{Phase voltage, } V_p = \frac{1{,}100}{\sqrt{3}} = 635.1 \textbf{ A}$$

$$\text{Full-load phase current, } I_p = I_L = \frac{100 \times 1{,}000}{\sqrt{3} \times 1{,}100} = 52.5 \text{ A}$$

$$\text{Armature dc resistance per phase, } R_{dc} = \frac{E_{dc}}{2 \times I_{dc}} = \frac{6}{2 \times 10} = 0.3\ \Omega$$

$\because$ dc voltage is connected across two phases

$$\text{Armature effective resistance per phase, } R_e = 1.667 \times 0.3 = 0.5\ \Omega$$

assuming 66.7 % of dc resistance for skin effect etc.

$$\text{Synchronous impedance per phase, } Z_s = \frac{\text{OC voltage per phase}}{\text{SC current per phase}} \text{ for same excitation}$$

$$= \frac{420/\sqrt{3}}{52.5} = 4.62\ \Omega$$

$$\text{Synchronous reactance per phase, } X_s = \sqrt{(4.62)^2 - (0.5)^2} = 4.59\ \Omega$$

$$\text{At 0.8 lagging power factor, } \cos\phi = 0.8 \text{ and } \sin\phi = 0.6$$

$$\text{Open-circuit voltage per phase, } E_{op} = \sqrt{\left(V_p \cos\phi + I_p R_e\right)^2 + \left(V_p \sin\phi + I_p X_s\right)^2}$$

$$= \sqrt{(635.1 \times 0.8 + 52.5 \times 0.5)^2 + (635.1 \times 0.6 + 52.5 \times 4.59)^2}$$

$$= 820 \text{ V}$$

$$\text{Percentage regulation} = \frac{820 - 635.1}{635.1} \times 100 = 29.11\ \% \text{ Ans.}$$

Example 5.15. (*a*) A 3-phase, star-connected, 5 kVA, 400 V, 50 Hz, 4-pole alternator has the following test data at rated speed :

(*i*) Exciting current (in amperes)	**0.5**	**1.0**	**1.5**	**2.0**	**2.5**	**3.0**	**3.5**	**4.0**	**4.5**	**5.0**	**6.0**	**8.0**
Per phase O C volts	**75**	**140**	**173**	**202**	**224**	**240**	**250**	**257**	**260**	**263**	**266**	**271**

(*ii*) Exciting current (amperes)	**1.0**	**2.0**	**3.0**
SC line current	**3.6**	**7.2**	**10.8**

(*iii*) Armature resistance per phase 2.0 Ω

Draw OC and SC characteristics on a squared graph paper and then determine unsaturated value of synchronous reactance per phase in ohms and in per unit.

(*b*) For the synchronous machine determine per cent voltage regulation at rated load at 0.8 pf lag and lead by synchronous impedance method under unsaturated condition. Draw relevant phasor diagrams.

Solution : OCC and SCC of the machine are shown in Fig. 5.37 alongwith the appropriate scales.

Unsaturated value of the synchronous reactance is the value of the reactance which is obtained for a linear portion of OCC.

For a field current of 1 A

Open-circuit line emf = 140 V

$$\text{Short-circuit current, } I_{sc} = 3.6 \text{ A}$$

$$\text{Synchronous impedance per phase, } Z_s = \frac{\text{OC voltage per phase}}{\text{SC phase current}} = \frac{140}{3.6} = 38.88\ \Omega$$

$$\text{Armature resistance per phase, } R_e = 2.0\ \Omega$$

$$\text{Synchronous reactance per phase, } X_s = \sqrt{(Z_s)^2 - R_e^2} = \sqrt{(38.88)^2 - (2)^2} = 38.84\ \Omega \textbf{ Ans.}$$

$$\text{Rated phase voltage, } V_p = \frac{V_L}{\sqrt{3}} = \frac{400}{\sqrt{3}} = 231 \text{ V}$$

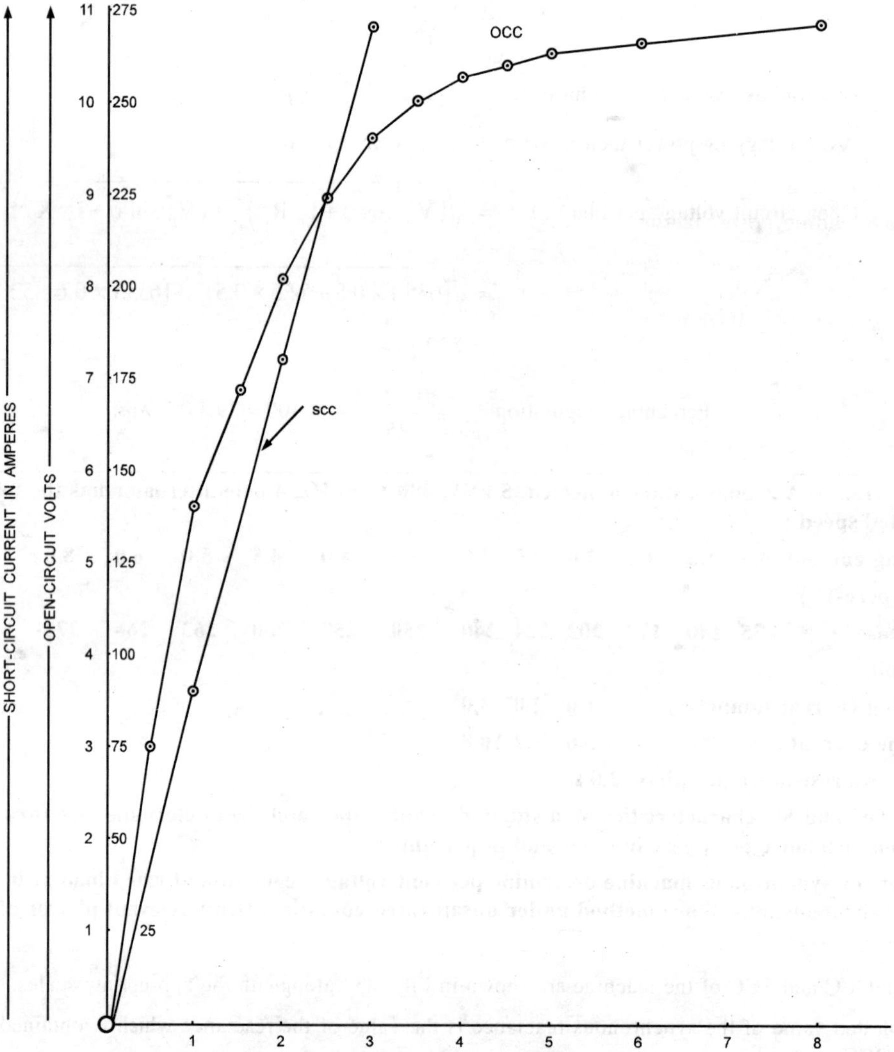

Fig. 5.37

$$\text{Phase current, I = Line current, } I_L = \frac{5 \times 1{,}000}{\sqrt{3} \times 400} = 7.217 \text{ A}$$

$$\text{Per unit value of synchronous reactance, } X_s = \frac{I X_s}{V_p} = \frac{7.217 \times 38.84}{231} = 1.213 \text{ pu } \textbf{Ans.}$$

For 0.8 lagging power factor

$$\cos \phi = 0.8 \text{ and } \sin \phi = 0.6$$

$$\text{Open-circuit voltage per phase, } E_{op} = \sqrt{\left(V_p \cos \phi + I R_e\right)^2 + \left(V_p \sin \phi + I X_s\right)^2}$$

$$= \sqrt{(231 \times 0.8 + 7.217 \times 2)^2 + (231 \times 0.6 + 7.217 \times 38.84)^2}$$

$$= 463.88 \text{ V}$$

$$\text{Percentage regulation} = \frac{E_{op} - V_p}{V_p} \times 100 = \frac{463.88 - 231}{231} \times 100 = 100.81\% \textbf{ Ans.}$$

For 0.8 leading power factor

$$\cos \phi = 0.8 \text{ and } \sin \phi = -0.6$$

$$\text{Open-circuit voltage per phase } E_{op} = \sqrt{(231 \times 0.8 + 7.217 \times 2)^2 + [231 \times (-0.6) + 7.217 \times 38.84]^2}$$

$$= 244.5 \text{ V}$$

$$\text{Voltage regulation} = \frac{E_{op} - V}{V} \times 100 = \frac{244.5 - 231}{231} \times 100 = 5.844 \% \textbf{ Ans.}$$

Phasor diagrams for 0.8 power factor lagging and leading are shown in Fig. 5.38 (*a*) and 5.38 (*b*) respectively.

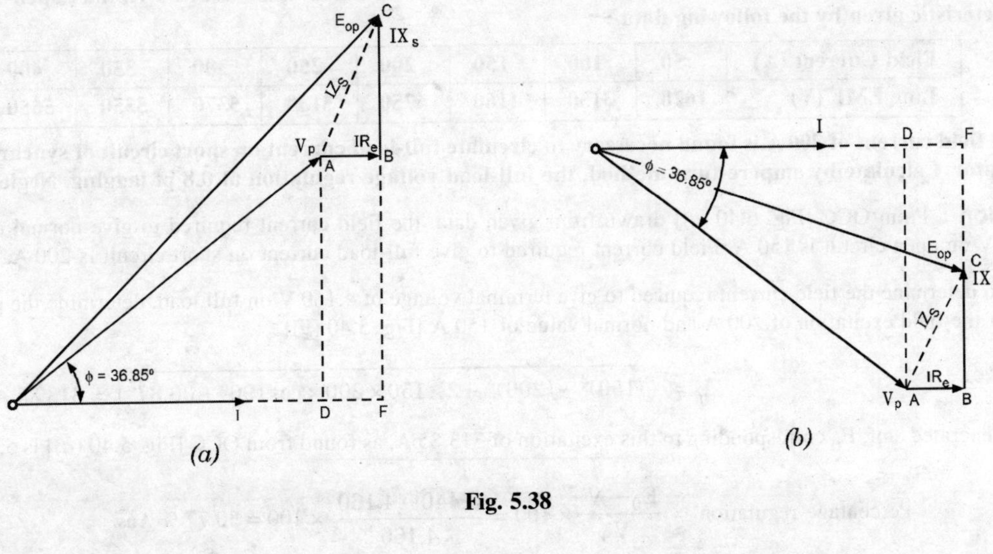

Fig. 5.38

Example 5.16. The no-load excitation of an alternator required to give rated voltage is 100 A. In a short-circuit test with full-load current flowing in the armature, the field excitation was 80 A. Determine the approximate excitation that will be required to give full load current at 0.8 pf lag and at the rated terminal voltage.

Solution : Field current required to give full-load rated voltage on open circuit,

$$I_{f1} = 100 \text{ A}$$

Field current required to give full-load current on short circuit in armature,

$$I_{f2} = 80 \text{ A}$$

To find the field current, required to give full-load current at 0.8 pf lag and at the rated terminal voltage draw OA equal to 100 A, then draw AB equal to 80 A making an angle (90° + ϕ) with OA where $\phi = \cos^{-1} 0.8$. The required field current will be OB, the phasor sum of OA and AB.

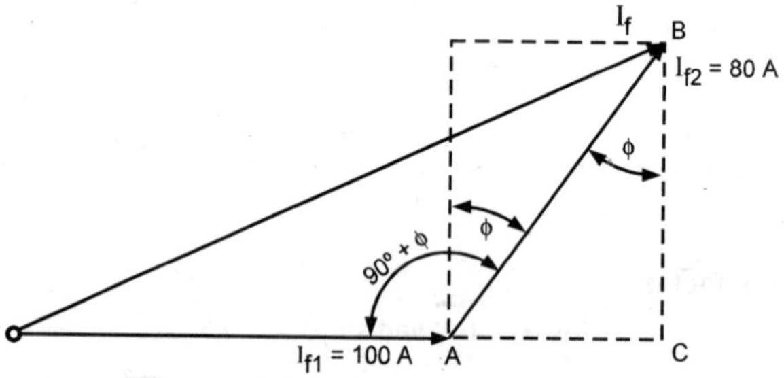

Fig. 5.39

The required field current I_f to give full-load current at 0.8 pf lag and at the rated terminal voltage

$$= OB = \sqrt{(OA + AC)^2 + (CB)^2} = \sqrt{(OA + AB \sin \phi)^2 + (AB \cos \phi)^2}$$

$$= \sqrt{(100 + 80 \times 0.6)^2 + (80 \times 0.8)^2} = 161.25 \text{ A } \textbf{Ans.}$$

Example 5.17. A 3.5 MVA Y-connected synchronous generator rated 4,160 V at 50 Hz has open-circuit characteristic given by the following data :

Field Current (A)	50	100	150	200	250	300	350	400
Line EMF (V)	1620	3150	4160	4750	5130	5370	5550	5650

A field current of 200 A is found necessary to circulate full-load current on short circuit of synchronous generator. Calculate by ampere-turn method, the full-load voltage regulation at 0.8 pf lagging. Neglect R_α.

Solution : From OCC [Fig. 5.40 (*a*)] drawn from given data, the field current required to give normal emf of 4,160 V on open circuit is 150 A. Field current required to give full-load current on short circuit is 200 A.

To determine the field current required to give terminal voltage of 4,160 V on full load, determine the phasor sum of the field excitation of 200 A and normal value of 150 A [Fig. 5.40 (*b*)]

$$I_f = \sqrt{(150)^2 + (200)^2 - 2 \times 150 \times 200 \times \cos(90° + 36.87°)} = 313.85 \text{ A}$$

Generated emf, E_0 corresponding to this excitation of 313.85 A, as found from OCC [Fig. 5.40 (*a*)] is 5,440 V

$$\text{Percentage regulation} = \frac{E_0 - V}{V} \times 100 = \frac{5,440 - 4,160}{4,160} \times 100 = 30.77 \% \textbf{ Ans.}$$

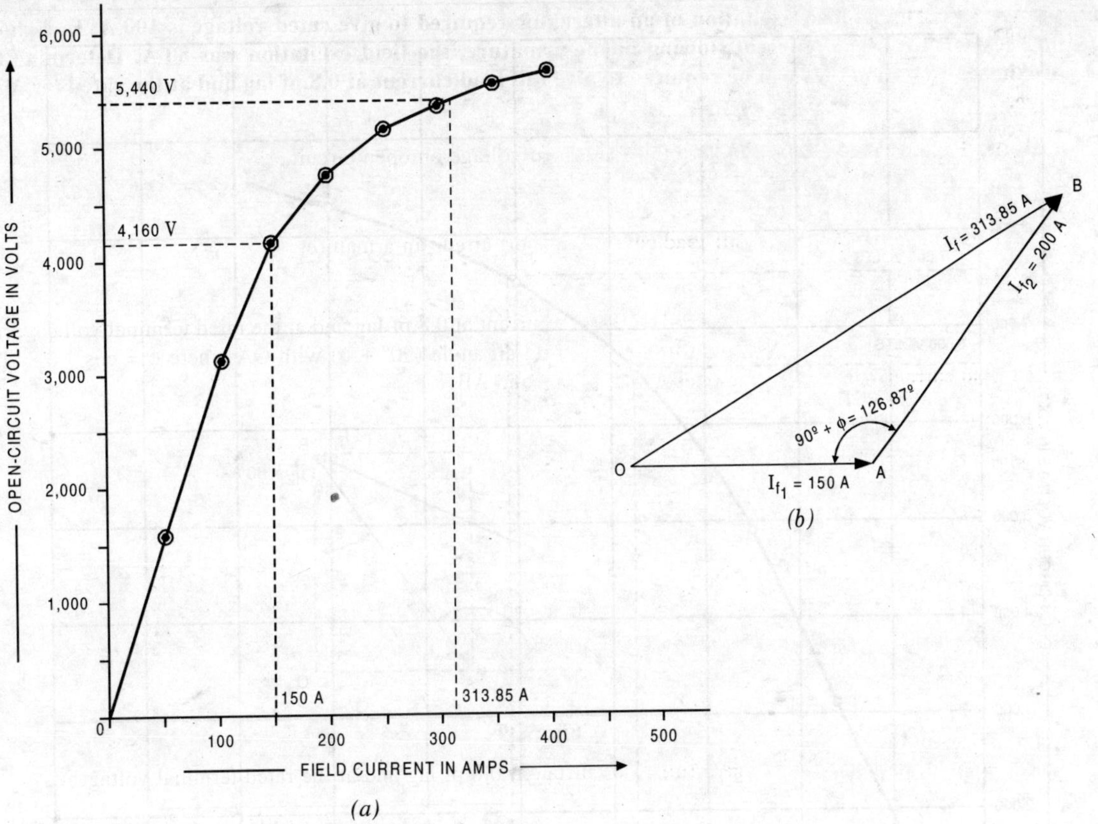

Fig. 5.40

Example 5.18. The following test results are obtained on a 6,600 V alternator.

Open-circuit voltage	**3,100**	**4,900**	**6,600**	**7,500**	**8,300**
Field current amperes	**16**	**25**	**37.5**	**50**	**70**

A field current of 20 A is found necessary to circulate full-load current on short circuit of the armature. Calculate by *(i)* the ampere-turn method and *(ii)* the synchronous impedance method the full-load regulation at 0.8 pf (lag). Neglect resistance and leakage reactance. State the drawbacks of each of these methods.

Solution : *(i)* Ampere-turn Method. From OCC (Fig. 5.41) drawn from given data the field current required to give normal emf of 6,600 volts on open circuit is 37.5 A.

Field current required to give full-load current on short circuit is 20 A.

To determine the field current required to give terminal pd of 6,600 V on full load, determine the phasor sum of the field excitation of 20 A and normal value of 37.5 A (Fig. 5.42)

$$I_f = \sqrt{(37.5)^2 + (20)^2 - 2 \times 37.5 \times 20 \times (-0.6)} = 52 \text{ A}$$

The generated emf E_0, corresponding to this excitation of 52 A, as found from OCC, (Fig. 5.42) is 7,600 V.

$$\text{Percentage regulation} = \frac{E_0 - V}{V} \times 100$$

$$= \frac{7,600 - 6,600}{6,600} \times 100 = 15.15\ \%\ \textbf{Ans.}$$

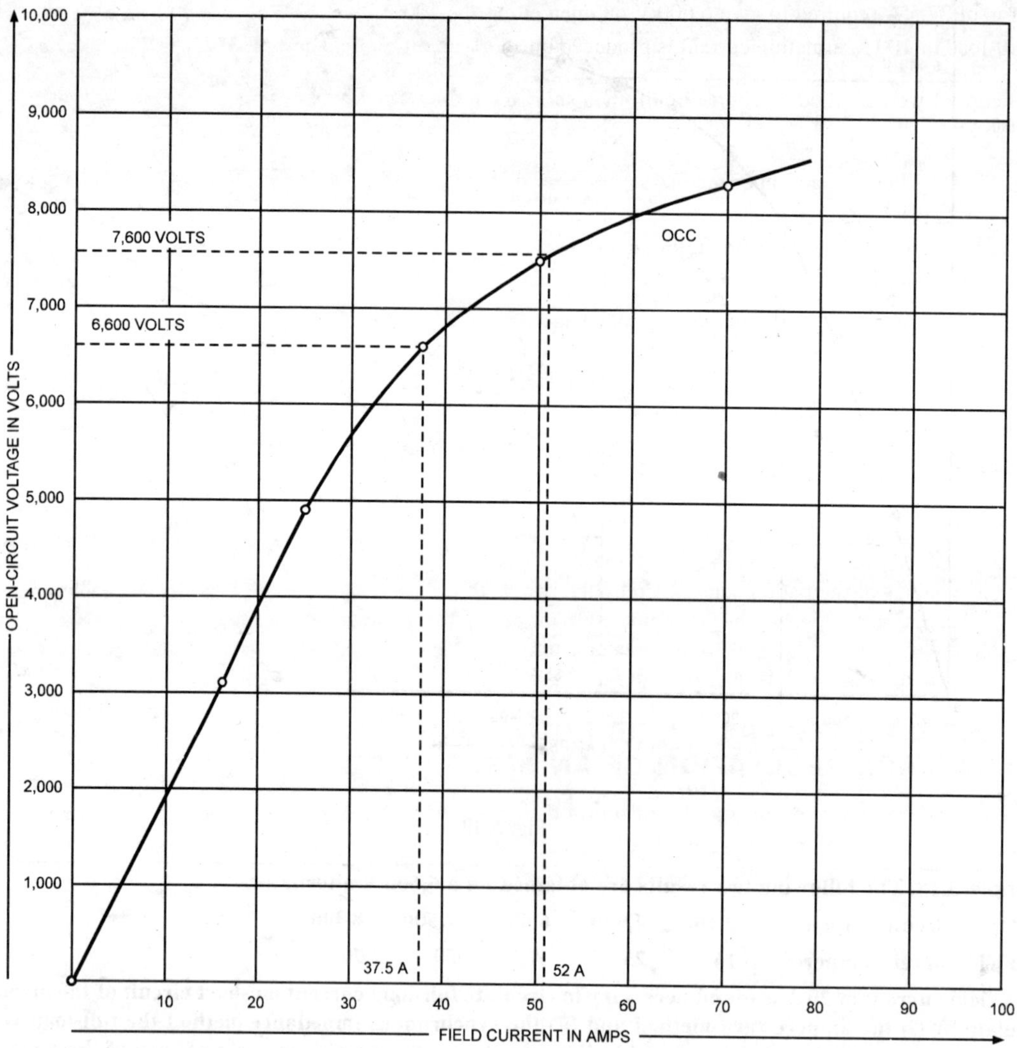

Fig. 5.41

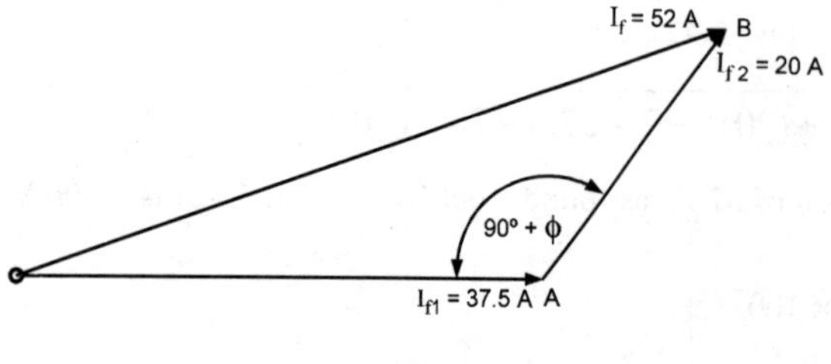

Fig. 5.42

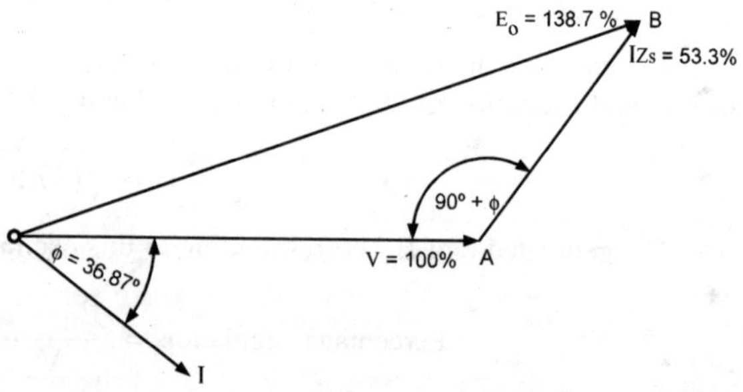

Fig. 5.43

***(ii)* Synchronous Impedance Method.** Let the voltage of 6,600 V be taken as 100 per cent and also let the excitation of 37.5 A required to give 6,600 V on open circuit be 100%.

Full-load or 100 % armature current is produced on short circuit by field current of 20 A. If 100 % *i.e.* 37.5 A of field current were applied on short circuit, then short-circuit current would be $100 \times \frac{37.5}{20} = 187.5$ %.

$$\text{Percentage impedance, } Z_s = \frac{\text{OC voltage}}{\text{SC current}} \text{ for given excitation} = \frac{100}{187.5} \times 100 = 53.3\ \%$$

Since armature resistance is negligible, therefore, impedance drop I Z_s = 53.3 % of the normal voltage will lead the load current by 90° or $\angle OAB = 90° + \phi$

The phasor sum of normal voltage and impedance drop will give open-circuit voltage, E_0

$$i.e.\ E_0 = \sqrt{(100)^2 + (53.3)^2 + 2 \times 100 \times 53.3 \times (90° + \phi)}$$

$$= \sqrt{(100)^2 + (53.3)^2 - 2 \times 100 \times 53.3 \cos(-0.6)} = 138.7\ \%$$

$$\text{Percentage regulation} = \frac{138.7 - 100}{100} \times 100 = 38.7\ \%\ \textbf{Ans.}$$

The two values of regulation determined by the two methods are found to differ widely from each other. The first method gives a little lower value as compared to the actual value while the other gives somewhat higher value. However, the first value is more likely to be nearer to the actual value because the second method employs synchronous impedance Z_s which is not constant but depends upon field excitation.

5.17. ZERO POWER FACTOR OR POTIER METHOD OF DETERMINATION OF VOLTAGE REGULATION OF AN ALTERNATOR

The regulation obtained by mmf and emf methods is based on the total synchronous reactance (the sum of reactance due to armature leakage flux and due to armature reaction effect). This method is based on the separation of reactances due to leakage flux and that due to armature reaction flux, therefore, it is more accurate.

For determination of voltage regulation by this method, the data required are *(i)* effective resistance of armature winding *(ii)* open-circuit characteristic *(iii)* field current to circulate full-load current in the stator. and *(iv)* zero power factor full-load voltage characteristic—*a curve between terminal voltage and excitation while the machine is being run on synchronous speed and delivering full-load current at zero power factor.*

The first three have already been discussed in Art. 5.14. The zero power factor characteristic curve can be obtained as follows.

The machine is run at rated synchronous speed by a prime mover. A purely inductive load (variable load reactors) is connected across the armature terminals and the excitation or field current is raised so as to cause flow of full-load armature current. The value of the reactance is then increased step by step in such a way that the excitation current is adjusted to a value that causes full-load rated

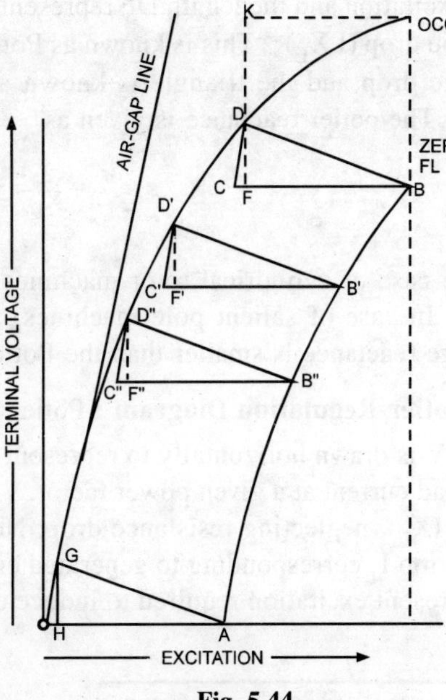

Fig. 5.44

armature current to flow. The armature terminal voltages are varied from 125 % to 25 % of the rated voltage in steps, maintaining the speed and rated armature current constant throughout the test. The armature terminal voltages and excitation currents are noted at each step. The curve drawn between terminal voltage and excitation current, as shown in Fig. 5.44, gives the *zero power factor (lagging) characteristic.*

When the zero power factor (lagging) characteristic is to be used only for obtaining the Potier reactance, it is sufficient to determine the point representing rated armature current and rated voltage. This is indicated by point B in Fig. 5.44.

The alternator can be loaded alternatively by an under-excited synchronous motor.

There is a definite relationship between the zero power factor (lagging) characteristic and an open-circuit characteristic of an alternator. The zero power factor characteristic curve is of exactly the same shape, as the OCC but it is shifted vertically downward by leakage reactance drop $I X_L$ and horizontally by the armature reaction mmf.

Zero power factor full-load voltage-excitation characteristic can be drawn by knowing two points A and B. Point A is obtained from a short-circuit test with full-load armature current. Hence OA represents excitation (field current) required to overcome demagnetising effect of armature reaction and to balance leakage reactance drop at full load. Point B is obtained when full-load current flows through the armature and wattmeter reads zero, as discussed above.

From point B, line BC is drawn equal and parallel to AO. Then a line is drawn through C parallel to initial straight part of OCC (parallel to OG), intersecting the OCC at D. BD is joined and a perpendicular DF is dropped on BC. The triangle BFD is imposed at various points OCC to obtain corresponding points on the zero factor curve.

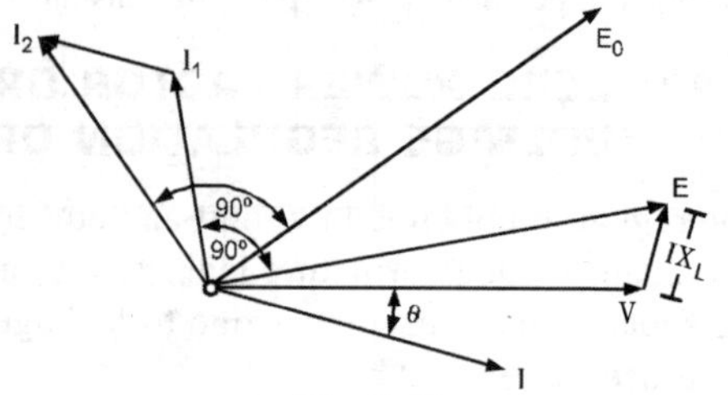

Fig. 5.45

In Δ BDF the length BF represents armature reaction excitation and the length DF represents leakage reactance drop ($I X_L$)*. This is known as Potier reactance voltage drop and the triangle is known as *Potier triangle.* The potier reactance is given as

$$X_p = \frac{\text{DF (voltage drop per phase)}}{\text{Zero power factor current per phase}}$$

In case of cylindrical rotor machines, Potier reactance is nearly equal to armature leakage reactance. In case of salient pole machines, the magnetising circuit is more saturated and the armature leakage reactance is smaller than the Potier reactance.

Potier Regulation Diagram : Potier regulation diagram is drawn as follows.

OV is drawn horizontally to represent terminal voltage, V on full load and OI is drawn to represent full load current at a given power factor. VE is drawn perpendicular to phasor OI and equal to reactance drop (IX_L), neglecting resistance drop. Now phasor OE represents generated emf E. From OCC field excitation I_1 corresponding to generated emf E is determined, OI_1 is drawn perpendicular to phasor OE to represent excitation required to induce emf OE on open circuit. I_1 I_2 is drawn parallel to load current

* The vertical distance FD represents the leakage reactance drop and DK represents the voltage drop due to armature reaction.

phasor OI to represent excitation equivalent to full-load armature reaction. OI_2 gives total excitation required. If the load is thrown off, then terminal voltage will be equal to generated emf corresponding to field excitation OI_2. Hence emf E_0 may be determined from OCC corresponding to field excitation OI_2. Phasor OE_0 will lag behind phasor OI_2 by 90º. EE_0 represents voltage drop due to armature reaction. Now regulation can be obtained from the relation

$$\% \text{ Regulation} = \frac{E_0 - V}{V} \times 100$$

Example 5.19. Tests on a 15,000 kVA, 11,000 volt, 3-phase, 50 Hz, star-connected alternator gave the following results:

Field AT per pole, thousands	**10**	**18**	**25**	**30**	**40**	**45**	**50**
Open-circuit line emf , kV	**5.0**	**8.4**	**10.0**	**11.2**	**12.8**	**13.3**	**13.65**
Full-load current, zero power factor test, line pd kV	**—**	**0**	**—**	**—**	**—**	**10.0**	**—**

Find the armature reaction ampere-turns, the leakage reactance and the regulation for full load at 0.8 pf lagging. Neglect resistance.

Solution : From the above datas first OCC is drawn between line voltage and field current. Full-load zero power factor curve is drawn, as discussed in Art. 5.17, point A (18, 0) and point B (45, 10.2) being known.

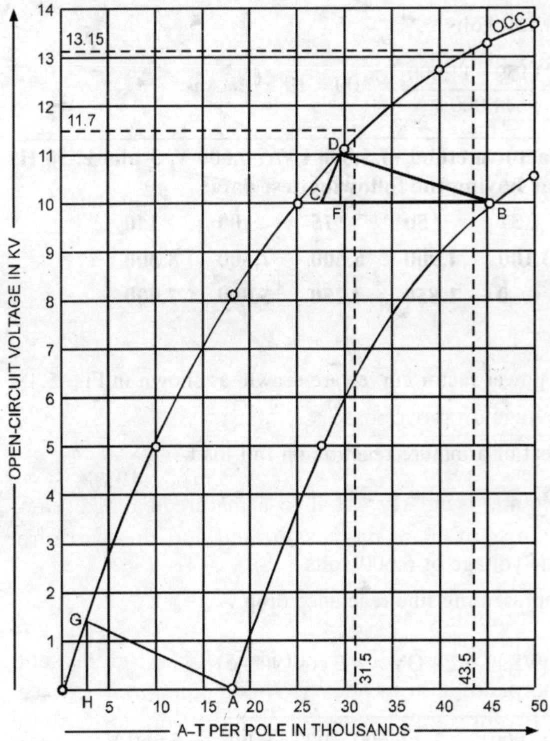

Fig. 5.46

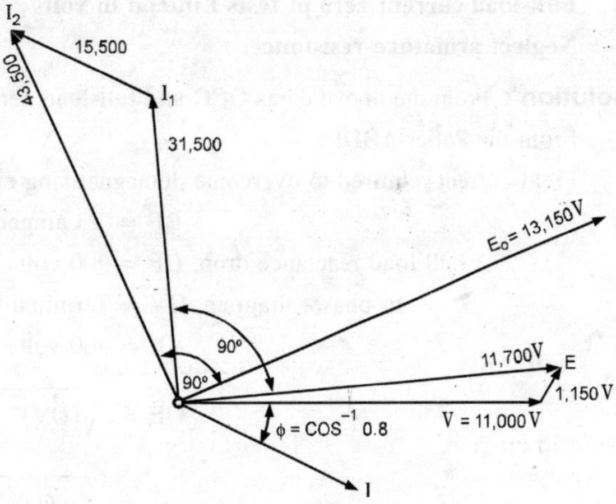

Fig. 5.47

From the Δ BDF (Fig. 5.46)

Armature reaction ampere turns = BF = 15,500 AT/pole **Ans.**

Full-load reactance drop = DF = 1.15 kV = 1,150 volts

$$\text{Leakage reactance drop per phase} = \frac{1,150}{\sqrt{3}} \text{ volts}$$

$$\text{Full-load current} = \frac{15,000 \times 1,000}{\sqrt{3} \times 11,000} = \frac{5,000\sqrt{3}}{11} \text{ A}$$

$$\text{Leakage reactance per phase} = \frac{1,150 \times 11}{\sqrt{3} \times 5,000 \times \sqrt{3}} = 0.843\ \Omega \ \textbf{Ans.}$$

In phasor diagram OV = Terminal line voltage = 11,000 volts

VE = 1,150 volts representing reactance drop

OE = 11,700 volts

Excitation corresponding to 11,700 volts = 31,500 AT.

$\therefore$ OI_1 = 31,500 AT

$I_1 I_2$ = Armature reaction AT = 15,500 AT

Total AT = OI_2 = 43,500

Induced emf corresponding to 43,500 AT, E_0 = 13,150 volts

$$\% \text{ Regulation} = \frac{13,150 - 11,000}{11,000} \times 100 = 19.55\ \% \ \textbf{Ans.}$$

Example 5.20. Find the regulation by the zero-power factor method of 5,000 kVA, 6,600 V, 3-phase, 50 Hz star-connected alternator at full load, unity power factor having the following test data:

Field current in amperes	**32**	**50**	**75**	**100**	**140**
Open-circuit terminal voltage in volts	**3,100**	**4,900**	**6,600**	**7,500**	**8,300**
Full-load current zero pf tests Line pd in volts	**0**	**1,850**	**4,250**	**5,800**	**7,000**

Neglect armature resistance.

Solution : From the above datas OCC and full-load zero power factor curves are drawn, as shown in Fig. 5.48.

From the Potier ΔBDF

Field current required to overcome demagnetising effect of armature reaction on full load,

BF = 23 amperes

Full load reactance drop, DF = 900 volts

In phasor diagram, OV = Terminal line voltage of 6,600 volts

VE = 900 volts representing line reactance drop

$$\therefore \ OE = \sqrt{(OV)^2 + (VE)^2 - 2 \times OV \times VE \cos(90° + \phi)}$$

$$= \sqrt{(6,600)^2 + (900)^2 - 2 \times 6,600 \times 900 \cos 90°} = 6,660 \text{ V}$$

As seen from OCC field current required to induce 6,660 line volts is 76 amperes. Hence phasor OI_1 = 76 A and is perpendicular to phasor OE and phasor $I_1 I_2$ = 23 A and is parallel to phasor OI.

Total field current, $OI_2 = 79.6$ A

From OCC it is found that line emf corresponding to this field current of 79.6 A is 6,820 volts.

$$\therefore E_0 = 6{,}820 \text{ volts}$$

$$\text{Percentage regulation} = \frac{E_0 - V}{V} \times 100 = \frac{6{,}820 - 6{,}600}{6{,}600} \times 100$$

$$= 3.33 \text{ \% } \textbf{Ans.}$$

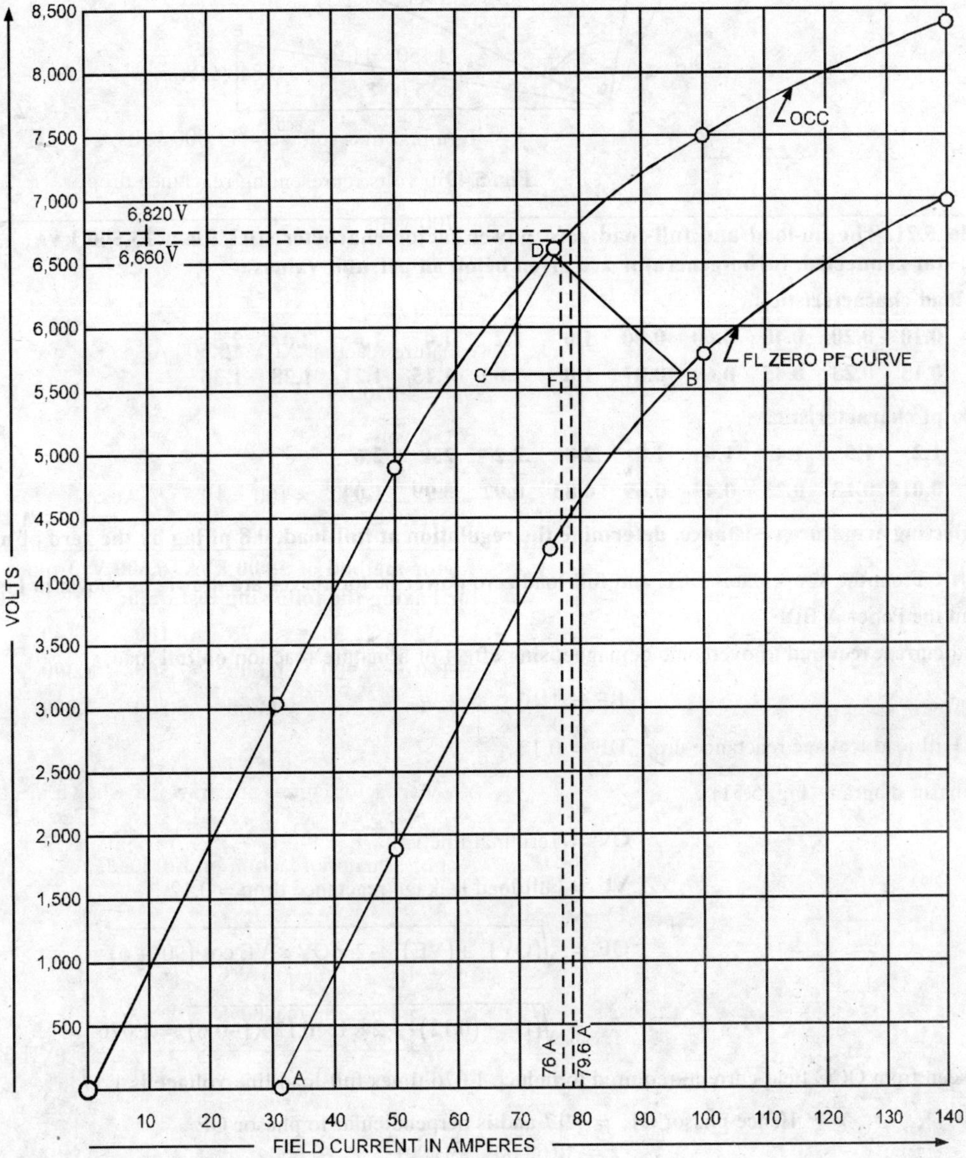

Fig. 5.48

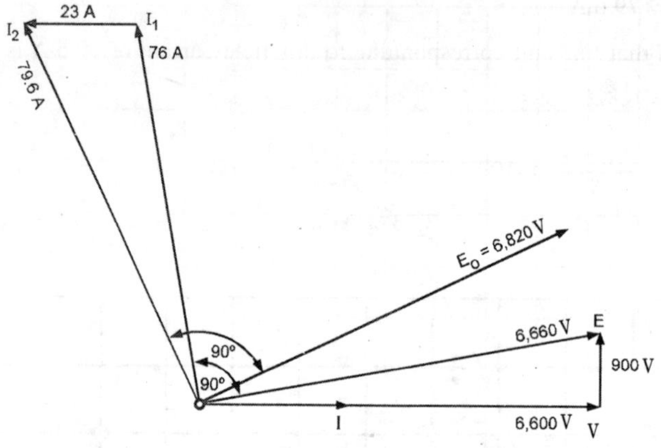

Fig. 5.49

Example 5.21. The no-load and full-load zero power factor characteristics for a 23,500 kVA, 13.8 kV, 3-phase, star-connected, turbo-generator are given below in per unit values :

No-load characteristic :

I_f	0.10	0.20	0.40	0.60	0.80	1.0	1.2	1.4	1.6	2.0	2.5
V	0.13	0.23	0.45	0.69	0.87	1.0	1.09	1.15	1.21	1.28	1.36

Zero pf characteristic:

I_f	1.2	1.3	1.4	1.6	1.8	2.0	2.2	2.4	2.6
V	0.015	0.13	0.25	0.49	0.69	0.83	0.92	0.99	1.04

Neglecting armature resistance, determine the regulation at full-load, 0.8 pf lag by the zero pf method.

Solution : From the above datas OCC and full-load zero power factor curves are drawn, as shown in Fig. 5.50.

From the Potier Δ BDF

Field current required to overcome demagnetising effect of armature reaction on full load,

$$BF = 1.10$$

Full load leakage reactance drop, DF = 0.12

In phasor diagram (Fig. 5.51)

OV = Terminal line voltage = 1

VE = Full load leakage reactance drop = 0.12

$$OE = \sqrt{(OV)^2 + (VE)^2 - 2 \times OV \times VE \cos(90^\circ + \phi)}$$

$$= \sqrt{(1)^2 + (0.12)^2 - 2 \times 1 \times 0.12 \times (-0.6)} = 1.076$$

As seen from OCC field current, required to induce 1.076 times full-load line voltage is 1.17.

Hence phasor OI_1 = 1.17 and is perpendicular to phasor OE

and phasor $I_1 I_2$ = 1.1 and is parallel to phasor OI

Total field current OI_2 = 2.00

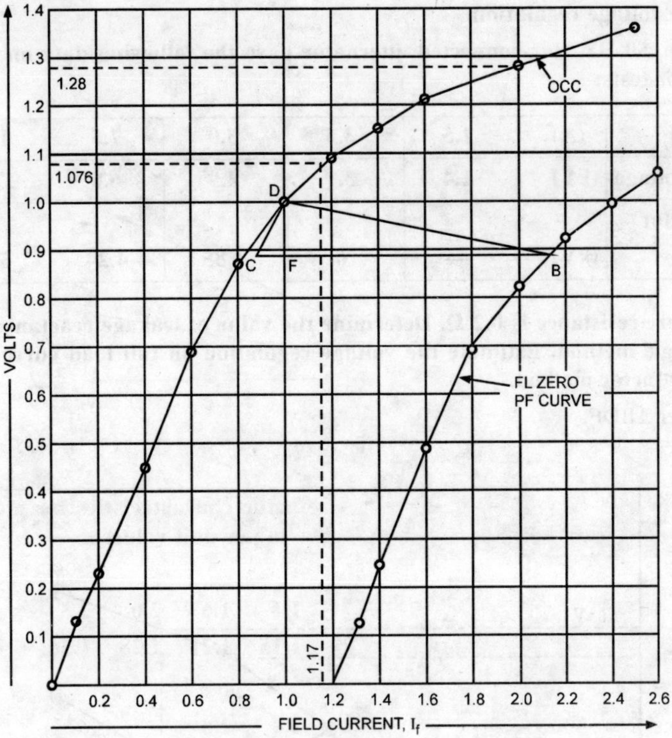

Fig. 5.50

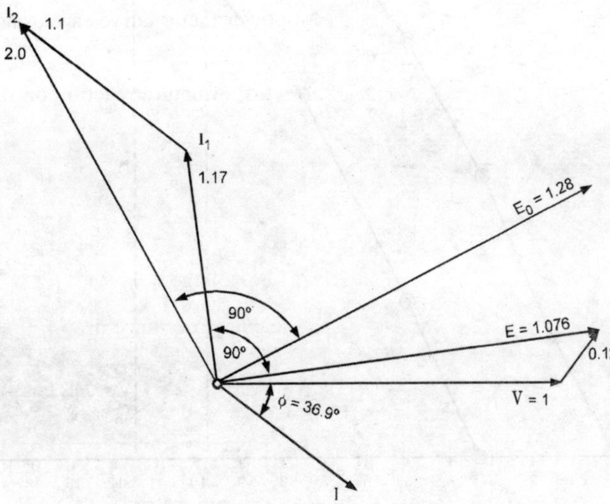

Fig. 5.51

From OCC it is found that line emf corresponding to field current of 2.00 is 1.28

$$\therefore E_0 = 1.28$$

$$\text{Regulation} = \frac{E_0 - V}{V} = \frac{1.28 - 1.00}{1.00} = 0.28 \text{ Ans.}$$

Example 5.22. Define voltage regulation.

A 6.6 kV, 3-phase, 50 Hz, star-connected alternator gave the following data on open-circuit and full load, zero power factor tests:

Field Current I_f	(A)	1.5	3.3	5.0	7.5	10.0	14.0
Open-circuit voltage	(kV)	1.4	3.1	4.9	6.6	7.5	8.24
Zero power factor terminal voltage	(kV)	—	0	1.85	4.24	5.78	7.0

Per phase armature resistance is 0.2 Ω. Determine the value of leakage reactance and armature reaction by Potier's triangle method. Estimate the voltage regulation on full load current of 500 A at 0.8 pf lagging by zero power factor method.

Solution : From Potier ΔBDF

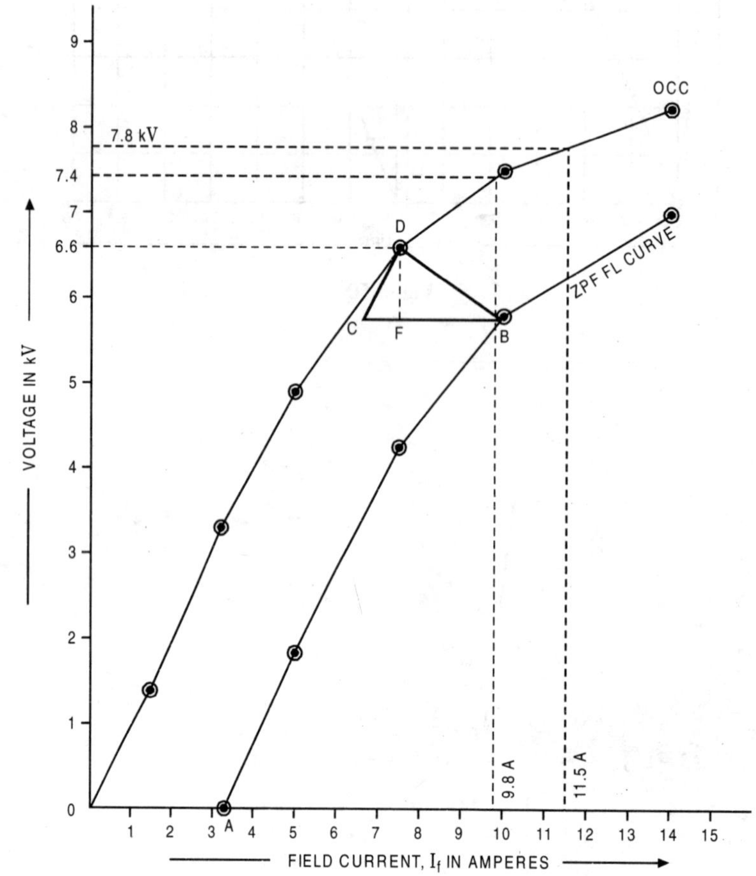

Fig. 5.52

Field current required to overcome demagnetising effect of armature reaction on full-load,

BF = 2.4 A **Ans.**

Full-load reactance drop, DF = 900 V

Full-load current, I = 500 A

$$\text{Leakage reactance per phase} = \frac{900}{\sqrt{3} \times 500} = 1.04\ \Omega\ \textbf{Ans.}$$

In phasor diagram shown in Fig. 5.53, $OV = V_L = 6.6$ kV

$$VA = \text{Armature resistance drop} = \sqrt{3} \times 500 \times 0.2 = 173.2\ \text{V}$$

$$AE = \text{Reactance drop} = 900\ \text{V}$$

$$OE = \text{Generated emf} = 7.4\ \text{kV}$$

$$\text{Excitation corresponding to 7.4 kV} = 9.8\ \text{A}$$

$$\text{So } OI_1 = 9.8\ \text{A}$$

$$I_1I_2 = \text{Armature reaction amperes} = 2.4\ \text{A}$$

$$\text{Total excitation} = OI_2 = 11.5\ \text{A}$$

Induced emf corresponding to excitation of 11.5 A, $E_0 = 7.8$ kV

$$\% \text{ Regulation} = \frac{E_0 - V}{V} \times 100 = \frac{7.8 - 6.6}{6.6} \times 100 = 18.182\ \%\ \textbf{Ans.}$$

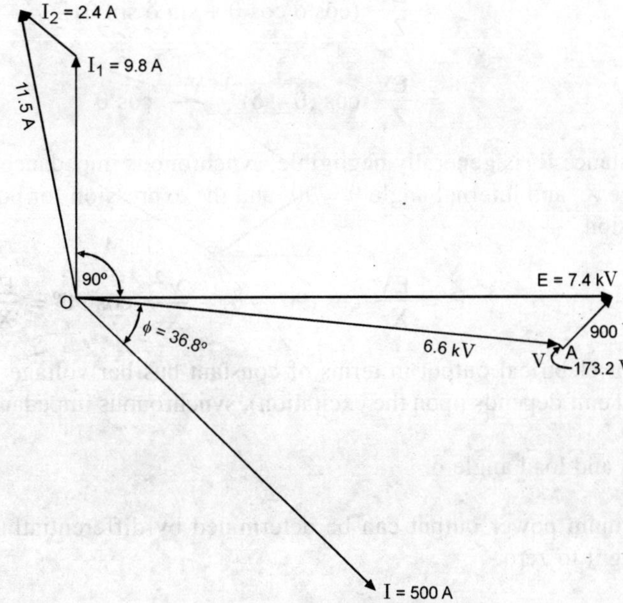

Fig. 5.53

5.18. POWER OUTPUT

Let us consider a 3-phase synchronous generator having effective resistance R_e ohms per phase and synchronous reactance X_s ohms per phase. Let it be connected to the bus-bars having phase voltage of V volts and supply current I amperes at power factor cos ϕ. Let the phase angle between induced emf E and bus-bar voltage V be δ.

The phasor diagram is drawn with voltage phasor as reference phasor (Fig. 5.54).

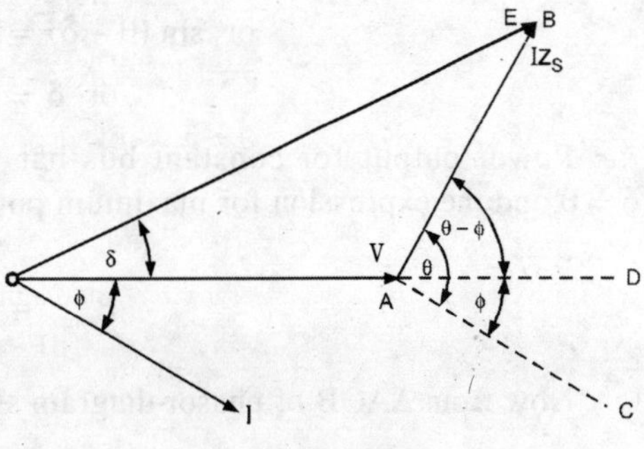

Fig. 5.54

Power output per phase, $P = V I \cos \phi$

$$= V I \cos [\theta - (\theta - \phi)]$$

$$= V I \frac{Z_s}{Z_s} [\cos \theta \cdot \cos (\theta - \phi) + \sin \theta \cdot \sin (\theta - \phi)]$$

$$= \frac{V}{Z_s} [I Z_s \cos (\theta - \phi) \cos \theta + I Z_s \sin (\theta - \phi) \sin \theta] \quad ...(5.17)$$

Now $I Z_s \cos (\theta - \phi) = E \cos \delta - V$

and $I Z_s \sin (\theta - \phi) = E \sin \delta$

Subtituting $I Z_s \cos (\theta - \phi) = (E \cos \delta - V)$ and $I Z_s \sin (\theta - \phi) = E \sin \delta$ in the Eq. (5.17) for power output we have

$$P = \frac{V}{Z_s} [(E \cos \delta - V) \cos \theta + E \sin \delta \sin \theta]$$

$$= \frac{EV}{Z_s} (\cos \delta \cos \theta + \sin \delta \sin \theta) - \frac{V^2}{Z_s} \cos \theta$$

$$= \frac{EV}{Z_s} \cos (\theta - \delta) - \frac{V^2}{Z_s} \cos \theta \quad ...(5.18)$$

Since effective resistance R_e is generally negligible, synchronous impedance Z_s can be taken equal to synchronous reactance X_s and internal angle $\theta = 90°$ and the expression for power output is reduced to the following expression

$$P = \frac{EV}{X_s} \cos (90° - \delta) - \frac{V^2}{X_s} \cos 90° = \frac{EV}{X_s} \sin \delta \quad ...(5.19)$$

The Eq. (5.18) is for electrical output in terms of constant bus-bar voltage V, induced emf E, (or excitation as the induced emf depends upon the excitation), synchronous impedance Z_s, internal angle θ being equal to $\tan^{-1} \frac{X_s}{R_e}$ and load angle δ.

Condition for maximum power output can be determined by differentiating the Eq. (5.18) with respect to δ and equating it to zero.

$$\frac{dP}{d\delta} = \frac{EV}{Z_s} \sin (\theta - \delta) = 0$$

or $\sin (\theta - \delta) = 0$

or $\delta = \theta$

i.e. Power output for constant bus-bar voltage V and fixed excitation will be maximum when $\delta = \theta$ and the expression for maximum power output per phase will become

$$P_{max} = \frac{EV}{Z_s} \cos (\theta - \theta) - \frac{V^2}{Z_s} \cos \theta = \frac{EV}{Z_s} - \frac{V^2}{Z_s} \cos \theta \quad ...(5.20)$$

Now from ΔAOB of phasor diagram shown in Fig. 5.54

$$I Z_s = \sqrt{V^2 + E^2 - 2VE\cos\delta}$$

$$\text{or } I = \frac{\sqrt{V^2 + E^2 - 2VE\cos\delta}}{Z_s}$$

Current corresponding to maximum power output *i.e.* when $\delta = \theta$

$$I_{max} = \frac{\sqrt{V^2 + E^2 - 2VE\cos\theta}}{Z_s}$$

and corresponding power factor, $$\cos\phi = \frac{P_{max}}{VI_{max}} = \frac{\frac{EV}{Z_s} - \frac{V^2}{Z_s}\cos\theta}{\frac{V}{Z_s}\sqrt{V^2 + E^2 - 2VE\cos\theta}}$$

$$= \frac{E - V\cos\theta}{\sqrt{V^2 + E^2 - 2VE\cos\theta}} \quad \text{...(5.21)}$$

Reactive power output per phase, $$Q = VI\sin\phi$$

$$= VI\sin[\theta - (\theta - \phi)]$$

$$= VI\frac{Z_S}{Z_S}[\sin\theta\cos(\theta - \phi) - \cos\theta \cdot \sin(\theta - \phi)]$$

$$= \frac{V}{Z_S}[IZ_S\cos(\theta - \phi)\sin\theta - IZ_S\sin(\theta - \phi)\cos\theta$$

Substituting $IZ_S\cos(\theta - \phi) = (E\cos\delta - V)$ and $IZ_S\sin(\theta - \phi) = E\sin\delta$ we have

$$Q = \frac{V}{Z_S}[(E\cos\delta - V)\sin\theta - E\sin\delta\cos\theta]$$

$$= \frac{EV}{Z_S}(\cos\delta\sin\theta - \sin\delta\cos\theta) - \frac{V^2}{Z_S}\sin\theta$$

$$= \frac{EV}{Z_S}\sin(\theta - \delta) - \frac{V^2}{Z_S}\sin\theta \quad \text{...(5.22)}$$

Since effective resistance R_e is generally negligible, therefore, synchronous reactance can be taken equal to synchronous reactance and internal angle $\theta = 90°$, and the expression for reactive power output becomes

$$Q = \frac{EV}{X_S}\sin(90° - \delta) - \frac{V^2}{X_S}\sin 90°$$

$$= -\frac{V^2}{X_S} + \frac{EV}{X_S}\cos\delta \quad \text{...(5.23)}$$

5.19. POWER ANGLE CHARACTERISTIC (OR POWER ANGLE CURVE)

The relationship of Eq. (5.19) is known as the *power angle characteristic* of the synchronous machine and is plotted in Fig. 5.55 for given E and V. The maximum power occurs at power angle $\delta = 90°$ beyond which the machine falls out of synchronism. The machine can be taken up to P_{max} only by gradually increasing the load. This is known as the *steady-state stability limit* of the machine.

The machine is normally operated at power angle δ much less than 90° in order to avoid the machine from going into an unstable region, where it will fall out of step during transient power swings.

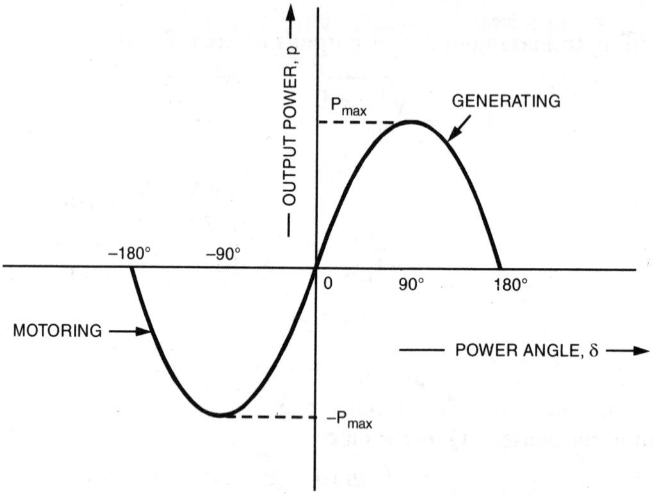

Fig. 5.55. ***Power Angle Characteristic***

Example 5.23. A 3-phase, 11,000 V, star-connected turbo-alternator having a synchronous reactance of 6 Ω per phase and negligible resistance has an armature current of 200 A at unity power factor when operating on constant frequency, constant voltage bus-bars. If the steam admission remains the same and the emf is raised by 25%, determine the new values of current and power factor.

Solution : Terminal voltage per phase,

$$V = \frac{11{,}000}{\sqrt{3}} = 6{,}350 \text{ V}$$

Reactance per phase, $X_s = 6\ \Omega$

Armature current, $I = 200$ A

Power factor, $\cos\phi = 1.0$

Phase angle, $\phi = \text{Cos}^{-1}\ 1.0 = 0°$

At unity power factor the reactance drop IX_s

$$= I_R X_s = 200 \times 6 = 1{,}200 \text{ V}$$

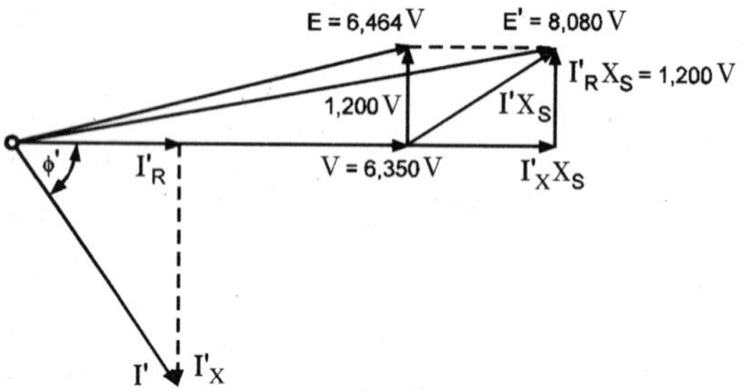

Fig. 5.56

The phasor diagram for this condition is shown in Fig. 5.56

$$\text{The emf generated per phase, } E = \sqrt{(6{,}350)^2 + (1{,}200)^2} = 6{,}464 \text{ V}$$

If the excitation is increased keeping the steam supply or power input same, the power output will not change. The result is to give the armature current a lagging reactive component which exerts a demagnetising effects on the main field and neutralise the increases in excitation. The phasor diagram is given in Fig. 5.56, which shows that the locus of the emf phasor is the dotted line perpendicular to $I'_R X_s$. I'_R is the power or active component of new armature current = 200 A as before and I'_X is the reactive component of new armature current.

With increased excitation the emf generated per phase

$$E' = 1.25\ E = 1.25 \times 6{,}464 = 8{,}080\ V$$

From phasor diagram shown in Fig. 5.56

$$(E')^2 = (V + I'_X X_s)^2 + (I'_R X_s)^2$$

$$\text{or } (8{,}080)^2 = (6{,}350 + 6\ I'_X)^2 + (200 \times 6)^2$$

$$\text{or } 6{,}350 + 6\ I'_X = \sqrt{(8{,}080)^2 - (1200)^2} = 7{,}990\ V$$

$$\text{or } I'_X = \frac{7{,}990 - 6350}{6} = 273.33\ A$$

New value of armature current, $I' = \sqrt{(I'_R)^2 + (I'_X)^2} = \sqrt{(200)^2 + (273.33)^2} = 338.7$ A **Ans.**

New power factor, $\cos\phi' = \frac{I'_R}{I'} = \frac{200}{338.7} = 0.59$ (lag) **Ans.**

Example 5.24. A 3-phase, 11 kV 10 MW, star-connected synchronous generator has a synchronous impedance of (0.8 + *j* 8.0) ohms per phase. If the excitation is such that the open-circuit voltage is 14 kV, determine (*i*) the maximum output of the generator (*ii*) the current and power factor at the maximum output.

Solution : Bus-bar voltage per phase, $V = \frac{11{,}000}{\sqrt{3}}$ volts

Open circuit voltage per phase, $E = \frac{14{,}000}{\sqrt{3}}$ volts

Synchronous impedance per phase, $Z_s = \sqrt{(0.8)^2 + (8)^2} = 8.04\ \Omega$

Internal angle $\theta = \text{Tan}^{-1}\frac{X_s}{R_e} = \text{Tan}^{-1}\frac{8}{0.8} = 84.3°$

$$\text{Maximum power output/phase} = \frac{EV}{Z_s} - \frac{V^2}{Z_s}\cos\theta$$

$$= \frac{\frac{14{,}000}{\sqrt{3}} \times \frac{11{,}000}{\sqrt{3}}}{8.04} - \frac{\left(\frac{11{,}000}{\sqrt{3}}\right)^2}{8.04}\cos 84.3°$$

$$= 63{,}84{,}743 - 4{,}99{,}168 = 58{,}85{,}575 \text{ watts or } 5.886 \text{ MW}$$

(*i*) Total maximum power output = 3 × 5.886 = 17.66 MW

(*ii*) Current corresponding to maximum output

$$I_{max} = \frac{\sqrt{V^2 + E^2 - 2VE\cos\theta}}{Z_s}$$

$$= \frac{\sqrt{\left(\frac{11000}{\sqrt{3}}\right)^2 + \left(\frac{14000}{\sqrt{3}}\right)^2 - 2\left(\frac{11000}{\sqrt{3}} \times \frac{14000}{\sqrt{3}}\right)\cos 84.3°}}{8.04} = 1{,}215 \text{ A } \textbf{Ans.}$$

and corresponding power factor, $\cos\phi = \frac{P_{max}}{VI_{max}} = \frac{58{,}85{,}575}{\frac{11{,}000}{\sqrt{3}} \times 1{,}215} = 0.763$ **Ans.**

5.20. TWO REACTANCE THEORY FOR SALIENT POLE SYNCHRONOUS MACHINES

A multipolar machine with cylindrical rotor has a uniform air gap and therefore, its reactance remains the same, irrespective of the spatial position of rotor. The effect of armature reaction, fluxes and voltages induced can, therefore, be treated in a simple way with concept of a synchronous reactance and taking it as constant for all positions of field poles with respect to the armature. However, a salient pole synchronous machine has non-uniform air gap due to which its reactance varies with the rotor position. Thus the salient pole machine possesses two axes of geometric symmetry (*i*) field pole axis, called *direct axis* or *d-axis* and (*ii*) axis passing through the centres of the interpolar space, called the *quadrature axis* or *q-axis*, as shown in Fig. 5.57 whereas in case of a cylindrical rotor machine there is only one axis of symmetry (pole axis or direct axis).

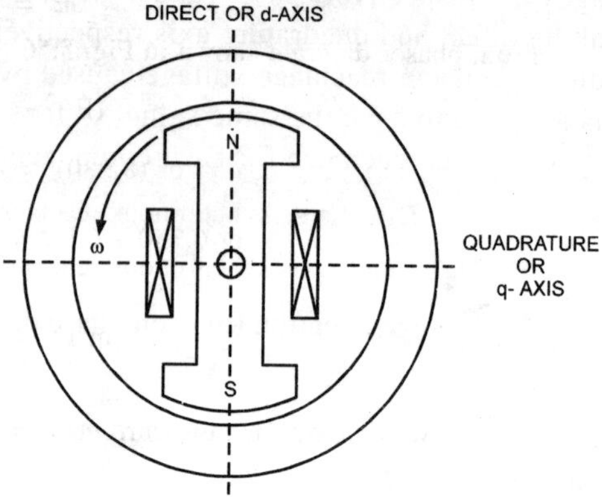

Fig. 5.57

In case of salient pole machine, the reluctance of the magnetic paths on which the emf acts are different along the direct axis and quadrature axis. The reluctance of the direct axis magnetic circuit is due to yoke and teeth of the stator, air gap, and pole and core of the rotor. In quadrature axis, the reluctance is mainly due to large air-gap in the interpolar space.

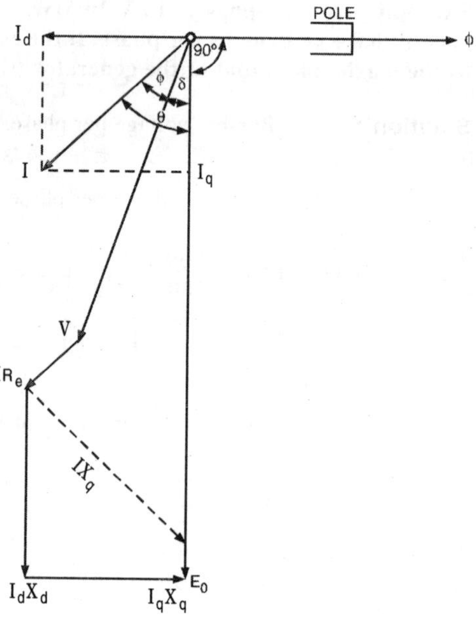

Fig. 5.58

Because of non-uniformity of the reluctance of the magnetic paths, the mmf of the armature is divided into two components namely (*i*) a direct acting component and (*ii*) a quadrature (or cross) component. It has been explained in Art. 5.11 that when armature current is in phase with the excitation voltage, the entire mmf of the armature acts at right angles to the axis of the salient poles and, therefore, all the armature mmf is in quadrature. On the other hand, if the armature current is in quadrature with the excitation voltage E_0, the entire mmf of the armature acts directly upon the magnetic paths through the salient poles and all of the armature mmf is direct acting, either directly opposing or directly aiding the mmf of the salient pole field windings. When the phase difference between armature current and excitation voltage is of some angle in between 0 and 90°, the armature mmf will have both a direct acting and a quadrature component. The direct-acting component is proportional to the sine of the phase angle between the armature current and excitation voltage whereas the quadrature (or cross) component is proportional to the cosine of the phase angle between the armature current and the excitation voltage. The two reactance concept is similar to the synchronous impedance concept in that the effect of armature reaction is taken into account by means of equivalent armature reactance voltage. However, owing to the difference in the reluctance of the magnetic paths upon which the two components of the armature mmf act, the value of the equivalent reactance for the direct component of armature mmf is greater than the value of the equivalent reactance for the quadrature component of the armature mmf. Thus the two reactance concept for salient-pole machines replaces the effect of armature reaction by two fictitious voltages. These reactance voltages are respectively $I_d X_{ad}$ and $I_q X_{aq}$, where I_d and I_q are the components of the armature current along direct and quadrature axis respectively. Each of these components of armature current also, produce a leakage-reactance voltage caused by the armature leakage flux. The armature leakage reactance is assumed to have the same value X_L for both components of the armature current, Therefore, there is synchronous reactance for each component of the armature mmf as follows :

Direct axis synchronous reactance, $X_d = X_{ad} + X_L$

Quadrature axis synchronous reactance, $X_q = X_{aq} + X_L$

The voltage equation for each phase of the armature based on the two-reactance concept is given as :

$$\mathbf{V} = \mathbf{E_0} - \mathbf{I}\,R_e - \mathbf{I}_d\,X_d - \mathbf{I}_q\,X_q \qquad ...(5.24)$$

In salient pole synchronous machines X_q = 0.6 to 0.7 times X_d whereas in cylindrical rotor machines $X_q = X_d$

5.20.1. Determination of X_d and X_q By Low Slip Test

The values of X_d and X_q are determined by applying a balanced reduced external voltage (say, V volts) to an unexcited machine at a speed a little less than the synchronous speed (the slip being less than 1%). Connection diagram is shown in Fig. 5.59. Applied voltage to the armature, armature current and the voltage induced in the field winding are measured by oscillographs. Due to voltage V applied to the stator terminals a current I will flow causing a stator mmf. This stator mmf moves slowly relative to the poles and induces an emf in the field circuit in a similar fashion to that of rotor in an induction motor at slip frequency. The effect will be that the stator mmf will move slowly relative to the poles.

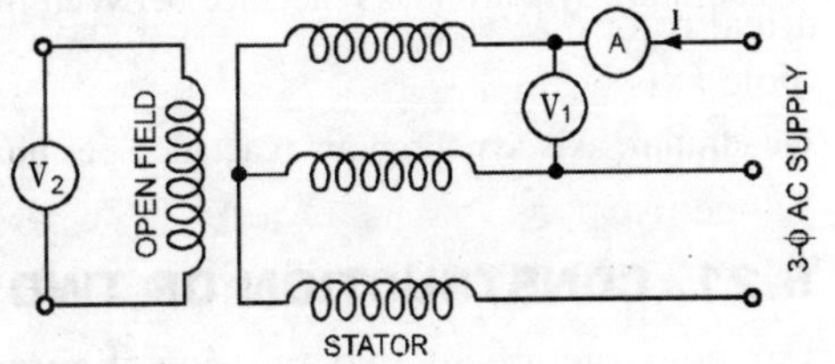

Fig. 5.59

The physical poles and the armature-reaction mmf are alternately in phase and out, the change occuring at slip frequency. When the axis of the poles and the axis of the armature reaction mmf wave coincide, the armature mmf acts through the field magnetic circuit. The voltage applied to the armature is then equal to drop caused by the direct component of armature reaction and leakage reactance. When the armature reaction mmf is in quadrature with the field poles, the applied voltage is equal to the leakage reactance drop plus the equivalent voltage drop of the cross-magnetising field component. From oscillograph record.

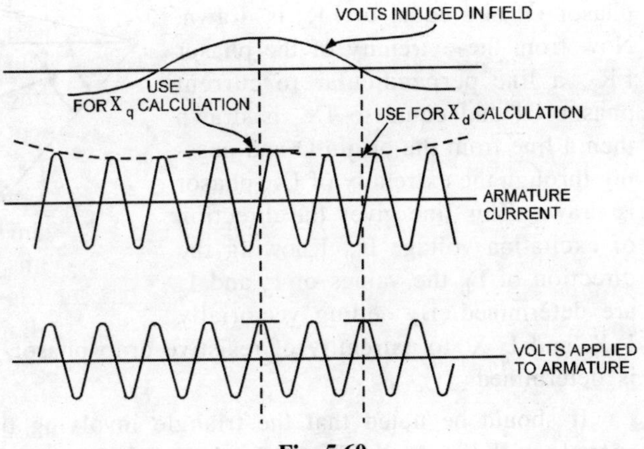

Fig. 5.60

$$X_d = \frac{\text{Maximum voltage}}{\text{Minimum current}} \qquad ...(5.25)$$

$$X_q = \frac{\text{Minimum voltage}}{\text{Maximum current}} \qquad ...(5.26)$$

Example 5.25. A 3-phase star-connected salient pole synchronous generator is driven at a speed near synchronous with the field circuit open, and the stator is supplied from a balanced 3-phase supply. Voltmeter connected across the line gave minimum and maximum readings of 2,800 and 2,820 volts. The line current fluctuated between 360 and 275 amperes. Find the direct and quadrature axis synchronous reactances per phase. Neglect armature resistance.

Solution :

Maximum voltage = 2,820 volts

Minimum voltage = 2,800 volts

Maximum current = 360 A

Minimum current = 275 A

$$\text{Direct-axis synchronous reactance between lines} = \frac{\text{Maximum voltage}}{\text{Minimum current}} = \frac{2{,}820}{275} = 10.25\ \Omega$$

$$\text{Direct-axis synchronous reactance per phase, } X_d = \frac{10.25}{\sqrt{3}} = 5.92\ \Omega. \textbf{ Ans.}$$

$$\text{Quadrature synchronous reactance between lines} = \frac{\text{Minimum voltage}}{\text{Maximum current}} = \frac{2{,}800}{360} = 7.778\ \Omega$$

$$\text{Quadrature axis synchronous reactance per phase, } X_q = \frac{7.778}{\sqrt{3}} = 4.5\ \Omega \textbf{ Ans.}$$

5.21. CONSTRUCTION OF TWO REACTION DIAGRAM

For construction of two reaction diagram, the values of X_d and X_q must be known.

From the load specified the voltage phasor and current phasor at the load power factor angle are drawn as shown in Fig. 5.61. From the extremity of the voltage phasor a line parallel to current phasor OI and equal to I R_e is drawn. Now from the extremity of the phasor I R_e, a line perpendicular to current phasor OI and equal to IX_q is drawn then a line from the origin O and passing through the extremity of IX_q phasor is drawn. This line gives the direction of excitation voltage E_0. Knowing the direction of E_0 the values of I_d and I_q are determined. By adding vectorially I_dX_d and $I_q X_q$ to extremity of resistive drop phasor IR_e, the magnitude of excitation voltage E_0 is determined.

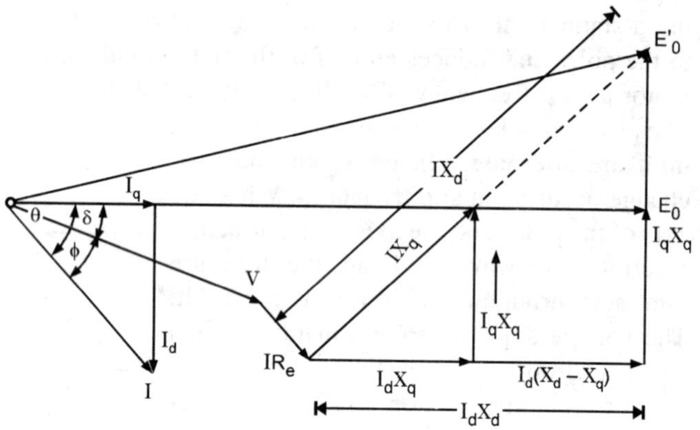

Fig. 5.61

It should be noted that the triangle involving the voltages IX_q and I_qX_q is similar to the triangle of the currents. Therefore, the left hand portion of the phasor $I_d X_d$ must be equal to $I_d X_q$. The portion of the components of $I_d X_q$ shown provides the ratio of X_q and X_d.

If the perpendicular drawn from the extremity of the phasor I R_e is produced so that its length is equal to IX_d – I X_q; the phasor E'_0, which is normally obtained from the synchronous impedance method, can be determined. In Fig. 5.61 the extremity of E'_0 is shown on the line perpendicular to $I_d X_d$ and it is slightly greater than E_0, making the regulation as obtained by the synchronous impedance method greater than that obtained by the two reaction method. The difference depends upon the saliency (ratio of pole pitch to pole arc) and increases with the increase in the value of saliency. One of the greatest difference between the two methods is the magnitude of the torque angle, which is much larger when determined from the synchronous impedance phasors than when found from the two reaction diagram.

The angle θ between E_0 and I is called the *internal power factor angle* whereas δ, the angle between E_0 and V is called the *load* or *power angle.*

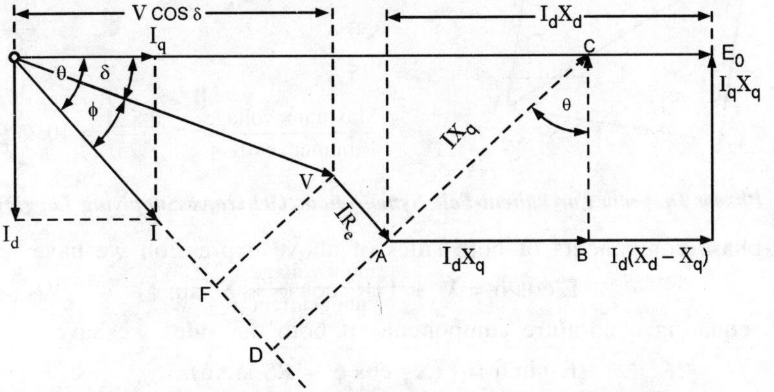

Fig. 5.62

The phasor diagram is redrawn in Fig. 5.62 taking E_0 in horizontal direction.

From Fig. 5.62

$$I_d = I \sin \theta$$
$$I_q = I \cos \theta$$

In ΔABC,

$\angle ACB = \theta$ as angle between the two lines is the same as between their perpendiculars. Line AC is perpendicular to phasor I and line CB perpendicular to phasor E_0.

From ΔODC

$$\text{Tan } \theta = \frac{DC}{OD} = \frac{DA + AC}{OF + FD} = \frac{V \sin \phi + IX_q}{V \cos \phi + IR_e} \qquad \text{...(5.27) generating}$$

$$= \frac{V \sin \phi - IX_q}{V \cos \phi - IR_e} \qquad \text{...(5.28) motoring}$$

The angle θ can be determined from the above equations.

Load angle δ = θ – φ in case of generating and δ = φ – θ in case of motoring.

Angle φ is taken + ve for lagging power factor and will be taken negative for leading power factor.

Power angle δ can be determined from the phasor diagram, of a salient pole synchronous generator supplying lagging current, drawn with phasor OV in horizontal direction (Fig. 5.63).

From phasor diagram shown in Fig. 5.63 we have

$$\mathbf{E} = \mathbf{V} + \mathbf{I}\,(R_e + j\,X_q)$$

$$\text{or } E\,(\cos \delta + j \sin \delta) = V\,(1 + j\,0) + I\,(\cos \phi - j \sin \phi)(R_e + j\,X_q)$$

$$\text{or } E\cos\delta + jE\sin\delta = V + I\,(R_e\cos\phi + X_q\sin\phi) + jI\,(X_q\cos\phi - R_e\sin\phi)$$

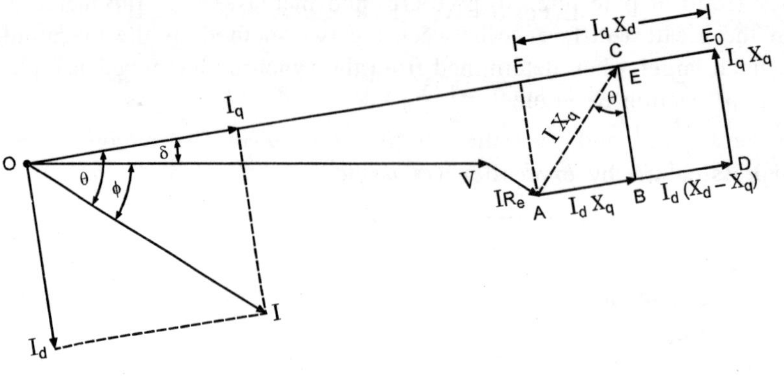

Fig. 5.63. *Phasor Diagram For Salient-Pole Synchronous Generator Supplying Lagging Current*

Equating in-phase components of both sides of above expression we have

$$E\cos\delta = V + I\,(R_e\cos\phi + X_q\sin\phi) \quad \text{...(i)}$$

Similarly on equating quadrature components of both the sides we have

$$E\sin\delta = I\,(X_q\cos\phi - R_e\sin\phi) \quad \text{...(ii)}$$

Dividing Eq. (*ii*) by Eq. (*i*) we have

$$\text{Tan}\,\delta = \frac{I(X_q\cos\phi - R_e\sin\phi)}{V + I(R_e\cos\phi + X_q\sin\phi)} \quad \text{...(5.29)}$$

For a salient-pole synchronous motor drawing lagging current with phasor OV in horizontal direction, phasor diagram is drawn, as illustrated in Fig. 5.64.

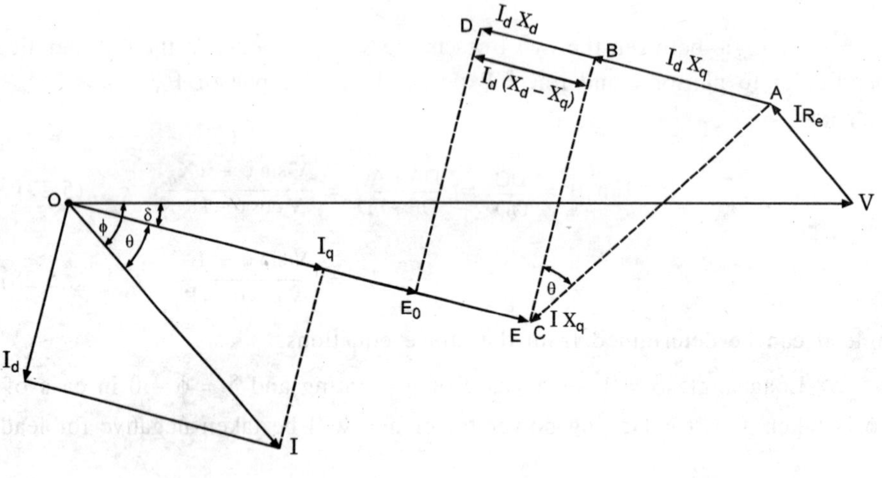

Fig. 5.64. *Phasor Diagram For Salient-Pole Synchronous Motor Drawing Lagging Current*

From phasor diagram shown in Fig. 5.64, we have

$$\mathbf{E} = \mathbf{V} - \mathbf{I}\,(R_e + jX_q)$$

$$\text{or } E\,(\cos\delta - j\sin\delta) = V\,(1 + j0) - I\,(\cos\phi - j\sin\phi)\,(R_e + jX_q)$$

$$\text{or } E\,(\cos\delta - j\sin\delta) = V - I\,(R_e\cos\phi + X_q\sin\phi) - jI\,(X_q\cos\phi - R_e\sin\phi)$$

From above equation we have

$$E \cos \delta = V - I(R_e \cos \phi + X_q \sin \phi) \quad ...(i),$$

considering in-phase components

$$\text{and } E \sin \delta = I(X_q \cos \phi - R_e \sin \phi) \quad ...(ii),$$

considering quadrature components

Dividing expression (*ii*) by expression (*i*) we have

$$\text{Tan } \delta = \frac{I(X_q \cos\phi - R_e \sin\phi)}{V - I(R_e \cos\phi + X_q \sin\phi)} \quad ...(5.30)$$

Thus power angle δ for a salient pole synchronous machine can be given by expression.

$$\text{Tan } \delta = \frac{I(X_q \cos\phi - R_e \sin\phi)}{V \pm I(R_e \cos\phi + X_q \sin\phi)} \quad ...(5.31)$$

+ *ve* sign for synchronous generator and – *ve* sign for synchronous motor

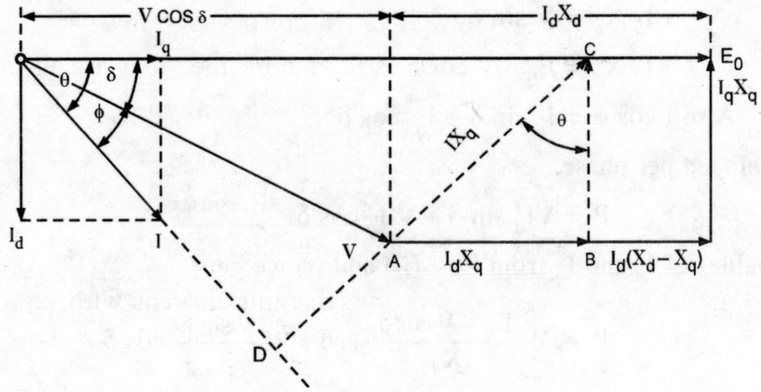

Fig. 5.65

If armature resistance is neglected, the phasor diagram becomes as shown in Fig. 5.65. The load angle δ can be determined directly as below :

$$\delta = \theta - \phi \text{ in case of generating}$$

$$\text{and } \delta = \phi - \theta \text{ in case of motoring}$$

$$\text{or } \theta = \phi \pm \delta \text{ in general} \quad ...(5.32)$$

$$\text{Direct-axis component, } I_d = I \sin \theta = I \sin (\phi \pm \delta) \quad ...(5.33)$$

$$\text{Quadrature axis component, } I_q = I \cos \theta = I \cos (\phi \pm \delta) \quad ...(5.34)$$

$$V \sin \delta = I_q X_q = I \cos \theta \, X_q = I X_q \cos (\phi \pm \delta)$$

$$= IX_q (\cos \phi \cos \delta \mp \sin \phi \sin \delta)$$

$$\text{or } V \sin \delta = IX_q \cos \phi \cos \delta \mp IX_q \sin \phi \sin \delta$$

$$\text{or } V = IX_q \cos \phi \cot \delta \mp IX_q \sin \phi$$

$$\text{or } I X_q \cos \phi \cot \delta = V \pm I X_q \sin \phi$$

$$\text{or } \tan \delta = \frac{IX_q \cos\phi}{V \pm IX_q \sin\phi} \quad ...(5.35)$$

+ ve sign for synchronous generator and – ve sign for synchronous motor

The same result is obtained if R_e is substituted equal to zero in Eq. (5.31).

When R_e is neglected then

$$E_0 = V \cos\delta \pm I_d X_d \qquad ...(5.36)$$

When R_e is to be taken into account

$$E_0 = V \cos\delta + I R_e \cos\theta + I_d X_d$$

$$= V \cos\delta + I_q R_e + I_d X_d \qquad ...(5.37) \quad ...\text{generating}$$

$$\text{and } E_0 = V \cos\delta - I_q R_e - I_d X_d \qquad ...(5.38) \quad ...\text{motoring}$$

5.22. POWER DEVELOPED BY A SYNCHRONOUS GENERATOR

Power developed per phase, P_d = Power output (P_{out}) per phase

= VI cos ϕ provided R or copper loss is negligible

Now from Fig. 5.65

$$I_q X_q = V \sin\delta \qquad ...(i)$$

$$I_d X_d = E_0 - V\cos\delta \qquad ...(ii)$$

Also $I \cos\phi = I_d \sin\delta + I_q \cos\delta$

So power developed per phase,

$$P = VI_d \sin\delta + V I_q \cos\delta$$

Substituting values of I_d and I_q from Eqs. (*ii*) and (*i*) we get

$$P = V\cdot\frac{E_0 - V\cos\delta}{X_d}\sin\delta + V\cdot\frac{V\sin\delta}{X_q}\cos\delta$$

$$= \frac{E_0 V}{X_d}\sin\delta - \frac{V^2}{X_d}\sin\delta\cos\delta + \frac{V^2\sin\delta\cos\delta}{X_q}$$

$$= \frac{E_0 V}{X_d}\sin\delta + V^2\left[\frac{1}{X_q} - \frac{1}{X_d}\right]\sin\delta\cos\delta$$

$$= \frac{E_0 V}{X_d}\sin\delta + \frac{V^2}{2}\left[\frac{1}{X_q} - \frac{1}{X_d}\right]\sin 2\delta \qquad ...(5.39)$$

Total power developed will be three times of the above *i.e* total power developed by a 3-phase synchronous generator

$$= \frac{3 E_0 V}{X_d}\sin\delta + \frac{3V^2}{2}\left[\frac{1}{X_q} - \frac{1}{X_d}\right]\sin 2\delta \qquad ...(5.40)$$

Above expression for power developed consists of two terms, the first term representing power due to field excitation and the second term gives the reluctance power *i.e.* power due to saliency. In case of a cylindrical rotor machine $X_d = X_q$ and so the second term becomes zero and the power is given by the first term only. In case there is no field exicitation *i.e.* E_0, the first term becomes zero and the power developed is given by the second term.

It may be noted that value of δ is positive in case of an alternator and negative for a motor.

Example 5.26. A salient pole, 3-phase, star-connected alternator at 5 MVA, 6.6 kV, has a resistance of 1.5 % and leakage reactance of 10 %. The short-circuit armature reaction for full load current is equivalent to 60 A field current. The armature cross-reaction per armature turn is half of the direct reaction. The open-circuit characteristic is as follows:

Field current in amps	**32**	**50**	**75**	**100**	**140**
Terminal pd in volts	**3,100**	**4,700**	**6,600**	**7,500**	**8,300**

Find the percentage regulation on full load at pf 0.8 (lagging)

Solution : The OCC is drawn as shown in Fig. 5.66 (*a*)

The phasor diagram for full load 0.8 power factor (lagging) is drawn to represent terminal voltage (6,600 V) taken as 100 % [Fig. 5.66 (*b*)]. Then to the same scale the resistance and leakage reactance drops of 1.5 % and 10 % are added in phase and in quadrature respectively with the current to give generated emf OE,

$$OE = 107.5 \% = 7{,}095 \text{ V}$$

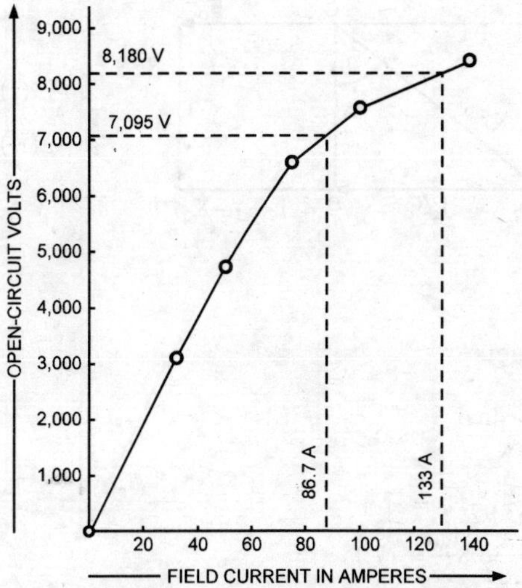

Fig. 5.66 (*a*)

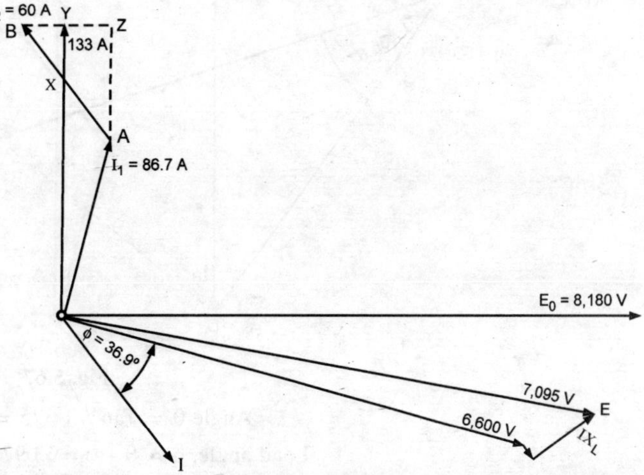

Fig. 5.66 (*b*)

From OCC the field excitation corresponding to 7,095 volts, I = 86.7 **A.** Now OI_1 is drawn perpendicular to phasor OE to represent excitation required to induce emf OE on open circuit. $I_1 I_2 = 60$ A is drawn parallel to load current phasor OI to represent excitation equivalent to full-load armature reaction on short circuit. I_1I_2 is now divided at X such that $I_1X = KI_1 I_2$ where K is the ratio of the cross-reaction to the direct reaction per ampere-turn. In this case K = 0.5 so that X bisects AB.

A line is now drawn through OX and produced; a perpendicular from I_2 is drawn to this line meeting it at Y. Then OY is the excitation required and is made up of the phasor sum of OI_1, I_1Z (to balance the direct reaction) and ZY (to balance the cross-reaction).

By measurement OY = 133 A

The emf generated corresponding to excitation of 133 amperes = 8,180 volts

$$\text{Percentage regulation} = \frac{8{,}180 - 6{,}600}{6{,}600} \times 100 = 23.94 \% \textbf{ Ans.}$$

Example 5.27. A synchronous generator has a direct-axis synchronous reactance 'X_d' of 0.8 pu and a quadrature-axis synchronous reactance 'X_q' of 0.5 pu, Determine the pu open-circuit voltage for full-load at a lagging pf of 0.8. Neglect saturation.

Solution : Direct-axis synchronous reactance, $X_d = 0.8$ pu

Quadrature-axis synchronous reactance, $X_q = 0.5$ pu

Power factor, $\cos\phi = 0.8$ lagging

Phase angle, $\phi = \text{Cos}^{-1}\, 0.8 = 36.87°$

$\text{Sin}\,\phi = \text{Sin}\, 36.87° = 0.6$

Terminal voltage, V = 1.0 pu

Armature current, I = 1.0 pu

$$\text{Tan}\,\phi = \frac{DC}{OD} = \frac{DA + AC}{OD} = \frac{V\sin\phi + IX_q}{V\cos\phi} = \frac{1\times 0.6 + 1\times 0.5}{1\times 0.8}$$

$$= 1.375$$

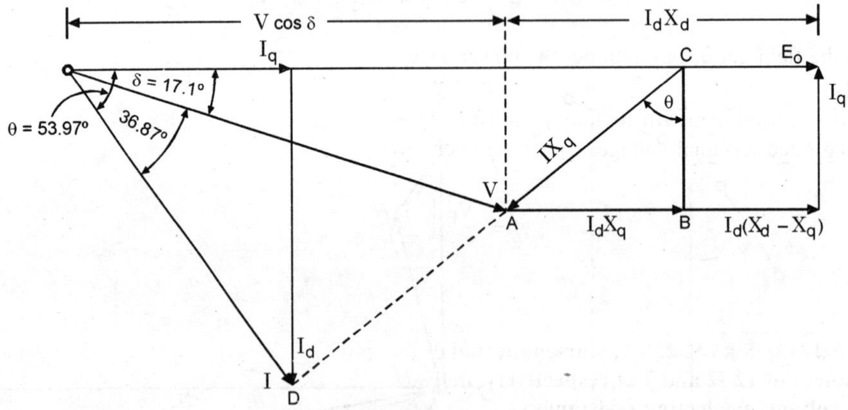

Fig. 5.67

Angle $\theta = \text{Tan}^{-1}\, 1.375 = 53.97°$

Load angle, $\delta = \theta - \phi = 53.97° - 36.87° = 17.1°$

Direct-axis component of current, $I_d = I \sin\theta = 1 \times \sin 53.97° = 0.8087$ pu

No-load emf, $E_0 = V\cos\delta + I_d X_d$

$= 1.0 \cos 17.1° + 0.8087 \times 0.8 = 1.60278$ pu **Ans.**

Example 5.28. A 3.5 MVA, slow speed, 3-phase synchronous generator rated at 6.6 kV has 32 poles. Its direct and quadrature axis synchronous reactances as measured by the slip test are 9.6 and 6 Ω respectively. Neglecting armature resistance, determine the regulation and excitation emf needed to maintain 6.6 kV at the terminals when supplying a load of 2.5 MW at 0.8 pf lagging. What maximum power can generator supply at the rated terminal voltage, if the field becomes open circuited?

Solution : Terminal voltage per phase, $V = \dfrac{6.6 \times 1{,}000}{\sqrt{3}} = 3{,}810.5$ V

Armature current, $I = \dfrac{2.5 \times 10^6}{\sqrt{3} \times 6{,}600 \times 0.8} = 273.37$ A

Load phase angle, $\phi = \text{Cos}^{-1}\, 0.8 = 36.87°$, $\text{Sin}\,\phi = \text{Sin}\, 36.87° = 0.6$

Direct-axis synchronous reactance per phase, $X_d = 9.6\ \Omega$

Quadrature axis synchronous reactance per phase, $X_q = 6\ \Omega$

From Eq. (5.35)

$$\text{Tan } \delta = \frac{I X_q \cos\phi}{V + I X_q \sin\phi} = \frac{273.37 \times 6 \times 0.8}{3{,}810.5 + 273.37 \times 6 \times 0.6} = 0.2737$$

$$\text{or } \delta = \text{Tan}^{-1}\ 0.2737 = 15.3°$$

$$\text{Angle } \theta = \phi + \delta = 36.87 + 15.3 = 52.17°$$

Direct-axis component of current, $I_d = I \sin\theta = 273.31 \sin 52.17° = 215.94$ A

Excitation voltage per phase, $E_0 = V \cos\delta + I_d X_d = 3{,}810.5 \times \cos 15.3° + 215.94 \times 9.6 = 5{,}748$ V

Excitation voltage (line-to-line) $= \sqrt{3} \times 5{,}748 = 9{,}956$ V **Ans.**

$$\text{Percentage regulation} = \frac{9{,}956 - 6{,}600}{6{,}600} \times 100 = 50.85\ \%\ \textbf{Ans.}$$

When the field get open circuited, the power developed $= \dfrac{V_L^2}{2}\left[\dfrac{1}{X_q} - \dfrac{1}{X_d}\right] \sin 2\delta$ Refer to Eq. (5.40)

The power developed will be maximum for $\sin 2\delta = 1$ and so the maximum power, that the generator can supply at the rated terminal voltage, with field open-circuited

$$= \frac{V_L^2}{2}\left[\frac{1}{X_q} - \frac{1}{X_d}\right] = \frac{(6{,}600)^2}{2}\left[\frac{1}{6} - \frac{1}{9.6}\right]$$

$$= 1{,}361{,}250 \text{ W or } 1.361 \text{ MW } \textbf{Ans.}$$

Example 5.29. A 5 kVA, 220 V, star-connected 3-phase salient pole alternator with direct and quadrature axis reactances of 12 Ω and 7 Ω respectively, delivers full-load current at unity power factor. Calculate the excitation voltage, neglecting resistance.

Solution : Direct-axis synchronous reactance per phase, $X_d = 12\ \Omega$

Quadrature-axis synchronous reactance per phase, $X_q = 7\ \Omega$

Power factor, $\cos\phi = 1.0$

Load phase angle, $\phi = \text{Cos}^{-1}\ 1.0 = 0°$

$\text{Sin } \phi = \text{Sin } 0° = 0$

Terminal voltage per phase, $V = \dfrac{220}{\sqrt{3}} = 127$ V

Armature current, $I = \dfrac{\text{KVA} \times 1{,}000}{\sqrt{3}\ V_L} = \dfrac{5 \times 1{,}000}{\sqrt{3} \times 220} = 13.122$ A

From Eq. (5.35)

$$\text{Tan } \delta = \frac{I X_q \cos\phi}{V + I X_q \sin\phi} = \frac{13.122 \times 7 \times 1.0}{127 + 13.122 \times 7 \times 0} = 1.3826$$

Load angle $\delta = \text{Tan}^{-1}\ 1.3826 = 54.123°$

Angle $\theta = \delta + \phi = 54.123° + 0 = 54.123°$

Direct-axis component of current, $I_d = I \sin\theta = 13.122 \times \sin 54.123° = 10.632$ A

Excitation voltage, $E_0 = V \cos\delta + I_d X_d = 127 \cos 54.123° + 10.632 \times 12$

$= 74.428 + 127.854 = 202.012$ V

Excitation line voltage, $E_{0L} = \sqrt{3} \times 202.012 = 349.9$ V **Ans.**

Example 5.30. A 10 kVA, 380 V, 50 Hz, 3-phase, star-connected salient pole alternator has direct axis and quadrature axis reactances of 12 Ω and 8 Ω respectively. The armature has a resistance of 1 Ω per phase. The generator delivers rated load at 0.8 pf lagging with the terminal voltage being maintained at rated value. If the load angle is 16.15°, determine (*a*) the direct axis and quadrature axis components of armature current, (*b*) excitation voltage of the generator.

Solution :

Armature resistance, $R_e = 1\ \Omega$

Direct-axis synchronous reactance, $X_d = 12\ \Omega$

Quadrature-axis synchronous reactance, $X_q = 8\ \Omega$

Power factor, $\cos\phi = 0.8$

$\phi = \text{Cos}^{-1}\ 0.8 = 36.87°$

Load angle, $\delta = 16.15°$

$\theta = \phi + \delta = 36.87° + 16.15° = 53.02°$

Terminal voltage per phase, $V = \dfrac{380}{\sqrt{3}} = 219.4\ V$

Armature current, $I = \dfrac{\text{kVA} \times 1{,}000}{\sqrt{3}\ V_L} = \dfrac{10 \times 1{,}000}{\sqrt{3} \times 380} = 15.2\ A$

Phasor diagram is shown in Fig. 5.68.

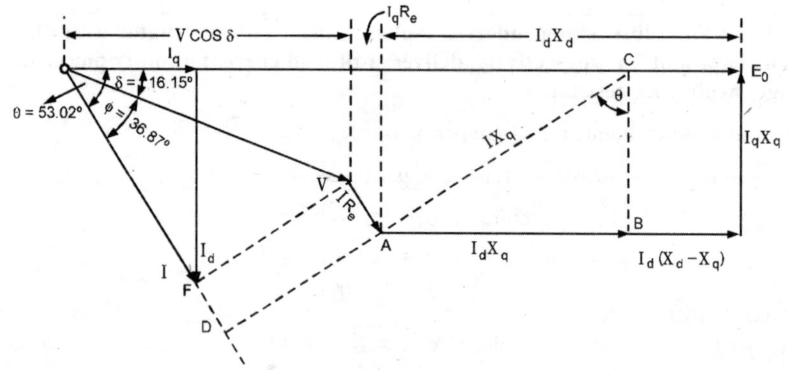

Fig. 5.68

Angle θ can also be determined as follows

$$\text{Tan}\ \theta = \frac{DC}{OD} = \frac{DA + AC}{OF + FD} = \frac{V\sin\phi + IX_q}{V\cos\phi + IR_e}$$

$$= \frac{219.4 \times 0.6 + 15.2 \times 8}{219.4 \times 0.8 + 15.2 \times 1} = 1.3278$$

or $\theta = \text{Tan}^{-1}\ 1.3278 = 53.02°$ The same as determined above.

Direct-axis component of armature current, $I_d = I\sin\theta = 15.2 \sin 53.02° = 12.14$ A **Ans.**

Quadrature-axis component of armature current, $I_q = I \cos \theta = 15.2 \cos 53.02° = 9.14$ A **Ans.**

Excitation voltage, $E_0 = V \cos \delta + I_q R_e + I_d X_d$

$= 219.4 \cos 16.15° + 9.14 \times 1 + 12.14 \times 12$

$= 210.74 + 9.14 + 145.68 = 365.56$ V

Excitation voltage (line-to-line) $= \sqrt{3} \times 365.56 = 633$ V **Ans.**

5.23. TRANSIENT CONDITIONS OF ALTERNATORS

In normal operation, the alternator may be subjected to transient conditions which cannot be explained from steady-state theory, discussed so far. These transients may occur from (*i*) switching, (*ii*) sudden changes of load, (*iii*) sudden short circuit (either line-to-line or line to neutral) or from a symmetrical short circuit (short circuit on all output terminals). These short-circuits may develop severe mechanical stresses on the armature coils and their end turns or large torques, which may damage the alternator or its prime mover. The analysis of the synchronous machine under conditions of such transients is useful in predicting the possible conditions that may result from abnormal operation resulting from short circuits.

The complete analysis of the transient conditions that may occur from short circuits is extensive and far beyond the scope of this book. Here the discussions will be limited to the phenomena that determine the short-circuit transient currents and the reactances that limit these currents.

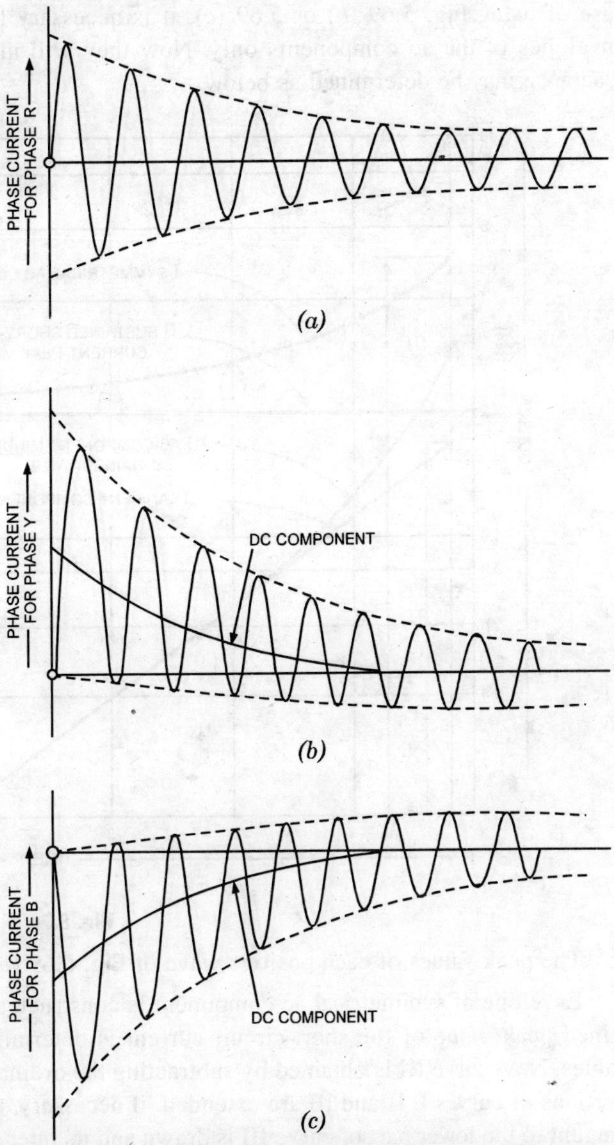

Fig. 5.69. ***Short-Circuit Current Waveforms in Three Phases of a Synchronous Generator***

5.23.1. Constant Flux Linkage Concept

Effects of short-circuit currents can be conveniently determined by using the concept of *constant flux-linkage,* which may be stated as below:

It is impossible to change the flux linkages in a closed electrical circuit that contains no resistance or capacitance (i.e. in a purely inductive circuit).

In case of a synchronous machine, the armature and field windings can be assumed almost purely inductive because they do not contain any capacitance and their resistances are almost negligible in comparison to their inductive reactances. So the flux linkages in the armature circuit and the field circuit cannot be changed suddenly by the application of short circuit to the armature winding. For maintaining these flux linkages constant, large changes of current may take place in both windings when the short circuit occurs in order to keep their respective flux linkages constant.

5.23.2. Transient Reactances

The characteristics of short-circuit currents of a 3-ϕ synchronous generator are shown in Fig. 5.69. These curves indicate that the short circuit occurred at the instant when the voltage of phase R was maximum (assuming that the machine was not delivering any current to the load prior to fault). The transient reactances can be obtained from any one of the three oscillograms by graphical methods. In case of using Fig. 5.69 (*b*) or 5.69 (*c*), it is necessary to eliminate the dc components and plot the envelopes of the ac components only. Now they will also appear as in Fig. 5.69 (*a*). The transient reactance may be determined as below:

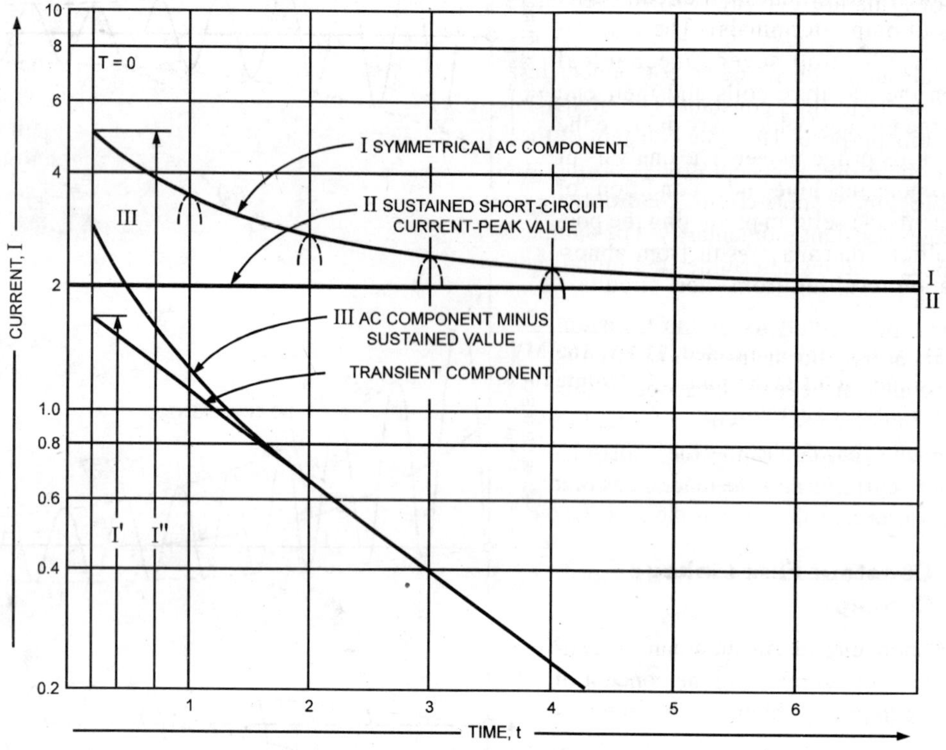

Fig. 5.70

The peak values of each positive wave of Fig. 5.69 (*a*) are plotted as shown in Fig. 5.70.

Envelope of symmetrical ac component is constructed as represented by curve I in Fig. 5.70. Sustained peak value of the short-circuit current is determined and curve II is drawn in terms of peak values. Now curve III is obtained by subtracting the ordinates of curve II from those of curve I. The left portions of curves I, II and III are extended, if necessary, to intersect the zero -time ordinate (T = 0). A tangent to the lower part of curve III is drawn and its intersection with zero time ordinate is determined.

If the terminal voltage of the machine before the fault taking place is E_0, the rated voltage is V and the rated rms value of the armature current is I, the direct transient reactance is given as

$$X'_d = \frac{E_0/V}{I'/1.414\,I} \text{ pu} \quad \text{or} \quad \frac{\sqrt{2}\,E_0}{I'} \text{ ohms} \qquad \text{...(5.41)}$$

The direct-axis subtransient reactance is given as

$$X''_d = \frac{E_0/V}{I''/1.414I} \text{ pu} \quad \text{or} \quad \frac{\sqrt{2}\,E_0}{I''} \text{ ohms} \qquad \text{...(5.42)}$$

Direct-axis subtransient reactance X''_d almost equals the leakage reactance of the machine.

The short-circuit produces an armature reaction tending to reduce the main field flux. As the flux of armature reaction is being established, a voltage is induced in the field circuit tending to increase the field current momentarily. This transient component of the field current is eventually dampened by the field resistance and the air-gap flux is gradually reduced at the rate, determined by the time constant of the field circuit. The rate of decay of this transient d-c current in the field and likewise the transient component of armature current is illustrated in Fig. 5.70. It is shown as a straight line tangent to curve III.

The initial rate of decay of ac component less the sustained value (curve III) is larger than that produced by the field time constant. The higher rate of decay is due to the action of damper winding provided for this purpose. The time constant of the damper winding is lower than that of the main field winding and, therefore, transient effects on the armature current are of short duration. The actions of damper winding under short-circuit condition are similar to those of field circuit in that it tends to keep its flux linkages constant momentarily. The damper winding current decays rapidly. Its action alongwith that of the field winding determines the initial rate of decay of the armature current at the instant that the short-circuit occurs.

Example 5.31. A 3-ϕ, star-connected, 11 kV, 165 MVA alternator has a synchronous reactance of 1 Ω and armature resistance of 0.1 Ω per phase. Calculate the short-circuit current at the instant of short circuit and sustained short-circuit current.

Solution : The magnitude of the short-circuit current at the instant of fault depends upon the magnitude of voltage at that instant and the value of sub-transient reactance. Since nothing is mentioned, it is assumed that the sub-transient reactance is the same as the synchronous reactance.

$$\text{Synchronous impedance, } Z_s = \sqrt{R_e^2 + X_s^2} = \sqrt{1^2 + 0.1^2} = 1.00\,5\ \Omega$$

$$\text{Short-circuit current at the instant of fault} = 2\sqrt{2}\,\frac{\text{Phase voltage}}{Z_s}$$

$$= 2\sqrt{2}\,\frac{11{,}000/\sqrt{3}}{1.005} = 17{,}874 \text{ A } \textbf{Ans.}$$

$$\text{Sustained short-circuit current} = \frac{\text{Phase voltage}}{Z_s} = \frac{11{,}000/\sqrt{3}}{1.005} = 6{,}319 \text{ A } \textbf{Ans.}$$

5.24. LOSSES AND EFFICIENCY

Just as in case of generators, the input to the alternators is not readily measured. The direct measurement of efficiency by actually loading is accompanied by the difficulties of providing the necessary power and finding suitable load. Efficiency is, therefore, determined by measuring the losses of the machine.

The losses in an alternator are :

1. Electrical losses that include field loss, armature winding loss and brush contact loss.

The copper loss in the field circuit is obtained by adding the $I^2_f R_f$ loss of field winding and the electrical loss of the field rheostat or, more simply, by multiplying the excitation voltage by the field current. This loss is constant for any given voltage and power factor, but varies with these two. The maintenance of rated voltage at low lagging pf needs comparatively large field current and gives the maximum field (or excitation) loss.

The armature copper loss in the stator windings is determined on the basis of the armature resistance. Whether the ohmic or the effective resistance should be used, depends upon the treatment of the stray load loss.

The electrical loss at the brush contacts, between the brushes and slip rings, is usually quite small and is often neglected.

2. Core loss is due to eddy currents and hysteresis caused by the flux resulting from the combined rotor and stator fields. Such loss occurs in pole faces and armature teeth and core. This loss is assumed to be independent of load, but varies with the excitation, if the excitation is to be varied with load and with the change in power factor, then this loss changes also. It is equal to the difference of power required to drive the alternator with and without the field excited.

3. Friction and windage loss includes ventilation loss (power required to circulate the cooling air), and loss due to bearing and brush friction. Since the speed is constant, so this loss for a particular machine remains constant. It is equal to the mechanical power required to drive the alternator at rated speed with no excitation.

4. Load or stray power loss is due to armature leakage flux which causes eddy currents and hysteresis in the iron surrounding the armature conductors in addition to the additional core loss caused by distortion of the magnetic field under load conditions. It may, however, be included in the efficiency calculation by using the effective value of the armature resistance instead of the dc resistance (See Art. 5.14).

After determining the foregoing losses, the efficiency may be determined from the following relation :

$$\text{Percent efficiency, } \eta = \left[1 - \frac{\text{kW losses}}{\text{kVA output} \times \text{pf} + \text{kW losses}}\right] \times 100 \qquad \ldots(5.43)$$

5.25. POWER FLOW IN AN ALTERNATOR

For generator operation the rotor of the machine is supplied with mechanical energy from its prime mover. A part of the mechanical energy is utilized or lost to overcome friction and windage, and hysteresis and eddy current losses (all losses except armature copper loss) and remainder of the energy supplied is converted into electrical energy in the armature circuit. Some of the electrical energy developed in the armature circuit is lost in overcoming armature copper loss. The remainder becomes the useful electrical output of the generator. The power flow diagram in case of a synchronous generator is given in Fig. 5.71.

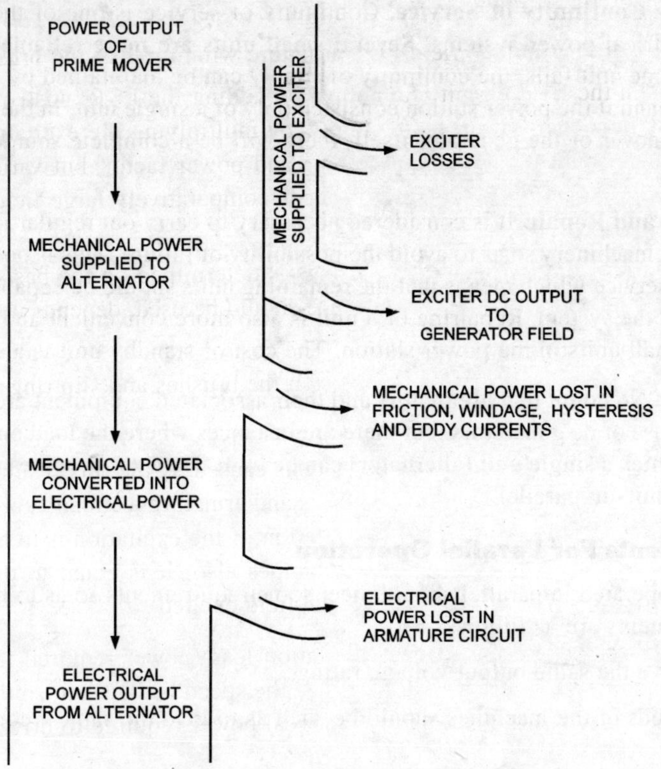

Fig. 5.71. *Power Flow Diagram For An Alternator*

5.26. PARALLEL OPERATION OF ALTERNATORS

In most power stations (whether dc or ac), it will be found that the power is supplied from several smaller units (generators) operating in parallel rather than from a single large unit capable of taking care of the maximum peak loads. There are a number of good reasons for this practice, even though in some instances it might be cheaper to install only one or two large units to take care of the system load (as a rule, large machines and their auxiliary equipment cost less per kW output than small machines). The demand for electrical energy continues to grow at a steady space, electrical utilities and others find it necessary to increase generating capacity at regular intervals. Since it is not economical to discard serviceable alternator (or alternators) in favour of larger ones, it becomes necessary to either operate alternators in parallel or else subdivide the load and feed it from multiple isolated systems. Of these two alternatives, parallel operation is much preferred because it does not require major changes in the existing system and provides greater reliability, improves efficiency, facilitates repairs and maintenance and makes the supply of load possible when it (load) exceeds the capacity of the largest single unit available.

1. **Efficiency.** The load on an electric power station fluctuates, usually having its peak value sometimes during the day and its minimum value during the night hours. Since generators operate most efficiently when operated on full load (or nearly full load), it is logical to operate a small unit delivering rated output when the load demand is light. Then as the load increases, a larger unit is substituted for the smaller one or another unit is connected to operate in parallel with the one already in operation. This keeps the machines loaded up to their rated capacity and increases the *efficiency of operation*.

2. **Reliability or Continuity of Service.** Continuity of service is one of the uttermost requirements of electrical power systems. Several small units are more reliable than a large single unit, since if one unit fails, the continuity of supply can be maintained by the remaining units. On the other hand if the power station consisted only of a single unit, in the event of breakdown of the prime mover or the generator itself, there will be a complete shutdown (failure of supply).

3. **Maintenance and Repair.** It is considered necessary to carry out regular inspection and maintenance of the machinery so as to avoid the possibility of failure. This is possible only when the unit is out of service which means that the remaining units should be capable to take care of the entire load on the system. Repairing of a unit is also more convenient and economical if there are several small units in the power station. The cost of standby unit is also small.

4. **Physical Size.** No doubt that alternators and their associated equipment are built in sizes much larger than those of dc generators, but there are instances where the load on the system exceeds the sizes to which a single unit (alternator) can be built. Hence it becomes necessary to operate two or more units in parallel.

5.26.1. Requirements For Parallel Operation

The alternators to be operated in parallel should meet some requirements so as to make their operation proper. These requirements are as follows:

1. They must have the same output voltage rating.
2. The rated speeds of the machines should be such as to give the same frequency (*f* being equal to $\frac{PN}{120}$).
3. The alternators should be of the same type so as to generate voltages of the same waveform. They may differ in their kVA ratings.
4. The prime movers of the alternators should have same speed-load characteristics, which of course must be drooping ones, so as to load the generators in proportion to their output ratings.
5. The alternators should have reactances in their armatures, otherwise they will not operate in parallel successfully.

5.27. SYNCHRONIZATION OF ALTERNATORS

It is imperative that changing from one unit to another or connecting an additional unit in parallel with the others (already operating in parallel) is carried out with the minimum possible disturbance to the existing system. Exactly how an alternator is taken off or put on a line without disturbing the voltage and frequency of the system will be discussed here.

The process of connecting an alternator in parallel with another alternator or with the common bus-bars (the bus-bars to which a number of alternators are connected) is called *synchronization*.

5.27.1. Conditions For Proper Synchronization

It will be logical to assume that only one alternator is to be placed in parallel with one or more alternators already in operation to meet an additional load. Those alternators, which are in operation and sharing the load, are called the *running machines*, while that which is to be introduced in the system is known as the *incoming machine*. Before the *incoming machine* is connected in parallel with the running machine or machines it is necessary that it meets certain conditions, which are given on next page.

1. The terminal voltage of the incoming machine must be exactly equal to that of the others, or of the bus-bars connecting them.
2. The speed of the incoming machine must be such that its frequency $\left(\text{being equal to } \frac{PN}{120}\right)$ equals bus-bar frequency.
3. The phase of the incoming machine voltage must be the same as that of the bus-bar voltage relative to the load *i.e.* the phase voltages of the incoming machine and the bus-bar should be in phase opposition. This implies that there will be no circulating current between the windings of the alternators already in operation (the bus-bars) and the incoming machine.

In case of 3-phase alternators, an additional requirement to be met is that the phase sequence of the incoming machine is the same as that of the bus-bars.

5.27.2. Synchronization of Single Phase Alternators

Before connecting an alternator, to the bus-bars or in parallel with another alternator already supplying the load, the three conditions mentioned in Art. 5.27.1. must be fulfilled.

The equality of the terminal voltages* can be determined readily by connecting a voltmeter first to the running alternator or the bus-bars and then to the incoming alternator. The remaining two conditions *i.e.* equality of frequency and phasing out can be determined with the help of two lamps connected across the ends of the double-pole paralleling switch and known as *synchronising lamps*.

Let alternator B is to be connected to bus-bars, to which alternator A is already connected. The method to be followed for doing so is given below.

The prime mover of alternator B (incoming alternator) is started and brought up close to the rated speed. The alternator is then excited and its voltage is raised, by increasing its excitation, to that of bus-bars or the running alternator. The synchronising lamps L_1 and L_2 are connected across the double pole paralleling switch of the incoming alternator B, as shown in Fig. 5.72. If the frequencies of the alternators A and B are exactly the same and their terminal voltages, with respect to local series circuit, are in exact phase opposition, no resultant voltage will act across the lamps L_1 and L_2 and, therefore, these lamps will remain dark. If the frequencies of the alternators A and B are not equal, the phase angle between their voltages will be continually changing, and therefore, the current through the synchronising lamps and through the local series circuit (shown dotted in the figure) will be changing, resulting in flickering of lamps. The frequency of flickering of lamps will be equal to the difference of frequencies of the two alternators *i.e.* greater the difference in frequencies, greater will be frequency of flickering and lesser the difference in frequencies, lesser will be the frequency of flickering. In the middle of the dark period the two voltages will be in phase opposition with respect to the local series circuit. Synchronising is done in the middle of the dark period. This is known as *dark lamp method*.

Hence at the time of synchronising of alternators the speed of the incoming machine is adjusted until the lamps go in and out very slowly, the terminal voltage of the incoming alternator is made equal to bus-bar voltage by adjusting the excitation of the incoming alternator and then switch of incoming alternator is closed in the middle of the dark period.

Since it is somewhat easier to judge the middle of bright than middle of dark period, therefore, some engineers prefer to synchronise in the middle of the bright period. This method is known as *bright*

* The equality of terminal voltages can also be determined by taking the readings of the voltmeters connected to the running alternator or the bus-bars and to the incoming alternator.

lamp method and necessitates the crossing-over of the lamp connections, as in the case of alternaor C in Fig. 5.72.

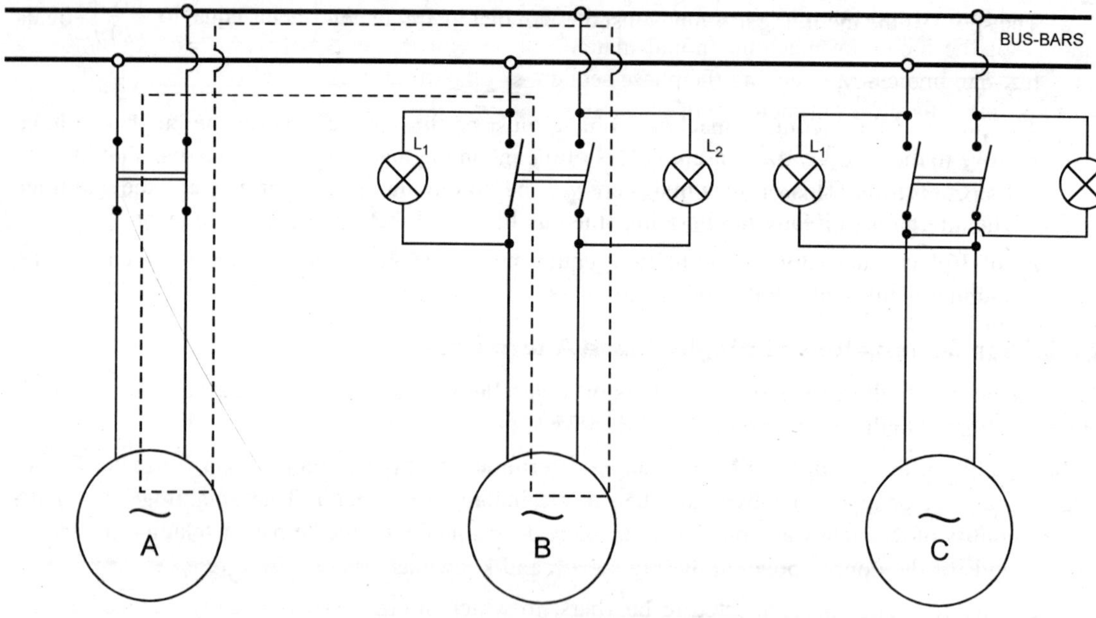

Fig. 5.72. ***Synchronising of Single Phase Alternators***

The maximum voltage across individual lamp in dark lamp method will occur when, during synchronising process alternator B is in phase with alternator A with respect to local series circuit. In case of bright lamp method the maximum voltage across individual lamp will occur when alternator C is in phase opposition with alternator A. The magnitude of maximum voltage that can exist across a lamp is double that of the voltage of one machine.

As it is obviously impossible to use lamps on any other except low voltages, small instrument transformers, known as potential transformers, are, therefore, necessary for reducing the voltage applied to the synchronising lamps.

After alternator B or C has been synchronised and is operating in parallel with alternator A, the incoming alternator may be made to deliver the load in part or in full, assuming of course, that its rating is sufficient to do so. The load is shifted from the running alternator to the incoming alternator by increasing the mechanical power input to the prime mover of the incoming alternator and simultaneously reducing the mechanical power input to the prime mover of existing alternator A. This may be readily done by admitting more steam (opening the steam valve) to the prime mover of the incoming alternator (if it is a steam machine) and simultaneously closing the steam valve of alternator A. When the incoming alternator carries the entire load, as indicated by the ammeters and wattmeters in the line circuits, the line circuit breaker and field breaker of alternator A are opened, the later unit is then shut down if desired.

The most important point, to be emphasized, is that load cannot be shifted from one machine to another by adjustment of excitation. The adjustment of excitation changes the voltage and power factor of the alternator.

5.27.3. Synchronization of 3-Phase Alternators

The same two methods of synchronising, as already discussed for synchronising of single phase alternators, in Art. 5.27.2, but with slight modifications are used for three phase alternators.

The additional condition to be met in case of synchronising of 3-phase alternators is that the phase sequence of the incoming machine should match with that of the system to which it is going to be connected. The process of checking the phase sequence and getting it correct is known as *'phasing out the alternator.'* The different methods of phasing out of alternators are given below:

I. Three Dark Lamp Method. Figure 5.73 shows an alternator A supplying a load, with alternator B all wired up and ready for start-up. If these are low voltage alternators, the alternator B can be phased out by connecting three lamps across the ends of the triple-pole paralleling switch as shown in Fig. 5.73. The prime mover of the incoming alternator B is started and brought up close to the rated speed. The alternator is then excited and its voltage is raised by increasing the excitation to that of the running alternator or the bus-bars and the lamps are watched closely. If the incoming alternator B is properly connected all the three lamps should become bright and dim together. If they bright and dim in sequence, it shows that incoming alternator B is not properly connected with the bus-bars, and the phase sequence of the incoming alternator B, therefore, must be reversed relative to the system. The usual way to do this is to interchange any two of the leads on either the alternator side or the line side of the switch. It is possible to achieve the same effect by reversing the direction of rotation of the alternator, but this is very seldom convenient.

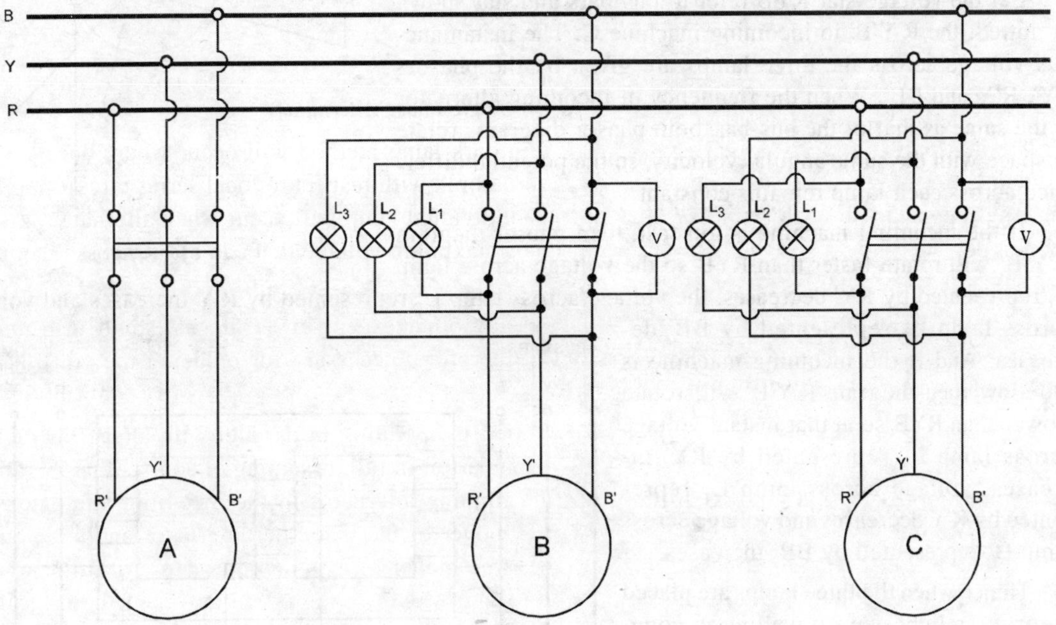

Fig. 5.73. ***Synchronising of 3-Phase Alternators***

Once the alternator B has been properly phased out, it will always be correct unless either the wiring or the rotation is altered so establishing the proper phase sequence is not a continual part of the synchronising process of the alternators.

The speed of the prime mover of the incoming machine is further adjusted until the lamps flicker at a very low rate (usually less than one dark period per second) and the voltage of the incoming machine is again adjusted and made equal to that of the running generator (or bus-bars) and the paralleling switch is closed at the instant all the three lamps are dark.

The disadvantage of this method is that lamps go dark at somewhat less than half their rated voltage, and so the paralleling switch might be closed when there is a considerable phase difference between the machines, with a high circulating current resulting in possible damage to the machines. Another disadvantage is that the lamp filaments might burn out. For this reason it is desirable to have the two frequencies nearly equal, so that the synchronising is not dependent on a constant dark period. A third disadvantage is that the flicker of lamps does not indicate whether the incoming alternator is slow or fast.

The above difficulties are eliminated in *two bright and one dark lamp method* in which two of the lamps are cross-connected, as shown in case of incoming alternator C in Fig. 5.73.

II. Two Bright and One Dark Lamp Method. When the incoming alternator is in synchronism, with alternator A or bus-bars, lamps L_1 and L_3 are bright and L_2 is dark.

Since near the point of synchronism, the brightness of one lamp increases and of another decreases so the instant at which the incoming machine is in synchronism with the bus-bars can be accurately determined and the paralleling switch should be closed at this instant.

Let the voltage star RYB refer to bus-bars and star shown in dotted, the R'Y'B' to incoming machine C. The instantaneous voltage across the three lamps are given by the phasors RY', R'Y and BB'. When the frequency of incoming alternator is the same as that of the bus-bar, both phasor diagrams rotate in space with the same angular velocity, so the potential difference across each lamp remains constant.

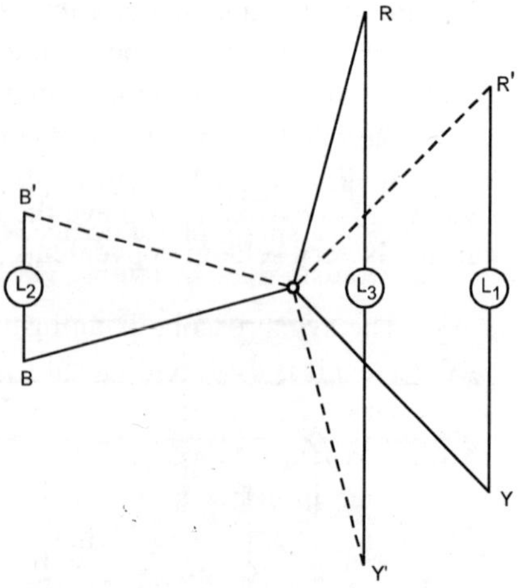

Fig. 5.74

If the incoming machine is too fast, then phasor diagram R′Y′B′ will rotate faster than RYB so the voltage across lamp L_3 represented by RY' decreases, the voltage across lamp L_1 represented by R'Y increases and voltage across lamp L_2 represented by BB' decreases. And if the incoming machine is too slow then diagram R'Y'B' will rotate slower than RYB so at that instant voltage across lamp L_3 represented by RY' increases, voltage across lamp L_1 represented by R'Y decreases and voltage across lamp L_2 represented by BB' increases.

Hence when the three lamps are placed in a ring, a light wave travelling in counter-clockwise direction indicates that the incoming machine is slow and light wave travelling in clockwise direction indicates that the incoming machine is fast, so by observing the sequences of brightness of the lamps, it can be determined whether the incoming machine is fast or slow.

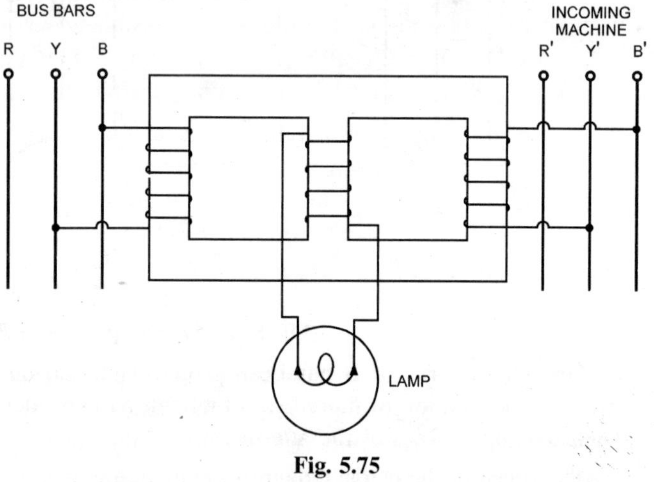

Fig. 5.75

The paralleling switch is closed when the changes in light are very slow and at the instant the lamp L_2 connected directly across one phase is dark.

In case of high-voltage generators, which is the usual case, direct connection of lamps is not practical. In such cases the lamps may be connected through potential transformers (the potential transformers directly across the switch and the lamps into the secondary circuits of the potential transformers). However, in most installations there will be potential transformers connected across the line on both sides of the paralleling switch.

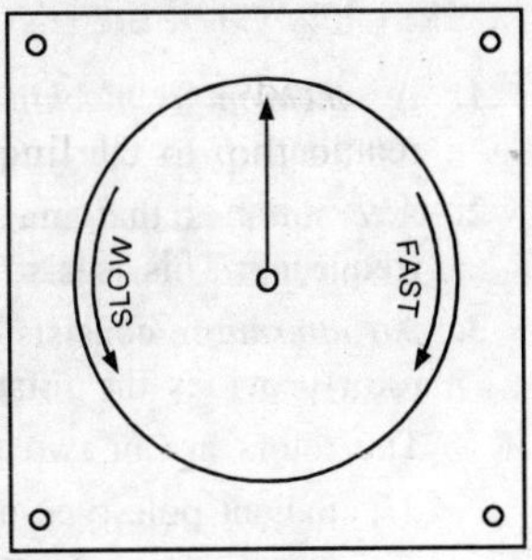

Fig. 5.76

Alternatively, the synchronising can be done with the help of one three limbed transformer and a lamp, the two primary windings on the outer limbs being connected to the bus-bars and the incoming alternator respectively, and the secondary winding on the central limb to a lamp (Fig. 5.75). The correct moment for closing the paralleling switch is the middle of the dark period if the primaries are each connected the same way round, and the middle of the bright period if the connections of one winding are reversed.

If the phase sequence is known to be correct, the best method of synchronising alternators is by means of a single phase device, called the *synchroscope*, which provides a more accurate indication of synchronism than do lamps. The dial markings of the synchroscope are shown in Fig. 5.76.

The synchroscope is an instrument for indicating difference of phase and frequency between two voltages. It is essentially a split-phase motor in which torque is developed if the frequencies of the two voltages differ. Voltages from corresponding phases of the incoming and running generators are applied to the synchroscope. A pointer, which is attached to the rotating part of the instrument, moves over the dial face in either a clockwise or counter-clockwise direction, depending on whether the incoming machine is fast or slow, when the frequencies of the two alternators are equal, no torque is exerted on the rotating part of the instrument and so the pointer stops. When the pointer stops in the vertical position, the frequencies are equal, the voltages are in phase, and the paralleling switch may be closed.

In practice it is sometimes difficult to adjust the speed of the incoming machine closely enough to stop the synchroscope pointer in the vertical position. In such a case, the speed of the incoming machine should be adjusted as closely as possible and the paralleling switch closed just before the pointer reaches the vertical position when travelling in the fast direction. This makes the incoming machine to take a small amount of load immediately after closure of switch and results in stable operation.

It is very important that the synchronising operation be done as accurately as possible. If the switch is closed when the alternator is out of phase with the system, a high circulating current will momentarily flow through the machine and pull it ahead (or back) to its correct position. However, this creates a voltage disturbance on the system, and sometimes the magnetic forces developed on the alternator windings will damage the coils.

If the synchronising operation has been correctly accomplished, the machine will now be "floating on the line," *i.e.*, it will neither contribute to nor draw anything from the system. To make the machine deliver load (kW) the driving torque is to be increased. This is done by raising the speed setting of the governor on the prime mover, but this tends to raise the frequency of the electrical system. The only way to prevent frequency rise is to lower the governor settings on the prime mover (s) of one or more other alternators on the system.

HIGHLIGHTS

1. A *synchronous machine* is an ac machine in which the rotor moves at a speed which bears a constant relationship to the frequency of currents in the armature winding.
2. *Alternator* is that machine which converts mechanical power into ac electrical power at a desired frequency. This is also known as a synchronous generator.
3. An *alternator* consists of a stator and a rotor. The stator provides the armature windings whereas the rotor provides the rotating magnetic field.

   The rotors are of two types :

   (*i*) Salient pole type used almost entirely for low and moderate speed alternators (hydro-generators or diesel-engine driven generators) being least expensive and providing ample space for field ampere-turns.

   (*ii*) Smooth cylindrical or non-salient pole type used in very high speed alternators (turbo-generators), being of smaller diameter and robust construction.
4. The alternators may be rotating armature type or rotating field type but practically all medium and large machines are always constructed with revolving field because of its superiority over the rotating armature type.
5. In a multipolar machine one pair of poles or one cycle of flux distribution equals 360 electrical degrees or 2π electrical radians. Since there P/2 complete wavelengths or cycles in one complete revolution, it follows that

$$\theta_{elec} = \frac{P}{2}\,\theta_{mech}$$

6. The frequency of induced emf in a coil in a multipolar machine is given by the expression

$$f = \frac{PN}{120}\ \text{Hz}$$

   where P is number of poles and N is the mechanical speed of rotation in rpm.
7. The radial frequency ω_{elec} of the wave of emf induced in P-pole machine is given as

$$\omega_{elec} = \frac{P}{2}\,\omega_{mech}$$

8. The expression for induced emf in a three phase alternator is given by expression

$$E_{rms}/\text{phase} = 4.44\ K_d\,K_p\,\phi f T \text{ volts}$$

   where ϕ is the useful flux per pole in webers, f is frequency in Hz and T is the number of series turns per phase and K_d and K_p are respectively the distribution factor and pitch factor and are given below:

   Distribution factor,

$$K_d = \frac{\text{Phasor sum of component emfs}}{\text{Arithmetic sum of component emfs}} = \frac{\text{Sin}\,\frac{m\beta}{2}}{m \sin\frac{\beta}{2}}$$

   For rth harmonic

$$\text{Distribution factor, } K_{dr} = \frac{\text{Sin}\,r\frac{m\beta}{2}}{m \sin\frac{r\beta}{2}}$$

   where $m\beta$ is known as the phase spread and is expressed in electrical degrees – m being the number of slots per pole per phase and β is the angular displacement between the slots $\left(\beta = \frac{180°}{\text{Number of slots per pole, } n}\right)$

$$\text{Pitch factor, } K_p = \frac{\text{Phasor sum of coil side emfs}}{\text{Arithmetic sum of coil side emfs}} = \text{Cos } \frac{\alpha}{2}$$

For *r*th harmonic

$$\text{Pitch factor, } K_{pr} = \text{Cos } \frac{r\alpha}{2}$$

where α is the chording angle in electrical degrees.

9. Winding factor, $K_w = K_d K_p$.
10. The sources of harmonics in the output voltage of an alternator are the non-sinusoidal waveform of the field flux and variation in reluctance of the air gap due to slotting of the stator core.

    The harmonics from the output voltage, however, can be eliminated by using (*i*) distributed winding (*ii*) short-pitched winding (*iii*) skewed slots (*iv*) fractional slot winding (*v*) larger length air gap.
11. The effect of armature flux on the main field flux is known as *armature reaction*. Armature reaction has distorting effect on unity power factor, wholly demagnetising at zero *pf* lagging and wholly magnetizing at zero power factor leading.
12. The sum of leakage reactance X_L and fictitious reactance X_a (representing armature reaction effect) is known as synchronous reactance X_s *i.e.*,

$$X_s = X_L + X_a$$

13. The phasor sum of effective resistance of armature R_e and synchronous reactance X_s is known as synchronous impedance Z_s *i.e.*,

$$\mathbf{Z}_s = R_e + j\,X_s$$

    Armature winding effective resistance R_e in alternators is usually very small in comparison to synchronous reactance X_s, and therefore, synchronous impedance may be assumed equal to the synchronous reactance for many purposes.
14. With the variation in load on an alternator, voltage drops due to (*i*) armature resistance (*ii*) armature leakage reactance and (*iii*) on account of armature reaction vary, so its terminal voltage varies.
15. For constant load current, excitation and speed, the terminal voltage of an alternator decreases when power factor is changed from leading to lagging.
16. Voltage regulation of an alternator is given by the expression

$$\text{Voltage regulation} = \frac{E_0 - V}{V} \text{ per unit}$$

$$\text{or } \frac{E_0 - V}{V} \times 100 \text{ in percentage}$$

    where E_0 is no-load terminal voltage and V is the full-load terminal voltage.

    The voltage regulation of an alternator is positive and increases with the decrease in lagging power factor. It may become negative for leading power factor.
17. Voltage regulation may be determined by any of the following methods:

    1. Synchronous impedance (or emf) method – pessimistic method.
    2. Ampere-turn (or mmf) method – optimistic method.
    3. Potier (or zero power factor) method.
18. Zero power factor characteristic of a synchronous machine is used for determination of (*i*) leakage reactance drop (*ii*) voltage drop due to armature reaction and (*iii*) armature reaction ampere-turns.
19. Power developed per phase by a 3-phase non-salient synchronous machine is given by

$$P = \frac{EV}{Z_S}\cos(\theta - \delta) - \frac{V^2}{Z_S}\cos\theta = \frac{EV}{X_S}\sin\delta$$

if stator resistance is negligible.

20. Reactive power developed per phase by a 3-phase non-salient synchronous machine is given by

$$Q = \frac{EV}{Z_S}\sin(\theta - \delta) - \frac{V^2}{Z_S}\sin\theta$$

$$= \frac{EV}{X_S}\cos\delta - \frac{V^2}{X_S} \qquad \text{if stator resistance is negligible.}$$

21. A multipolar machine with cylindrical rotor has a uniform air gap and, therefore, its reactance remains the same, irrespective of the spatial position of rotor. But in case of salient pole machines, the air gap is not uniform and its reactance varies with the rotor position. Because of non-uniformity of the reluctance of the magnetic paths, the mmf of the armature is divided into two components viz. (*i*) a direct-acting component along the field pole axis, called the direct-axis and (*ii*) quadrature-component along the axis passing through centre of the interpolar space, called the quadrature-axis. These facts form the basis of the two-reactance theory applied to the salient pole machines.

22. Power developed by a 3-phase salient pole synchronous machine is given by

$$\frac{3E_0V}{X_d}\sin\delta + \frac{3V^2}{2}\left[\frac{1}{X_q} - \frac{1}{X_d}\right]\sin 2\delta \quad \text{where } E_0 \text{ and V are phase values.}$$

Above expression for power developed consists of two terms, the first term representing power due to field excitation and the second term gives the reluctance power *i.e.* power due to saliency. In case of a cylindrical rotor machine $X_d = X_q$ and so the second term becomes zero and the power is given by the first term only. In case there is no field excitation *i.e.* E_0, the first term becomes zero and the power developed is given by the second term.

It may be noted that value of δ is positive in case of an alternator and negative for a motor.

23. Electrical losses (including armature winding loss, brush contact loss and field loss), core loss, friction and windage loss and stray *power loss occur in an alternator.*

Efficiency of a synchronous generator is given by

$$\eta = 1 - \frac{\text{kW losses}}{\text{kVA output} \times \text{pf} + \text{kW losses}} \times 100$$

24. Requirements of parallel operation:

(*i*) Same output voltage rating.

(*ii*) The rated speeds of the machines should be such as to give the same frequency.

(*iii*) The alternators should be of the same type so as to generate voltages of the same waveform.

(*iv*) The prime movers of the alternators should have same speed-load characteristics, which of course must be drooping ones.

(*v*) The alternators should have reactances in their armatures.

25. Conditions for proper synchronizing:

(*i*) Terminal voltage of incoming machine must be exactly equal to that of other machines or bus-bar.

(*ii*) The frequency of the incoming machine must be equal to bus-bar frequency.

(*iii*) The phase of the incoming machine must be same as that of the bus-bar voltage.

(*iv*) In case of 3-phase synchronous generators, phase sequence of the incoming machine and that of the bus-bar must be same.

EXERCISES

1. What are the different types of ac generators in use. Explain the essential differences in their construction.

2. Discuss the salient constructional features of ac generators driven by (*i*) diesel engines (*ii*) steam engines.

3. Explain the essential difference between cylindrical and salient pole rotors used in large alternators. Also suggest various methods of excitation system for large alternators.
4. Explain the essential difference between cylindrical (smooth) and salient pole rotors used in large alternators. What types of rotor would you expect in (*i*) a 2-pole machine (*ii*) a 12-pole machine.
[M.D. Univ. Electromechanical Energy Conversion May 2004]
5. Give the constructional details of salient pole synchronous machine.
[M.D. Univ. Electromechanical Energy Conversion December 2005]
6. Explain the difference between salient and cylindrical pole type of rotor used in alternators.
[M.D. Univ. Electromechanical Energy Conversion December 2006]
7. Explain the constructional features of synchronous generators. What are two types of generators? Give the field of their application. Derive an expression for the emf generated in the armature winding of a synchronous machine.
8. What are the advantages of using distributed winding and short pitch coil ? Derive, from the fundamental, the expressions for winding factor and pitch factor.
9. Explain the terms coil span factor and distribution factor in connection with alternator armature windings and deduce the emf equation of an alternator incorporating effects of these factors.
10. Derive, from the fundamental, generalized equation of the EMF generated in the non-salient pole synchronous generator, taking into account the effect of 5th and 7th harmonic components.
11. Derive an expression for the emf generated in the armature winding of a synchronous machine. Also explain how harmonics in the emf wave are eliminated by short-pitch winding.
12. Explain the principle of operation of synchronous generator. Derive the emf equation.
[M.D. Univ. May 2007]
13. Derive an expression for induced emf of an alternator. **[M.D. Univ. December 2006]**
14. Write a short note on “armature reaction in alternator”. **[M.D. Univ. Electrical Machines-II May 2009]**
15. What is the distribution factor ? What is its effect ? Derive an expression for distribution factor of a winding having β slots per pole per phase and a slot angle β.
[M.D. Univ. Electrical Machines-II, December 2010]
16. Explain how the armature reaction influences the field distribution of an alternator for varying power factor.
17. Define armature reaction. How it affects the main field under unity, 90º lagging and 90º leading power factor conditions. Explain with the help of suitable diagrams.
18. Discuss operation of an alternator on load. Define synchronous reactance and develop phasor diagram of a loaded alternator. **[M.D. Univ. May 2006, May 2008]**
19. Draw and explain the phasor diagram of a loaded synchronous generator under lagging, leading and unity power factor conditions.
20. Explain synchronous impedance of an alternator.
21. Draw phasor diagram of a non-salient pole machine supplying full load at (*i*) lagging power factor and (*ii*) leading power factor.
22. Sketch and explain the open-circuit and short-circuit characteristics of a synchronous machine.
23. Discuss about the ‘open-circuit’ and ‘short-circuit’ characteristics of a synchronous generator. Draw the phasor diagram under short-circuit condition. What do you understand by the term ‘short-circuit ratio’? Discuss how the short-circuit ratio can be calculated from the two characteristic curves.
24. In an alternator, explain why short-circuit characteristic is a straight line whereas open-circuit characteristic is a curve.
25. Discuss the synchronous impedance method of calculating the voltage regulation of an alternator, at different load power factors. Give phasor diagrams.

26. Explain the mmf method of determining the voltage regulation of alternator.

[M.D. Univ. Electrical Machines-II, May 2009]

27. Compare synchronous impedance method and ampere-turn method of predetermining regulation of alternators.

28. Explain an experimental method of separating stator leakage reactance drop and drop due to armature reaction, when an alternator is loaded.

29. Explain the determination of per unit regulation of an alternator at full load unity power factor, 0.8 pf leading and 0.8 pf lagging using mmf method. Use suitable example.

30. Using suitable phasor diagram of alternator, explain why Potier test is conducted at zero pf lagging load.

31. Derive the expression for power in the case of synchronous generator from the fundamentals. Determine also the maximum power and corresponding power angle δ.

32. Explain the Potier triangle method of determining voltage regulation of an alternator.

[M.D. Univ. Electrical Machines-II, December 2010]

33. Derive the condition for maximum power output of a synchronous generator connected to infinite bus-bars and operating at constant excitation.

34. Mention effect of saliency of alternator pole. Also discuss two reactance concept for salient pole synchronous machine. **[M.D. Univ. December 2007]**

35. Explain the two reaction theory applicable to salient pole synchronous machine.

[M.D. Univ. Electrical Machines-II, December 2010]

36. Discuss "Two Reaction Theory" as applied to salient pole synchronous machines.

Derive an expression of power developed in a cylindrical rotor alternator in terms of power angle and synchronous reactance and therefrom write the value of maximum power developed. Ignore stator resistance.

37. Explain how to determine the direct and quadrature axis reactances of a salient pole synchronous machine.

38. Explain the two reaction theory of salient pole synchronous machines.

Describe a method of determining direct and quadrature axis reactances (X_d and X_q) of salient pole alternator.

39. Describe the slip test method for the measurement of X_d and X_q of synchronous machines.

40. Draw and explain the phasor diagram of a salient pole alternator supplying a lagging pf load.

41. For salient pole synchronous machine, neglecting the effect of armature resistance, derive an expression for power developed as a function of load angle.

42. Explain two reaction theory applicable to salient pole synchronous machine. For a salient pole synchronous machine show that the power output per phase is given by

$$P = \frac{E_f V_t}{X_d} \sin \delta + \frac{V_t^2}{2}\left[\frac{1}{X_d} - \frac{1}{X_q}\right] \sin 2\delta$$

where E_f = Excitation emf and V_t = Terminal voltage.

43. Show that the expression of power developed (P_d) for a salient synchronous machine is

$$P_d = \frac{E_0 V}{X_d} \sin \delta + \frac{V^2}{2} \cdot \frac{X_d - X_q}{X_d X_q} \sin 2\delta$$

where
E_0 = Excitation voltage
V = Bus-bar voltage
X_d = *d*-axis synchronous reactance
X_q = *q*-axis synchronous reactance
δ = Torque angle.

Draw also power-torque angle characteristics for both cylindrical and salient pole machines. Explain, what is meant by reluctance power and why salient pole machines are considered more stable.

44. Discuss the phenomenon of sudden 3-phase short-circuit at the armature terminals of an alternator. Draw a typical wave shape of current and mark the different regions. Write an expression for the current. **[M.D. Univ. Electrical Machines-II, December 2010]**

45. List the various losses that incur in an alternator. Describe a suitable method of determining losses and efficiency of alternator.

46. What is the necessity of parallel operation of alternators ? State the conditions necessary for parallel operation. **[M.D. Univ. Electrical Machines-II, December 2010]**

47. Mention the need for parallel operation of alternators. State the conditions to be satisfied before connecting an alternator to the infinite bus-bars.

48. What do you mean by synchronising of alternators ? Discuss two bright and one dark lamp synchronising method.

49. What are the conditions of parallel operation of synchronous generators ? Write, in sequence, the steps needed to connect a synchronous generator with infinite bus by dark bright lamp method.

50. What do you mean by synchronising of alternators ? Describe any one method of synchronising. **[M.D. Univ. Electrical Machines-II, May 2009]**

SHORT ANSWER TYPE QUESTIONS WITH ANSWERS

Q. 1. What are the essential elements for generating emf in alternators ?

Ans. The essential elements for generating emf in alternators are prime mover, stator, rotor and exciter.

Q. 2. Why is it possible to design alternators to generate much higher voltages than dc generators ?

Ans. In case of an alternator, field system is moving and armature is stationary and it is easier to insulate stationary armature winding for very high voltage because insulation of stationary armature is not subjected to mechanical stresses due to centrifugal action and more space is available on the armature for providing more insulation as the stator is outside the rotor. In case of dc generators field is stationary and armature is moving.

Q. 3. What type of rotor is adopted for high speed alternators ?

Ans. Cylindrical or non-salient pole rotor construction is used for high speed alternators.

Q. 4. What type of alternators will be used (*a*) for hydraulic turbines (*b*) for steam turbines as prime movers ?

Ans. (*a*) Salient pole type (*b*) smooth cylindrical rotor type.

Q. 5. Name the two types of alternators depending on the rotor construction.

Ans. Salient pole alternators and cylindrical rotor alternators.

Q. 6. Name the types of alternators depending on the type of prime mover.

Ans. (*i*) Turbo-generators (*ii*) hydro-generators (or water-wheel generators) and (*iii*) diesel-engine driven generators.

Q. 7. What will be the number of poles of a 50 Hz alternator if it runs at its greatest speed ?

Ans. Two.

Q. 8. What is the maximum speed of a 50 Hz alternator ?

Ans. 3,000 rpm.

Q. 9. Why are salient-pole alternators more suitable for low speed and non-salient pole for high speed operation ?

Ans. The salient pole type field structure is used almost entirely for low speed alternators, since it is least expensive and provides ample space for the field ampere-turns but it cannot be used for high speed alternators on account of high peripheral speed and the difficulty of obtaining sufficient mechanical strength. So non-salient pole field structure is used for high speed operation.

Q. 10. Which one of the cylindrical and salient pole alternators is more stable ?

Ans. Salient-pole alternator is more stable.

Q. 11. Do we use laminations for all the iron parts of an electrical machine ?

Ans. All the iron parts of an electrical machine are laminated except the end covers and outer stator frame. The laminations are used to reduce the eddy current loss.

Q. 12. Why are the poles and pole shoes laminated ?

Ans. For reducing eddy current loss.

Q. 13. Why machines having long core length are provided with radial ventilating ducts while machines having deep cores are provided with axial cooling.

Ans. Radial ventilating ducts provided with long core machines take away heat at faster rate. Similarly in case of deep cores axial ventilating ducts do. Otherwise the central portion of long core and the inner portion of deep core may attain a temperature higher than the permissible value.

Q. 14. What is the role of damper winding in (*a*) synchronous generator and (*b*) synchronous motor ?

Ans. (*a*) In a synchronous generator damper winding is provided to suppress the negative sequence field and to damp the oscillations when the generator starts hunting.

(*b*) In a synchronous motor, damper winding is provided to develop starting torque so as to make the motor self starting and to develop damping torque when the machine starts hunting.

Q. 15. Why rotating machines with aluminium armature coils have increased leakage reactance ?

Ans. Rotating machines with aluminium armature coils need more space to accommodate conductors and so large size of stator slots resulting into increased leakage reactance.

Q. 16. At what voltage is the field of an alternator usually excited ?

Ans. 125 or 250 V dc.

Q. 17. What is an exciter ?

Ans. An exciter is a small dc generator (usually flat compound wound type) to supply dc power to the field magnet system or rotor of the alternator.

Q. 18. Why are distributed windings preferred over concentrated windings ?

Ans. Distributed windings are preferred over concentrated windings as they (*i*) improve waveform of induced emf (*ii*) provide better utilization of core (*iii*) diminish armature reaction and (*iv*) facilitate cooling.

Q. 19. What is the necessity for chording in the armature winding of a synchronous machine ?

Ans. The armature winding is chorded in order to (*i*) improve the waveform of induced emf (*ii*) save copper in the coil ends due to less span and (*iii*) reduce inductance of the winding.

Q. 20. A full pitch coil in an alternator has a span of 18 slots. What should be the span if the third harmonic component of the emf is to be eliminated ?

Ans. 12 slots.

Q. 21. What is the other name for distribution factor ?

Ans. Breadth factor.

Q. 22. Is it possible to make a balanced 3-ϕ, 6-pole winding with 48 slots.

Ans. Number of slots per pole per phase, $m = \frac{48}{3 \times 6} = \frac{8}{3}$. As the denominator of m is not divisible by 3, it is not possible to have balanced winding.

Q. 23. What is the value of distribution factor for a uniformly distributed 3-phase winding.

Ans. Distribution factor for a uniformly distributed 3-phase winding with 60° phase spread is 0.955.

Q. 24. What is meant by "armature reaction" of a synchronous machine ?

Ans. The effect of armature flux on the main field flux is known as armature reaction. Armature reaction has distorting effect on unity power factor, wholly demagnetising at zero power factor lagging and wholly magnetising at zero power factor leading.

Q. 25. What parameter of load influences the armature reaction of an alternator ?

Ans. Power factor of load.

Q. 26. Does the armature mmf of an alternator delivering a capacitive load magnetise or demagnetise the main field.

Ans. Magnetises.

Q. 27. When the load on an alternator is varied, its terminal voltage is also found to vary. Why ?

Ans. With the variation in load on an alternator, voltage drops due to (*i*) armature resistance (*ii*) armature leakage reactance and (*iii*) on account of armature reaction vary, so its terminal voltage varies

Q. 28. By which tests synchronous reactance of a synchronous machine is determined ?

Ans. By open-circuit and short-circuit tests.

Q. 29. Why is the short-circuit characteristic of an alternator linear?

Ans. The short-circuit characteristic is normally a straight line through the origin because the net excitation is so small that there is no saturation in the magnetic circuit.

Q. 30. What do you mean by the short-circuit characteristic of an alternator ?

Ans. The short-circuit characteristic of an alternator is a curve (usually a straight line passing through origin) plotted between short-circuit current and field current while the alternator being driven at its rated speed.

Q. 31. What are the effects on voltage regulation of an alternator of power factor (*a*) at lagging power factor and (*b*) leading power factor.

Ans. The regulation of an alternator is positive and increases with the decrease in lagging power factor. The regulation of an alternator improves first with the decrease in leading power factor, becomes unity and further decrease in leading power factor makes the regulation negative and increasing.

Q. 32. Why voltage regulation of an alternator is negative for leading power factor ?

Ans. When the power factor is leading, the effect of armature flux is to assist the main flux, hence to generate more emf and so to increase the terminal voltage when the alternator is loaded. Thus the terminal voltage of an alternator decreases when the load of leading pf (i.e capacitive load) is thrown off and voltage regulation is –ve.

Q. 33. Why does synchronous impedance method give a poorer voltage regulation ?

Ans. In synchronous impedance method of determination of voltage regulation synchronous reactance is assumed to be constant while actually it varies with the saturation (at low saturation its value is higher because the effect of armature reaction is larger than that at high saturation). Now under short-circuit conditions, saturation is very low and, therefore, the value of synchronous impedance measured is higher than that in actual operating conditions and the regulation determined is higher than actual one.

Q. 34. Which method of regulation is called pessimistic and why ?

Ans. Synchronous impedance or emf method of determination of voltage regulation of an alternator is pessimistic method. For reasons see answer to above question.

Q. 35. What is the use of Potier triangle ?

Ans. Potier triangle is used in determination of voltage regulation of an alternator.

Q. 36. Which quantities are usually determined using the zero power factor characteristic of a synchronous machine ?

Ans. Leakage reactance drop, voltage drop due to armature reaction and armature reaction ampere-turns.

Q. 37. What are the characteristics required for Potier method of computing voltage regulation ?

Ans. OCC, SCC and zero-power factor full load voltage characteristic.

Q. 38. What is an air-gap line ?

Ans. Tangent to OCC is called the air-gap line.

Q. 39. At what power angle a synchronous generator will develop maximum power ?

Ans. Synchronous generator will develop maximum power when power angle $\delta = \theta$, internal angle where $\theta = \text{Tan}^{-1} \dfrac{X_s}{R_e}$.

Q. 40. Two reactance theory is applied only to salient pole machines. State the reason.

Ans. A multipolar machine with cylindrical rotor has a uniform air gap and, therefore, its reactance remains the same, irrespective of the spatial position of rotor. But in case of salient pole machines, the air gap is not uniform and its reactance varies with the rotor position. Because of non-uniformity of the reluctance of the magnetic paths, the mmf of the armature is divided into two components viz. (*i*) a direct-acting component along the field pole axis, called the direct-axis and (*ii*) quadrature-component along the axis passing through centre of the interpolar space, called the quadrature-axis. These facts form the basis of the two reactance theory applied to the salient pole machines.

Q. 41. State the use of slip test on an alternator.

Ans. The slip test is performed on an alternator to determine X_d and X_q.

Q. 42. Upon what factor does the power angle 'δ' depend ?

Ans. Power angle δ depends upon supply voltage, armature current, power factor of load and quadrature component of synchronous reactance.

Q. 43. Why an alternator with low value of SCR has lower limit stability ?

Ans. An alternator with low value of SCR has high value of $X_d \left(X_d = \dfrac{1}{\text{SCR}} \right)$ and, therefore, has lower stability limit because maximum power output of a machine is inversely proportional to X_d.

Q. 44. An unloaded three phase alternator is suddenly symmetrically short circuited. The peak value of the alternating component of current at the instant of short circuit is I_1 and the peak value of steady-state current is I_2 while E is the rms voltage per phase before short circuit. What is the value of direct-axis sub-transient reactance ?

Ans. $\sqrt{2}\ E/I_1$.

Q. 45. What are the losses that take place in an alternator ?

Ans. 1. Electrical losses including armature winding loss, brush contact loss and field loss, 2. core loss 3. Friction and windage loss and 4. Stray power loss.

Q. 46. Why are alternators put in parallel ?

Ans. For high efficiency of operation, reliability, convenience and economy in maintenance and repair and possibility of additions to plant capacity with the growth in load on the power station.

Q. 47. What is meant by synchronising of alternators ?

Ans. The process of connecting an alternator in parallel with another alternator or with the common bus-bars is called the synchronising.

Q. 48. What are the conditions for parallel operation of alternators ?

Ans. Terminal voltage, frequency and the phase of the incoming machine must be the same as that of the bus-bars (or other alternators already operating in parallel).

In case of 3-phase alternators, an additional requirement to be met is that the phase sequence of the incoming machine must be the same as that of the bus-bars.

Q. 49. What are the two methods by which two alternators are put in parallel ?

Ans. Synchronising lamp method and synchroscope method.

Q. 50. A 3-phase, 60-pole alternator is to be synchronised to an infinite bus-bar using two bright one dark lamp method. If the lamps glow 60 times per minute in a particular sequence, determine the speed of the alternator.

Ans. 98 rpm or 102 rpm.

PROBLEMS

1. An 8-pole ac generator is running at 750 rpm. What is the frequency. At what speed must the generator be run so that frequency shall be 25 Hz ? [**Ans.** 50 Hz; 375 rpm]
2. Calcultate the pitch factor of a winding having 4/5 full pitch. [**Ans.** 0.951]
3. Find the distribution factor in a 4-pole, 3-phase, 36 slots alternator. [**Ans.** 0.96]
4. For a 3-phase winding with 4 slots per pole per phase and with the coil span of 10 slot pitch, calculate the values of the distribution factor and coil span factor. [**Ans.** 0.958; 0.966]
5. The armature of a l-phase alternator is completely wound with T single turn coils distributed uniformly. The induced emf in each turn is 1.6 V. Determine the emf induced for the whole winding with all T coils in series. $\left[\textbf{Ans.}\ \frac{3.2\,T}{\pi}\ \text{volts}\right]$
6. A 3-phase, 16-pole altenator has the following data:

   Number of slots = 192, conductors per slot = 8; coil span 10 slots; speed of alternator = 375 rpm; flux per pole = 55 mwb. Calculate the phase and line voltage. [**Ans.** 2,852 V; 5.010 V]
7. The stator of a 3-phase, 20-pole alternator has 120 slots, and there are 4 conductors per slot accommodated in two layers. If the speed of the alternator is 300 rpm, calculate the emf induced per phase. Resultant flux in the air-gap is 0.05 Wb per pole. Assume the coil span as 160º electrical. [**Ans.** 844.7 V]
8. A star-connected, 3-phase, 6-pole alternator has a stator with 90 slots and 8 conductors per slot. The rotor revolves at 1,000 rpm. The flux per pole is 4×10^{-2} Wb. Calculate the emf generatcd if all the conductors in each phase are in series. Assume sinusoidal flux distribution and full-pitched coils.

   [**Ans.** 1,765.7 V (line emf)]
9. A 16 pole, 3-phase alternator has a star-connected winding with 144 slots and 10 conductors per slot. The flux per pole is 0.03 Wb distributed sinusoidally and the speed is 375 rpm. Find the line voltage, if the coil span is 150º elec. [**Ans.** 2,567 V]
10. Find the no-load phase and line voltage of star-connected 3-phase, 6-pole alternator which runs at 1,200 rpm, having flux per pole of 0.1 Wb sinusoidally distributed. Its stator has 54 slots having double layer winding. Each coil has 8 turns and the coil is chorded by 1 slot. [**Ans.** 1,813 V; 3,141 V]
11. The stator of a 3-phase, 16-pole alternator has 144 slots and there are 4 conductors per slot connected in two layers and the conductors of each phase are connected in series. If the speed of the alternator is 375 rpm, calculate the emf induced per phase. Resultant flux in the air gap is 5×10^{-2} Wb/pole, sinusoidally distributed. Assume the coil span as 150º electrical. [**Ans.** 988 V]
12. Calculate the rms value of induced emf per phase of a 10-pole, 3-phase, 50 Hz alternator with 2 slots/pole/phase and 4 conductors per slot. Assume coil span of 150º. Flux per pole has a fundamental component of 0.12 Wb and a third harmonic component of 20 % of the fundamental. What is the value of line-to-line emf ? [**Ans.** 1,000; 1,722 V]

    [**Hint:** Since phase emfs induced by the 3rd, 9th and 15th harmonics are eliminated from the line voltages, the line voltage for star connection = $994.4 \times \sqrt{3}$ = 1,722 V]
13. A 500 kVA, 1,100 V, 50 Hz star-connected 3-phase alternator has armature resistance per phase of 0.1 Ω and synchronous reactance per phase 1.5 Ω. Find its voltage for (*a*) unity pf (*b*) 0.9 lagging and (*c*) 0.8 leading pf. Also, calculate voltage regulation in each case.

    [**Ans.** (*a*) 1,333 V; 21.187% (*b*) 1,556 V; 41.447% (c) 925.7 V; – 15.845%]

14. A 500 V, 50 kVA single phase alternator has an effective resistance of 0.2 Ω. A field current of 10 A produces an armature current of 200 A on short circuit and an emf of 450 V on open circuit. Calculate the full-load regulation. [**Ans.** 13.247% at unity power factor]

15. Determine the voltage regulation of a 2,000 V, 1-phase alternator giving a current of 100 A at (*i*) 0.8 power factor leading and (*ii*) 0.707 power factor lagging. Use the test results given follows.

Full load current of 100 A is produced on short circuit by a field excitation of 2.5 A; an emf of 500 V is generated on open circuit by the same excitation. The armature resistance is 0.8 Ω.

[**Ans.** (*i*) – 8.9 % (*ii*) 21.15 %]

16. A full load current of 100 A requires an excitation current of 3 A, when a single phase 1,200 V alternator is short circuited by an ammeter of resistance 0.01 Ω. On open circuit the same excitation produces 350 V. The resistance of the armature is 0.5 Ω. Calculate the regulation of the alternator at power factor of 0.8 lagging. [**Ans.** 22.39 %]

$$\left[\textbf{Hint.} \text{ Synchronous reactance } X_s = \sqrt{\left(\frac{350}{100}\right)^2 - (0.5+0.01)^2}\right] = 3.4626\ \Omega$$

17. A 3-phase, 50 Hz, 60 kVA, 600 V star-connected alternator has an effective resistance of 0.2 Ω. A field current of 10 A produces 480 V on open circuit and a field current of 5 A gives armature current of 105 A. Calculate the voltage regulation of this alternator at 0.8 pf lagging at full-load.

[**Hint :** A field current of 10 A will produce a short-circuit current twice of that produced by 5 A field current] [**Ans.** 16.733 %]

18. A 3.3 kV, 3-ϕ star-connected alternator has a full-load current of 100 A. Under short-circuit condition it takes 5A field current to produce full-load short-circuit current. The emf on open circuit for the same excitation is 900 V (line to line). The armature resistance was 0.9 Ω / phase. Determine synchronous reactance per phase and regulation for (*i*) 0.8 power factor lagging (*ii*) 0.8 power factor leading.

[**Ans.** 5.118 Ω; 21.34 %; – 9.024 %]

19. Find the synchronous impedance and reactance in an alternator in which a given field current produces an armature current of 250 A on short circuit and a generated emf of 1,500 V on open circuit. The armature resistance is 2 Ω. Hence calculate the terminal pd when a load of 250 A at 6,000 V and lagging power factor 0.8 is switched off. [**Ans.** 6 Ω, 5.657 Ω, 7,892 V]

20. A 415 V, 30 kVA, 50 Hz, 3-phase star-connected turbo-alternator has the following open-circuit test results.

I_f (A)	6	12	18	24	28
Line emf (V)	282	408	435	459	574

An excitation of 8 A produced full-load current in the armature on short circuit. If R_a = 0.5 Ω per phase, calculate the voltage regulation at full-load and at 0.707 pf lagging. [**Ans.** 73.42%; –19.57%]

21. The OC and SC test readings for a 3-phase, 50 Hz, 1,000 kVA, 2,000 V, star-connected alternator are as follows:

Field current (A)	10	20	25	30	40	50
OC terminal voltage (V)	800	1,500	1,760	2,000	2,350	2,600
SC armature current (A)	–	200	250	300	–	–

The effective armature resistance per phase is 0.2 Ω. Estimate the full-load percentage regulation at 0.8 pf lagging and 0.6 pf leading. Use mmf method. [**Ans.** 34.6%; – 6.4%] [**Ans.** 21.25 %]

22. The effective resistance of a 1,200 kVA, 3.3 kV, 50 Hz, 3-phase, star-connected alternator is 0.25 Ω/ phase. A field current of 35 A produces a current of 200 A on short circuit and 1.1 kV (line to line) on open circuit. Calculate the power angle and per unit change in magnitude of the terminal voltage when the full load of 1,200 kVA at 0.8 pf (lag) is thrown off. Draw the corresponding phasor diagram.

[**Ans.** 0.258; 12.81°]

23. An 11 kV, 3-phase alternator has X_d and X_q respectively as 1 pu and 0.6 pu. Armature resistance is negligible. Compute the generated emf when the machine delivers rated kVA at 0.8 pf lag.

[**Ans.** 19.525 kV]

24. The direct and quadrature axis reactances of a salient pole synchronous generator are 1.0 pu and 0.6 pu respectively. The armature resistance is negligible. Calculate the excitation voltage in pu when the generator delivers rated kVA at 0.8 pf lagging and rated voltage. [**Ans.** 1.775 pu]

25. A 3-phase alternator has a direct-axis synchronous reactance of 0.8 pu and a quadrature-axis synchronous reactance of 0.45 pu. Draw the phasor diagram for full load of 0.8 pf lagging and obtain therefrom (*i*) the load angle and (*ii*) the open-circuit voltage in pu. [**Ans** (*i*) 15.826° (*iii*) 1.5984 pu]

26. A 5 kVA, 240 V, star-connected, 3-phase salient pole alternator with direct -axis and quadrature-axis synchronous reactance of 12 Ω and 7 Ω respectively delivers full load current at unity pf. Calculate the excitation voltage and reluctance power. [**Ans.** 334.93 V; 1,521.5 watts]

27. A 3-phase star-connected synchronous generator supplies current of 10 A having phase angle of 20° lagging at 400 V. Find the load angle and the components of armature current I_d and I_q if X_d = 10 Ω and X_q = 6.5 Ω. Assume armature resistance to be negligible. [**Ans.** 8.23°; 4.73 A; 8.81 A]

28. A salient pole synchronous generator has the following parameters: X_d = 1.2, X_q = 0.8, r_a = 0.025. Compute the excitation voltage E on a per unit basis when the alternator is delivering at rated voltage for 0.8 pf lagging. [**Ans.** 1.9696 pu]

❑❑❑

SECTION D

SYNCHRONOUS MOTOR

CONTENTS OF SECTION-D

6

CHAPTER

Synchronous Motors

INSIDE THIS CHAPTER

6.1. INTRODUCTION

A synchronous motor has the same relationship to an alternator as a dc motor has to a dc generator *i.e.* if an alternator is supplied ac power it is capable of rotating as a motor and doing mechanical work. If the mechanical power supplied to a rotating alternator is removed while the dc field remains energized, and an ac supply is then connected across the armature terminals, torque will be developed and the alternator will continue to rotate at a speed determined by the ac supply frequency and number of poles on the synchronous machine. Changes in mechanical load within the machine's rating will not cause a change in speed.

In case of a dc motor the field and armature, both, are supplied from dc supply mains, since both require direct current. But in case of a synchronous motor the field structure is to be energized by direct current, as in case of an alternator, whereas the armature winding is connected to a 3-phase ac supply mains. Such a machine, therefore, requires two sources of supply, viz., an ac source to supply the power for driving the armature and a dc source to excite the field. DC excitation may be provided either from an exciting plant dc system or, if no direct current is available, from separate dc generators. Direct-connected exciters are frequently found on high speed motors. These are overhung from specially constructed motor brackets with the exciter armature mounted on an extension of the motor shaft. On low-speed machines, belted exciters are sometimes employed, notably on air compressor drives. Separate motor-generator sets are widely used for excitation of low-speed synchronous motors. Where a group of synchronous motors is installed, it is sometimes economical to use a single motor-generator set for exciting the entire group, although greater flexibility is obtained when each motor is provided with its own exciter. Synchronous motor with a directly connected exciter is shown in Fig. 6.1.

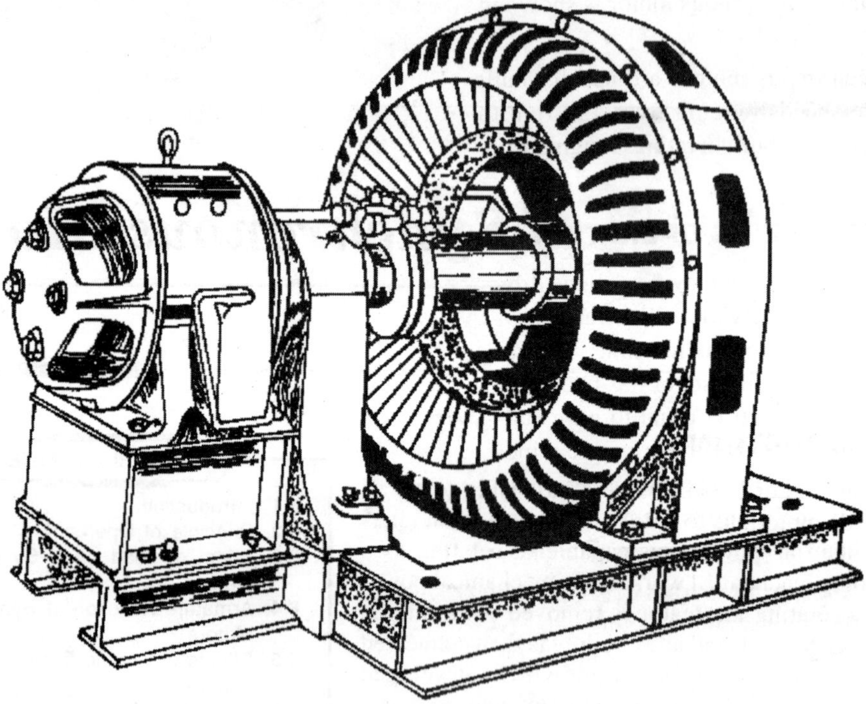

Fig. 6.1. ***Synchronous Motor With Directly Connected Exciter***

Essentially, the construction of a polyphase synchronous motor is the same as that of an alternator. Certain features of its design may be varied to suit the particular requirements. A synchronous motor may have either a revolving armature or a revolving field, but the revolving field construction is the more common.

The essential parts of a three-phase synchronous motor are (*i*) laminated stator core with 3-phase armature winding (*ii*) rotating field structure complete with damper winding and slip rings (*iii*) brushes and brush holders and (*iv*) two end shields to house the bearings that support the rotor shaft.

The stator core and windings of a synchronous motor are similar to those of a 3-phase alternator. The leads from the stator windings terminate in a terminal box usually mounted on the side of the motor frame. The rotor of the synchronous motor usually has salient field poles connected to give alternate polarity. The number of rotor field poles must be the same as that of stator field poles. The field circuit leads are brought out to two slip rings mounted on the shaft. Carbon brushes mounted in brush holders make contact with the two slip rings. The terminals of the field circuit are brought out from the brush holders to a second terminal box mounted on the motor frame.

In the most common type of field construction, the poles are somewhat different from those of the alternator. In addition to the exciting winding, a synchronous motor often is provided with a squirrel cage winding on the rotor. The slots of this winding are in the pole faces and are parallel to the shaft. The ends of the copper or bronze bars embedded in the slots are short circuited at both ends by conducting rings. This short-circuited winding on the rotor is known as the *cage* winding, the *damper* winding, the *amortisseur* winding, or sometimes, to describe its function, the *starting* winding. A typical rotor for a polyphase synchronous motor is shown in Fig. 6.2.

The synchronous motors are usually built for moderate speeds ranging from 150 rpm to 1,500 rpm. They are built in ratings ranging from 150 kW to 15 MW.

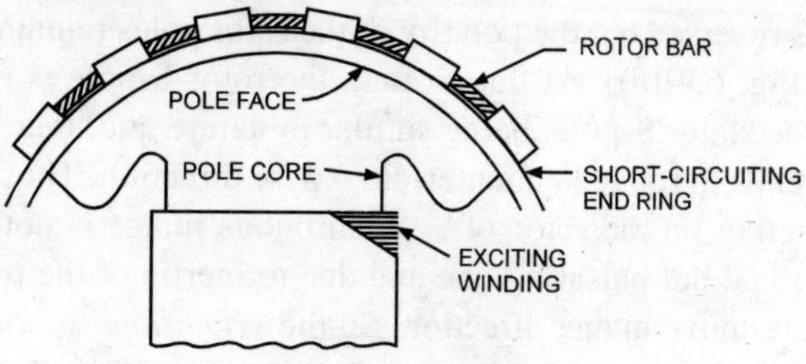

Fig. 6.2. ***Rotor Pole***

6.2. PRINCIPLE OF OPERATION

The action of the motor may be followed by referring to Fig. 6.3, which illustrates a 3-phase, 2-pole synchronous machine. The rotor has two poles and the stator has two poles per phase. In practice it is usual to have salient poles on the rotor, but the armature winding is housed in slots in the concave periphery of the stator. In the figure, salient type stator poles have been shown for convenience only.

As the rotor is excited from dc supply, so the poles of the rotor retain the same polarity throughout but the polarity of the stator poles changes as it is connected to an ac supply mains.

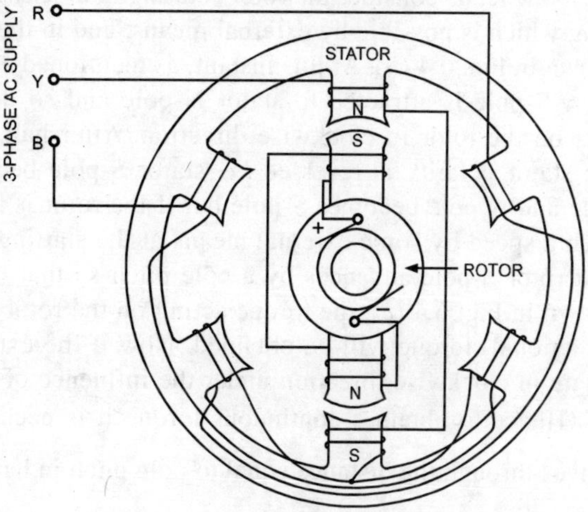

Fig. 6.3. ***Three-Phase, Two-Pole Synchronous Motor***

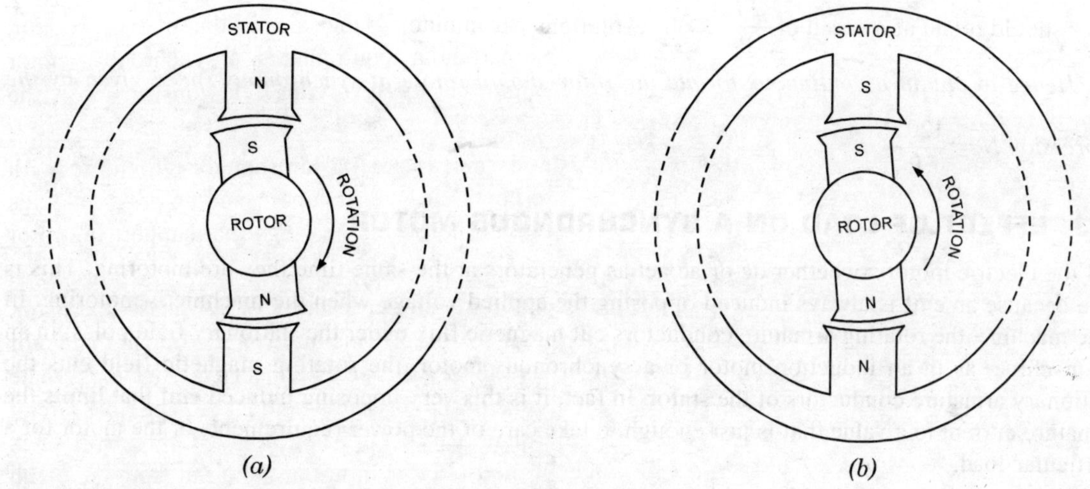

Fig. 6.4

First of all let us consider the rotor as stationary and in the position shown in Fig. 6.4 (*a*), At this instant the rotor S-pole is attracted to the stator N-pole and, therefore, the rotor tends to rotate in clockwise direction. After half a period $\left(i.e. \text{ after } \dfrac{T}{2} \text{ second whereas } T = \dfrac{1}{f}\right)$ the polarity of stator poles is reversed but the polarity of the rotor poles remains the same [Fig. 6.4 (*b*)]. At this instant, the rotor S-pole is repelled by the stator S-pole, being similar in nature and, therefore, rotor tends to rotate in counter-clockwise direction. Thus the torque acting on the rotor of a synchronous motor is not uni-directional but pulsating one and due to inertia of the rotor, it will not move in any direction. So the *synchronous motor has got no self-starting torque.*

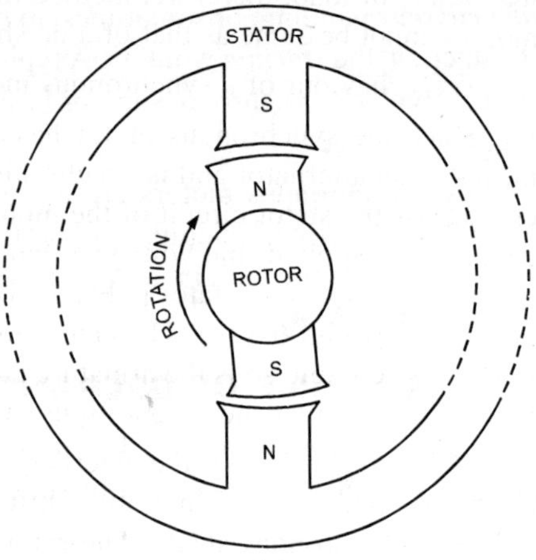

(*c*)

Fig. 6.4

Now let us consider the rotor rotating in clockwise direction, which is possible by external means, and in the position shown in Fig. 6.4 (*a*). At this instant, as mentioned above, the rotor S-pole is attracted to stator N-pole and so the torque acts on the rotor in clockwise direction. After half a period the stator polarity is reversed *i.e.* stator S-pole becomes N-pole and N-pole becomes S-pole but if the rotor is rotated at such a speed by some external means at the starting moment that rotor S-pole advances by a pole pitch so that it is again under the influence of stator N-pole, as shown in Fig. 6.4 (*c*), the torque acting on the rotor will be again clockwise. Hence a continuous (uni-directional) torque will be obtained. Now if the external means is removed the rotor will continue to rotate in clockwise direction under the influence of clockwise continuous torque acting on the rotor.

Hence to obtain a continuous torque it is necessary that the rotor rotates at such a speed that it moves through the distance equal to pole pitch in half the period *i.e.* in $\frac{T}{2}$ or $\frac{1}{2f}$ second where f is the supply frequency. If there are P poles then to complete one revolution it should take $\frac{P}{2f}$ second or the rotor should rotate at a speed of $\frac{2f}{P} \times 60$ revolutions per minute.

Hence to obtain a continuous torque the rotor should rotate at synchronous speed given by the expression $N_s = \frac{120f}{P}$.

6.3. EFFECT OF LOAD ON A SYNCHRONOUS MOTOR

All the electric motors, whether dc or ac act as generators at the same time they are motoring. This is true because an emf is always induced opposing the applied voltage when the machine is motoring. In a dc machine, the rotating armature conductors cut magnetic flux under the stationary field poles; in an ac machine, as in an induction motor or a synchronous motor, the rotating magnetic field cuts the stationary armature conductors of the stator. In fact, it is this very opposing induced emf that limits the armature current to a value that is just enough to take care of the power requirements of the motor for a particular load.

When the load on a dc shunt motor is increased, speed of motor falls slightly. This reduces the induced emf, called the back emf, so that resultant voltage across the armature ($V - E_b$) increases, which in turn increases the armature current, since armature current is given by the expression $I_a = \left(\frac{V - E_b}{R_a}\right)$, to supply extra torque and power required by the increased load. In the case of synchronous motor, however, the speed, as we have already discussed, remains constant irrespective of magnitude of load, therefore, method of drawing increased current from supply mains on synchronous motor cannot be same as that of a dc shunt motor.

The behaviour of a synchronous motor on load is discussed below:

Since the synchronous motor has got no self starting torque, first it is driven by some external means as an alternator and is synchronised with the supply. At the instant of synchronising the induced emf across the stator circuit of the motor (also called the back or counter-emf or excitation voltage) is equal to the applied line voltage V but opposite in direction, as shown in Fig. 6.6 (*a*) vectorially. The relative positions of a stator pole and dc field pole of a synchronous motor at the synchronising instant are shown in Fig. 6.5 (*a*). The centre lines of the two poles coincide. Since V = E, hence the resultant voltage is zero and so is the armature current. Motor intake is zero, as there is neither load nor losses to be met by it. Under these conditions the motor is said to be *floating*.

After synchronising, the external means is removed and the motor tends to slow down. There is no change in speed; there is merely a shift in relative positions of the two poles, both of which continue to rotate at synchronous speed. The rotor pole falls back a little relative to the stator pole, as shown in Fig. 6.5 (*b*). The angular displacement between the rotor and stator poles δ is called the *torque angle*. Now the applied voltage and induced emf are no longer in direct opposition since induced emf at this instant lags behind the induced emf at the synchronising instant, say by an angle δ_0, as shown in Fig. 6.6 (*b*). The resultant voltage E_R (the resultant of applied voltage **V** and induced emf **E**) causes a current $I_0 = \frac{E_R}{Z_s}$ to flow in the stator winding; where Z_s is the synchronous impedance per phase of the armature circuit. I_0 lags E_R by an angle θ, called the *internal angle*, where $\theta = \tan^{-1} \frac{X_s}{R_e}$; X_s and R_e being the stator synchronous reactance and effective resistance respectively. Hence on no load motor draws power equal to $VI_0 \cos \phi_0$ (for one phase) from the supply mains which is sufficient to make the motor running continuously at synchronous speed.

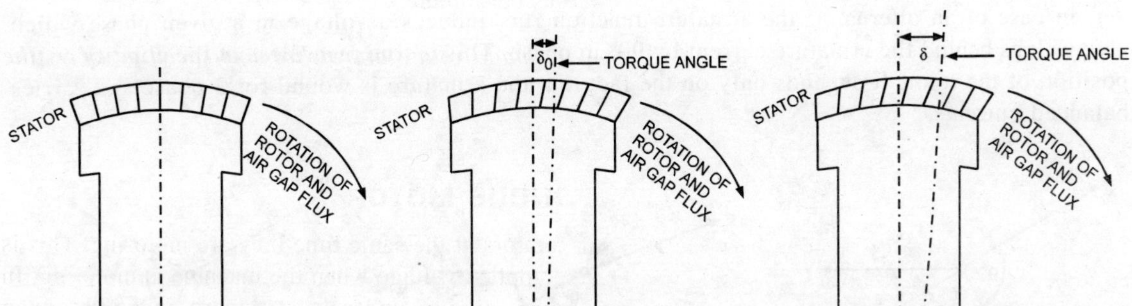

Fig. 6.5

At no load the torque angle δ_0 is very small, so the resultant voltage E_R acting across the armature circuit is small, due to which a small current is drawn from the supply mains. At no load, the torque angle δ_0 is so small that it can be assumed zero (neglecting the motor internal losses).

Now when the motor is loaded, it slows down momentarily to adjust itself to the change in loading conditions, so the rotor pole falls back a little more relative to the stator pole, as shown in Fig. 6.5 (*c*). Hence the torque angle increases with the increase in load. Due to increase in load (or torque) angle δ, the resultant voltage E_R across the armature circuit increases, and, therefore, current drawn from the supply mains increases, as illustrated in Fig. 6.6 (*c*).

Thus a synchronous motor is able to supply increased mechanical load, not by reduction in speed, but by shift in relative positions of the rotor and rotating magnetic field. From Fig. 6.6 it is obvious that for increasing load with a constant value of induced emf E the phase angle φ increases in lagging direction.

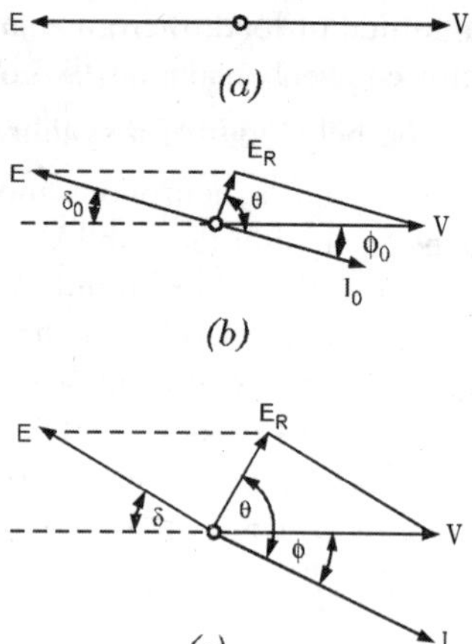

Fig. 6.6

If too great mechanical load is applied to a synchronous motor, the rotor will pull out of synchronism, after which it will come to standstill. The maximum value of torque that a motor can develop without loosing synchronism is called the *pull-out torque*, like pull-in torque. The motor may be designed for widely varying values, depending on the applications of the motor. These values range from 125% to 350% of full-load torque values.

6.4. ARMATURE REACTION IN SYNCHRONOUS MOTORS

The polyphase synchronous motor stator, as already mentioned, is identical to that of the polyphase synchronous generator. The balanced currents in the armature, therefore, set up a magnetic field which, for a given load, has a constant magnitude and always rotates at synchronous speed. Moreover, in case of an alternator, the armature reaction flux induces a voltage in a given phase which always lags behind the armature current by 90° in phase. This is true regardless of the polarity or the position of the rotor. It depends only on the fact that the armature is wound for 3-phase and carries balanced currents.

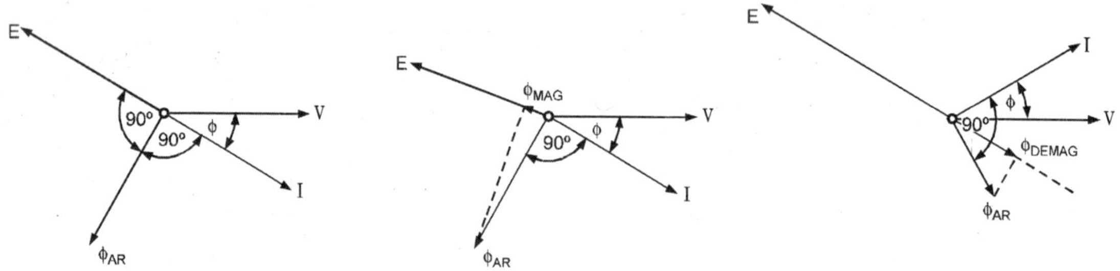

(*a*) *Cross-Magnetising* (*b*) *Magnetising* (*c*) *Demagnetising*

Fig. 6.7

In a synchronous motor too, the flux created by the armature current lags the respective armature currents by 90°. But now since the generated or excitation voltage E is more than 90° out of phase with the armature current, armature reaction flux has a different effect on the rotor flux than it has in the case of an alternator, except when the armature current I and excitation voltage E are exactly 180° apart as illustrated in Fig. 6.7 (*a*). Under this condition, the armature reaction is cross-magnetizing, as it is when the generated or excitation voltage of an alternator is in phase with the armature current. When the motor operates at lagging power factor such that the armature flux leads the excitation voltage by less than 90°, as shown in Fig. 6.7 (*b*), the armature reaction has a magnetising component as well as cross-magnetising one. With a leading power factor on the motor, the armature reaction flux has a demagnetising as well as cross-magnetising component, as shown in Fig. 6.7 (*c*). In Fig. 6.7, ϕ_{AR} represents the armature reaction flux. The component along the axis of the rotor poles is shown along the line of excitation voltage E, assisting the rotor flux for the lagging power factor and opposing it in the case of the leading power factor.

It is thus seen that the armature reaction is one which opposes the action of the main field flux. When the dc field current is reduced so as to draw the current from the ac supply mains at lagging power factor, the armature current drawn from the ac supply mains acts to strengthen the field. On the other hand, if the field current is increased so as to draw a current at leading power factor, the armature current now sets up a flux tending to reduce the flux created by the main field. In any case, the motor tends to operate at a nearly constant flux or generated voltage, as does the transformer.

6.5. EFFECTS OF VARYING EXCITATION ON ARMATURE CURRENT AND POWER FACTOR

The change in field excitation neither affects the speed of the motor nor the output of the motor but does affect the power factor and consequently armature current for constant supply voltage and constant input power.

The operation of a synchronous motor at a constant supply voltage demands nearly constant flux, as mentioned in the foregoing section. Both the dc source and ac source cooperate in establishing this resultant flux. When the dc excitation is weak, the flux produced by it is weak and needs stengthening, so the stator draws the current lagging behind the supply voltage by 90° so as to magnetise the air gap to the extent needed to provide the necessary resultant flux. The magnetising component of current being a large part of the total input current results in low lagging power factor. For constant applied voltage and constant input power, the active component of current drawn from the supply mains should be constant and so because of large wattless (magnetising) current armature current I is much more. With the increase in excitation to 100%, the magnetising current drawn from the 3-phase ac supply mains is reduced *i.e.* the magnetising component becomes a smaller part of the total input current and so the power factor improves. This results in reduction of armature current I drawn from 3-ϕ ac supply mains. If the excitation is further increased such that power factor becomes unity, the stator draws only energy or active current, and the dc field circuit supplies all the current necessary to magnetize the rotor field. For unity power factor, the armature current drawn will be minimum. On increasing the dc excitation further, the field produced by it becomes stronger than it is required and needs demagnetization. So the motor draws the leading current so as to demagnetise the main field. This makes the power factor leading and lesser than unity and so the armature current increases.

The effect of variation in field excitation on armature current and power factor can also be explained as below.

Consider a synchronous motor loaded with a constant mechanical load and operating at a constant supply voltage. When the excitation is 100% *i.e.* when E = V, armature current I lags behind the resultant voltage E_R (phasor sum of V and E) by a fixed angle θ, given as $\theta = \text{Tan}^{-1} \dfrac{X_s}{R_e}$ and lags behind the applied voltage by an angle ϕ, power factor angle [Fig. 6.8 (*a*)] where X_s is the synchronous reactance per phase, R_e is the effective resistance of armature per phase, θ is the *internal angle* and δ is the load or torque angle. The power drawn per phase from the 3-ϕ ac supply mains is VI cos ϕ. Since applied voltage and input power or load is constant so the active component of the current drawn from the supply mains, I cos ϕ remains constant for constant mechanical load (neglecting internal losses). As the speed of the motor remains constant so increase in dc excitation causes increase in excitation voltage E and decrease in excitation causes reduction in excitation voltage E.

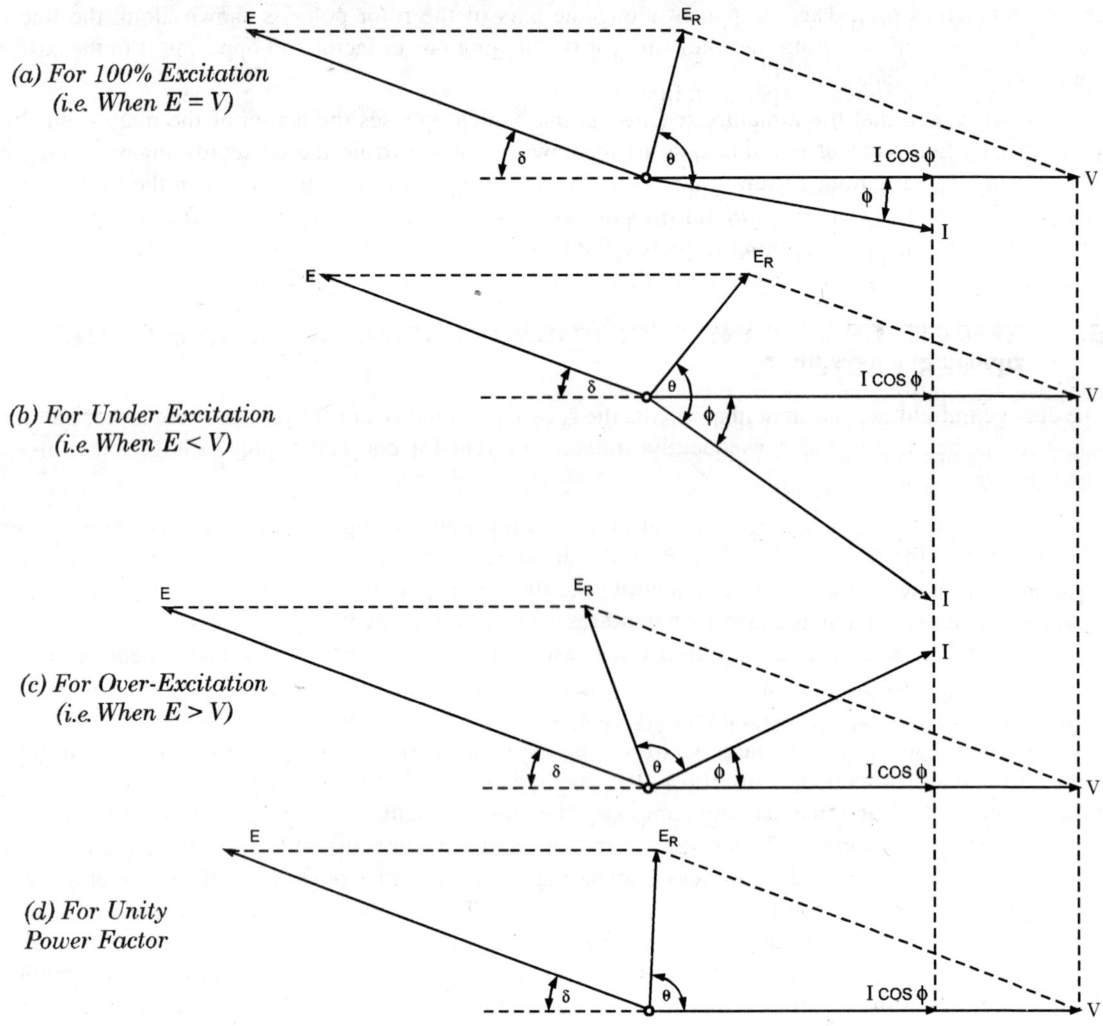

Fig. 6.8

When the excitation is reduced the excitation voltage E decreases in magnitude to the value shown in Fig. 6.8 (*b*) which shifts the resultant voltage E_R in clockwise direction and since the phase angle θ between the resultant voltage E_R and current I, being equal to $\tan^{-1}\dfrac{X_s}{R_e}$, is fixed, so the armature current phasor is also shifted in clockwise direction. Thus power factor angle ϕ increases, power factor decreases and armature current increases because I cos ϕ is constant. *Hence with reduction in dc excitation synchronous motor draws more current from the supply mains at lower lagging power factor.*

When the excitation is increased, the excitation voltage E increases in magnitude to a value shown in Fig. 6.8 (*c*) which shifts the resultant voltage E_R in counter-clockwise direction. This results in change in position of phasor of current I with respect to terminal voltage so that the motor then operates at a leading power factor. For a certain value of excitation phase angle ϕ between voltage V and current I is zero *i.e.* the current I will be in phase with voltage V and at this instant the value of current drawn from supply mains, I will be minimum, as shown in Fig. 6.8 (*d*). Now if the excitation is further increased, it draws current with leading power factor, as shown in Fig. 6.8 (*c*).

Hence when the excitation of a synchronous motor is increased above 100 per cent, first power factor improves until it becomes unity; at this instant the current drawn from the supply mains is minimum and is in phase with supply voltage. With further increase in field current the power factor becomes leading one and decreases and current drawn from supply main increases.

The variation of current and power factor of a synchronous motor with a variation of field current (excitation) and for a constant load is shown in Fig. 6.9. Because of their shape, excitation-armature current curves are called the *V-curves* of a synchronous motor.

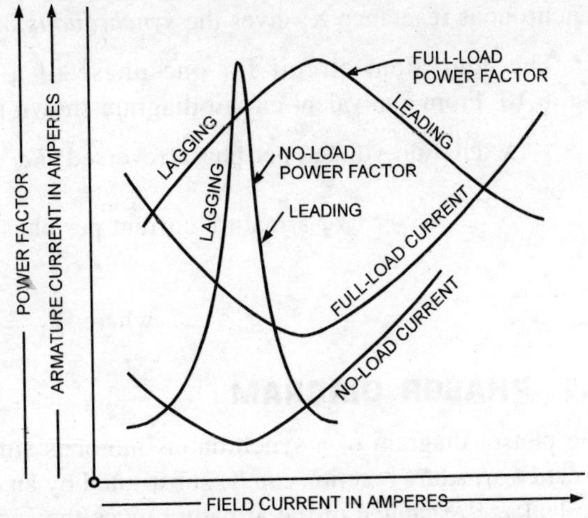

Fig. 6.9

6.6. EQUIVALENT CIRCUIT

The synchronous motor, as already mentioned in Art. 6.1, is fed from two sources of supply; field structure from dc source and armature winding from a 3-phase ac supply source.

Under normal operating conditions, no emf is induced in the rotor due to stator field because the rotor revolves at the same speed as the stator field (*i.e.* there is no relative motion between rotor and stator revolving field). Only the direct current flows through the rotor winding and ohmic resistance of the field winding and that of field rheostat is the only opposition to it, as shown in Fig. 6.10.

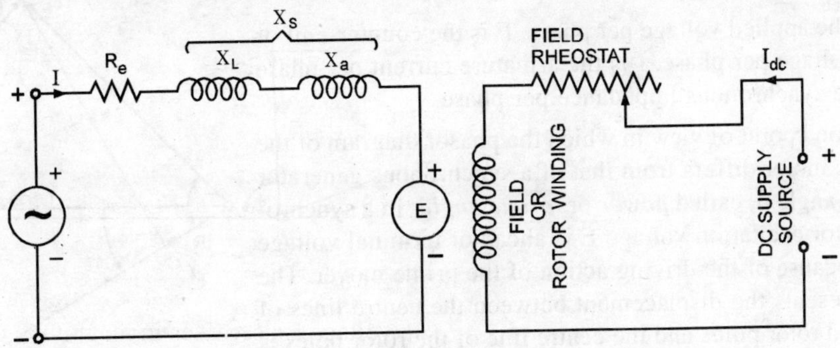

Fig. 6.10

In stator winding, two effects are to be considered; the effect of the rotor field cutting the stator conductors at synchronous speed and the effect of stator revolving field. A voltage is induced in the stator winding due to the field rotating at synchronous speed. This induced emf E is known as *excitation voltage* or *back emf* or *counter-emf* and opposes the applied voltage, V, to the stator. The magnitude of this induced emf depends upon the rotor flux per pole or strength of excitation current I_{dc} because the motor operates at a constant speed.

The effect of stator revolving field on the stator (or armature) conductors is accounted for by including an inductive reactance in the armature (or stator) winding. This is called the *synchronous reactance* X_s. An effective resistance R_e is considered to be in series with this reactance to account for copper loss in the armature winding and stray power loss. This resistance R_e when combined with synchronous reactance X_s gives the *synchronous impedance* Z_s of the machine.

The equivalent circuit for one phase of a star-connected synchronous motor is shown in Fig. 6.10. From equivalent circuit diagram shown in Fig. 6.10 we have

Excitation voltage per phase(reversed) *i.e.* $-\mathbf{E} = \mathbf{V} - \mathbf{I}\,\mathbf{Z}_s$

$$\text{or armature current per phase, } \mathbf{I} = \frac{\mathbf{V}+\mathbf{E}}{\mathbf{Z}_s} = \frac{\mathbf{E}_R}{\mathbf{Z}_s} \qquad \text{...(6.1)}$$

where $\mathbf{Z}_s = R_e + j\,X_s$ and is equal to $\sqrt{R_e^2 + X_s^2}$ in magnitude.

6.7. PHASOR DIAGRAM

The phasor diagram of a synchronous motor is similar to that of an alternator in many respects. The effect of armature reaction can be substituted by an equivalent reactance X_a which when combined with the leakage reactance of the armature gives the synchronous reactance X_s, as illustrated in equivalent circuit diagrams (Fig. 6.10). The synchronous reactance X_s combined with armature effective resistance* R_e gives the synchronous impedance, Z_s.

The phasor diagram of a synchronous motor differs from that of a synchronous generator in two respects. First one is the voltage equation for the stator circuit. In case of synchronous generator the excitation voltage E plays the role of a source voltage and is equal to the phasor sum of terminal voltage V and the synchronous impedance drop $\mathbf{I}\,\mathbf{Z}_s$. But in the synchronous motor supply voltage V is the source voltage applied to the stator winding of the synchronous motor and E is a reaction or counter-emf which is internally generated. The supply voltage V has to overcome the synchronous impedance drop $\mathbf{I}\,\mathbf{Z}_s$ and counter-emf E. Therefore, the voltage equation for a synchronous motor is given as

$$\mathbf{V} = -\mathbf{E} + \mathbf{I}\,\mathbf{Z}_s$$

where V is the applied voltage per phase, E is the counter-emf or excitation voltage per phase, I is the armature current per phase and Z_s is the synchronous impedance per phase.

The second point of view in which the phasor diagram of the synchronous motor differs from that of a synchronous generator involves the angle δ, called *power* or *torque angle*. In a synchronous generator excitation voltage E is ahead of terminal voltage V in time because of the driving action of the prime mover. The angle δ represents the displacement between the centre lines of the stator and rotor poles and the centre line of the rotor poles is ahead of the centre line of the stator poles by an electrical angle δ in case of a synchronous generator. But in case of a synchro-

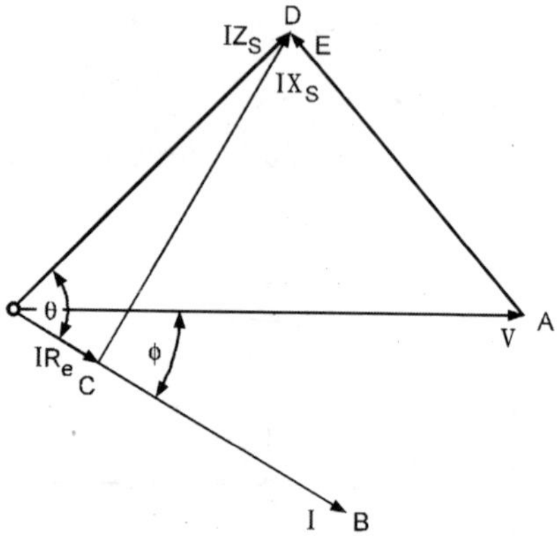

(a) When ϕ is – Ve

Fig. 6.11

* In synchronous machines the armature effective resistance is usually very small in comparision to synchronous reactance $\left(R_e \text{ is usually } \frac{1}{100} \text{ to } \frac{1}{10} \text{th of } X_s\right)$

nous motor the situation is reverse of that in a synchronous generator *i.e.* the rotor falls behind the stator and this is because of load on the motor shaft. Thus in a synchronous motor the phasor of applied voltage V leads the phasor of excitation voltage E.

On the basis of the modifications mentioned above the phasor diagrams for a synchronous motor with armature current (*a*) lagging behind (*b*) in phase with and (*c*) leading the applied voltage are shown in Figs. 6.11 (*a*), (*b*) and (*c*) respectively. In Fig. 6.11. phasors OA, OB, OC, CD, OD and AD represent the applied voltage V, load current I, resistive drop IR_e, reactive drop IX_s, resultant voltage across the stator winding $\mathbf{E_R = I\,Z_s}$ and excitation voltage E (all phase values) respectively.

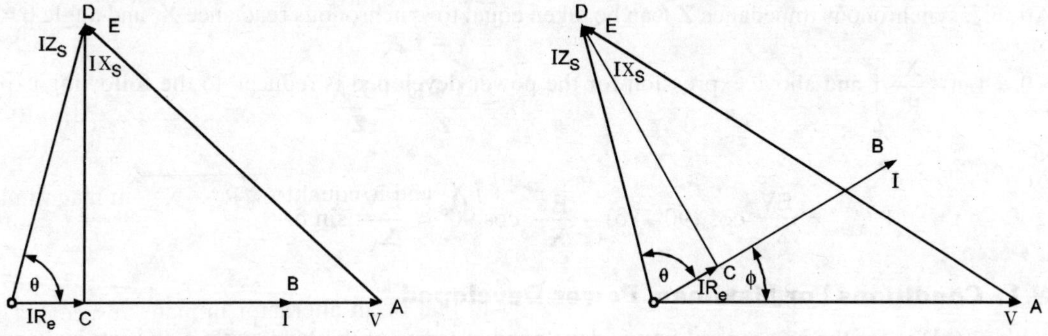

(*b*) *When* $\phi = 0$ (*c*) *When* ϕ *is + Ve*

Fig. 6.11

6.8. POWER DEVELOPED IN A SYNCHRONOUS MOTOR

Let us consider a 3-phase synchronous motor having effective resistance R_e ohms/phase and synchronous reactance X_s ohms/phase. Let the motor be connected across 3-phase ac supply having voltage per phase equal to V volts and have induced emf (or excitation voltage) per phase of E volts and load angle δ electrical degrees.

Taking induced voltage E horizontally, resolving V and E horizontally and vertically and equating the algebraic sum of their horizontal and vertical components separately with the horizontal and vertical components of resultant voltage E_R respectively we have

$$E_R \cos\gamma = E - V\cos\delta \quad \text{...(6.2)}$$

$$\text{and } E_R \sin\gamma = V\sin\delta \quad \text{...(6.3)}$$

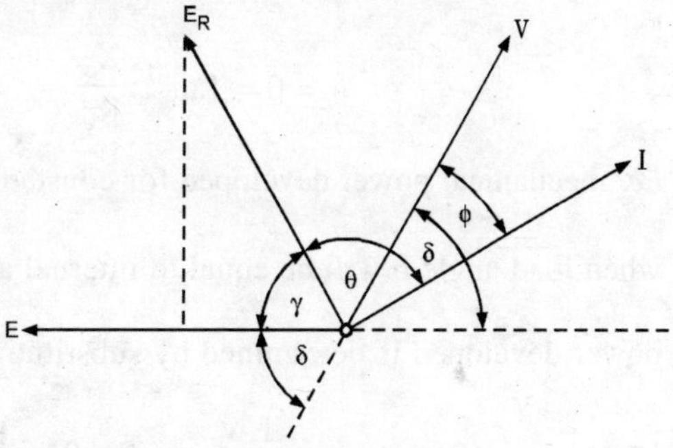

Fig. 6.12

Mechanical power developed per phase,

P_{mech} = EMF induced/phase × armature current per phase × cosine of the angle between the two (*i.e.* angle between I and E reversed)

$$= E\,I\cos[180° - (\theta + \gamma)] = -EI\cos(\theta + \gamma)$$

$$= -E \times \frac{E_R}{Z_s}(\cos\theta\cos\gamma - \sin\theta\sin\gamma)$$

$$\because I = \frac{E_R}{Z_s}$$

$$= \frac{-E}{Z_s}(\cos\theta\, E_R\cos\gamma - \sin\theta\, E_R\sin\gamma)$$

Substituting $E_R \cos \gamma = (E - V \cos \delta)$ from Eq. (6.2) and $E_R \sin \gamma = V \sin \delta$ from Eq. (6.3) in above equation we have

$$P_{mech} = \frac{-E}{Z_s} [\cos \theta (E - V \cos \delta) - \sin \theta V \sin \delta]$$

$$= \frac{EV}{Z_s} (\cos \delta \cos \theta + \sin \theta \sin \delta) - \frac{E^2}{Z_s} \cos \theta = \frac{E V}{Z_s} \cos (\theta - \delta) - \frac{E^2}{Z_s} \cos \theta \quad ...(6.4)$$

Since effective resistance R_e of the synchronous motor is usually negligible, as already mentioned in Art. 6.7, synchronous impedance Z_s can be taken equal to synchronous reactance X_s and angle $\theta = 90°$ $\left(\because \theta = \text{Tan}^{-1} \frac{X_s}{R_e}\right)$ and above expression for the power developed is reduced to the following expression

$$P_{mech} = \frac{EV}{X_s} \cos (90° - \delta) - \frac{E^2}{X_s} \cos 90° = \frac{E V}{X_s} \sin \delta \quad ...(6.5)$$

6.8.1. Conditions For Maximum Power Developed

Equation (6.4) is for the mechanical power developed in terms of the load angle δ and internal angle θ of the motor for a constant supply voltage V and induced emf E (or excitation as the induced emf depends upon the excitation).

Now following two cases may be considered

(*i*) *Fixed V, E, and* θ $\left(\text{or } R_e \text{ and } X_s \text{ because } \theta = \text{Tan}^{-1} \frac{X_s}{R_e}\right)$. Under these conditions the power developed P_{mech} will vary with the load angle δ and the maximum power developed can be determined by differentiating equation (6.4) with respect to load angle δ and then equating it to zero.

$$\therefore \quad \frac{dP_{mech}}{d\delta} = \frac{E V}{Z_s} \sin (\theta - \delta) = 0$$

$$\text{or} \quad \sin (\theta - \delta) = 0$$

$$\text{or} \quad \delta = \theta = \text{Tan}^{-1} \frac{X_s}{R_e} \quad ...(6.6)$$

i.e. mechanical power developed for constant supply voltage and constant excitation will be maximum when load angle δ will be equal to internal angle θ (or equal to $\tan^{-1} \frac{X_s}{R_e}$) and the value of maximum power developed is determined by substituting $\delta = \theta$ in equation (6.4) so we get

$$(P_{mech})_{max} = \frac{E V}{Z_s} \cos (\theta - \theta) - \frac{E^2}{Z_s} \cos \theta = \frac{E V}{Z_s} - \frac{E^2}{Z_s} \cos \theta \quad ...(6.7)$$

$$\text{or} \quad (P_{mech})_{max} = \frac{E V}{Z_s} - \frac{E^2}{Z_s} \times \frac{R_e}{Z_s} = \frac{E V}{Z_s} - \frac{E^2}{Z_s^2} R_e \quad ...(6.8)$$

The above expression shows that the maximum mechanical power developed or torque (since speed is constant) depends upon the applied voltage V, induced emf E (or excitation) and internal angle, θ. Maximum value of θ (and hence δ) is 90° and, therefore, maximum mechanical power developed

$$(P_{mech})_{max} = \frac{EV}{Z_s} \qquad ...(6.9)$$

$(P_{mech})_{max}$ is the value of power at which the motor falls out of step.

Solving for E from equation (6.8), we get

$$\text{Excitation voltage,} \quad E = \frac{Z_s}{2R_e}\left[V \pm \sqrt{V^2 - 4R_e (P_{mech})_{max}}\right] \qquad ...(6.10)$$

The two values of E so obtained give the excitation limits for any load.

(ii) Fixed V and θ (or R_e and X_s). For determining the value of excitation voltage E to give maximum power developed, differentiate equation (6.8) with respect to E and equate to zero

$$\frac{d(P_{mech})_{max}}{dE} = \frac{V}{Z_s} - \frac{2E}{Z_s^2} R_e = 0$$

$$\text{or} \quad E = \frac{VZ_s}{2R_e}$$

Substituting $E = \frac{VZ_s}{2R_e}$ in equation (6.8) we have

$$(P_{mech})_{max} = \frac{V^2}{2R_e} - \frac{V^2}{4R_e}$$

$$= \frac{V^2}{4R_e} \qquad ...(6.11)$$

The value of excitation voltage to develop maximum power, given by $\frac{VZ_s}{2R_e}$ would be very large, because effective resistance of the stator winding is very small. Usually it is physically impossible to achieve these maximum conditions because of saturation and practical limits on excitation. Thus this equation is of theoretical interest only because the assumptions on which it is based are not in sufficient agreement with the physical facts.

6.9. POWER FLOW IN A SYNCHRONOUS MOTOR

In motor operation, electrical energy from some source outside the machine is supplied to the armature circuit. Some of this supplied energy is utilised to overcome the armature copper loss $I^2 R_e$ and remainder of it is converted into mechanical energy through the

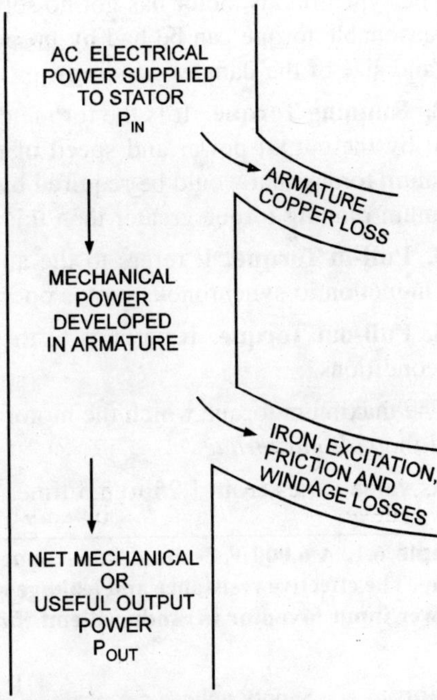

Fig. 6.13. ***Power Flow Diagram For a Synchronous Motor***

electromagnetic motor action. Some of this energy which is converted into mechanical energy is lost through friction, windage, and hysteresis and eddy-current phenomena. The remainder of the mechanical energy is available for useful work. The power flow diagram is shown in Fig. 6.13.

Power supplied to stator of the synchronous motor = $\sqrt{3}\ V_L\ I_L \cos\phi$ watts

where V_L is the line voltage across the stator terminals, I_L is the line current supplied to the stator and $\cos\phi$ is the power factor (cos of the phase angle between phase voltage V and phase current I).

$$\text{Armature copper losses for 3-phases} = 3\ I^2\ R_e \text{ watts}$$

$$\text{Mechanical power developed} = \sqrt{3}\ V_L\ I_L \cos\phi - 3\ I^2\ R_e$$

The mechanical power developed per phase also

= EMF induced/phase × armature current per phase × cosine of the angle between E and I

$= -EI \cos(\theta + \gamma)$ Refer to Fig. 6.12.

6.10. DIFFERENT TORQUES OF A SYNCHRONOUS MOTOR

The torque required to operate the driven machine at every moment between the initial breakaway and the final shutdown is important in determination of the motor characteristics. The various torques associated with synchronous motors are termed starting torque, running torque, pull-in torque, and pull-out torque.

1. Starting Torque. It pertains to the ability of the motor to accelerate the load. The starting torque, sometimes also called the *breakaway torque*, required by the driven machine may be as low as 10%, as in case of centrifugal pumps, and as high as 200 or 250% of full-load torque, as in case of loaded reciprocating two-cylinder compressors.

The synchronous motor has got no self-starting torque, but in modern synchronous motors, almost any reasonable torque can be had by proper design of the damper windings (by changes in the resistance and size of the damper winding).

2. Running Torque. It is the torque developed by the motor under running conditions. It is determined by the output power and speed of the driven machine. The peak output power determines the maximum torque that would be required by the driven machine. The motor must have a breakdown or a maximum running torque greater than this value so as to avoid stalling of the machine.

3. Pull-in Torque. It refers to the ability of the motor to pull into synchronism when changing from induction to synchronous motor operation.

4. Pull-out Torque. It pertains to the ability of the motor to remain in synchronism under rated load conditions.

The maximum torque which the motor will develop without pulling out of step (or synchronism) is called the *pull-out torque*.

Its value varies from 1.25 to 3.5 times the full-load torque.

Example 6.1. A 6,000 V, 3-phase, star-connected synchronous motor has an input current of 40 A at 0.8 pf lagging. The effective resistance and leakage reactance per phase are 0.5Ω and 10 Ω. respectively. Determine: (*i*) Power input to motor (*ii*) Induced emf (*iii*) Internal mechanical power developed.

Solution : Supply voltage per phase, $V = \dfrac{V_L}{\sqrt{3}} = \dfrac{6{,}000}{\sqrt{3}} = 3{,}464$ V (*i*) Power supplied to motor, $P = \sqrt{3}\, V_L I_L \cos\phi$

$$= \sqrt{3} \times 6{,}000 \times 40 \times 0.8 = 3{,}32{,}550 \text{ W or } 332.55 \text{ kW } \textbf{Ans.}$$

$$\text{Total copper loss} = 3\, I^2 R_e = 3 \times 40^2 \times 0.5 = 2{,}400 \text{ W or } 2.4 \text{ kW}$$

(*iii*) Internal mechanical power developed,

$$P_{mech} = P - 3I^2R_e = 332.55 - 2.4 = 330.15 \text{ kW } \textbf{Ans.}$$

$$\text{Internal angle, } \theta = \text{Tan}^{-1} \frac{X}{R_e} = \text{Tan}^{-1} \frac{10}{0.5} = 87.14^\circ \text{ (lagging)}$$

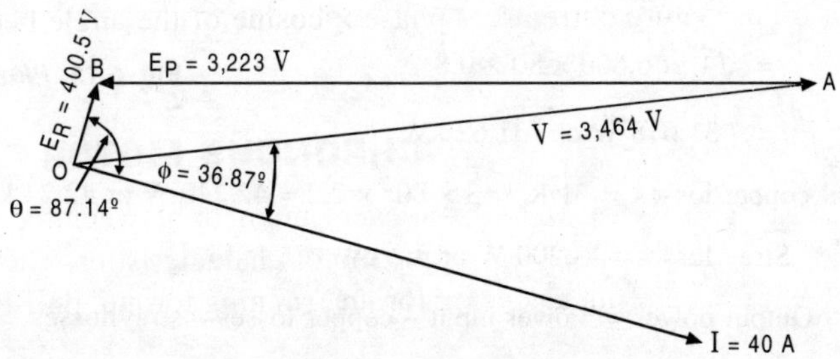

Fig. 6.14. ***Phasor Diagram***

$$\text{Power factor angle, } \phi = \text{Cos}^{-1}\, 0.8 = 36.87^\circ \text{ (lagging)}$$

$$\text{Impedance per phase, } Z = \sqrt{R_e^2 + X^2} = \sqrt{(0.5)^2 + (10)^2} = 10.0125\ \Omega$$

$$\text{Impedance drop per phase, } E_R = IZ = 40 \times 10.0125 = 400.5 \text{ V}$$

From phasor diagram shown in Fig. 6.14, we have

$$(ii) \text{ Induced emf per phase, } E_P = \sqrt{V^2 + E_R^2 - 2VE_R \cos(\theta - \phi)}$$

$$= \sqrt{(3{,}464)^2 + (400.5)^2 - 2 \times 3{,}464 \times 400.5 \cos(87.14^\circ - 36.87^\circ)}$$

$$= 3{,}223 \text{ V}$$

$$\text{Line value of induced emf, } E_L = \sqrt{3} \times 3{,}223 = 5{,}582 \text{ V } \textbf{Ans.}$$

Example 6.2. A star-connected synchronous motor rated at 187 kVA, 3-φ, 2,300 V, 47 A, 50 Hz, 187.5 rpm has an effective resistance of 1.5 Ω and a synchronous reactance of 20 Ω per phase. Determine the internal power developed by the motor when it is operating at rated current and 0.8 power factor leading.

Solution : Line voltage, $V_L = 2{,}300$ volts

Line current, $I_L = 47$ A

Power factor, $\cos\phi = 0.8$ (leading)

$$\text{Power supplied to the motor, } P = \sqrt{3}\, V_L\, I_L \cos\phi$$

$$= \sqrt{3} \times 2{,}300 \times 47 \times 0.8 = 1{,}49{,}780 \text{ watts or } 149.78 \text{ kW}$$

$$\text{Total copper loss} = 3\,I^2\,R = 3 \times (47)^2 \times 1.5 = 9{,}940 \text{ watts or } 9.94 \text{ kW}$$

$$\text{Internal power developed} = P - 3\,I^2\,R = 149.78 - 9.94 = 139.84 \text{ kW } \textbf{Ans.}$$

Example 6.3. A 6600 V, 3-phase, star-connected synchronous motor draws a full-load current of 80 A at 0.8 power factor leading. The per phase armature resistance is 2.2 Ω and synchronous reactance is 22 Ω. If the stray losses of the machine are 3,200 W, calculate the induced emf, output power and efficiency.

Solution : Line voltage, $V_L = 6{,}600$ V

Line current, $I_L = 80$ A

Power factor, $\cos\phi = 0.8$ (leading)

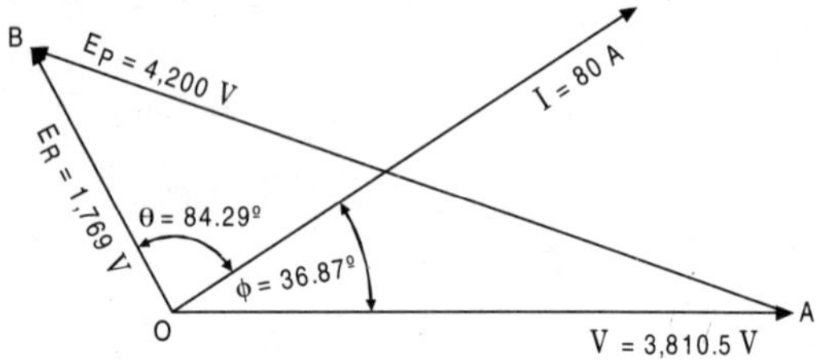

Fig. 6.15. ***Phasor Diagram***

$$\text{Power input to motor, } P = \sqrt{3}\, V_L I_L \cos\phi$$

$$= \sqrt{3} \times 6{,}600 \times 80 \times 0.8$$

$$= 7{,}31{,}618 \text{ W or } 731.62 \text{ kW}$$

$$\text{Total copper losses} = 3I^2R_e = 3 \times 80^2 \times 2.2 = 42{,}240 \text{ W or } 42.24 \text{ kW}$$

$$\text{Stray losses} = 3{,}200 \text{ W or } 3.2 \text{ kW}$$

$$\text{Output power} = \text{Power input} - \text{copper losses} - \text{stray losses}$$

$$= 731.62 - 42.24 - 3.2 = 686.18 \text{ kW } \textbf{Ans.}$$

$$\text{Efficiency, } \eta = \frac{\text{Output}}{\text{Input}} \times 100 = \frac{686.18}{731.62} \times 100 = 93.79\,\% \textbf{ Ans.}$$

$$\text{Supply voltage per phase, } V = \frac{6{,}600}{\sqrt{3}} = 3{,}810.5 \text{ V}$$

$$\text{Internal angle } \theta = \text{Tan}^{-1}\frac{X_s}{R_e} = \text{Tan}^{-1}\frac{22}{2.2} = 84.29^\circ \text{ (lagging)}$$

$$\text{Phase angle, } \phi = \text{Cos}^{-1}\,0.8 = 36.87^\circ \text{ (leading)}$$

$$\text{Synchronous impedance per phase, } Z_s = \sqrt{R_e^2 + X_s^2} = \sqrt{(2.2)^2 + (22)^2} = 22.11\ \Omega$$

$$\text{Impedance drop per phase, } E_R = I_L Z_s = 80 \times 22.11 = 1{,}769 \text{ V}$$

From phasor diagram shown in Fig. 6.15.

$$\text{Induced emf per phase, } E_P = \sqrt{V^2 + E_R^2 - 2VE_R\cos(\phi + \theta)}$$

$$= \sqrt{(3{,}810.5)^2 + (1{,}769)^2 - 2 \times 3{,}810.5 \times 1{,}769 \cos(36.87^\circ + 84.29^\circ)}$$

$$= 4{,}200 \text{ V}$$

$$\text{Line value of induced emf, } E_L = 4{,}200 \times \sqrt{3} = 7{,}276 \text{ V } \textbf{Ans.}$$

Example 6.4. A 2.3 kV, 3-ϕ, star-connected synchronous motor has Z_s = (0.2 + *j* 2.2) ohms per phase. The motor is operating at 0.5 power factor leading with a line current of 200 A. Determine the generated emf per phase.

Solution : Input current, I = 200 A

Supply voltage per phase, $V = \frac{2.3 \times 1{,}000}{\sqrt{3}} = 1,328$ V

Internal angle, $\theta = \text{Tan}^{-1} \frac{X_s}{R_e}$

$= \text{Tan}^{-1} \frac{2.2}{0.2} = 84.8°$

Power factor angle, $\phi = \text{Cos}^{-1}\ 0.5 = 60°$

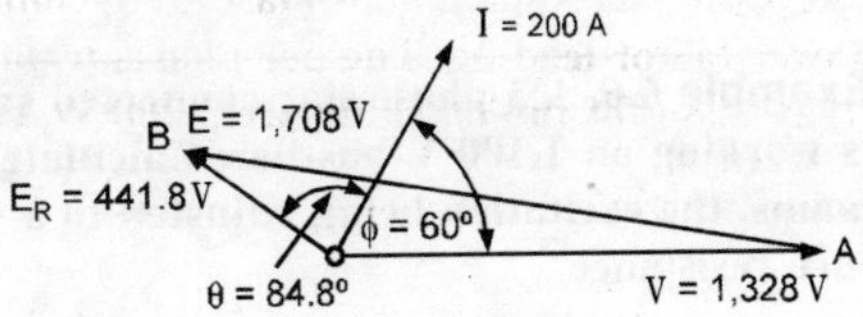

Fig. 6.16

Synchronous impedance per phase,

$$Z_s = \sqrt{R_e^2 + X_s^2} = \sqrt{0.2^2 + 2.2^2} = 2.209\ \Omega$$

Impedance drop/phase, $E_R = I Z_s = 200 \times 2.209 = 441.8$ V

From phasor diagram shown in Fig. 6.16

$$\text{Induced emf per phase, } E = \sqrt{V^2 + E_R^2 - 2 V E_R \cos(\phi + \theta)}$$

$$= \sqrt{(1{,}328)^2 + (441.8)^2 - 2 \times 1{,}328 \times 441.8 \times \cos(60° + 84.8°)}$$

= 1,708 volts **Ans.**

Example 6.5. A 75 kW, 400 V, 4-pole, 3-phase, star-connected synchronous motor has a resistance and synchronous reactance per phase of 0.04 Ω and 0.4 Ω respectively. Compute for full load 0.8 pf lead the open-circuit emf per phase and gross mechanical power developed. Assume an efficiency of 92.5%.

Solution : Motor input $= \frac{\text{Motor output}}{\eta} = \frac{75}{0.925} = 81.081$ kW or 81,081 watts

Armature current, $I = \frac{\text{Motor input}}{\sqrt{3}\, V_L \cos\phi} = \frac{81{,}081}{\sqrt{3} \times 400 \times 0.8} = 146.3$ A

Supply voltage/phase, $V = \frac{400}{\sqrt{3}} = 231$ V

Resultant voltage, $E_R = I Z_s = 146.3 \sqrt{(0.04)^2 + 0.4^2}$

= 58.81 V

Internal phase angle, $\theta = \text{Tan}^{-1} \frac{X_s}{R_e} = \text{Tan}^{-1} \frac{0.4}{0.04}$

= 84.3°

From phasor diagram shown in Fig. 6.17.

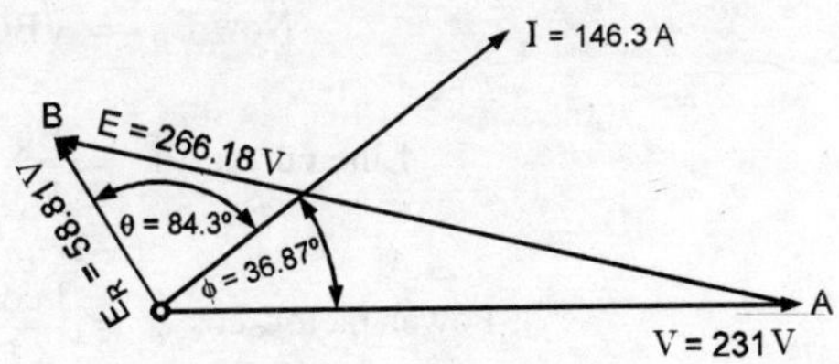

Fig. 6.17

$$E = \sqrt{V^2 + E_R^2 - 2 V E_R \cos(\theta + \phi)}$$

$$= \sqrt{(231)^2 + (58.81)^2 - 2 \times 231 \times 58.81 \times \cos(84.3° + 36.87°)} = \sqrt{70{,}855} = 266.18\text{ V}$$

So open-circuit emf/phase, E = 266.18 V **Ans.**

Gross mechanical power developed for all 3 phases,

$$P_m = P_{in} - 3\,I^2 R_e = 81{,}081 - 3 \times (146.3)^2 \times 0.04$$
$$= 78{,}512 \text{ watts or } 78.512 \text{ kW } \textbf{Ans.}$$

Example 6.6. A 3-phase star-connected synchronous motor has a synchronous reactance of 4 Ω/ phase and is working on 1,100 V bus-bar. Calculate the power factor of this machine when taking 90 kW from the mains, the excitation being adjusted to a value corresponding to an induced emf of 1,200 V. Neglect armature resistance.

Solution : Supply voltage per phase, $V = \dfrac{1{,}100}{\sqrt{3}} = 635 \text{ V}$

Induced emf per phase, $E = \dfrac{1{,}200}{\sqrt{3}} = 692.8 \text{ V}$

Since induced emf is greater than supply voltage, the motor must operate with a leading power factor.

Since power input $= \sqrt{3}\ V_L\ I_L \cos\phi = \sqrt{3}\ V_L\ I \cos\phi$ ∵ in star- connected synchronous motor, phase current I, = Line current I

$$\text{or } I\cos\phi = \frac{\text{Power input}}{\sqrt{3}\,V_L} = \frac{90 \times 1{,}000}{\sqrt{3} \times 1{,}100} = 47.24 \text{ A}$$

Synchronous reactance/phase, $X_s = 4\ \Omega$

Since resistance is negligible, synchronous impedance is equal to synchronous reactance *i.e.* 4 Ω and internal angle $\theta = 90°$

Impedance drop/phase, $E_R = I X_s = 4 I$

In ΔABC of phasor diagram shown in Fig. 6.18 we have

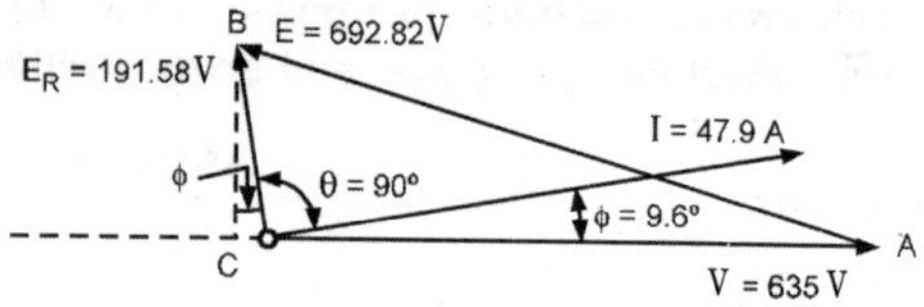

Fig. 6.18

$$AB^2 = BC^2 + AC^2$$
$$E^2 = (E_R \cos\phi)^2 + AC^2$$
$$\text{or } AC = \sqrt{E^2 - (E_R \cos\phi)^2}$$
$$= \sqrt{(692.82)^2 - (4I\cos\phi)^2} = \sqrt{(692.8)^2 - (4 \times 47.24)^2} = 666.6 \text{ V}$$
$$OC = AC - AO = 666.6 - 635 = 31.6 \text{ V}$$
$$\text{Now } E_R = \sqrt{BC^2 + OC^2} = \sqrt{(4 \times 47.24)^2 + (31.6)^2} = 191.58 \text{ V}$$
$$\text{Line current, } I = \frac{E_R}{Z_s} = \frac{191.58}{4} = 47.9 \text{ A}$$
$$\text{Power factor, } \cos\phi = \frac{I\cos\phi}{I} = \frac{47.24}{47.9} = 0.986 \text{ (lead) } \textbf{Ans.}$$

Example 6.7. A 2,000 V, 3-phase, star-connected synchronous motor has an effective resistance and synchronous reactance of 0.2 Ω and 2.2 Ω respectively. The input is 800 kW at normal voltage and the generated line emf is 2,500 V. Calculate the line current and power factor.

Solution : Supply voltage/phase, $V = \dfrac{2{,}000}{\sqrt{3}} = 1{,}155$ volts

$$\text{Induced emf/phase, } E = \frac{2{,}500}{\sqrt{3}} = 1{,}443 \text{ V}$$

$$\text{Internal phase angle of the motor, } \theta = \text{Tan}^{-1} \frac{X_s}{R_e} = \text{Tan}^{-1} \frac{2.2}{0.2} = 84.8°$$

Since induced emf is greater than supply voltage, the motor must be operating with a leading power factor. If the motor current is I, then in phase (or active component) is I cos ϕ and reactive (or wattless) component is I sin ϕ

$$\text{or } I = (I \cos \phi + j\, I \sin \phi)$$

$$\text{Since power input} = \sqrt{3}\ V_L\ I_L \cos \phi$$

$$\text{or } I \cos \phi = \frac{\text{Power input}}{\sqrt{3}\ V_L}$$

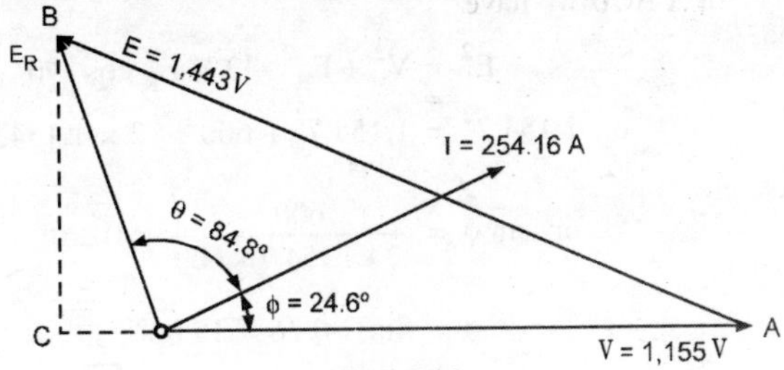

Fig. 6.19

$$= \frac{800 \times 1{,}000}{\sqrt{3} \times 2{,}000} = 231 \text{ A} \quad \because \text{ in star-connected motor phase current I = Line current } I_L$$

$$\text{Impedance drop per phase, } \mathbf{E_R} = \mathbf{I\, Z_s} = (I \cos \phi + j\, I \sin \phi)\ (0.2 + j\, 2.2)$$

$$= (231 + j\, I \sin \phi)\ (0.2 + j\, 2.2)$$

$$= (46.2 - 2.2\, I \sin \phi) + j\, (508.2 + 0.2\, I \sin \phi)$$

Now from Fig. 6.19.

$$OA = 1{,}155 \text{ V; } AB = 1{,}443 \text{ V; } \theta = 84.8°$$

$$OC = 46.2 - 2.2\, I \sin \phi;$$

$$BC = 508.2 + 0.2\, I \sin \phi$$

From the right angled Δ ABC we have

$$AB^2 = BC^2 + AC^2 = BC^2 + (AO + OC)^2$$

$$\text{or } (1{,}443)^2 = (508.2 + 0.2\, I \sin \phi)^2 + [1{,}155 + (46.2 - 2.2\, I \sin \phi)]^2$$

$$\text{or } I \sin \phi = 106.0 \text{ A}$$

$$\text{Line current, } I_L = I = \sqrt{(I \cos\phi)^2 + (I \sin\phi)^2} = \sqrt{(231)^2 + (106)^2} = 254.16 \text{ A } \textbf{Ans.}$$

$$\text{Power factor, } \cos \phi = \frac{I \cos\phi}{I} = \frac{231}{254.16} = 0.909 \text{ (leading) } \textbf{Ans.}$$

Example 6.8. A 2,000 V, 3-phase, 4-pole, Y-connected synchronous motor runs at 1,500 rpm. The excitation is constant and corresponds to an open-circuit voltage of 2,000 V. The resistance is negligible as compared to synchronous reactance of 3 Ω/phase. Determine the power input, power factor and torque developed for an armature current of 200 A.

Solution : Supply voltage per phase, $V = \frac{2{,}000}{\sqrt{3}} = 1{,}154.7$ V

$$\text{Induced emf per phase, } E = \frac{2{,}000}{\sqrt{3}} = 1{,}154.7 \text{ V}$$

$$\text{Input line current, } I_L = \text{Input phase current, } I = 200 \text{ A}$$

Impedance drop per phase, $E_R = I Z_s = 200 \times 3 = 600$ V

Internal angle, $\theta = 90°$ $\because$ resistance is negligible

Assuming armature current lagging behind the supply voltage by an angle ϕ, as shown in phasor diagram (Fig. 6.20)

In Δ AOB we have

$$E^2 = V^2 + E_R^2 - 2 V E_R \cos(90° - \phi)$$

or $$1{,}154.7^2 = 1{,}154.7^2 + 600^2 - 2 \times 1{,}154.7 \times 600 \sin\phi$$

or $$\sin\phi = \frac{600^2}{2 \times 1{,}154.7 \times 600} = 0.26$$

$$\phi = \text{Sin}^{-1}\, 0.26 = 15.06°$$

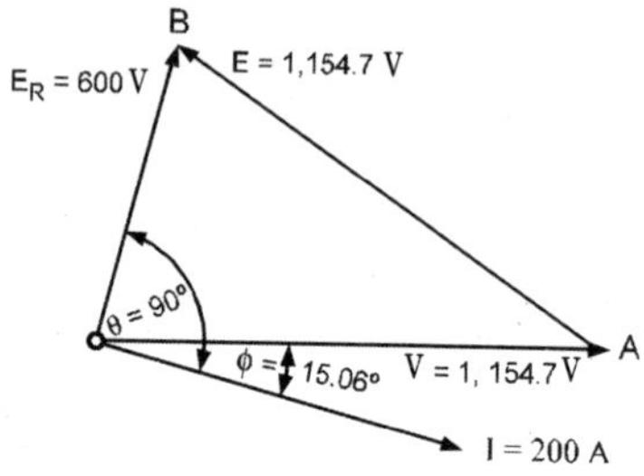

Fig. 6.20

Power factor = Cos 15.06° = 0.966 (lag) **Ans.**

$$\text{Power input} = \sqrt{3}\, V_L I_L \cos\phi$$

$$= \sqrt{3} \times 2{,}000 \times 200 \times 0.966$$

= 6,69,029 W or 669.029 kW **Ans.**

$$\text{Torque developed} = \frac{\text{Power input} - \text{copper losses in armature}}{2\pi N_s / 60}$$

$$= \frac{6{,}69{,}029}{2\pi \times 1{,}500/60} = 4{,}259 \text{ Nm}$$ **Ans.** neglecting armature copper losses.

Example 6.9. A 400 V, 10 hp, 3-phase synchronous motor has negligible armature resistance and synchronous reactance of 10 Ω per phase. Determine the minimum current and the corresponding induced emf for full load condition. Assume an efficiency of 85%.

Solution : $$\text{Motor input} = \frac{\text{Output}}{\eta} = \frac{10 \times 735.5}{0.85} = 8{,}653 \text{ watts}$$

Since motor input $= \sqrt{3}\, V_L I_L \cos\phi$,

$$I\cos\phi = \frac{\text{Motor input in watts}}{\sqrt{3}\, V_L}$$

$$= \frac{8{,}653}{\sqrt{3} \times 400} = 12.5 \text{ A}$$

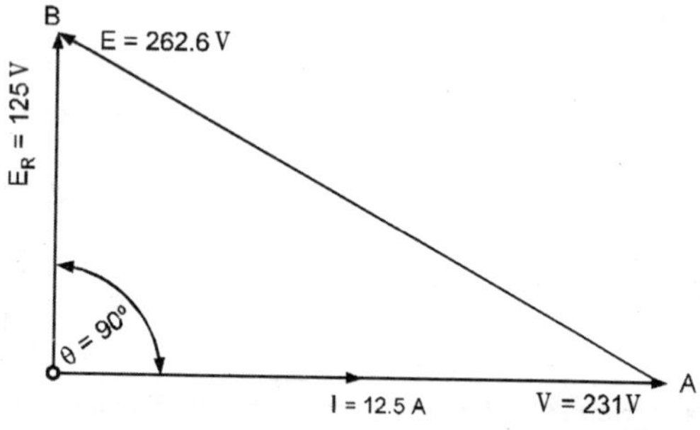

Fig. 6.21

The current is minimum when power factor, cos ϕ is unity and in such a condition

$$I = I\cos\phi = 12.5 \text{ A}$$

Impedance drop, $E_R = I Z_s = 12.5 \times 10 = 125$ V

$$\text{Supply voltage/phase}, V = \frac{400}{\sqrt{3}} = 231 \text{ V}$$

$$\text{Induced emf/phase}, E = \sqrt{V^2 + E_R^2} = \sqrt{(231)^2 + (125)^2} = 262.6 \text{ V}$$

Line induced emf = $\sqrt{3} \times 262.6 = 455$ V **Ans.**

Example 6.10. A 2200 V, 400 kW, 3-phase star-connected synchronous motor has a synchronous impedance of 0.3 + *j* 3Ω/ph. Determine the excitation emf (back emf) and the power angle if the motor works at rated load at 0.8 pf leading with an efficiency of 94 %.

Solution : Motor output = 400 kW

$$\text{Efficiency, } \eta = 94\% = 0.94$$

$$\text{Motor input} = \frac{\text{Motor output}}{\eta} = \frac{400}{0.94} = 425.532 \text{ kW}$$

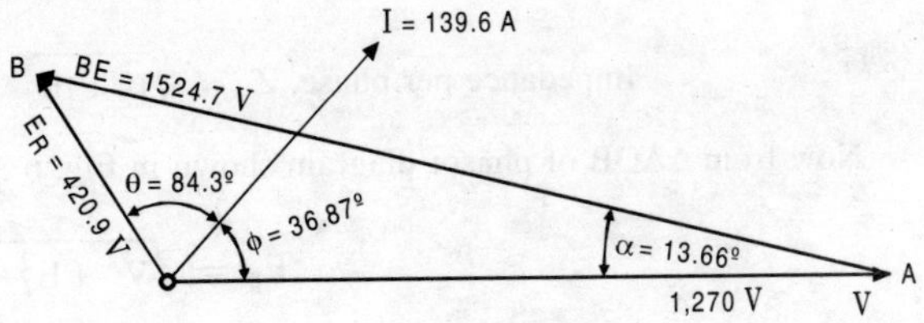

Fig. 6.22

$$\text{Motor input current, I} = \frac{\text{Motor input in watts}}{\sqrt{3}\, V_L \cos\phi} = \frac{425.532 \times 1{,}000}{\sqrt{3} \times 2{,}200 \times 0.8} = 139.6 \text{ A}$$

$$\text{Synchronous impedance per phase, } Z_s = \sqrt{(R_e)^2 + (X_s)^2} = \sqrt{(0.3)^2 + (3)^2} = 3.015\ \Omega$$

$$\text{Internal angle, } \theta = \text{Tan}^{-1}\frac{X_s}{R_e} = \text{Tan}^{-1}\frac{3}{0.3} = \text{Tan}^{-1} 10 = 84.3°$$

$$\text{Power factor angle, } \phi = \text{Cos}^{-1} 0.8 = 36.87°$$

$$\text{Voltage per phase, V} = \frac{2{,}200}{\sqrt{3}} = 1{,}270 \text{ V}$$

$$\text{Impedance drop per phase, } E_R = IZ_s = 139.6 \times 3.015 = 420.9 \text{ V}$$

In ΔOAB of phasor diagram shown in Fig. 6.22, we have

$$E = \sqrt{V^2 + E_R^2 - 2VE_R \cos(\theta + \phi)}$$

$$= \sqrt{(1{,}270)^2 + (420.9)^2 - 2 \times 1{,}270 \times 420.9 \cos(84.3° + 36.87°)}$$

$$= 1{,}524.7 \text{ V}$$

$$\text{Line induced emf} = \sqrt{3} \times 1{,}524.7 = 2{,}640 \text{ V}$$ **Ans.**

$$\text{Also } \frac{E_R}{\text{Sin}\,\delta} = \frac{E}{\text{Sin}(\theta + \phi)}$$

$$\text{or } \sin\delta = \frac{E_R \sin(\theta + \phi)}{E} = \frac{420.9 \times \sin(84.3° + 36.87°)}{1{,}524.7} = 0.2362$$

$$\text{or Power angle, } \delta = \text{Sin}^{-1} 0.2362 = 13.66°$$ **Ans.**

Example 6.11. A 20-pole, 30 kW, 660 V, 50 Hz, 3-ϕ, star-connected synchronous motor is operating with its generated voltage per phase exactly equal to the phase voltage applied to its armature. At loaded condition the motor is retarded by 5 mechanical degrees from its synchronous position. The synchronous reactance and armature resistance are 10 Ω and 1 Ω per phase. Calculate (*i*) armature current per phase (*ii*) the power per phase and the total power drawn by the motor from the bus and (*iii*) the developed power.

Solution : $$\text{Supply voltage per phase, V} = \frac{660}{\sqrt{3}} = 381 \text{ V}$$

$$\text{Induced emf per phase, E} = V = 381 \text{ V}$$

$$\text{Load angle, } \delta = \frac{P}{2} \times \text{angle of retardation in mechanical degrees}$$

$$= \frac{20}{2} \times 5 = 50^\circ \text{ electrical}$$

$$\text{Impedance per phase, } Z_s = \sqrt{1^2 + 10^2} = 10.05\ \Omega$$

Now from ΔAOB of phasor diagram shown in Fig. 6.23

$$E_R = \sqrt{V^2 + E^2 - 2VE\cos\delta}$$

$$= \sqrt{(381)^2 + (381)^2 - 2\times 381\times 381\cos 50^\circ} = 322\text{ V}$$

$$(i)\text{ Armature current per phase, } I = \frac{E_R}{Z_s} = \frac{322}{10.05} = 32\text{ A } \textbf{Ans.}$$

From Δ AOB of phasor diagram shown in Fig. 6.23

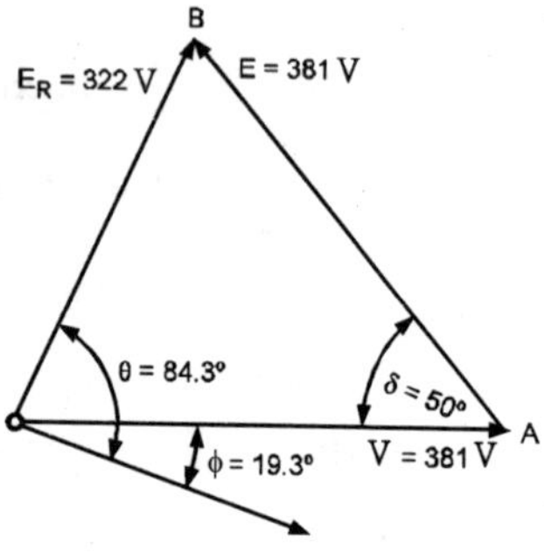

Fig. 6.23

$$\frac{AB}{\sin\angle AOB} = \frac{OB}{\sin\angle OAB}$$

$$\frac{381}{\sin(\theta+\phi)} = \frac{322}{\sin 50^\circ}$$

$$\text{or } \sin(\theta+\phi) = \frac{381}{322}\sin 50^\circ = 0.9064$$

$$\text{or } \theta + \phi = 65^\circ$$

$$\text{and } \theta = \text{Tan}^{-1}\frac{X_s}{R_e} = \text{Tan}^{-1}\frac{10}{1} = 84.3^\circ$$

$$\phi = 65^\circ - \theta = 65 - 84.3^\circ = -19.3^\circ \textit{ i.e. } 19.3^\circ \text{ lagging}$$

$$(iii)\text{ Power per phase} = VI\cos\phi = 381 \times 32 \times \cos 19.3^\circ = 11{,}507 \text{ watts or } 11.5\text{ kW } \textbf{Ans.}$$

$$\text{Total power drawn} = 3 \times 11.5 = 34.5\text{ kW } \textbf{Ans.}$$

$$\text{Power developed} = \text{Total power drawn} - 3I^2R_e = 34.5 - \frac{3\times 32^2\times 1}{1{,}000} = 31.43\text{ kW } \textbf{Ans.}$$

Example 6.12. A 3-phase, 415 V, 6-pole, 50 Hz star-connected synchronous motor has emf of 520 V (L – L). The stator winding has a synchronous reactance of 2 Ω per phase, and the motor develops a torque of 220 Nm. The motor is operating at 415 V, 50 Hz bus (*a*) Calculate the current drawn from the supply and its power factor (*b*) draw the phasor diagram showing all the relevant quantities.

Solution : Supply voltage per phase, $V = \frac{415}{\sqrt{3}} = 239.6\text{ V}$

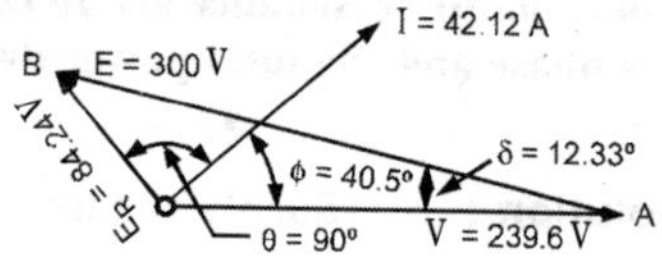

Fig. 6.24

$$\text{Induced emf per phase, } E = \frac{520}{\sqrt{3}} = 300\text{ V}$$

$$\text{Synchronous speed of motor, } N_s = \frac{120f}{P} = \frac{120\times 50}{6} = 1{,}000\text{ rpm}$$

$$\text{Torque developed, T} = 220 \text{ Nm}$$

$$\text{Total power developed, } 3\,P_{mech} = \frac{T \times 2\pi N_s}{60} = \frac{220 \times 2\pi \times 1{,}000}{60} = 23{,}038 \text{ watts}$$

$$\text{Power developed per phase} = \frac{23{,}038}{3} = 7{,}679 \text{ watts}$$

$$\text{Synchronous impedance/phase, } \mathbf{Z}_s = 2\ \angle 90^\circ\ \Omega$$

Power developed per phase is given as

$$P_{mech} = \frac{EV}{Z_s}\cos(\theta - \delta) - \frac{E^2}{Z_s}\cos\theta = \frac{EV}{Z_s}\sin\delta \qquad \because \theta = 90^\circ$$

$$\text{So } \frac{EV}{Z_s}\sin\delta = 7{,}679$$

$$\text{or } \sin\delta = \frac{7{,}679 \times Z_s}{EV} = \frac{7{,}679 \times 2}{300 \times 239.6} = 0.2135$$

$$\text{or } \delta = \text{Sin}^{-1}\ 0.2135 = 12.33^\circ$$

From phasor diagram shown in Fig. 6.24.

$$E_R = \sqrt{V^2 + E^2 - 2VE\cos\delta}$$

$$= \sqrt{(239.6)^2 + (300)^2 - 2 \times 239.6 \times 300 \times \cos 12.33^\circ} = 84.24 \text{ V}$$

$$\text{Current drawn per phase, I} = \frac{E_R}{Z_s} = \frac{84.24}{2} = 42.12 \text{ A } \textbf{Ans.}$$

Again from phasor diagram

$$\text{Sin}(\phi + \theta) = \frac{E}{E_R}\sin\delta = \frac{300}{84.24}\sin 12.33^\circ = 0.76$$

$$\text{or } \phi + \theta = \text{Sin}^{-1}\ 0.76 = 130.5^\circ$$

$$\text{or } \phi = 130.5^\circ - 90^\circ = 40.5^\circ$$

$$\text{Power factor} = \text{Cos}\,\phi = \text{Cos}\ 40.5^\circ = 0.76 \text{ (leading) } \textbf{Ans.}$$

Example 6.13. A 500 V, 6-pole, 3-phase, 50 Hz, star-connected synchronous motor has a resistance and synchronous reactance of 0.3 Ω and 3 Ω per phase respectively. The open-circuit voltage is 600V. If the friction and core losses total 1 kW, calculate the line current and power factor when the motor output is 100 hp.

Solution : Supply voltage per phase, $V = \frac{500}{\sqrt{3}} = 288.7$ V

$$\text{Induced emf per phase, E} = \frac{600}{\sqrt{3}} = 346.41 \text{ V}$$

$$\text{Total power developed, } 3\,P_{mech} = 100 \times 0.7355 + 1$$

$$= 74.55 \text{ kW}$$

$$\text{Power developed per phase} = \frac{74.55}{3}$$

$$= 24.85 \text{ kW}$$

or 24,850 watts

$$\text{Internal angle, } \theta = \text{Tan}^{-1} \frac{X_s}{R_e}$$

$$= \text{Tan}^{-1} \frac{3}{0.3} = 84.3°$$

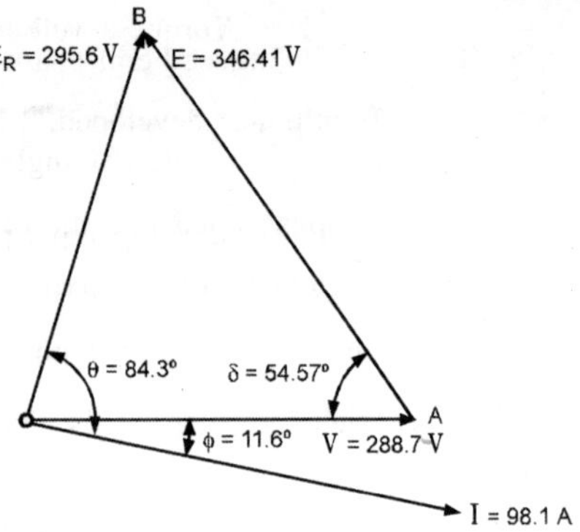

Fig. 6.25

$$\text{Synchronous impedance/phase, } Z_s = \sqrt{0.3^2 + 3^2} = 3.015 \ \Omega$$

Power developed per phase is given as

$$P_{mech} = \frac{EV}{Z_s} \cos(\theta - \delta) - \frac{E^2}{Z_s} \cos \theta$$

$$\text{or } 24{,}850 = \frac{288.7 \times 346.41}{3.015} \cos(84.3° - \delta) - \frac{(346.41)^2}{3.015} \cos 84.3°$$

$$\text{or } \text{Cos}(84.3° - \delta) = 0.8684$$

$$\text{or } 84.3° - \delta = \text{Cos}^{-1} 0.8684 = 29.73°$$

$$\text{or } \delta = 84.3° - 29.73° = 54.57°$$

From phasor diagram shown in Fig. 6.25.

$$E_R = \sqrt{V^2 + E^2 - 2VE \cos\delta}$$

$$= \sqrt{288.7^2 + 346.41^2 - 2 \times 288.7 \times 346.41 \cos 54.57°} = 295.6 \text{ V}$$

$$\text{Line current} = \text{Phase current, } I = \frac{E_R}{Z_s} = \frac{295.6}{3.015} = 98.1 \text{ A } \textbf{Ans.}$$

Again from phasor diagram

$$\text{Sin}(\theta + \phi) = \frac{E}{E_R} \sin \delta = \frac{346.41}{295.6} \sin 54.57° = 0.9542$$

$$\text{or } \phi + \theta = \sin^{-1} 0.9542 = 72.7°$$

$$\text{or } \phi = 72.7° - 84.3° = -11.6° = 11.6° \text{ (lag)}$$

$$\text{Power factor} = \text{Cos } 11.6° = 0.98 \text{ (lag) } \textbf{Ans.}$$

Example 6.14. A 50 Hz, 4-pole, 3-phase, star-connected synchronous motor has a synchronous reactance of 12.0 Ω/phase and negligible armature resistance. The excitation is such as to give an open-circuit voltage of 13.2 kV. The motor is connected to 11.5 kV, 50 Hz supply.

What maximum load can the motor supply before loosing synchronism ? What is the corresponding motor torque, line current and power factor ?

Solution : Supply voltage/phase, $V = \frac{11.5 \times 1{,}000}{\sqrt{3}} = 6{,}640$ V

$$\text{Induced emf/phase, } E = \frac{13.2 \times 1{,}000}{\sqrt{3}} = 7{,}621 \text{ V}$$

$$\text{Internal angle, } \theta = 90^\circ \qquad \because \text{ armature resistance is negligible}$$

$$\text{Synchronous impedance/phase, } Z_s = X_s = 12.0\ \Omega$$

Power developed will be maximum when

$$\text{load angle, } \delta = \theta = 90^\circ$$

$$\text{Impedance drop per phase, } E_R = V^2 + E^2 - 2\text{ VE} \cos\delta$$

$$= \sqrt{(6{,}640)^2 + (7{,}621)^2 - 2 \times 6{,}640 \times 7{,}621 \times \cos 90^\circ} = 10{,}108 \text{ V}$$

$$\text{Line current, } I_L = \text{Phase current, } I = \frac{E_R}{Z_s} = \frac{10{,}108}{12} = 842.3 \text{ A } \textbf{Ans.}$$

The maximum load that motor can deliver,

$$(P_{mech})_{max} = 3\left[\frac{EV}{Z_s} - \frac{E^2}{Z_s}\cos\theta\right] = 3\left[\frac{6{,}640 \times 7{,}621}{12} - \frac{(7{,}621)^2}{12}\cos 90^\circ\right]$$

$$= 1{,}26{,}50{,}000 \text{ watts or } 12.65 \text{ MW } \textbf{Ans.}$$

$$\text{Power supplied to motor, } P_{IN} = P_{mech} + \text{armature copper loss}$$

$$= P_{mech} = 12.65 \text{ MW} \qquad \because \text{ armature resistance is negligible}$$

$$\text{So } \sqrt{3}\, V_L I_L \cos\phi = 12.65 \times 10^6$$

$$\text{or Power factor, } \cos\phi = \frac{12.65 \times 10^6}{\sqrt{3} \times 11{,}500 \times 842.3} = 0.754 \textbf{ Ans.}$$

$$\text{Synchronous speed, } N_s = \frac{120 f}{P} = \frac{120 \times 50}{4} = 1{,}500 \text{ rpm}$$

Torque developed corresponding to maximum power developed,

$$(T_{mech})_{max} = \frac{(P_{mech})_{max}}{2\pi N_s/60} = \frac{12.65 \times 10^6}{2\pi \times 1{,}500/60} = 80{,}532 \text{ Nm } \textbf{Ans.}$$

Example 6.15. The excitation of a 415 V, 3-phase, mesh connected synchronous motor is such that the induced emf is 520 V. The impedance per phase is (0.5 + *j* 4.0) ohms. If the friction and iron losses are constant at 1,000 watts, calculate the power output, line current, power factor and efficiency for maximum power output.

Solution : Supply voltage/phase, $V = 415$ V $\qquad \because$ motor is mesh connected

$$\text{Induced emf/phase, } E = 520 \text{ V}$$

$$\text{Synchronous impedance/phase, } Z_s = \sqrt{(0.5)^2 + 4^2} = 4.03\ \Omega$$

$$\text{Internal angle, } \theta = \text{Tan}^{-1}\frac{X_s}{R_e} = \text{Tan}^{-1}\frac{4}{0.5} = 82.9^\circ$$

For a constant value of supply voltage, excitation (or induced emf), effective resistance R_e and synchronous reactance X_s maximum power will be developed when load angle

$$\delta = \theta = 82.9° \qquad ...(Art. 6.8.1)$$

Now from phasor diagram ΔOAB (Fig. 6.26) we have

$$E_R = \sqrt{V^2 + E^2 - 2VE\cos\delta}$$

$$= \sqrt{(415)^2 + (520)^2 - 2 \times 415 \times 520 \cos 82.9°}$$

$$= 625 \text{ V/phase}$$

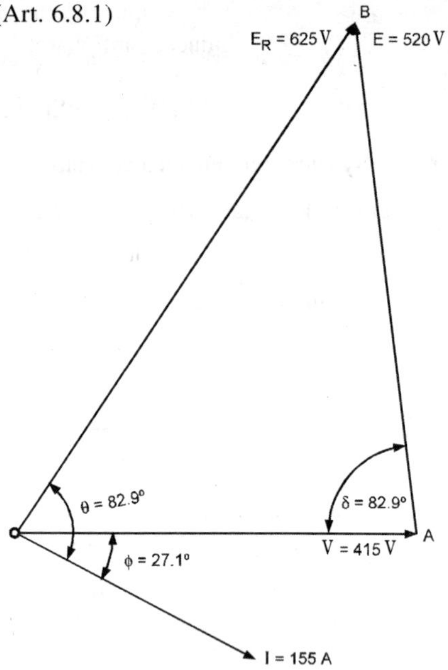

Fig. 6.26

$$\text{Load current, } I = \frac{E_R}{Z_s} = \frac{625}{4.03} = 155 \text{ A/phase}$$

Line current, $I_L = \sqrt{3}\, I = \sqrt{3} \times 155 = 268.2$ A **Ans.**

Maximum power developed per phase,

$$(P_{mech})_{max} = \frac{EV - E^2 \cos\theta}{Z_s}$$

$$= \frac{520 \times 415 - 520^2 \times \cos 82.9°}{4.03} = 45{,}255 \text{ watts}$$

Maximum power for 3-phase,

$$= 3 \times 45{,}255 = 1{,}35{,}765 \text{ watts}$$

Output power = Power developed – iron and friction losses

= 1,35,765 – 1,000

= 1,34,765 watts or 134.765 kW **Ans.**

Total copper losses = $3 I^2 R_e = 3 \times (155)^2 \times 0.5 = 36{,}000$ watts

Input to motor = Power developed + copper losses = 1,35,765 + 36,000 = 1,71,765 watts

$$\text{Power factor, } \cos\phi = \frac{\text{Input to motor in watts}}{\sqrt{3}\, V_L I_L} = \frac{1{,}71{,}765}{\sqrt{3} \times 415 \times 268.2} = 0.89 \textbf{ Ans.}$$

$$\text{Efficiency, } \eta = \frac{\text{Output}}{\text{Input}} \times 100 = \frac{1{,}34{,}765}{1{,}71{,}765} \times 100 = 78.46\% \textbf{ Ans.}$$

Example 6.16. A 400 V, 6-pole, 3-ϕ, 50 Hz, star-connected synchronous motor has a resistance and synchronous impedance of 0.5 Ω and 4 Ω per phase respectively. It takes a current of 15 A at unity power factor when operating with a certain field current. If the torque load is increased until the line current is increased to 60 A, the field current remaining the same determine the gross torque developed and the new power factor.

Solution :

$$\text{Supply voltage, } V = \frac{400}{\sqrt{3}} = 231 \text{ V}$$

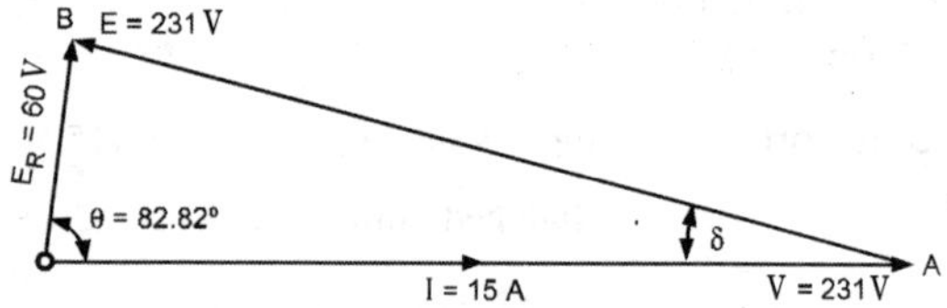

Fig. 6.27

Effective resistance/phase, $R_e = 0.5\ \Omega$

Synchronous impedance/phase, $Z_s = 4\ \Omega$

$$\text{Synchronous reactance/phase, } X_s = \sqrt{(4)^2 - (0.5)^2}$$

$$= 3.968\ \Omega.$$

$$\text{Internal angle, } \theta = \text{Cos}^{-1}\frac{R_e}{Z_s} = \text{Cos}^{-1}\frac{0.5}{4} = 82.82°$$

When input current, $I = 15$ A

and $\cos\phi$ = unity

Impedance drop, $E_R = I Z_s = 15 \times 4 = 60$ V

Induced emf/phase, $E = \sqrt{V^2 + E_R^2 - 2 V E_R \cos\theta}$

$= \sqrt{(231)^2 + (60)^2 - 2 \times 231 \times 60 \cos 82.82^\circ} = 231$ V

Now when the load on the motor is increased, the angle of retardation δ increases. The phasor diagram is shown in Fig. 6.27.

Input current, $I' = 60$ A

Voltage per phase, $V = 231$ V, as before

Induced emf per phase, $E = 231$ V, as before

Since excitation is constant

Impedance drop/phase, $E'_R = I' Z_s = 60 \times 4 = 240$ V

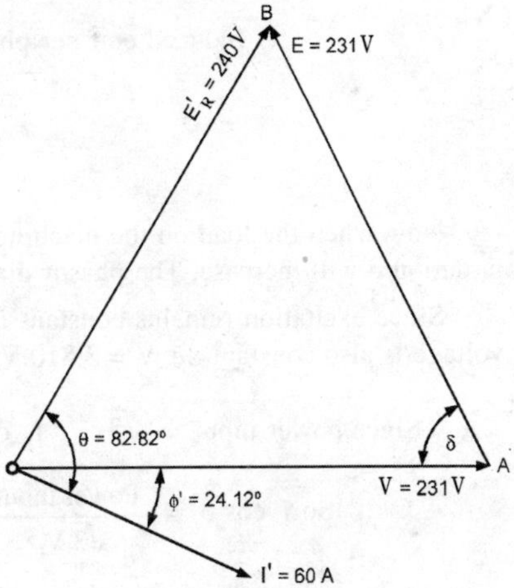

Fig. 6.28

Now in Δ OAB of phasor diagram (Fig. 6.28) we have

$$E^2 = V^2 + E_R'^2 - 2 V E'_R \cos \angle AOB$$

or $231^2 = 231^2 + 240^2 - 2 \times 231 \times 240 \cos \angle AOB$

or $\angle AOB = \cos^{-1} 0.5195 = 58.7^\circ$

Internal angle, $\theta = 82.82^\circ$

$\therefore$ Power factor angle, $\phi' = \theta - \angle AOB$

$= 82.82^\circ - 58.7^\circ$

$= 24.12^\circ$ (lag)

New power factor, $\cos\phi' = \cos 24.12^\circ$

$= 0.9127$ (lag) **Ans.**

New motor input $= \sqrt{3}\, V_L I'_L \cos\phi'$

$= \sqrt{3} \times 400 \times 60 \times 0.9127 = 38{,}000$ W or 38 kW

Total armature copper loss $= 3 I'^2 R_e = 3 \times (60)^2 \times 0.5 = 5{,}400$ W or 5.4 kW

Mechanical power developed $= 38 - 5.4 = 32.6$ kW

Synchronous speed, $N_s = \dfrac{120 f}{P} = \dfrac{120 \times 50}{6} = 1{,}000$ rpm

Gross torque developed, $T_g = \dfrac{\text{Power developed in watts}}{2\pi N_s/60} = \dfrac{32.6 \times 1{,}000}{2\pi \times \dfrac{1{,}000}{60}} = 311$ N-m **Ans.**

Example 6.17. A 6.6 kV star-connected, 3-phase synchronous motor works at constant voltage and constant excitation. Its synchronous reactance is 20 Ω per phase. When the input is 1,000 kW, the power factor is 0.8 leading. Find the power factor when the input is increased to 1,500 kW. Neglect armature resistance.

Solution : Supply voltage per phase, $V = \dfrac{6.6 \times 1{,}000}{\sqrt{3}} = 3{,}810$ V

Synchronous impedance per phase, $Z_s = X_s = 20\ \Omega$ $\because$ armature resistance is negligible

Internal angle, $\theta = 90^\circ$

When the input is 1,000 kW at 0.8 pf leading

$$\text{Armature current/phase, } I = \frac{1{,}000 \times 1{,}000}{\sqrt{3} \times 6{,}600 \times 0.8} = 109.35 \text{ A}$$

$$\text{Power factor angle, } \phi = \text{Cos}^{-1}\ 0.8 = 36.87^\circ$$

$$\text{Impedance drop per phase, } E_R = IZ_s = 109.35 \times 20 = 2{,}187 \text{ V}$$

$$\text{Induced emf per phase, } E = \sqrt{V^2 + E_R^2 - 2VE_R \cos(\theta + \phi)}$$

$$= \sqrt{(3{,}810)^2 + (2{,}187)^2 - 2 \times 3{,}810 \times 2{,}187 \cos(90^\circ + 36.87^\circ)}$$

$$= 5{,}413 \text{ V.}$$

Now when the load on the machine is increased, the angle of retardation δ will increase. The phasor diagram is shown in Fig. 6.29.

Since excitation remains constant *i.e.* 5,413 V per phase. Supply voltage is also constant *i.e.* V = 3,810 V per phase.

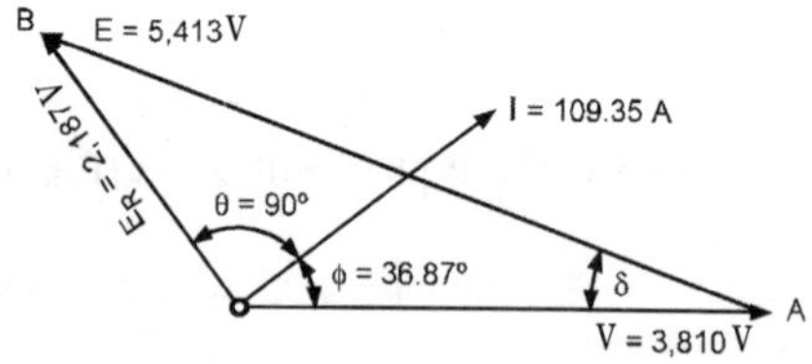

Fig. 6.29

$$\text{Since power input} = \sqrt{3}\ V_L I_L \cos\phi$$

$$\text{So } I' \cos\phi' = \frac{\text{Power input}}{\sqrt{3}\, V_L} = \frac{1{,}500 \times 1{,}000}{\sqrt{3} \times 6{,}600} = 131.2 \text{ A}$$

Impedance drop per phase,

$$E'_R = I'X = 20\, I$$

In Δ ABC of phasor diagram shown in Fig. 6.30 we have

$$AB^2 = BC^2 + AC^2$$

$$\text{or } AC = \sqrt{AB^2 - BC^2} = \sqrt{E^2 - (E'_R \cos\phi')^2}$$

$$= \sqrt{(5{,}413)^2 - (20 \times 131.2)^2} = 4{,}734.3 \text{ V}$$

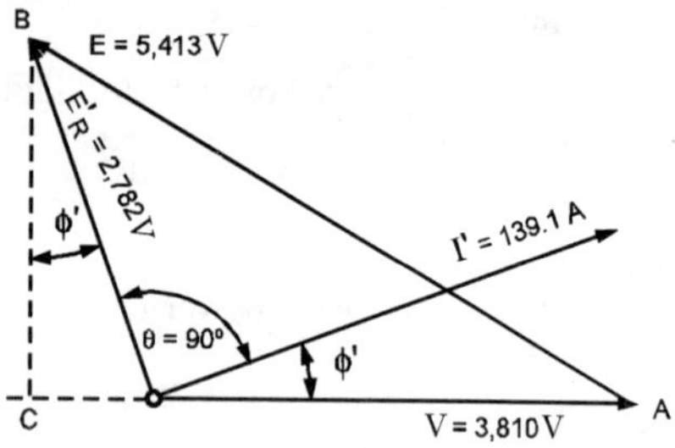

Fig. 6.30

$$OC = AC - AO = 4{,}734.3 - 3{,}810 = 924.3 \text{ V}$$

$$E'_R = \sqrt{BC^2 + OC^2} = \sqrt{(20 \times 131.2)^2 + (924.3)^2} = 2{,}782 \text{ V}$$

$$\text{Line current, } I' = \frac{E'_R}{Z_s} = \frac{2{,}782}{20} = 139.1 \text{ A}$$

$$\text{Power factor, } \cos\phi' = \frac{I'\cos\phi'}{I'} = \frac{131.2}{139.1} = 0.9432 \text{ (leading)}$$ **Ans.**

Example 6.18. A 3-phase, Y-connected synchronous motor takes 48 kW at 693 V (line), the pf being 0.8 lag. The induced emf is now increased by 30%, the power input being the same. Find the new current and pf. Z_s equals (0 + *j* 2) ohm/phase.

Solution : Supply voltage per phase, $V = \dfrac{693}{\sqrt{3}} = 400$ V

Synchronous impedance per phase, $Z_s = X_s = 2\ \Omega$ $\because R_e = 0$

Internal angle $\theta = 90°$

At normal excitation

Power factor, $\cos\phi = 0.8$ (lag)

Power factor angle, $\phi = \cos^{-1} 0.8 = -36.87°$

$$\text{Armature current/phase, } I = \frac{\text{kW input} \times 1{,}000}{\sqrt{3}\, V_L \cos\phi} = \frac{48 \times 1{,}000}{\sqrt{3} \times 693 \times 0.8} = 50 \text{ A}$$

Impedance drop per phase, $E_R = I Z_s = 50 \times 2 = 100$ V

$$\text{Induced emf per phase, } E = \sqrt{V^2 + E_R^2 - 2 V E_R \cos(\theta + \phi)}$$

$$= \sqrt{400^2 + 100^2 - 2 \times 400 \times 100 \cos[90° + (-36.87°)]}$$

$$= 349.3 \text{ V}$$

When the excitation is increased by 30% the induced emf for phase E′ will become 1.3 × 349.3 = 454 V.

Since power input and supply voltage is constant, active component of current drawn remains unchanged *i.e.*

$$I' \cos\phi' = 50 \times 0.8 = 40 \text{ A}$$

The phasor diagram is shown in Fig. 6.31 assuming that the current drawn from supply mains is leading one now.

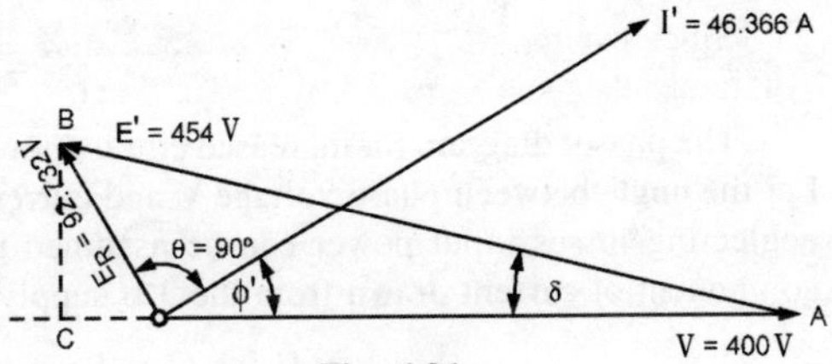

Fig. 6.31

In Δ ABC of the phasor diagram shown in Fig. 6.31

$$AC = \sqrt{AB^2 - BC^2} = \sqrt{E'^2 - (E'_R \cos\phi')^2}$$

$$= \sqrt{454^2 - (I' Z_s \cos\phi')^2}$$

$$= \sqrt{454^2 - (40 \times 2)^2} = 446.9 \text{ V}$$

$$OC = AC - OA = 446.9 - 400 = 46.9 \text{ V}$$

$$E'_R = \sqrt{BC^2 + OC^2} = \sqrt{(40 \times 2)^2 + (46.9)^2} = 92.732 \text{ V}$$

$$\text{Current, } I' = \frac{E'_R}{Z_s} = \frac{92.732}{2} = 46.366 \text{ A } \textbf{Ans.}$$

$$\cos\phi' = \frac{I' \cos\phi'}{I'} = \frac{40}{46.366} = 0.863 \text{ (lead) } \textbf{Ans.}$$

Example 6.19. A 20 hp, 3-phase, 400 V synchronous motor operating on full load from infinite bus-bar, has its excitation so adjusted that power factor is 0.87 lagging. Load being kept constant, excitation is now increased by 20%. Find the new power factor. Synchronous reactance may be taken as 1.0 per unit.

Solution : Assuming synchronous motor star-connected

$$\text{Supply voltage per phase, } V = \frac{400}{\sqrt{3}} = 231 \text{ volts}$$

Synchronous reactance = 1.0 per unit

$$\text{Since per unit reactance} = \frac{IX_s}{V}$$

$\therefore$ Synchronous reactance drop, IX_s = V × per unit reactance

$$= V \times 1 = 231 \text{ volts}$$

Neglecting effective resistance

Impedance drop per phase, $E_R = IZ_s = IX_s = 231$ V

and internal angle $\theta = 90°$

Power factor, $\cos\phi = 0.87$ (lagging)

$$\phi = \text{Cos}^{-1}\, 0.87 = 29.54°$$

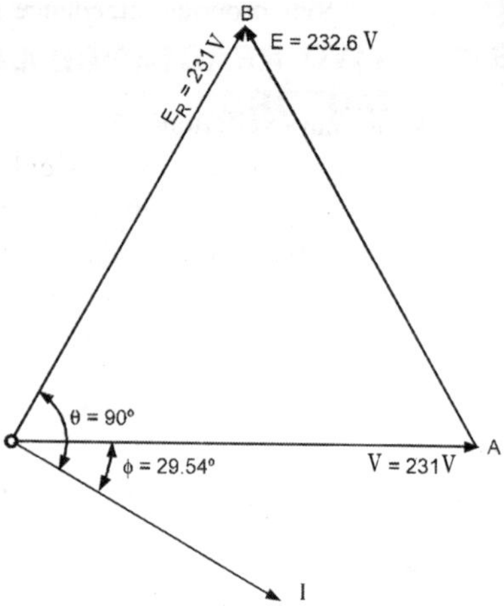

Fig. 6.32

From phasor diagram shown in Fig. 6.32

$$E = \sqrt{V^2 + E_R^2 - 2VE_R\cos(\theta - \phi)}$$

$$= \sqrt{(231)^2 + (231)^2 - 2 \times 231 \times 231 \cos(90° - 29.54°)} = 232.6 \text{ V}$$

The phasor diagram for increased emf is shown in Fig. 6.33. Now induced emf, $E' = 1.2\,E = 1.2 \times 232.6 = 279$ V. Let the angle between phase voltage V and current I be ϕ'. Let the new current be I'. As load remains the same so neglecting losses input power can be assumed to remain the same and since supply voltage is same so active component of current drawn from the 3-φ supply main would remain the same.

$$\therefore\ I'\cos\phi' = I\cos\phi = 0.87\, I$$

and impedance drop/phase $= E'_R = I'Z_s = I'X_s$

and $E'_R \cos\phi' = I'X_s\cos\phi'$

$= I\cos\phi \times X_s$

$= 0.87\, I X_s$

$= 0.87 \times 231$

$= 201$ V

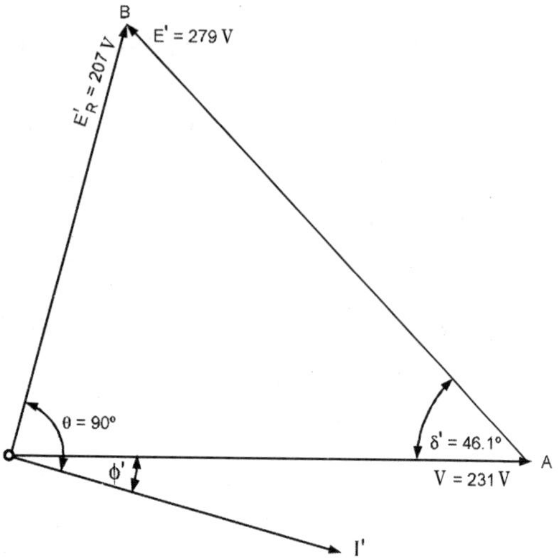

Fig. 6.33

From Δ AOB of phasor diagram shown in Fig. 6.33.

$$\frac{E'}{\sin(90° - \phi')} = \frac{E'_R}{\sin\delta'}$$

$$\text{or } \sin\delta' = \frac{E'_R\cos\phi'}{E'} = \frac{201}{279}$$

or $\delta' = 46.1°$

Again from ΔAOB, we get

$$E'_R = \sqrt{V^2 + (E')^2 - 2VE'\cos\delta'}$$

$$= \sqrt{(231)^2 + (279)^2 - 2 \times 231 \times 279 \cos 46.1°} = 207 \text{ volts}$$

New power factor, $\cos\phi' = \dfrac{E'_R\cos\phi'}{E'_R} = \dfrac{201}{279} = 0.971$ (lagging) **Ans.**

6.11. TWO REACTANCE CONCEPT FOR SALIENT POLE SYNCHRONOUS MOTORS

The analysis of smooth cylindrical (or non-salient pole) synchronous machine is much easier than that of a salient pole synchronous motor. It is due to the fact that in a smooth cylindrical rotor synchronous motor the air gap is uniform and, therefore, the effect of armature reaction, fluxes and voltages induced can be treated in a simple way with synchronous reactance concept. But in case of a salient pole synchronous motor, the air gap is not uniform—air gap is much larger in the inter polar space than along the field pole axis. The two-reactance theory, applicable to the salient pole motor*, will now be considered.

The salient pole synchronous motor has two axis—field pole axis, called the *direct axis* or *d-axis* and the axis passing through the centre of interpolar space, called the *quadrature axis* or *q-axis*, as in case of an alternator. The armature current of the motor can also be resolved along *d*-axis and *q*-axis called I_d and I_q components as in case of an alternator.

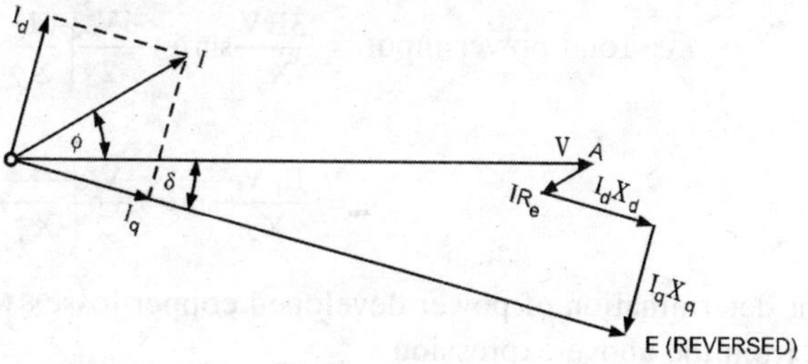

Fig. 6.34. ***Phasor Diagram With Leading Current For Synchronous Motor***

The voltage equation for each phase of the armature based on the two reactance concept is given below:

Induced emf for phase (reversed) *i.e.* $-\mathbf{E} = \mathbf{V} - \mathbf{I}\,\mathbf{R}_e - \mathbf{I}_d\ \mathbf{X}_d - \mathbf{I}_q\,\mathbf{X}_q$ ...(6.12)

The complete phasor diagram of a salient pole synchronous motor, for a leading power factor is given in Fig. 6.34.

The phasor diagram of a salient pole synchronous motor for a leading power factor becomes, as shown in Fig. 6.35, when the effect of armature resistance is neglected.

From phasor diagram shown in Fig. 6.35 using E as reference phasor, we have

$$V \cos \delta = E - I_d X_d \quad ...(6.13)$$

$$\text{and } V \sin \delta = I_q X_q \quad ...(6.14)$$

$$\text{So } I_d = \frac{E - V\cos\delta}{X_d} \quad ...(6.15)$$

$$\text{and } I_q = \frac{V\sin\delta}{X_q} \quad ...(6.16)$$

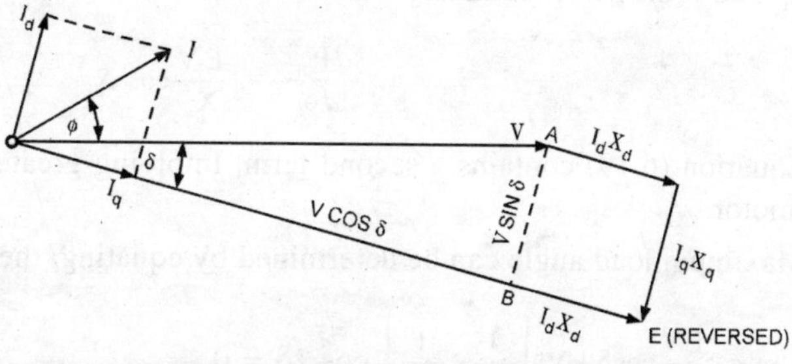

Fig. 6.35. ***Phasor Diagram With Leading Current For Synchronous Motor (Neglecting Effect of Armature Resistance)***

Power input being equal to the product of the in-phase components of current and voltage plus the product of the quadrature components, regardless of the axis of reference, is given as

$$P_{in} = I_q V \cos \delta + I_d V \sin \delta$$

* Synchronous reactance concept is reasonably accurate in predicting the steady-state performance of salient pole synchronous motors and two-reaction theory is required only when very high degree of accuracy is required or when problems concerning transients or power stability are to be handled.

Substituting the values of I_d and I_q from Eqs. (6.15) and (6.16) we have

$$P_{in} = \frac{V^2}{X_q} \cos \delta \sin \delta + \frac{E - V\cos\delta}{X_d} V \sin \delta$$

$$= \frac{EV}{X_d} \sin \delta + \frac{V^2}{2}\left[\frac{1}{X_q} - \frac{1}{X_d}\right] \sin 2\delta \quad ...(6.17)$$

Total power input will be 3 times of the above

$$\text{i.e. Total power input} = \frac{3EV}{X_d}\sin\delta + \frac{3V^2}{2}\left(\frac{1}{X_q} - \frac{1}{X_d}\right) \sin 2\delta$$

$$= \frac{E_L V_L}{X_d}\sin\delta + \frac{V_L^2}{2}\left(\frac{1}{X_q} - \frac{1}{X_d}\right) \sin 2\delta \quad ...(6.18)$$

For determination of power developed copper losses will be deducted from the power input determined from the above expression.

6.12. STABILITY AND MAXIMUM LOAD ANGLE

Differentiating the Eq. (6.17) with respect to load angle δ we have

$$\frac{dP_{in}}{d\delta} = \frac{EV}{X_d}\cos\delta + V^2\left[\frac{1}{X_q} - \frac{1}{X_d}\right] \cos 2\delta \quad ...(6.19)$$

The above equation gives the rate of change of power as a function of the load angle δ. This is called the *stability factor*, *rigidity factor*, or simply the *stiffness of coupling*. This is approximately because it is based on power input rather than power developed.

In case of smooth cylindrical rotor motor the corresponding expression is obtained by differentiating equation (6.5) w.r.t. load angle δ so

$$\frac{dP_{in}}{d\delta} = \frac{EV}{X_s}\cos \delta$$

Equation (6.19) contains a second term, implying greater stiffness of the coupling for the salient pole motor.

Maximum load angle can be determined by equating, the RHS of the equation (6.19) equal to zero

$$\text{i.e. } \frac{EV}{X_d}\cos\delta + V^2\left[\frac{1}{X_q} - \frac{1}{X_d}\right] \cos 2\delta = 0$$

Solving above equation for cos δ we have

$$\text{Cos } \delta \simeq \frac{-E}{4V\dfrac{X_d - X_q}{X_q}} + \sqrt{\frac{1}{2} + \left[\frac{EX_q}{4V(X_d - X_q)}\right]^2} \quad ...(6.20)$$

This is the cosine of the maximum load angle when the effect of armature resistance is neglected.

Example 6.20. A salient pole synchronous motor has $X_d = 0.85$ pu and $X_q = 0.55$ pu. It is connected to bus-bar of 1.0 pu voltage while its excitation is adjusted to 1.2 pu. Calculate the maximum power the motor can deliver without loss of synchronism. What minimum pu excitation will be required for the machine to remain in synchronism while delivering full-load torque.

Solution : For maximum power to be delivered by a salient pole synchronous motor, according to Eq. (6.20)

$$\text{Cos}\,\delta = \frac{-E \times X_q}{4V(X_d - X_q)} + \sqrt{\frac{1}{2} + \left[\frac{E X_q}{4V(X_d - X_q)}\right]^2}$$

$$= \frac{-1.2 \times 0.55}{4 \times 1(0.85 - 0.55)} + \sqrt{\frac{1}{2} + \left[\frac{1.2 \times 0.55}{4 \times 1(0.85 - 0.55)}\right]^2} = -0.55 + 0.896 = 0.346$$

or Load angle $\delta_{(max)} = \text{Cos}^{-1}\, 0.346 = 69.8°$

Now maximum power is given by Eq. (6.18)

$$i.e.\ P_{max} = \frac{EV}{X_d} \sin\delta + \frac{V^2}{2}\left[\frac{1}{X_q} - \frac{1}{X_d}\right] \sin 2\delta$$

$$= \frac{1.2 \times 1}{0.85} \sin 69.8° + \frac{1}{2}\left[\frac{1}{0.55} - \frac{1}{0.85}\right] \sin(2 \times 69.8°) = 1.533 \text{ pu } \textbf{Ans.}$$

The power delivered due to excitation is given as

$$P = \frac{EV}{X_d} \sin\delta$$

Substituting $V = 1.0$, $X_d = 0.85$, $\delta = 69.8°$ and $P = 1.533$ in the above equation we have

$$1.533 = \frac{E \times 1.0}{0.85} \sin 69.8°$$

$$\text{or } E = \frac{1.533 \times 0.85}{\sin 69.8°} = 1.388 \text{ pu } \textbf{Ans.}$$

i.e., the minimum pu excitation required for the machine to remain in synchronous while delivering full-load torque = 1.388 pu **Ans.**

6.13. V- AND INVERTED V-CURVES

It has already been discussed how the variation in field current affects the power factor at which a synchronous motor operates. In case of a synchronous motor driving a constant mechanical load variation in field or dc excitation current will not only affect the power factor but also the current drawn by the motor. Except for the change in copper losses due to variation in armature current and a slight change in core loss due to variation in flux, the power input to the motor is almost constant for a constant load.

The power drawn by a 3-phase synchronous motor is given by, $P = \sqrt{3}\, V_L I_L \cos\phi$ where V_L is the line voltage, I_L is the armature current and $\cos\phi$ is the power factor. Since input power P and supply voltage V is constant, so decrease in power factor causes increase in armature current and vice versa.

Hence variation in excitation or in field current causes the variation in armature current and curves drawn between armature current and field current for different constant loads are known as "V-curves" due to their shape similar to English letter V. *The V-curves of a synchronous motor give relation between armature current and field current for different power inputs.*

Similarly the variation of power factor with a variation in field current (dc excitation) for a constant load gives inverted V curve, shown in Fig. 6.9.

The connections for determining the V-curves experimentally are shown in Fig. 6.36.

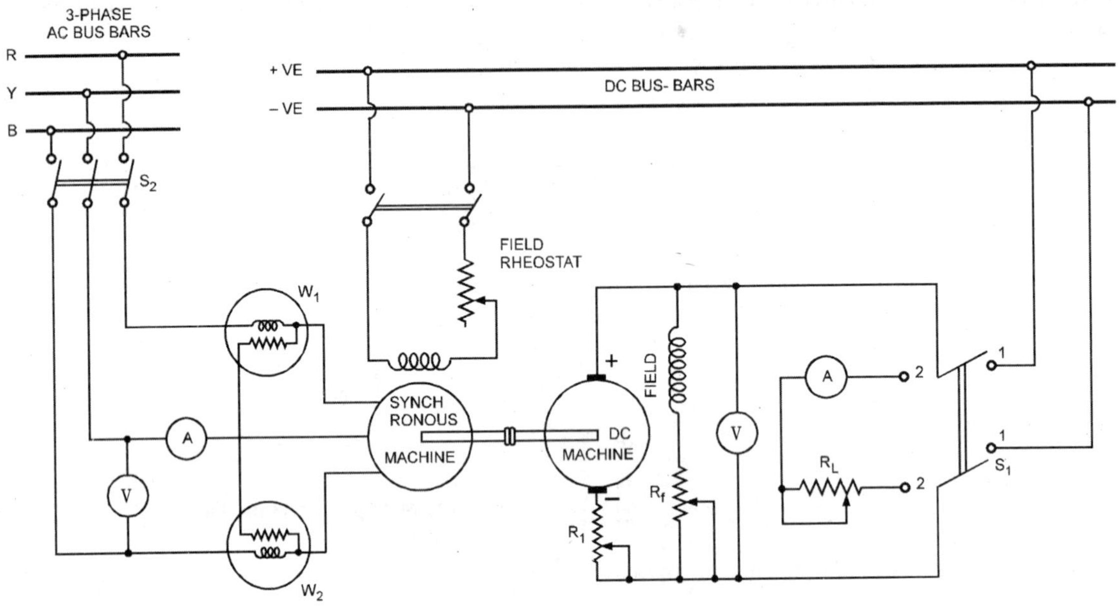

Fig. 6.36. ***Connections For Obtaining V-Curves of a Synchronous Motor***

The synchronous motor is coupled to a shunt wound dc machine which is used as a dc motor for driving the synchronous machine up to the time of synchronising. After synchronising the dc supply to the shunt machine is interrupted by throwing double pole double throw (DPDT) switch S_1 to the terminals 2-2. Now synchronous machine operates as a motor and drives the dc shunt machine, now working as a generator and supplying power to load resistance R_L.

First of all synchronous motor is started by the dc shunt motor and brought to its rated speed. Resistances R_1 and R_f are employed respectively for the purpose of starting and controlling the field current to dc shunt machine. During starting field control resistance R_f is kept zero.

When the synchronous motor attains nearly the synchronous speed, its field is energised and voltage of the machine is built up to the bus-bar voltage (ac). Now the synchronous motor is synchronised with the bus-bars, and dc supply to the dc shunt machine is interrupted by opening the switch S_1. Now the synchronous motor is running on no load and it can drive the dc shunt machine working as a dc shunt generator. DC shunt machine is made to operate as a dc shunt generator by making resistance R_1 in the armature circuit as zero and exciting it and then loading it.

(*i*) Keeping the synchronous motor on no load, take the readings of ammeter, voltmeter and wattmeters on ac side for different values of excitation. The excitation is varied with the help of rheostat provided in the field circuit of the synchronous motor.

Now the dc generator is loaded so that the motor is loaded to about one-fourth of its full load. Keeping this load constant again note the readings of ammeter, voltmeter and wattmeters on ac side for different excitations. Repeat this process for one-half full load, full load and 1.25 times full load on synchronous motor.

The V-curves for no-load, half load, full-load and 1.25 times full load are shown in Fig. 6.37.

By the aid of V-curves, the power factor for any value of armature current I and given input power can be determined from the relation power factor, $\cos\phi = \frac{I_{min}}{I}$ where I_{min} is the minimum current drawn by the motor at given load.

From V–curves for different power inputs it is observed that with low value of field current, the armature current is large and lagging. As the field current is increased the power factor increases and armature current decreases until it reaches its minimum value. When the armature current is minimum, the power factor is unity and the corresponding field current is known as *normal field current or excitation* of the motor for that particular load. The point at which unity power factor occurs is shown by dotted curve. The lines drawn through points of equal power factor at different loads are known as *compounding curves*. The region in which field current is less than its normal value is known as *region of under excitation or region of lag*. If the field current is further increased, the power factor becomes leading and begins to decrease, so armature current begins to increase. This region, in which field current is more than normal value of field current or armature current leads, is known as *region of over excitation or region of lead*.

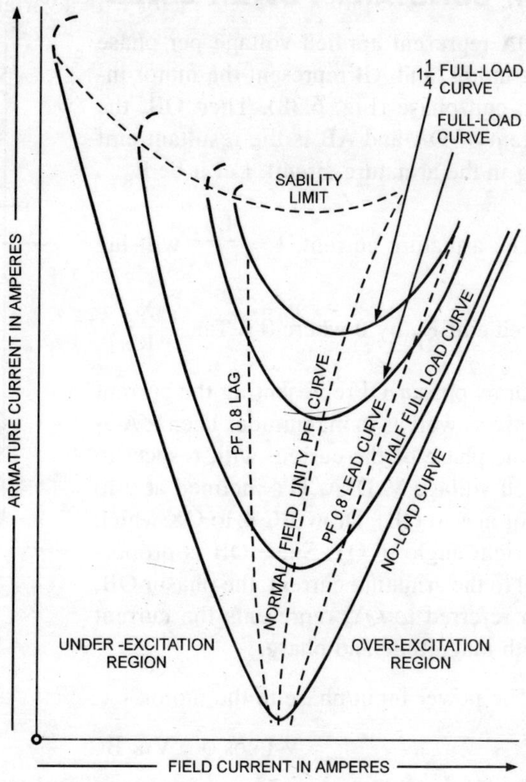

Fig. 6.37. *V-Curves of a Synchronous Motor*

From V–curves it is also to be noted that the no-load armature current rises sharply, with the change in field current, on either side of its unity power factor point. A synchronous motor when operated on no load with an unexcited field, but at synchronous speed, may draw as large as 150% of the rated current or even more. A large change of excitation makes relatively less difference in the armature current at full load. With the reduction in excitation, the limit of stability (for the given load) is reached; at this point the motor pulls out of synchronism. Under light load, say, upto 30% of rated, the excitation may be reduced to zero in most of the motors without causing them to loose synchronism.

In operation, the excitation current of the motor is usually adjusted at the rated value (value of dc excitation mentioned on the name plate), and is kept at this value at all loads. Thus by keeping the excitation at rated value, full pull-out torque is maintained and at the same time maximum amount of reactive kVA are available for power factor correction. If the motor is to be operated for long duration of time at a considerably reduced load, it may be desirable, from the point of view of efficiency, to reduce the excitation to a value below that marked on the name plate to give power factor mentioned on the name plate for the actual reduced load point of operation.

6.14. CONSTANT POWER LINES

Let OA represent applied voltage per phase of the motor, and AB represent the motor induced emf/phase (Fig. 6.38). Then OB, the resultant of OA and AB is the resultant emf acting in the armature circuit. Let it be E_R.

The armature current, $I = \frac{E_R}{Z_s}$ will lag induced emf E_R by θ where $\theta = \text{Tan}^{-1} \frac{X_s}{R_e}$.

Draw phasor OI representing the current in phase as well as in magnitude. Then $\angle AOI = \phi$, the phase of the current with respect to applied voltage V. Draw OY inclined at θ to OA (or at ϕ to OB). Draw BC $\perp$ to OX which is at right angle to OY. Since OB is proportional to the armature current, the phasor OB, when referred to OY, represents the current in both magnitude and phase.

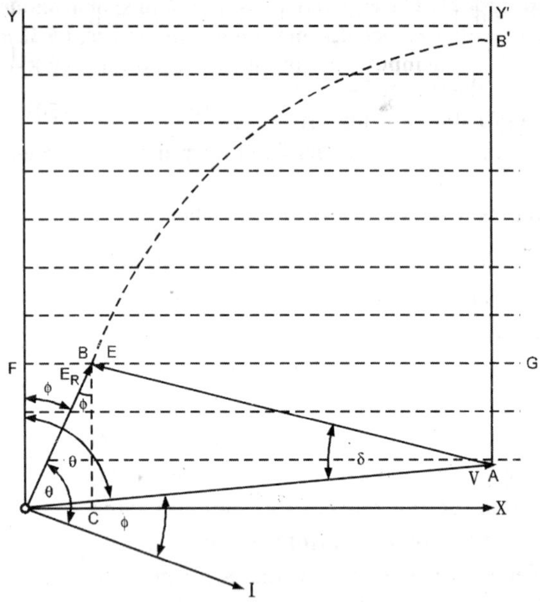

Fig. 6.38

The power input/phase to the motor

$$= V I \cos \phi = V \times BC.$$

As V is constant, power input is proportional to BC. Thus if the motor is operating with constant input power, then locus of the working point B is a straight line passing through point B and parallel to OX (or perpendicular to OY) since this line is at a constant distance from OX. Hence FG represents a constant power input line for a given voltage but varying excitation (or varying induced emf). Similarly, a series of such parallel lines can be drawn, each representing a definite input power of the motor.

The important points regarding constant power lines are:

(*i*) For equal increments of input power, the power lines are equally spaced.

(*ii*) The perpendicular distance from working point B to line OX represents the motor input.

(*iii*) The line OX represents zero power line.

(*iv*) If the excitation is constant, the magnitude of induced emf, E (or AB) is constant and the locus of the working point B is a circle shown dotted in Fig. 6.38 with A as centre and AB as radius. With the increase in load, the load angle δ increases, and so operating point B goes on to line of higher power point till B' is reached at which the input power is obviously a maximum. If the load is further increased, the operating point B will travel on to a line of smaller input power. Such a condition is impossible since an increase in load naturally demands the increased input power, and the whole of the diagram to the right of AY', therefore, represents unstable operation of the motor. For a given voltage and excitation AB' represents the maximum input power and if the motor losses are deducted from it and the resulting output is divided by the speed, the pull-out torque is obtained.

Example 6.21. The effective resistance and synchronous reactance per phase of a 50 Hz, 6,600 V, 1,000 kVA, 3-phase, star-connected synchronous motor are 0.5 Ω and 12 Ω respectively. Plot the V-curve for this machine when the input is maintained constant at 250 kW. No-load saturation curve:

I_f	50	100	150	175	200
V	3,400	5,700	7,200	7,900	8,400

Solution : The problem is solved graphcially as follows:

$$\text{Motor input} = 250 \text{ kW} = 2{,}50{,}000 \text{ watts}$$

$$\text{Supply line voltage, } V_L = 6{,}600 \text{ V}$$

$$\text{Supply current, I} = \frac{2{,}50{,}000}{\sqrt{3}\, V_L \cos\phi} = \frac{2{,}50{,}000}{\sqrt{3} \times 6{,}600 \cos\phi} = \frac{21{,}9\,\text{A}}{\cos\phi}$$

$$\text{or I} \cos\phi = 21.9 \text{ A}$$

$$\text{Impedance, } Z_s = \sqrt{(0.5)^2 + (12)^2} = 12.01\ \Omega$$

$$\text{Internal angle, } \theta = \text{Tan}^{-1}\frac{X_s}{R_e} = \text{Tan}^{-1}\frac{12}{0.5} = 87.6^\circ$$

$$\text{Line internal drop, } E_R = \sqrt{3}\ I Z_s = \sqrt{3} \times \frac{21.9}{\cos\phi} \times 12.01 = \frac{455}{\cos\phi} \text{ volts}$$

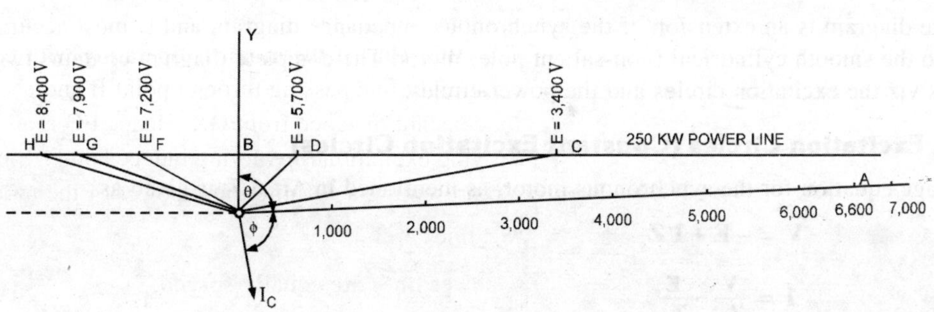

Fig. 6.39

Referring to Fig. 6.39, OA is drawn to represent the applied potential difference of 6,600 volts. Line OY is drawn at an angle θ = 87.6° from OA. Make OB to represent 455 volts and draw a line perpendicular to line OY and passing through point B. This line represents the power of 250 kW, and is the locus of the end of the impedance drop phasor since any line joining point O to any point on this line has a length representing $\frac{455}{\cos\phi}$ volts, where φ is the angle between this line and OY. φ is also the angle between OA and the current phasor drawn at angle of 87.6° from impedance drop phasor *e.g.* when φ = ∠YOC, the impedance drop = OC then OI_c lagging OC by 87.6° represents the current and $\angle AOI_c = \phi$.

Now with the centre A and radii representing 3400, 5700, 7200, 7900 and 8400 volts respectively are drawn to cut power line of 250 kW at C, D, F, G and H. These points are joined to O and OC, OD, OF, OG and OH represent the impedance drops for the condition when the induced (line) emfs are AC, AD, AF, AG and AH *i.e.* 3400, 5700, 7200, 7900 and 8400 volts respectively.

Now armature current,

$$I = \frac{\text{Impedance drop}}{\sqrt{3}\,Z_s}$$

$$= \frac{\text{Impedance drop}}{\sqrt{3}\times 12.01}$$

$$= \frac{\text{Impedance drop}}{20.8}$$

The results from the diagram (Fig. 6.39) are tabulated as below

Field current in amps	50	100	150	175	200
Open-circuit emf (line) in volts	3,400	5,700	7,200	7,900	8,400
Impedance drop (line) in volts	3,400	1,100	700	1,300	1,750
Armature current, I in amperes	164	53	33.7	62.5	84.3

From the above table V-curve is drawn, as shown in Fig. 6.40.

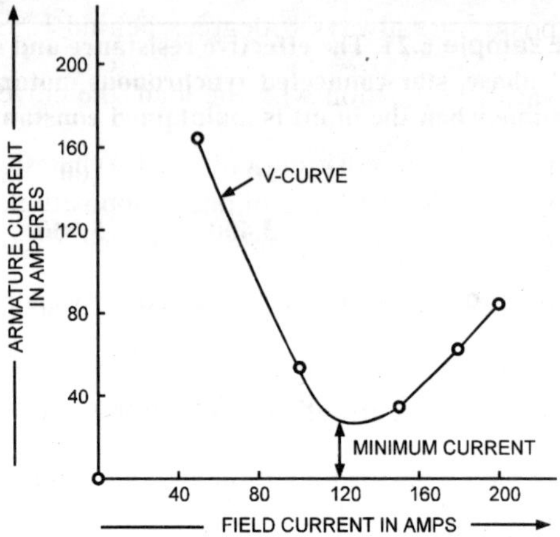

Fig. 6.40

6.15. CIRCLE DIAGRAMS

The circle diagram is an extension of the synchronous impedance diagram and is most accurate when applied to the smooth cylindrical (non-salient pole) motor. The complete diagram contains two groups of circles viz the excitation circles and the power circles.

6.15.1. Excitation Circles (Constant Excitation Circles)

The voltage equation for the synchronous motor, as mentioned in Art 6.7, is given as

$$\mathbf{V} = -\,\mathbf{E} + \mathbf{I}\,\mathbf{Z}_s$$

or

$$\mathbf{I} = \frac{\mathbf{V}}{\mathbf{Z}_s} + \frac{\mathbf{E}}{\mathbf{Z}_s} \qquad \text{...(6.21)}$$

Taking V as the reference, the phasor E, although constant in magnitude for the given condition of constant excitation, varies in phase with respect to V in accordance with the necessary variations in load angle δ. Since constant E traces out a circular locus (Fig. 6.38), the same must be true of $\frac{E}{Z_s}$.

The right-hand components of the equation (6.21) gives values of two current phasors and their phasor sum gives armature current I. The first component of armature current is produced by applied voltage V and its limiting impedance is Z_s. This component of current $\frac{V}{Z_s}$ must lag V by an angle θ where $\theta = \text{Tan}^{-1}\frac{X_s}{R_e}$. The second component of armature current produced by induced emf E must likewise lag E by an angle θ. The applied voltage V is assumed fixed in position and so the current $\frac{V}{Z_s}$ will be constant and fixed in position w.r.t. V. The second component of the armature current is constant in magnitude for a given excitation, but its phase position is always θ degrees behind E. Since E varies in position with the load, the current $\frac{E}{Z_s}$ must vary in position with load. When these two voltages V and E are in phase opposition, the armature current I is the arithmetic difference of the two components, $\frac{V}{Z_s}$ and $\frac{E}{Z_s}$. For any load angle δ the current $\frac{E}{Z_s}$ is retarded in time phase from the no-load position and the value of armature current is determined by the phasor sum of the two components, as expressed by Eq. (6.21).

From the relations mentioned above the extremity of the phasor $\frac{E}{Z_s}$ and likewise the extremity of the phasor I must follow the path of a circle as the load is varied on the motor. The locus is known as an *excitation circle*. It defines the magnitude and power factor of the armature current for the fixed excitation and varying load. The centre of this circle regardless of the excitation will be on a line θ degrees behind the voltage V phasor and at a distance from $\frac{V}{Z_s}$ amperes.

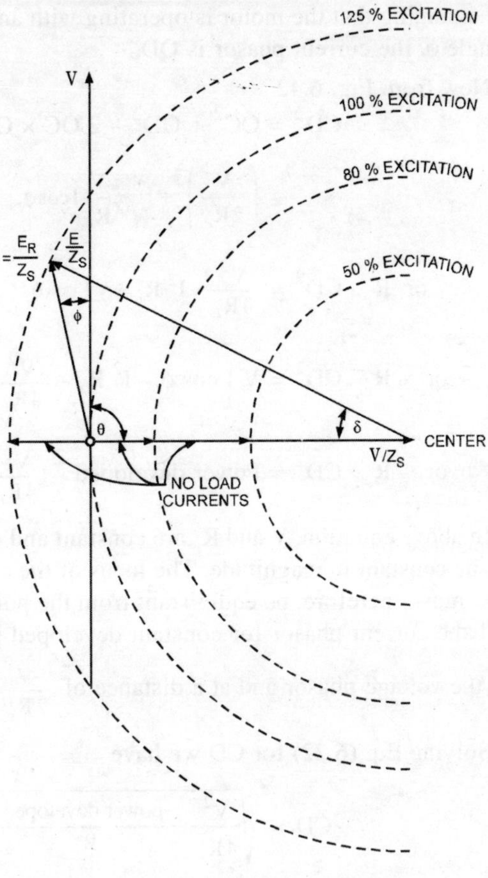

Fig. 6.41. *Constant Excitation Circle Diagram*

For 100% excitation the no-load current is zero. For other excitations the no-load current is determined by the upper intersection of the particular excitation circle. Group of excitation circles for different degrees of excitation is given in Fig. 6.41.

6.15.2. Constant Power Circle

Consider a reproduced portion diagram of Fig. 6.41, as illustrated in Fig. 6.42. The distance OA represents the current component $\frac{V}{Z_s}$. Draw a perpendicular to phasor OA from point A intersecting voltage line at point B, and

$$OB = \frac{OA}{\cos\theta} = \frac{V/Z_s}{R_e/Z_s} = \frac{V}{R_e} \text{ amperes}$$

Making $OC = \frac{1}{2}$ OB we have

$$OC = \frac{V}{2R_e}$$

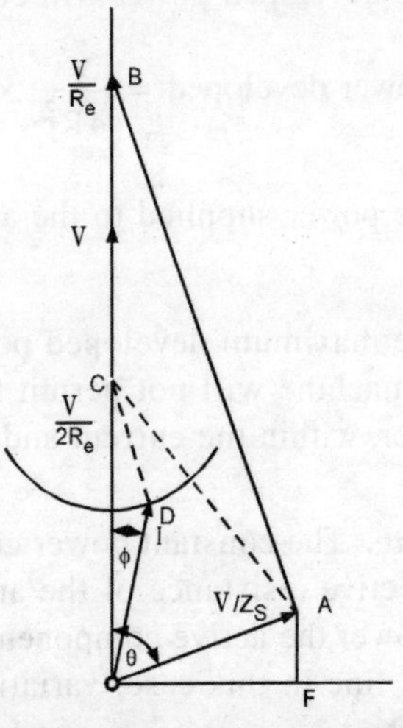

Fig. 6.42. *Constant Power Circle Diagram*

Now assuming that the motor is operating with an input current I lagging behind the applied voltage by an angle ϕ, the current phasor is OD.

Now from Fig. 6.42

$$CD^2 = OC^2 + OD^2 - 2\,OC \times OD \cos\phi$$

$$= \left(\frac{V}{2R_e}\right)^2 + I^2 - \frac{V}{R_e} I \cos\phi$$

$$\text{or } R_e \cdot CD^2 = \frac{V^2}{4R_e} + I^2 R_e - VI\cos\phi$$

$$\text{or } -R_e \cdot CD^2 = V\,I\cos\phi - I^2 R_e - \frac{V^2}{4R_e}$$

$$\text{or } -R_e \cdot CD^2 = \text{Power developed} - \frac{V^2}{4R_e} \quad ...(6.22)$$

In above equation, V and R_e are constant and so if the power developed is constant, the distance CD must be constant in magnitude. The locus of the extremity of the current phasor for constant developed power must, therefore, be equidistant from the point C for various excitations. The locus of the extremity of the current phasor for constant developed power is a circle with its centre on a line coincident with the voltage phasor and at a distance of $\frac{V}{2R_e}$ amperes from the origin O.

Solving Eq. (6.22) for CD we have

$$CD = \sqrt{\frac{V^2}{4R_e^2} - \frac{\text{power developed}}{R_e}} \text{ amperes} \quad ...(6.23)$$

From above expression it is evident that for zero developed power CD = $\frac{V}{2R_e}$ and since CA = OC, the no-load circle will pass through O and A. Circles of smaller radii give greater power. The maximum possible developed power will coincide points D and C and for this condition

$$\text{Power developed} = \frac{V^2}{4R_e^2} \times R_e = \frac{V^2}{4R_e} \quad ...(6.24)$$

The power supplied to the armature under these conditions will be equal to $\frac{V}{2R_e} \times V$ *i.e.* $\frac{V^2}{2R_e}$ watts.

The maximum developed power given by Eq. (6.24) is an imaginary one because the saturation of the machine will not permit the degree of excitation required to give this particular adjustment. However, within the current and power factor range of the motor the circle diagram will yield useful results.

Note. The constant power circle is not of great practical interest because the radius is so large. If the effective resistance of the armature is negligible, the radius is infinite since for constant developed power the active component of current I cos ϕ will be constant. The current locus is obviously a straight line in this case, variation of field current causing the kVAR to vary, with kW maintained constant.

Example 6.22. A 1,000 hp, 6000 V, 3-phase, star-connected synchronous motor has a synchronous impedance of (1.5 + *j* 16) Ω per phase. It is excited to develop an open-circuit emf of 5 kV. Draw the locus diagram of the current for loads up to 1,250 hp with constant excitation. Determine the maximum value of the power factor.

Solution : Supply voltage per phase, $V = \frac{6,000}{\sqrt{3}} = 3,464$ V

Induced emf per phase, $E = \frac{5 \times 1,000}{\sqrt{3}} = 2,887$ V

Impedance, $Z_s = \sqrt{(1.5)^2 + 16^2}$
$= 16.07\ \Omega$

Internal angle, $\theta = \text{Tan}^{-1} \frac{X_s}{R_e}$

$= \text{Tan}^{-1} \frac{16}{1.5}$

$= 84.6°$

Current due to supply voltage, $\frac{V}{Z_s} = \frac{3,464}{16.07}$

$= 215.6$ A

Current due to induced emf, $\frac{E}{Z_s} = \frac{2,887}{16.07}$

$= 179.6$ A

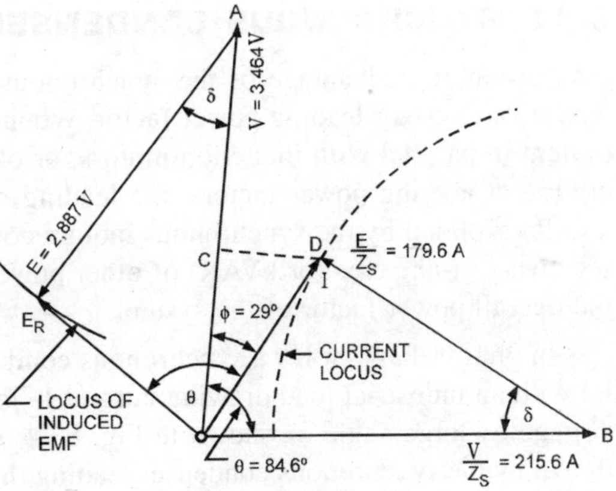

Fig. 6.43

Since power input $= \sqrt{3}\ V_L\ I \cos \phi$

So $I \cos \phi = \frac{\text{Power input}}{\sqrt{3}\ V_L} = \frac{1,250 \times 735.5}{\sqrt{3} \times 6,000} = 88.5$ A

Locii of induced emf of 5 kV and current for loads up to 1,250 hp (neglecting losses) are shown in Fig. 6.43.

Figure 6.43 is drawn with OA = 3,464 V as reference phasor. Locus of E is an arc with centre A and radius 2,887 V; OB = 215.6 A making an angle θ = 84.6°; CD a line perpendicular to phasor V with OC = I cos ϕ = 88.5 A. The arc of the circle with centre B and radius of current 179.6 A between origin O and point D is the locus of the stator current. Maximum value of power factor = Cos ∠COD = Cos 29° = 0.875 (lag) **Ans.**

6.16. MAIN CHARACTERISTICS OF SYNCHRONOUS MOTORS

1. Synchronous motor operates only at one speed, known as *synchronous speed*, given by the expression $N_s = \frac{120 f}{P}$. The speed is independent of load and can be varied either by varying the supply frequency or by changing the number of poles.
2. It is, inherently, not a self starting motor and, therefore, before connecting it to an ac supply it is run up to the synchronous speed by some external means.
3. It can be operated under a wide range of power factor both leading and lagging by controlling its excitation current.

4. On no load synchronous motor draws very little current from the supply mains to meet the internal losses. With fixed excitation the input current increases with the increase in load. After the input current reaches maximum no further increase in load is possible. If the motor is further loaded, the motor will stop.
5. If field excitation is small, the stator inductance will be more so the motor draws lagging current and, therefore, power factor is low. If the excitation is increased the stator flux is neutralized by the rotor flux and, therefore, power factor becomes unity. In this position, the current and voltage are in phase and motor current will be minimum.

If excitation is further increased rotor flux increases. It neutralizes the stator flux and causes leading current. This property makes the motor extremely useful as a power factor corrector. By taking leading current it compensates the lagging current drawn by other equipment from the supply mains and thus brings the total current drawn from the supply mains more nearly in phase with the voltage.

6.17. SYNCHRONOUS CONDENSERS

An outstanding advantage of the synchronous motors is that they can operate at unity or leading power factor. When operated in an electrical system in parallel with induction motors, or other power apparatus operating at lagging power factors, the leading reactive volt-amperes (or kVAR) supplied by the synchronous motors compensate for the lagging reactive volt-amperes (or kVAR) of other power apparatus and improve the overall power factor of the system.

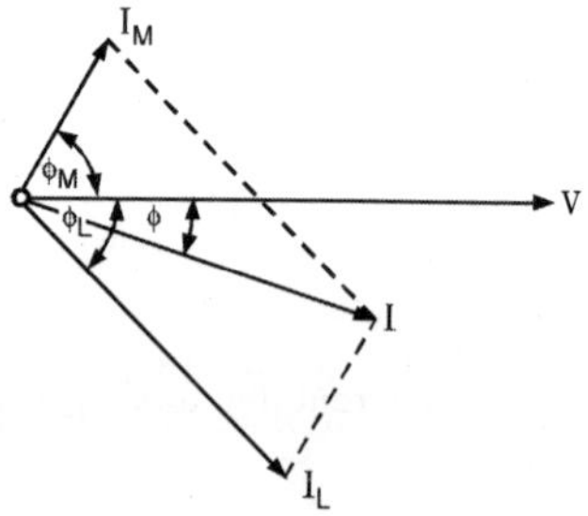

Fig. 6.44

In phasor diagram for a synchronous condenser connected in parallel with an industrial load drawing current I_L lagging behind the supply voltage by large angle ϕ_L shown in Fig. 6.44, I_M represents the current drawn by the synchronous condenser leading the supply voltage V by an angle ϕ_M. The resultant current I is the phasor sum of I_L and I_M and we see that its angle of lag ϕ is much smaller than ϕ_L. Thus overall power factor is improved from cos ϕ_L to cos ϕ by the use of synchronous condenser. In this way the power factor can be made unity even.

In large industrial plants, which have a low lagging power-factor load, it is often found economical to install a synchronous motor, even though the motor is not required to drive a load. The motor is operated overexcited at no load so that the current drawn by it leads the voltage by nearly 90°. A synchronous motor used in this way is said to "float"on the line because it has no mechanical output. Since the motor operating at no load acts in the same manner as a static capacitor and when operated in this manner it is called the *synchronous condenser* or a *synchronous capacitor.*

When a synchronous motor is operated on no load, it still takes power for its own rotational losses (*i.e.* friction, windage and iron losses). Although the inphase component of armature current is usually comparatively small under no-load operation, the total current may be rather large, as it depends upon dc excitation. When a synchronous condenser is carrying full-rated leading kVA, the current leads the supply voltage by 88 to 89°, depending on the size of the machine. The angles correspond to power factors of 3.5 and 1.7 per cent respectively.

The synchronous condenser is especially designed so that practically all its rated kVA are available for power factor correction. The design of a synchronous capacitor is characterized by its large synchronous reactance and field windings, because its function is to develop zero-power factor currents with the least expenditure of power in losses. In comparison to a synchronous motor with equal armature voltage and current ratings, a synchronous condenser needs more copper in the field winding to carry the increased field current. The synchronous condenser does not require so large shaft and bearings as the synchronous motor because no shaft torque is required.

The advantage of synchronous condensers over static capacitors is that power factor can very easily be controlled by variation of the field excitation. The disadvantages of synchronous condensors are (*i*) except in size above 500 kVA, the cost is higher than that of static capacitors of the same rating, (*ii*) the possibility of their falling out of synchronism with resulting interruption of supply, and (*iii*) increase of short-circuit currents when the fault occurs near the synchronous condensers.

So the synchronous condensers are employed only when power factor correction is required on a large scale.

The application of synchronous condensers in connection with power factor improvement is illustrated with the help of solved examples as below.

Example 6.23. A 500 V, 50 Hz, 3-phase circuit takes 20 A at a lagging power factor of 0.8. A synchronous motor is used to raise the power factor to unity. Calculate the kVA input to the motor, and its power factor when driving a mechanical load of 7.5 kW. The motor has an efficiency of 85%.

Solution : kVAR drawn by the 3-phase circuit, $Q = \sqrt{3}\, V_L I_L \sin\phi \times 10^{-3}$

$$= \sqrt{3} \times 500 \times 20 \times 0.6 \times 10^{-3} = 10.3923 \text{ kVAR}$$

$$\text{Power supplied by the motor, } P = \frac{\text{Output in kW}}{\eta} = \frac{7.5}{0.85} = 8.8235 \text{ kW}$$

Power factor will be raised to unity when kVAR (leading) drawn by a 3-phase synchronous motor will become equal to the kVAR drawn by 3-ϕ ac circuit.

i.e. kVAR drawn by synchronous motor, Q = 10.3923 (leading)

KW drawn synchronous motor, P = 8.8235

$$\text{KVA input to the motor, } S = \sqrt{P^2 + Q^2} = \sqrt{(8.8235)^2 + (10.3923)^2} = 13.63 \text{ kVA } \textbf{Ans.}$$

$$\text{Power factor, } \cos\phi = \frac{P}{S} = \frac{8.8235}{13.63} = 0.6472 \text{ (leading) } \textbf{Ans.}$$

Example 6.24. An alternator supplying 500 kW at 0.7 power factor lagging has its power factor raised to unity by means of an overexcited synchronous motor. At a constant armature current how much input power is required for the synchronous motor ? Find the power factor of the syncrhonous motor.

Solution : Load, P = 500 kW

$$\text{Load kVA, } S = \frac{P}{\text{Cos}\,\phi} = \frac{500}{0.7} = 714.3$$

$$\text{Power factor angle, } \phi = \text{Cos}^{-1}\, 0.7 = 45.57°$$

$$\text{Load kVAR} = S \times \sin\phi = 714.3 \sin 45.57° = 714.3 \times 0.714 = 510$$

Thus a synchronous motor when connected to improve power factor under overexcited condition will operate as a synchronous condenser. The power required to be supplied by synchronous condenser, therefore, is 510 kVAR and the power factor of the synchronous motor will be zero (leading).

Example 6.25. A substation operating at its full load of 1,000 kVA supplies a load of power factor 0.71 lagging. Calculate the permissible additional load at this power factor and the rating of synchronous condenser to raise the substation power factor to 0.87 lagging.

Solution : Rating of power plant, S = 1,000 kVA

Load supplied at 0.71 pf = 0.71 S = 0.71 × 1,000 = 710 kW

Load supplied at 0.87 pf = 0.87 S = 0.87 × 1,000 = 870 kW

Increase in load supplied = 870 – 710 = 160 kW

Additional load supplied at 0.71 pf (lagging) = $\frac{160}{0.71}$ = 225.35 kVA **Ans.**

Reactive power drawn by load = New load in kW × tan ϕ_1 = 870 tan cos^{-1} 0.71 = 863 kVAR

Reactive power supplied by the plant = S sin ϕ_2

= S sin cos^{-1} 0.87 = 1,000 × sin 29.54° = 493 kVAR

Rating of ideal synchronous condenser = Reactive power drawn by load – reactive power supplied by the plant

= 863 – 493 = 370 kVAR **Ans.**

Example 6.26. A 3-phase synchronous motor absorbing 60 kW is connected in parallel with a factory load of 240 kW having a lagging power factor of 0.8. If the combined load has a pf of 0.9, what is the value of the leading kVAR supplied by the motor and at what power factor is it working ?

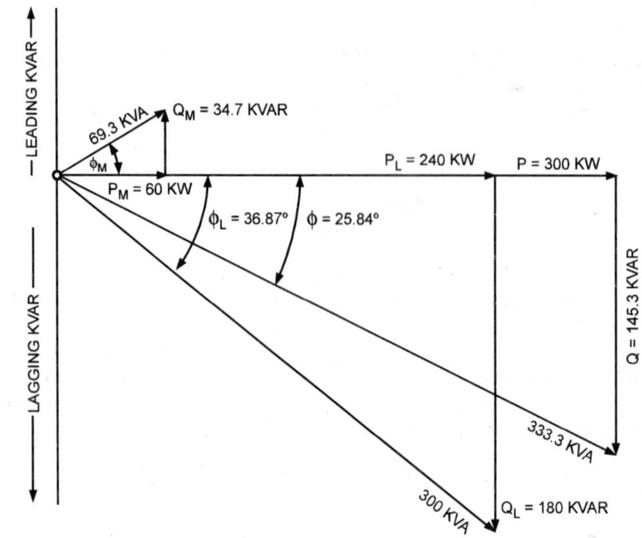

Fig. 6.45

Solution :

Factory load, P_L = 240 kW

Power factor, cos ϕ_L = 0.8 lagging

Load kVAR, Q_L = 240 tan (cos^{-1} 0.8) = 240 × 0.75 = 180 kVAR

Synchronous motor load, P_M = 60 kW

Total load, P = $P_L + P_M$ = 240 + 60 = 300 kW

Combined power factor, cos ϕ = 0.9 (lagging)

Phase angle, ϕ = cos^{-1} 0.9 = 25.84°

Combined kVAR, Q = P tan ϕ = P tan 25.84° = 300 × 0.4843 = 145.3 (lag)

Leading kVAR supplied by the motor, $Q_M = Q_L - Q = 180 - 145.3 = 34.7$ kVAR **Ans.**

$$\text{KVA supplied by the motor, } S_M = \sqrt{P_M^2 + Q_M^2} = \sqrt{60^2 + (34.7)^2} = 69.3$$

$$\text{Power factor of the motor, } \cos\phi = \frac{P_M}{S_M} = \frac{60}{69.3} = 0.866 \text{ (leading) } \textbf{Ans.}$$

Example 6.27. What is a synchronous condenser ? Give its practical utility.

A small industrial load of 500 kW at 0.6 pf lagging is supplied from a 3300 V, 3-phase, 50 Hz system. It is desired to raise the pf of the entire system to 0.8 lagging by means of a synchronous motor, which is also drawing a pump so that the synchronous motor takes 100 kW from the line. Determine the rating of the synchronous motor.

Solution :

Industrial load, $P_1 = 500$ kW

Power factor, $\cos\phi_1 = 0.6$ (lagging)

$\phi_1 = \text{Cos}^{-1}\, 0.6 = 53.13°$ (lagging)

Load kVAR, $Q_1 = P_1 \tan\phi_1 = 500 \tan 53.13° = 666.667$ (lagging)

Input to synchronous motor, $P_2 = 100$ kW

Total load, $P = P_1 + P_2 = 500 + 100 = 600$ kW

Power factor of combined load, $\cos\phi = 0.8$ (lagging)

Power factor of combined load, $\phi = \text{Cos}^{-1}\, 0.8 = 36.87°$ (lagging)

Combined kVAR, $Q = P \tan\phi = 600 \tan 36.87° = 450$ kVAR

kVAR supplied by the synchronous motor, $Q_2 = Q_1 - Q = 666.667 - 450 = 216.667$ (leading)

$$\text{Rating of motor, } S_2 = \sqrt{P_2^2 + Q_2^2} = \sqrt{100^2 + (216.667)^2} = 238.63 \text{ kVA } \textbf{Ans.}$$

6.18. HUNTING AND DAMPING

We have seen that when a synchronous motor is loaded the rotor poles fall back by a certain angle, called the load angle and represented by δ behind the poles of the forward rotating magnetic field. If the load is suddenly thrown off, the rotor poles are pulled into almost exact position to the poles of the forward field, but due to inertia of rotor the rotor poles travel too far. They are then pulled back again, and so on. Thus an oscillation is set up about the equilibrium position corresponding to new load. The oscillation of the rotor about its equilibrium position is called the *hunting*. The hunting is caused by change in load, change in excitation, or change in some other condition of the system. In some cases, when the mass of the rotating parts has an oscillating period that is the same as, or some multiple of, the hunting period, mechanical resonance is set up and the amplitude of the swing of the rotor poles relative to the poles of rotating field becomes too great that either the motor circuit gets opened by the circuit protective devices, or the motor falls out of synchronism and stall. Hunting is, of course, an objectionable characteristic of all synchronous motors, as it produces severe mechanical stresses as well as large variations in current and power drawn by the motor.

Hunting is suppressed by providing a damper winding, also known as the *squirrel cage* winding or *damper grids*. The damper winding consists of short-circuited copper or bronze bars embedded in the faces of the field poles of the synchronous motors (Fig. 6.2). When the machine is operating at exactly synchronous speed, there is no relative motion between the rotor and air-gap flux and so no voltage is induced in this damper or squirrel cage winding. Thus at synchronous speed of the motor no current flows through this additional winding and operation of the machine is not affected. But when the speed of the machine deviates from the synchronous speed or hunting takes place, there is a relative motion between the rotor (or damper winding) and the air-gap flux and, therefore, eddy currents are set up in this winding in such a way as to suppress the oscillations. The effectiveness of the damping winding in suppressing the rotor oscillations, however, depends upon the resistance of this winding; the lower the resistance, the stronger is the damping action. Fortunately this winding also serves to make the synchronous motor self-starting and a high squirrel cage resistance is necessary for developing good starting torque, it is customary to use a compromise value for the cage. The term *amortisseur* is generally used for the squirrel cage in connection with its damping action.

The dampers cannot prevent hunting completely because their operation depends upon the presence of oscillating motion.

6.19. STARTING METHODS OF SYNCHRONOUS MOTORS

As already stated the synchronous motor has got no self starting torque and the motor must first be brought up to synchronous speed by some external means before it is capable of operating with load angle δ in the manner previously described. Various methods of starting of synchronous motors are described as follows:

1. From DC Source. If the dc supply and dc compound motor is available, the synchronous motor is coupled and started by means of a dc compound motor. The speed of dc motor is adjusted by the speed regulator. The synchronous motor is then excited and synchronised with ac supply mains. At the moment of synchronising the synchronous motor is switched on with the ac mains, and either the dc motor is disconnected from the dc supply mains or the field of the dc machine is strengthened until it begins to function as a generator. Now the synchronous machine is operating, as a motor, from ac supply mains and dc machine acts as load on it.

The synchronous motor can also be started by the exciter mounted on an overhung synchronous motor bracket and shaft extension. Here again, an available dc source operates the exciter as a motor during the starting period; then, after the synchronous machine is brought up to speed and synchronised, the exciter assumes its normal function.

2. By Means of AC Motor. A small direct-coupled induction motor, called the *pony motor*, may be used for starting the synchronous motor unless the motor is required to start against full-load torque. The induction motor frequently has two poles less than the synchronous motor and so is capable of raising the speed of the latter to synchronous speed. Before switching the ac supply to the synchronous motor, it must be synchronised with the bus-bars. After normal operation is established, the pony motor is sometimes uncoupled from the synchronous motor. This method is not very satisfactory and not suited to industrial needs. Modern machines are usually of the self starting type and are arranged to start as induction motors.

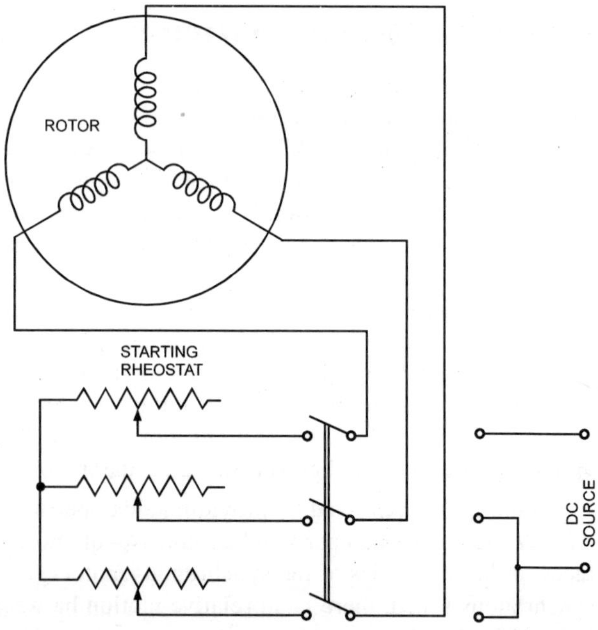

Fig. 6.46

3. By Means of Damper Grids in The Pole Faces. The synchronous motor is made self starting by providing a special winding on the rotor poles, known as *damper winding* or *squirrel cage winding*. The damper winding consists of short-circuited copper bars embedded in the face of the field poles. AC supply given to the stator produces a rotating magnetic field which causes the rotor to rotate, therefore, in the beginning synchronous motor provided with damper winding starts as a *squirrel cage induction motor*. The exciter moves along the rotor. When the motor attains about 95% of synchronous speed, the rotor winding is connected to exciter terminals and the rotor is magnetically locked by the rotating field of the stator and the motor runs as a synchronous motor.

Advantages. When the motor is overloaded it does not stop, continues running as a squirrel cage induction motor. When the speed falls slightly below synchronism due to overloading an emf is induced in the rotor short-circuited winding called *damper winding*. Due to this emf a torque is produced which keeps the motor running even on no load.

Disadvantages. (*i*) Since damper winding resistance is low so at start it takes large current from supply mains.

To avoid large starting current drawn from the supply mains the synchronous motor is started usually as slip-ring induction motor. The rotor winding is connected in series with triplex rheostat at the time of starting by throw-over switch. Rotor circuit is completed and an emf is induced in the rotor circuit as in the rotor of a slip-ring induction motor.

As the motor attains synchronous speed the throw-over switch is put on the side of the exciter terminals and the dc supply passes through the rotor winding. While running, the rotor acts like rotor of synchronous motor. Hence the motor runs as a synchronous motor. If the starting and excitation is done automatically the motor is known as *auto-synchronous motor*.

(*ii*) Field windings meant for 110 or 250 volts, the exciting voltage generally being this much, should be highly insulated for withstanding high induced emf due to induction at starting.

Precautions. Under normal running conditions, there is no relative motion between the rotating magnetic field and the field winding of a synchronous motor and, therefore, no voltage is induced in the field winding by the air-gap flux. However, if the voltage is applied to the stator winding and the motor is at standstill with the field winding unexcited, the revolving field produced by the polyphase armature (or stator) currents will cut across the field windings at synchronous speed and thereby induce a high voltage in the field winding. This induced voltage would be dangerous and often would result in breakdown of the insulation of the field winding. *Therefore, in starting a synchronous motor the field winding must be shorted through a suitable resistance so that the induced voltage is distributed throughout the whole winding and no part can be subjected to the high voltage induced in the entire winding.*

Another method of eliminating the hazard of breakdown of insulation of the field winding due to induced high voltage in it is by means of dividing the field winding into several sections during the starting period. While emf is still induced in the field coils, the number of coils in series is reduced, so that the magnitude of emf induced in any section is kept within safe limits. Sectionalizing of field winding may be accomplished by a centrifugally operated switch that keeps the field circuit sectionalized until the rotor revolves at near synchronous speed.

6.19.1. Starting of Synchronous Motor Under Load

Synchronous motor with the usual low resistance type of damper has a comparatively low starting torque. The starting torque of a synchronous motor may be increased by increasing the resistance of the rotor winding. Since in a synchronous motor the squirrel cage windings are not effective during normal operation, (as the high resistance starting winding gives at running speed high slip and low efficiency in case of induction motors) therefore, the resistance of squirrel cage can be made sufficiently high to give high starting torque and the motor can be started under considerable load. The high resistance of the squirrel cage may produce such a large slip that locking of rotor with the synchronously rotating stator poles may become impossible. This difficulty is overcome by short-circuiting the field winding near the synchronous speed of the motor.

For the purpose, where the motor is required to be started under considerable loads, synchronous motors with phase wound dampers are employed. These motors consist of usual salient poles but the damper bars, instead of being connected to end rings are phase connected and brought out to the resistors through slip rings. Thus such motors have five slip rings, two for conducting current to dc field winding and other three to be connected to the terminals of the 3 phase rotor winding. The resistors are put in the circuit at start and are taken out as the motor attains the nearly synchronous speed, just as done with slip-ring induction motors. When the three slip rings are short circuited, the winding acts as a damper winding.

This type of motor is used in heavy duty machinery such as conveyors, ball mills and crushers etc.

6.20. MERITS AND DEMERITS OF SYNCHRONOUS MOTORS

Merits. (*i*) The ease with which the power factor can be controlled. An overexcited synchronous motor having a leading power factor can be operated in parallel with induction motors and other power apparatus operating at lagging power factors, thereby improving the power factor of the supply system.

(*ii*) The speed is constant and independent of load. This characteristic is mainly of use when the motor is required to drive another alternator to generate a supply at a different frequency as in frequency changers.

(*iii*) Electromagnetic power varies linearly with the voltage.

(*iv*) These motors can be constructed with wider air gaps than induction motors, which make them better mechanically.

(*v*) These motors usually operate at higher efficiencies, especially in the low speed unity pf ranges.

Demerits (*i*) The cost per kW output is generaly higher than that of an induction motor.

(*ii*) It requires dc excitation which must be supplied from external source.

(*iii*) The synchronous motor is inherently not self-starting motor and needs some arrangement for its starting and synchronizing.

(*iv*) It cannot be used for variable speed jobs as there is no possibility of speed adjustment.

(*v*) It cannot be started under load. Its starting torque is zero.

(*vi*) It has a tendency to hunt.

(*vii*) It may fall out of synchronism and stop when overloaded.

(*viii*) Collector rings and brushes are required.

(*ix*) For some purposes synchronous motors are not desirable as for driving shafts in small workshops having no other power available for starting and in cases where frequent starting or strong starting torque is required.

6.21. APPLICATIONS OF SYNCHRONOUS MOTORS

Synchronous motors are rarely used below 40 kW output in the medium speed range because of their much higher initial cost in comparison to that of induction motors. In addition they need a dc excitation source, and the starting and control devices are usually more expensive—especially where automatic operation is required. However, there is a kW output and speed range where the disadvantage of higher initial cost vanishes, even to the point of putting the synchronous motors to the advantage. Where low speeds and high kW outputs are involved the induction motors are no longer cheaper because they need large amounts of iron in order to restrict the air-gap flux density to 0.7 T. In the synchronous machines, on the other hand, a value twice this figure is permissible because of the separate excitation.

The various classes of service for which synchronous motors are employed may be classified as (*i*) power factor correction (*ii*) voltage regulation and (*iii*) constant speed, constant-load drives.

1. They are used in power houses and substations in parallel to the bus-bars to improve the power factor. For this purpose they are run without mechanical load on them and overexcited.
2. In factories having a large number of induction motors, or other power apparatus operating at lagging power, they are employed to improve the power factor.
3. Such motors are also used to regulate the voltage at the end of transmission lines.
4. Because of the higher efficiency possible with synchronous motors, they can be employed advantageously for the loads where constant speed is required. Typical applications of high-speed synchronous motors (above 500 rpm) are such drives as fans, blowers, dc generators, line shafts, centrifugal pumps and compressors, reciprocating pumps and compressors, constant speed frequency changers, rubber and paper mills etc.

The fields of applications of low-speed synchronous motors (below 500 rpm) are drives such as reciprocating compressors when started unloaded, dc generators, centrifugal and screw type pumps, vacuum pumps, electroplating generators, line shafts, rubber and band mills, ball and tube mills, chippers, metal rolling mills etc. Flywheel is used for pulsating loads.

High power electronic convertors generating very low frequencies made the operation of synchronous motors at ultra low speeds. Synchronous motors in very large sizes (as high as 10 MW size) operating at ultra low speeds are employed to drive crushers, rotary kilns and variable speed ball mills.

HIGHLIGHTS

1. *Synchronous motor* is identical in construction with the alternator and converts ac electrical power into mechanical power.
2. When 3-phase supply is given to the stator of a 3-phase synchronous motor, a revolving field is produced. This field tries to drag the rotor with it but could not do so owing to rotor inertia. Hence starting torque is not developed. Thus a synchronous motor is inherently not self-starting and, therefore, it is necessary to provide some external means to start it and bring its speed near synchronous speed.
3. Synchronous motor operates only at one speed known as synchronous speed given as $N_s = 120f/P$. It can be operated under a wide range of power factor both leading and lagging by controlling its excitation current.
4. A synchronous motor is able to supply increased mechanical load, not by reduction in speed, but by shift in relative positions of the rotor and rotating magnetic field. For increased load with constant excitation, phase angle ϕ increases in lagging direction.
5. In synchronous motors, the armature reaction is one which opposes the action of the main field flux. When the dc field current is reduced so as to draw the current from the ac supply mains at lagging power factor, the armature current drawn from the ac supply mains acts to strengthen the field. On the other hand, if the field current is increased so as to draw a current at leading power factor, the armature current now sets up a flux tending to reduce the flux created by the main field. In any case, the motor tends to operate at a nearly constant flux or generated voltage, as does the transformer.
6. Power developed in a synchronous motor is given by the expression

$$P_{mech} = \frac{EV}{X_s} \sin \delta$$

where E is excitation or induced emf, V is terminal voltage, X_s is synchronous reactance (all phase values) and δ is the load angle in electrical degrees.

7. Maximum value of mechanical power developed is given by expression

$$(P_{mech})_{max} = \frac{EV}{X_s} \qquad \because \text{ Load angle corresponding to maximum power, } \delta = \theta = 90°$$

8. Mechanical power developed is given by expression.

$$\text{Mechanical power developed} = \sqrt{3}\, V_L I_L \cos\phi - 3\, I^2 R_e$$

Mechanical power developed per phase is also given by expression

$$P_{mech} = -\,E\, I \cos(\theta + \gamma)$$

where E is induced emf per phase, I is armature current per phase and $(\theta + \gamma)$ is the angle between E and I.

9. Total power input in a 3-phase salient synchronous motor is given by

$$P_{in} = \frac{E_L V_L}{X_d} \sin\delta + \frac{V_L^2}{2}\left(\frac{1}{X_q} - \frac{1}{X_d}\right) \sin 2\delta$$

Power developed will be equal to power input less copper losses.

10. The maximum value of load angle δ, assuming negligible, stator resistance is given by

$$\text{Cos}\,\delta = \frac{-E}{4V\,\dfrac{X_d - X_q}{X_q}} + \sqrt{\frac{1}{2} + \left[\frac{EX_q}{4V(X_d - X_q)}\right]^2}$$

11. The V-curves of a synchronous motor give relation between armature current and field current for different power inputs.

    Similarly variation of power factor with a variation in field current (dc excitation) for a constant load gives inverted V-curve.

12. Synchronous condenser is a specially designed synchronous motor for operation without mechanical load with its field over-excited so as to draw leading armature current. It is used for power factor correction.

13. Synchronous condensers are sometimes operated at power factors ranging from lagging through unity to leading for voltage control. When operated in this manner a synchronous condenser is called a synchronous *phase modifier* or *reactor*.

14. The oscillation of synchronous motor rotor about its equilibrium position is known as *hunting*. It is minimised by providing damper winding on rotor.

15. Synchronous motors are rarely used in small ratings (below 40 kW output) in the medium speed range because of being expensive in comparison to induction motors. Where low speeds and high kW outputs are involved the synchronous motors prove economical.

    The various services for which synchronous motors are used may be classified as (*i*) power factor correction (*ii*) voltage regulation and (*iii*) constant speed constant load drives.

EXERCISES

1. Explain the construction and principle of operation of a synchronous motor.
2. Explain the working principle of a synchronous motor. **[M.D. Univ. December 2004; May 2005]**
3. Explain the principle of operation of a 3-phase synchronous motor. Why is starting torque not produced in this motor?
4. Explain why the speed of a 3-phase synchronous motor remains constant at various loads when fed from a constant frequency.

5. Explain with neat sketches the principle of operation of a 3-phase synchronous motor. Also explain why it will not run at other than synchronous speed.

6. Explain the mechanism of torque production in a synchronous motor.

7. Explain how in a synchronous motor by changing the field excitation it is possible to operate the motor from lagging to leading power factor, while keeping the load constant. **[M.D. Univ. May 2006; December 2007]**

8. Describe briefly the effect of varying excitation upon armature current and power factor of a synchronous motor when input power to the motor is maintained constant.

[M.D. Univ. Electrical Machines-II, December 2010]

9. Explain the effect of varying excitation on the armature current and power factor in a synchronous motor. Draw V-curves and state the significance. **[M.D. Univ. Electrical Machines-II, May 2009]**

10. How V-curves of a synchronous motor can be obtained? Discuss also their importance.

[M.D. Univ. December 2005, December 2006]

11. Write a technical note on V-curves of synchronous motors. **[M.D. Univ. May 2007]**

12. Explain V-curves for synchronous motor. **[M.D. Univ. December 2010]**

13. The field excitation of a 3-phase synchronous motor operating on a fixed load is increased from a very low value. With the help of a V-curve or otherwise explain how the magnitude and power factor of the armature current will vary.

14. Draw a curve between field current and armature current of a synchronous motor, when power input to the motor and supply voltage are constant. What is the shape of the curve ? Show the region of lagging and leading pf. Why is power input at unity pf minimum ?

15. Draw a phasor diagram of a synchronous motor. Explain the effect of (*i*) change in excitation if load is constant (*ii*) change in load if excitation is constant.

16. Draw the families of curves of synchronous motor at no load and full load showing the relation between (*i*) armature current and field current (*ii*) power factor and field current. Using these curves explain how the motor may be overloaded with no load connected to its shaft.

17. Explain with the aid of a simplified phasor diagram how the power factor at which a 3-phase synchronous motor operates may be varied when the motor is developing a constant torque.

18. Draw and explain equivalent circuit of a synchronous motor.

19. Draw and explain the phasor diagrams of a 3-phase synchronous motor when (*i*) it is overexcited (*ii*) it is underexcited.

20. Draw the phasor diagrams of synchronous motor for lagging, leading and unity power factor conditions. Name all the phasors.

21. Derive the commonly used expression for the power developed by a synchronous motor.

22. Derive an expression for the power developed in a 3-phase synchronous motor.

[M.D. Univ. Electrical Machines-II May 2009]

23. Assume two-axis model and draw the phasor diagram of a synchronous motor drawing leading current.

24. A 3-phase salient pole synchronous motor is connected to an infinite bus. Derive an expression for the electromechanical power developed. What will happen if the excitation is reduced to zero ? Also comment on the stability of the machine as compared to that of a cylindrical rotor machine.

25. Explain V curves as applied to synchronous motors.

26. Explain the ability of a synchronous motor to maintain a constant excitation regardless of the power factor when connected to constant voltage mains.

27. What is meant by constant power circle for synchronous motor ? How is it derived ?

28. Show that the locus of stator current for a constant output of 3-ϕ synchronous motor connected to a constant voltage, constant frequency bus-bars is a circle.

29. Can a synchronous machine act as a capacitor ? If so under what conditions ? Explain it with phasor diagrams.

30. What is a synchronous condenser ? Show the region of operation of the condenser on V-curves. Where are synchronous condensers used ?

31. Explain the operation of a synchronous motor under different load and excitation. Discuss how a synchronous motor can function as synchronous capacitor.

32. Why does a static condenser differ from synchronous condenser ? Discuss the basic philosophy of synchronous condenser with the help of phasor diagram.

33. What is the nature of power factor in a power system ? Explain how a synchronous motor can be used for power factor improvement, with proper vector diagram.

34. What is a synchronous condenser ? Explain with the help of phasor diagram its operation. What are its applications ? **[M.D. Univ. Electrical Machines-II, December 2010]**

35. Explain hunting of a synchronous machine. What is the purpose of damper windings in a synchronous machine ? **[M.D. Univ. Electrical Machines-II, December 2010]**

36. Briefly describe the phenomenon of 'hunting' in a synchronous machine. How is it remedied ?

37. Why a synchronous motor is not self starting. Give its starting methods.

38. Explain that a 3-phase synchronous motor is not self starting. Disuss the use of damper winding in a synchronous motor.

39. Why does a synchronous motor not have a starting torque ? Explain briefly. Describe, with the help of neat diagram, the method of starting a synchronous motor.

40. Explain two important functions served by damper winding in a synchronous motor. State applications of synchronous motor.

41. A synchronous motor will operate at a constant speed on every load, why ? Explain in detail. Make a list of applications of a synchronous motor.

42. State the applications of synchronous motors. **[M.D. Univ. May 2004]**

SHORT ANSWER TYPE QUESTIONS WITH ANSWERS

Q. 1. What factors determine the number of poles of a synchronous motor ?

Ans. The number of poles of a synchronous motor is determined by supply frequency and speed.

Q. 2. A synchronous motor runs at N_S rpm at full load. What will be its speed if the load is reduced to half the full-load ?

Ans. N_s rpm.

Q. 3. Why a 3-phase synchronous motor will always run at synchronous speed ?

Ans. The synchronous motor always runs at synchronous speed because there is an interlocking action between stator and rotor fields of this motor.

Q. 4. In a 3-phase synchronous motor field mmf lags the air-gap flux, then what is the position of air-gap flux vis-a-vis armature mmf ?

Ans. It also lags.

Q. 5. Will the air-gap flux in a 3-phase synchronous motor lag or lead the armature mmf ?

Ans. Air-gap flux will lag behind the armature mmf.

Q. 6. How the power factor of a synchronous motor is changed keeping the shaft load undisturbed ?

Ans. By varying the field excitation.

Q. 7. Does the change in excitation affect the synchronous motor speed and power factor ?

Ans. Change in excitation will affect the power factor (not the speed) of the synchronous motor.

Q. 8. What is the effect of increase in excitation of a synchronous motor ?

Ans. When the excitation of a synchronous motor is increased, first power factor improves until it becomes unity and with the further increase in excitation the power factor becomes leading one and decreases.

Q. 9. If V and E_0 be respectively the terminal voltage/phase and open-circuit emf/phase and θ is the internal angle of a 3-phase synchronous motor, what would be the relationship among them if the pf is to be lagging ?

Ans. $E_0^2 = V^2 + E_R^2 - 2VE_R \cos(\theta + \phi)$

Q. 10. What should be the relationship between power angle δ and internal angle θ for stable operation of a 3-phase synchronous motor working with constant excitation. If excitation is increased, what will happen to the power angle ?

Ans. Power angle δ should not exceed the internal angle θ for stable operation of a 3-phase synchronous motor.

With the increase in excitation power angle will decrease.

Q. 11. A 3-phase synchronous motor driving a constant load torque draws power from infinite bus at leading power factor. How power angle and power factor will change if the excitation is increased ?

Ans. Both power angle and power factor will decrease.

Q. 12. A synchronous motor develops some mechanical power, even if the field is unexcited. Is it cylindrical or salient pole machine ?

Ans. Salient pole.

Q. 13. Compare the salient pole rotor and cylindrical pole rotor synchronous motor on stability point of view.

Ans. Salient pole rotor machine is more stable than cylindrical pole rotor machine.

Q. 14. What is meant by V curves of synchronous motor ?

Ans. The curves drawn between armature current and field current for different constant loads are known as V-curves.

Q. 15. What is meant by synchronous condenser ?

Ans. Synchronous condenser is specially designed synchronous motor for operation without mechanical load with its field overexcited so as to draw leading armature current.

Q. 16. In what operating condition is a synchronous motor referred to as synchronous condenser ?

Ans. A synchronous motor operated without mechanical load with its field overexcited is referred to as a synchronous condenser.

Q. 17. What is the use of synchronous condenser ?

Ans. Power factor correction.

Q. 18. An alternator can be used as a synchronous condenser. Under what conditions it should operate ?

Ans. Overexcited synchronous motor.

Q. 19. What is meant by hunting of synchronous motor ?

Ans. The oscillation of synchronous motor rotor about its equilibrium position is called the hunting.

Q. 20. How will you minimize hunting ?

Ans. Hunting is minimised by providing damper winding on the rotor.

Q. 21. What is the purpose of damper windings ?

Ans. The purpose of damper windings is to minimise hunting and make the synchronous motor self starting.

Q. 22. What is the common starting method used for synchronous motor ?

Ans. By means of damper grids in the pole shoes.

Q. 23. How is the speed of a synchronous motor varied ?

Ans. By varying the frequency of supply source.

Q. 24. What are the advantages of synchronous motor ?

Ans. The advantages of synchronous motor are (*i*) constant operation speed, (*ii*) higher efficiency, especially in the low-speed unity pf ranges and (*iii*) capability of being operated under a wide range of power factor both lagging and leading.

Q. 25. A synchronous motor is connected to an infinite bus and supplying constant torque and operates at unity power factor. If the field current is increased, explain what happens to the following:

(*a*) The magnitude of resultant flux wave.
(*b*) The magnitude of armature current and its power factor.
(*c*) Rotor current away from resultant mmf wave or towards it.
(*d*) Space angle between armature mmf and resultant air-gap mmf.

Ans. (*a*) The motor will become overexcited, the field mmf will increase but increase in field mmf will be neutralised by the increase in counter armature mmf and the magnitude of result flux wave will remain almost constant.(Refer Art. 6.5)

(*b*) The armature current will increase and power factor will drop.

(*c*) Rotor current will move towards the resultant mmf because the resultant flux wave will move towards the leading power factor side causing reduction in angle between rotor current and mmf wave.

(*d*) Space angle between armature mmf and resultant air gap mmf will also decrease because resultant flux wave will move towards the air-gap mmf.

PROBLEMS

1. The input to an 11,000 V, 3-ϕ, star-connected synchronous motor is 60A. The effective resistance and synchronous reactance per phase are respectively 1Ω and 30 Ω . Find (*i*) the power supplied to the motor, and (*ii*) the induced emf for a power factor of 0.8 leading. [**Ans.** 914.52 kW; 13,040 V]

2. A 400 V, 50 Hz, 3-phase, 37.3 kW star-connected synchronous motor has a full-load efficiency of 88%. The synchronous impedance of the motor is (0.2 + *j* 1.6) Ω per phase. If the excitation of the motor is adjusted to give a leading pf of 0.9, calculate for full load the induced emf. [**Ans.** 495 V; 39.612 kW]

3. A 400 V, 3-ϕ, star-connected synchronous motor has a synchronous reactance of 1Ω per phase. The input to motor is 100 kW at normal voltage and induced line emf is 600 V. Calculate line current and power factor. Neglect armature resistance. [**Ans.** 167A; 0.8645 (lead)]

4. A 1,000 kVA, 11,000 V, 3-ϕ, star-connected synchronous motor has an armature resistance and reactance per phase of 3.5 Ω and 40 Ω respectively. Determine the induced emf and angular retardation of the motor when fully loaded at (*a*) unity pf (*b*) 0.8 pf lagging (*c*) 0.8 pf leading. [**Ans.** (*a*) 11,284 V; 18.8° (*b*) 8,985 V; 17.6° (*c*) 13,295 V; 13.48°]

5. The synchronous reactance per phase of a 3-phase star-connected 6,600 V synchronous motor is 10 Ω. For a certain load the input is 900 kW and the induced line emf is 8,900 V (line to line). Evaluate the line current. Neglect armature resistance. [**Ans.** 149.23 A]

6. Consider a 3,300 V delta-connected synchronous motor having a synchronous reactance per phase of 18 Ω. It operates at a leading pf of 0.707 when drawing 800 kW from mains. Calculate the excitation emf and the rotor angle (δ), explaining the latter term. [**Ans.** 4,972 V; 17°]

7. A 3-phase, 8-pole, 3,300 V, 50-Hz, star-connected synchronous motor has armature resistance of 0.5 Ω / phase and synchronous reactance of 5 Ω / phase. While running on no load, the excitation has been adjusted so as to make the emf numerically equal to and anti-phase with terminal voltage. With a certain load torque applied the rotor gets retarded by 3 mechanical degrees, calculate the armature current and power factor of the motor. [**Ans.** 79.6A; 0.9992 (lag)]

8. A 400 V, 6-pole, 3-ϕ, 50 Hz, star-connected synchronous motor has negligible armature resistance and 4Ω synchronous reactance per phase. It takes a current of 15 A at unity power factor when operating with a certain field current.

If the load torque is gradually increased until the line current is 60 A, the field current remaining unchanged, find the gross torque developed and the new power factor. **[Ans.** 346.64 Nm; 0.8732 (lag)]

9. A 400 V, 50-Hz, 6-pole, 3-ϕ, λ-connected synchronous motor has a synchronous reactance of 4 Ω / phase and resistance of 0.5 Ω / phase. On full load, the excitation is adjusted so that machine takes an armature current of 60 A at 0.866 pf leading. Keeping the excitation unchanged, find the maximum power output. Excitation, friction, windage and iron losses total 2 kW. **[Ans.** 51.42 kW]

10. A 3-ϕ, 3,300 V, λ-connected synchronous motor has an effective resistance and synchronous reactance of 2.0 and 18.0 Ω per phase respectively. If the open-circuit generated emf is 3,800 V between lines, calculate (*i*) the maximum total mechanical power that the motor can develop and (*ii*) the current and pf at the maximum mechanical power.

[Ans. 604.35 kW; 151.4 A; 0.8572 (lead)]

11. Construct curves showing the relation (*a*) between armature and field current, (*b*) between pf and field current, for a star-connected synchronous motor with a synchronous reactance of 8.25 Ω/phase and negligible resistance. The machine takes a constant power input of 800 kW at 6.6 kV. The open-circuit characteristic is as follows:

Terminal voltage, kV	3	4	5	6	6.6	7	7.5	8
Field current, Amps	16	23	31	41	50	56	69	85

12. A synchronous motor absorbing 50 kW is connected in parallel with a factory load of 200 kW having a lagging pf of 0.8. If the combination has a pf of 0.9 lagging, how many leading kVAR are to be supplied by the motor ? What is its power factor ? **[Ans.** 29 kVAR (leading); 0.865 (leading)]

13. The load in a factory is 800 kVA at 0.85 pf lagging and in addition, there is a 3-phase synchronous motor having an input of 200 kW. Determine the input to the synchronous motor in kVA and the power factor at which it must operate in order that the power factor of the combined load comes to be 0.95 lagging. **[Ans.** 239.7 kVA; 0.834 (leading)]

14. A manufacturing plant takes 360 kW, at a lagging power factor of 0.6. Calculate (*a*) the rating of the synchronous motor which would deliver 140 kW, and at the same time make it possible for the plant to operate without overloading its 600 kVA supply transformer (*b*) the resultant pf of the system and (*c*) the power factor of the synchronous motor.

[Ans. (*a*) 204 kVA (*b*) 0.8333 (lag) (*c*) 0.687 (lead)]

❑❑❑

MODEL TEST PAPER–1

ELECTRICAL MACHINES-II

Time : 3 Hours *Maximum Marks : 100*

Note : *(i) Question No. 1 is compulsory. (ii) Attempt any **four** questions from remaining four parts by selecting **one** question from every part. Combining (i) and (ii) attempt **five** questions in total.*

1. (*a*) How is the mechanical power developed represented in the circuit model of an induction motor ? **2½**

(*b*) Why do you require starter for starting of 3-phase induction motors ? Name the various types of starters used for starting of squirrel cage induction motors. **2½**

(*c*) Is the torque developed by a single phase induction motor at synchronous speed zero ? **2½**
What is the effect of increasing rotor resistance in a single phase induction motor ?

(*d*) In the design of resistance-start single phase induction motor the starting winding is disconnected once the slip reaches 0.25 or less. Explain. **2½**

(*e*) Why are salient-pole alternators more suitable for low speed and non-salient pole for high speed operation ? **2½**

(*f*) What are the effects of power factor on voltage regulation of an alternator (*a*) at lagging power factor and (*b*) at leading power factor ? **2½**

(*g*) What is meant by hunting of a synchronous motor and how is it minimized ? **2½**

(*h*) Give the applications of synchronous motors. **2½**

SECTION–A

2. (*a*) Give constructional features and working principle of a squirrel cage induction motor. Give with reasoning if this motor can run at synchronous speed. **10**

(*b*) A delta- connected, 6-pole, 50 Hz, 3-phase induction motor has a rotor resistance of 0.15 Ω per phase and exerts maximum torque at 880 rpm. Calculate the percentage maximum torque that would be exerted (*i*) at standstill and (*ii*) at 940 rpm. **10**

3. (*a*) What are different methods of starting 3-ϕ induction motor ? Describe, with the help of neat diagram, the method of starting slip-ring induction motor and squirrel cage induction motor. **10**

(*b*) Describe any two methods of speed control of 3-phase squirrel cage induction motors. **10**

SECTION–B

4. (*a*) Explain cross-field theory as applied to single phase induction motor. **10**

(*b*) Derive the equivalent circuit of a single phase induction motor with the help of double revolving field theory. **10**

5. (*a*) Compare various types of single phase induction motors in terms of performance and explain where they are employed. Why a single winding single-phase induction motor does not have starting torque ? **10**

(*b*) A 230 V, 4-pole, 50 Hz, 1-phase induction motor has an effective rotor resistance and leakage reactance of 0.5 Ω and 5 Ω respectively. It is running at a speed of 1,350 rpm. Determine :

(*i*) frequencies of forward and backward rotor current components (*ii*) relative magnitudes of forward and backward fluxes.

Neglect magnetizing current and stator impedance. **10**

SECTION–C

6. (*a*) What are the advantages of using distributed winding and short pitch coil ? Derive, from the fundamentals, the expressions for winding factor and pitch factor. **10**

(*b*) A 550 V, 55 kVA, single-phase synchronous generator has an effective resistance of 0.2 Ω. A field current of 10 amps produced an armature current of 200 A on short circuit and an emf of 450 V on open circuit. Calculate synchronous impedance and voltage regulation at full load with pf 0.8 (lagging). **10**

7. (*a*) Discuss "Two Reactance Theory" as applied to salient pole synchronous machines.

Derive an expression of power developed in a cylindrical rotor alternator in terms of power angle and synchronous reactance and there from write the value of maximum power developed. Ignore stator resistance. **10**

(*b*) What are the conditions of parallel operation of synchronous generators ? Write, in sequence, the steps needed to connect a synchronous generator with infinite bus by dark bright lamp method. **10**

SECTION–D

8. (*a*) Explain with neat sketches the principle of operation of a 3-phase synchronous motor. Also explain why it will not run at other than synchronous speed. **10**

(*b*) Draw the families of curves of synchronous motor at no load and full load showing the relation between (*i*) armature current and field current (*ii*) power factor and field current. Using these curves explain how the motor may be overloaded with no load connected to its shaft. **10**

9. (*a*) What is a synchronous condenser ? Show the region of operation of the condenser on V-curves. Where are synchronous condensers used ? **10**

(*b*) Why does a synchronous motor not have a starting torque ? Explain briefly. Describe, with the help of neat diagram, the method of starting a synchronous motor. **10**

MODEL TEST PAPER–2

ELECTRICAL MACHINES-II

Time : 3 Hours *Maximum Marks : 100*

Note : *(i) Question No. 1 is compulsory. (ii) Attempt any* ***four*** *questions from remaining four parts by selecting* ***one*** *question from every part. Combining (i) and (ii) attempt* ***five*** *questions in total.*

1. (*a*) What is an induction generator and what are its aplications ? 2½
   (*b*) On what factors does the speed of an induction motor depend ? Name the various methods of speed control of induction motor. 2½
   (*c*) Name the two theories regarding single phase induction motor. State the principle that the double revolving field theory makes use of. 2½
   (*d*) How does a capacitor-start motor differ from a resistance-start motor ? 2½
   (*e*) What are the essential elements for generating emf in alternators ? Why is it possible to design alternators to generate much higher voltages than dc generators ? 2½
   (*f*) Why does synchronous impedance method give a poorer voltage regulation ? 2½
   (*g*) What factors determine the number of poles on a synchronous motor ? A synchronous motor is running at N rpm on full load. What will be the speed if load is reduced to one-fourth ? 2½
   (*h*) What is meant by synchronous condenser and for what is it used ? 2½

SECTION–A

2. (*a*) Draw and explain the slip-torque characteristic of a typical induction motor. Mark the starting torque and maximum torque on the diagram so drawn. How do starting and maximum torques vary with the rotor resistance ? 10
   (*b*) Derive the exact equivalent circuit of a 3-phase induction motor. What is the difference between the exact and approximate equivalents ? Discuss. From the approximate equivalent circuit, find the (*i*) rotor output (*ii*) output power (*iii*) output torque. Also find the slip at maximum torque. 10
3. (*a*) Describe speed control of a 3-phase slip-ring induction motor by injecting voltage in the rotor circuit. 10
   (*b*) A 12 kW, 3–ϕ, 6-pole, 50 Hz, 400 V, delta-connected induction motor runs at 960 rpm on full-load. If it takes 85 A on direct starting, find the ratio of the starting torque to full-load torque with a star-delta starter. Full-load efficiency and power factor are 88 % and 0.85 respectively. 10

SECTION–B

4. (*a*) Explain "double revolving field theory" and show that a single-phase induction motor is not self-starting. Draw the torque-slip curves based on this theory. 10
   (*b*) State the reasons for the inferior performance of single phase induction motors compared to three phase induction motors. 10
5. (*a*) Explain with neat diagrams the following types of single phase induction motors. 12
   (*i*) Split- phase induction motor (*ii*) Capacitor- start induction-run motor.
   Also draw their torque-speed characteristics.

(*b*) What will happen if a resistance start induction motor is started with a heavy overload and fails to accelerate to a sufficient speed to cause the centrifugal switch contacts to open ? **8**

SECTION–C

6. (*a*) Define armature reaction. How does it affect the main field under unity, 90º lagging and 90º leading power factor conditions. Explain with the help of suitable diagrams. **10**

(*b*) A 3-phase, 16-pole, star-connected alternator has 144 slots and 6 conductors per slot. The flux per pole is 0.03 Wb sinusoidally distributed and the speed is 375 rpm. If the coil span is 160º, determine: (*i*) Frequency (*ii*) Pitch factor (*iii*) Distribution factor (*iv*) Phase and line emfs. **10**

7. (*a*) Draw and explain the phasor diagram of a salient pole alternator supplying a lagging pf load. **10**

(*b*) Mention the need for parallel operation of alternators. State the conditions to be satisfied before connecting an alternator to the infinite bus-bars. **10**

SECTION–D

8. (*a*) Draw a curve between field current and armature current of a synchronous motor, when power input to the motor and supply voltage are constant. What is the shape of the curve ? Show the region of lagging and leading pf. Why is power input at unity pf minimum ? **10**

(*b*) Derive the commonly used expression for the power developed by a synchronous motor. **10**

9. (*a*) A star-connected synchronous motor rated at 187 kVA, 3-ϕ, 2,300 V, 47 A, 50 Hz, 187.5 rpm has an effective resistance of 1.5 Ω and a synchronous reactance of 20 Ω per phase. Determine the internal power developed by the motor when it is operating at rated current and 0.8 power factor leading. **10**

(*b*) A synchronous motor will operate at a constant speed on every load, why ? Explain in detail. Make a list of applications of a synchronous motor. **10**

❑❑❑

B.TECH.

FIFTH SEMESTER (ELECTRICAL ENGINEERING)

EXAMINATION, DECEMBER, 2011

Electrical Machines-II

Time : 3 hours *Maximum Marks : 100*

Before answering the question, candidates should ensure that they have been supplied the correct and complete question paper. No complaint in this regard, will be entertained after examination.

Note : (*i*) Question No. **1** is ***compulsory*** from **Section-A.**

(*ii*) Attempt ***four*** questions from remaining four sections selecting ***one*** question from each section.

(*iii*) Use of non-programmable calculator is allowed.

SECTION-A

1. (*i*) Explain the plugging of a 3-phase induction motor. **4**

(*ii*) How the direction of a capacitor start induction run motor can be reversed ? **4**

(*iii*) Explain the potier triangle method of voltage regulation for alternators. **4**

(*iv*) Explain the meaning of synchronizing power in alternators. **4**

(*v*) A synchronous motor start as usual but fails to develop its full torque. What could be the reasons ? **4**

SECTION-B

2. Derive the equation for torque under running condition for 3-phase induction motor. A 4-pole, 50 Hz induction motor has full-load slip of 5%. Each rotor phase has a resistance of 0.3 ohms and a standstill resistance of 1.2 ohms. Find the ratio of maximum torque to full-load torque and the speed at which maximum torque occurs. **20**

3. Draw and explain the equivalent circuit of 3-phase induction motor. Also explain with the help of mathematical equations, the requirement of starter in the starting of 3-phase induction motor. **20**

SECTION-C

4. Draw and explain the equivalent circuit of 1-phase induction motor. A single-phase 230 V, 50 Hz, 6 pole induction motor has the following parameters :

$r_1 = 0.12\ \Omega,\ r'_2 = 0.14\ \Omega,\ x_1 = x'_2 = 0.25\ \Omega,\ x_m = 15\ \Omega.$

If the core loss is 250 W and friction and windage losses are 500 W, determine the efficiency and torque at $s = 0.05$. **20**

5. Prove that single-phase induction motor is not self starting. How it is made self starting ? Explain the methods. **20**

SECTION-D

6. Explain the terms pitch factor and distribution factor in synchronous machines. A 4-pole, 50 Hz, 3-phase, Y-connected alternator has a single layer, full-pitched winding with 21 slots per pole and two conductors per slot. The fundamental flux is 0.6 wb and air-gap flux contains a third harmonic of 5% amplitude. Find the rms values of the phase emf due to fundamental and third harmonic flux and the total induced emf. **20**

7. Define the voltage regulation of alternators. A 100 kVA, 3000 V, 50 Hz 3-phase star-connected alternator has effective armature resistance of 0.25 Ω. The field current of 38 A produces short-circuit current of 195 A and an open-circuit emf of 1,025 V as line voltage. Calculate the full-load voltage regulation at 0.85 pf lagging and 0.8 pf leading. Also draw the phasor diagrams for both the cases. **20**

SECTION-E

8. Explain the effect of excitation on armature current and power factor on the working of synchronous motor. Also explain the methods of starting of synchronous motor. **20**

9. Define the term coupling angle in synchronous motor. The input to an 11 kV, 3-phase, Y-connected synchronous motor is 60 A. The effective resistance and synchronous reactance per phase are 1 Ω and 30 Ω respectively. Find (*i*) power supplied to the motor and (*ii*) induced emf for a power factor of 0.8 leading. **20**

ANSWERS/SOLUTION

1. (*i*) Plugging can be achieved in an induction motor merely by reversing any two of the three phases which causes a reversal of the direction of the rotating magnetic field. At the instant of switching the motor to the plugging position the motor runs in the opposite direction to that of the field and the relative speed is approximately twice [(2 – *s*) times] of synchronous speed *i.e.* the slip is very nearly equal to two, being equal to (2 – *s*). So voltage induced in the rotor will be twice of normally induced voltage at standstill and the winding must be provided with the additional insulation to withstand this much voltage.

During plugging period, the motor acts as a brake and it absorbs kinetic energy from the still revolving load causing its speed to fall. The associated power P_{mech} is dissipated as heat in the motor. At the same time, the rotor also continues to draw power from the stator which is also dissipated as heat. Due to increased rotor frequency there are additional iron losses in the rotor. Thus, the heat developed in the rotor during braking period are about three times the heat developed during starting period (or under blocked condition of rotor).

In case of a squirrel cage motor, energy is dissipated wholly within the machine; whereas in case of wound rotor motor this energy is dissipated also in the external resistance added in the rotor circuit for this purpose.

(*ii*) The direction of rotation of a capacitor-start induction motor can be reversed simply by reversing the line connections of either the main winding or the auxiliary winding. The noteworthy point is that reversal of a capacitor-start induction motor takes place only when it is at rest and the starting switch is in its normally closed de-energized position. If reversal is attemped under normal operating conditions, nothing will happen because under normal operating conditions, the starting switch is in open position and the motor is running as a pure single-phase induction motor and developing torque in the direction of rotation.

(*iii*) Refer to Art. 5.17.

(*iv*) Once a synchronous machine is synchronised, it will tend to remain in synchronism with the other alternators. Any tendency to depart from the condition of synchronism is opposed by a synchronising torque produced due to circulating current flowing through the alternators.

When two alternators are in exact synchronism, the two alternators have equal induced emfs which are in exact phase opposition, as shown in Fig. 1(*a*). No circulating current flows round the local circuit.

When the induced emfs of the two alternators are equal in magnitude but not in exact phase opposition, as shown in Fig. 1 (*b*), their resultant emf acts round the local circuit and causes flow of current called the *synchronising current*, I_{sy}.

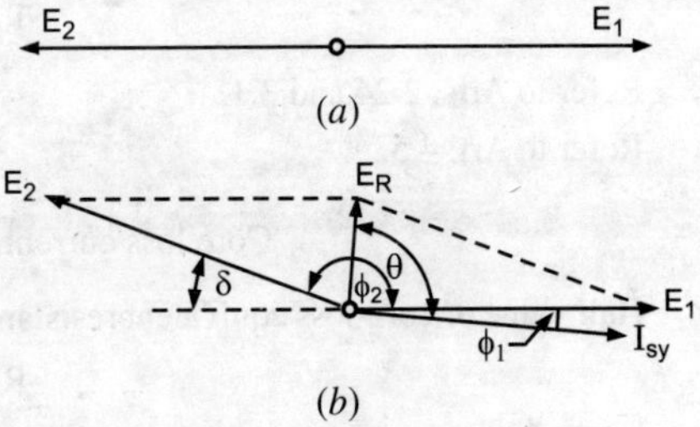

Fig. 1

If any alternator (say, the second alternator) due to some disturbance tends to retard, E_2 falls back by a phase angle δ electrical degrees, as shown in Fig. 1 (*b*). Now though their induced emfs E_1 and E_2 are equal in magnitude but have a phase difference of 180° – δ.

Synchronising current is given by $$I_{sy} = \frac{E\delta}{Z_s}$$

where Z_s is the combined synchronous impedance per phase of the two alternators (or of one alternator only if it is connected to infinite bus-bars).

In above case, machine no. 1 supplies power equal to $E_1 I_{sy} \cos \phi_1$ and the machine no. 2 receives power equal to $E_2 I_{sy} \cos (180° - \phi_2)$.

Power supplied by machine no. 1 = Power supplied to machine no. 2 + copper losses.

. The power supplied by machine no. 1 is called the *synchronising power* and is given by the expression

$$P_{sy} = E_1 I_{sy} \cos \phi_1$$

$$= E_1 I_{sy} = E \times \frac{E\delta}{X_s} = \frac{\delta E^2}{X_s} \quad \because E_1 = E \text{ and } \phi_1 \text{ is very small}$$

$$\text{Total synchronising power for 3-phases} = 3P_{sy} = \frac{3\,\delta E^2}{X_s}.$$

2. Refer to Art. 2.23.

Rotor resistance per phase, $R_2 = 0.3\ \Omega$

Rotor standstill reactance per phase, $X_2 = 1.2\ \Omega$

$$\text{Synchronous speed, } N_s = \frac{120f}{P} = \frac{120 \times 50}{4} = 1{,}500 \text{ rpm}$$

Full-load slip, $s_f = 5\% = 0.05$ pu

Ratio of rotor resistance to rotor standstill reactance,

$$a = \frac{R_2}{X_2} = \frac{0.3}{1.2} = 0.25$$

Slip corresponding to maximum torque,

$$s_{max\,T} = \frac{R_2}{X_2} = \frac{0.3}{1.2} = 0.25$$

Speed corresponding to maximum torque

$$N_{max\,T} = N_s(1 - s_{max\,T})$$

$$= 1{,}500\,(1 - 0.25) = 1{,}125 \text{ rpm} \quad \textbf{Ans.}$$

Ratio of maximum to full load torque,

$$\frac{T_{max}}{T_f} = \frac{a^2 + s_f^2}{2as_f} = \frac{0.25^2 + 0.05^2}{2 \times 0.25 \times 0.05} = 2.6 \quad \textbf{Ans.}$$

3. Refer to Arts. 2.24 and 3.1.

4. Refer to Art. 4.5.

$$\text{Core loss current, } I_e = \frac{250}{230} = 1.087 \text{ A}$$

Half value of core-loss equivalent resistance,

$$\frac{R_m}{2} = \frac{230}{1.087} = 211.6\ \Omega$$

For Forward Field [Fig. 4.9 (*c*)]

$$\text{Conductance of core-loss branch} = \frac{1}{R_m/2} = \frac{1}{211.6} = 0.004726 \text{ S}$$

$$\text{Susceptance of magnetising branch} = \frac{-j}{\frac{X_m}{2}} = \frac{-j}{15/2} = -j0.13333$$

$$\text{Rotor admittance} = \frac{1}{\frac{r_2'}{2s} + \frac{jx_2'}{2}} = \frac{1}{\frac{0.14}{2 \times 0.05} + j\frac{0.25}{2}} = \frac{1}{1.4 + j0.125}$$

$$= (0.70864 - j0.0327) \text{ S}$$

$$\text{Admittance, } \mathbf{Y}_f = (0.004726 - j0.13333 + 0.70864 - j0.0327) \text{ S} = 0.71366 - j0.166033 \text{ S}$$

$$\text{Impedance, } \mathbf{Z}_f = \frac{1}{\mathbf{Y}_f} = \frac{1}{0.71366 - j0.166033} = (1.3293 + j0.30926)\ \Omega$$
$$= 1.3648 \angle 13.097°\ \Omega$$

For Backward Field

$$\text{Rotor admittance} = \frac{1}{\frac{r_2'}{2(2-s)} + \frac{jx_2'}{2}} = \frac{1}{\frac{0.14}{2(2-0.05)} + \frac{j0.25}{2}}$$
$$= \frac{1}{0.0359 + j0.125} = (2.12255 - j7.39)\ \text{S}$$
$$\text{Admittance, } \mathbf{Y}_b = (0.004726 - j0.133333 + 2.12255 - j7.39) = (2.127276 - j7.52333)\ \text{S}$$
$$\text{Impedance, } \mathbf{Z}_b = \frac{1}{\mathbf{Y}_b} = \frac{1}{2.127276 - j7.52333} = (0.0348 + j0.12308)\ \Omega$$
$$= 0.1279 \angle 74.21°\ \Omega$$
$$\text{Equivalent impedance, } \mathbf{Z}_{eq} = \mathbf{Z}_1 + \mathbf{Z}_f + \mathbf{Z}_b$$
$$= (0.12 + j0.25) + (1.3293 + j0.30926) + (0.0348 + j0.12308)\ \Omega$$
$$= (1.4841 + j0.68234)\ \Omega = 1.6334 \angle 24.69°\ \Omega$$
$$\text{Input current, } \text{I}_1 = \frac{\text{V}}{\text{Z}_{eq}} = \frac{230}{1.6334} = 140.8\ \text{A}$$
$$\text{Power input} = \text{V}_1\text{I}_1 \cos\phi = 230 \times 140 \cos 24.69° = 29{,}256\ \text{W}$$
$$\text{E}_{mf} = \text{I}_1 \times \text{Z}_f = 140.8 \times 1.3648 = 192.173\ \text{V}$$
$$\text{E}_{mb} = \text{I}_1 \times \text{Z}_b = 140.8 \times 0.1279 = 18\ \text{V}$$
$$\text{I}_{2f}' = \frac{\text{E}_{mf}}{\sqrt{\left(\frac{r_2'}{2s}\right)^2 + \left(\frac{x_2'}{2}\right)^2}} = \frac{192.173}{\sqrt{(1.4)^2 + (0.125)^2}} = 136.72\ \text{A}$$
$$\text{I}_{2b}' = \frac{\text{E}_{mb}}{\sqrt{\left(\frac{r_2'}{2(2-s)}\right)^2 + \left(\frac{x_2'}{2}\right)^2}} = \frac{18}{\sqrt{0.0359^2 + 0.125^2}} = 138.4\ \text{A}$$
$$\text{Air-gap power for forward field, P}_{gf} = (\text{I}_{2f}')^2 \times \frac{r_2'}{2s} = (136.72)^2 \times \frac{0.14}{2 \times 0.05} = 26{,}170\ \text{W}$$
$$\text{Air-gap power for backward field, P}_{gb} = (\text{I}_{2b}')^2 \times \frac{r_2'}{2(2-s)} = (138.4)^2 \times \frac{0.14}{2 \times 1.95} = 688\ \text{W}$$
$$\text{Power developed, P}_{\text{mech}} = (1-s)(\text{P}_{gf} - \text{P}_{gb}) = (1 - 0.05)(26{,}170 - 688) = 2{,}4208\ \text{W}$$
$$\text{Shaft power} = \text{P}_{\text{mech}} - \text{friction and windage losses} = 2{,}4208 - 500 = 23{,}708\ \text{W}$$
$$\text{Efficiency, } \eta = \frac{\text{Shaft power}}{\text{Power input}} \times 100 = \frac{23{,}708}{29{,}256} \times 100 = 81\%\ \textbf{Ans.}$$
$$\text{Torque, T} = \frac{\text{Shaft power}}{2\pi\text{N}/60} = \frac{23{,}708 \times 60}{2\pi \times \frac{120 \times 50}{6}(1 - 0.05)} = \frac{23{,}708 \times 60}{2\pi \times 950}$$
$$= 238.3\ \text{N-m}\ \textbf{Ans.}$$

5. Refer to Arts. 4.2.2, 4.8, 4.8.1, 4.9, 4.10, 4.11, 4.12, etc.

6. Refer to Arts. 5.6.1 and 5.6.2.

$$\text{Number of slots per pole, } n = 21$$
$$\text{Number of slots per pole per phase, } m = \frac{n}{3} = \frac{21}{3} = 7$$

Number of conductors connected in series per phase,

$$Z_p = \frac{n \times p \times 2}{3} = \frac{21 \times 4 \times 2}{3} = 56$$

$$\text{Number of series turns per phase, } T = \frac{Z_p}{2} = \frac{56}{2} = 28$$

$$\text{Angular displacement between the slots, } \beta = \frac{180°}{21} = \frac{60°}{7}$$

$$\text{Pitch factor, } K_p = 1 \qquad \because \text{ winding is full-pitched}$$

$$\text{Distribution factor, } K_d = \frac{\sin m \times \frac{\beta}{2}}{m \sin \frac{\beta}{2}} = \frac{\sin 7 \times \frac{60°}{7 \times 2}}{7 \sin \frac{60°}{7 \times 2}} = 0.9558$$

Induced emf per phase (fundamental component),

$$E_1 = 4.44\, K_d K_p \phi f T = 4.44 \times 0.9558 \times 1 \times 0.6 \times 50 \times 28 = 3{,}564.8 \text{ V } \textbf{Ans.}$$

For Third Harmonic Flux

$$\text{Distribution factor, } K_{d_3} = \frac{\sin mr\frac{\beta}{2}}{m \sin \frac{r\beta}{2}} = \frac{\sin 7 \times 3 \times \frac{60°}{7 \times 2}}{7 \sin \frac{3 \times 60°}{7 \times 2}} = 0.642$$

$$\text{Coil span factor, } K_{p_3} = \text{Cos } 3 \times \frac{180° - 180°}{2} = \text{Cos } 0° = 1.0$$

$$\text{Frequency, } f = 3 \times f = 3 \times 50 = 150$$

$$\text{Flux per pole, } \phi_3 = \frac{1}{3} \times 0.6 \times \frac{5}{100} = 0.01 \text{ Wb}$$

Induced emf per phase (third harmonic component),

$$E_3 = 4.44\, K_d K_p \phi_3 f_3 T$$

$$= 4.44 \times 0.642 \times 1 \times 0.01 \times 150 \times 28 = 119.72 \text{ V } \textbf{Ans.}$$

$$\text{Total induced emf per phase, } E_p = \sqrt{E_1^2 + E_3^2} = \sqrt{3{,}564.8^2 + 119.72^2} = 3{,}567 \text{ V } \textbf{Ans.}$$

7. Voltage regulation of an alternator is defined as the increase in terminal voltage, expressed as percentage (or per unit) of the rated terminal voltage, when the load at a given power factor is thrown off, with speed and field current remaining the same.

$$\textit{i.e.}, \text{ voltage regulation} = \frac{E_0 - V}{V} \text{ per unit} = \frac{E_0 - V}{V} \times 100 \text{ in percentage}$$

where E_0 is no-load terminal voltage and V is the full load terminal voltage.

Note : In case of leading power factor E_0 is less than V, so voltage regulation will be negative.

$$\text{Load, } Q = 100 \text{ kVA}$$

$$\text{Phase voltage, } V_P = \frac{V_L}{\sqrt{3}} = \frac{3{,}000}{\sqrt{3}} = 1{,}732 \text{ V}$$

$$\text{Load current, } I = \frac{100 \times 1{,}000}{\sqrt{3} \times 3{,}000} = 19.245 \text{ A}$$

$$\text{Effective resistance per phase, } R_e = 0.25\ \Omega$$

$$\text{Synchronous impedance per phase, } Z_s = \frac{\text{Open-circuit emf per phase}}{\text{Short-circuit current}} \text{ for a given excitation}$$

$$= \frac{1{,}025/\sqrt{3}}{195} = 3.03\ \Omega$$

Synchronous reactance per phase, $X_s = \sqrt{Z_s^2 - R_C^2} = \sqrt{(3.03)^2 - (0.25)^2} = 3.02\ \Omega$

At power factor 0.85 lagging $\quad \text{Cos}\,\phi = 0.85$

$$\text{Sin}\,\phi = \sqrt{1-0.85^2} = 0.5268$$

Open-circuit voltage per phase,

$$E_{op} = \sqrt{(V_P \cos\phi + IR_e)^2 + (V_P \sin\phi + IX_s)^2}$$

$$= \sqrt{(1{,}732\times 0.85 + 19.245\times 0.25)^2 + (1{,}732\times 0.5268 + 19.245\times 3.02)^2}$$

$$= 1{,}767\ \text{V}$$

$$\text{Full-load regulation} = \frac{E_{op} - V_P}{V_P}\times 100 = \frac{1{,}767 - 1{,}732}{1{,}732}\times 100 = 2.02\%\ \textbf{Ans.}$$

At power factor 0.8 leading $\quad \text{Cos}\,\phi = 0.8$

$$\text{Sin}\,\phi = \sqrt{1-0.8^2} = -0.6 \qquad \text{minus sign for leading pf}$$

$$E_{op} = \sqrt{(1{,}732\times 0.8 + 19.245\times 0.25)^2 + (1{,}732\times(-0.6) + 19.245\times 3.02)^2}$$

$$= 1{,}701.7\ \text{V}$$

$$\text{Full-load regulation} = \frac{1701.7 - 1{,}732}{1{,}732}\times 100 = -1.75\%\ \textbf{Ans.}$$

Phasor diagrams may be drawn as shown in Fig. 5.38 (*a*) and (*b*).

8. Refer to Arts. 6.5 and 6.19.

9. The angular displacement between the rotor and stator poles, δ is called the torque or coupling angle in a synchronous motor

$$\text{Supply voltage per phase, } V_P = \frac{V_L}{\sqrt{3}} = \frac{11{,}000}{\sqrt{3}} = 6{,}351\ \text{V}$$

$$(i)\ \text{Power supplied to motor, } P = \sqrt{3}V_L I_L \cos\phi$$

$$= \sqrt{3}\times 11{,}000\times 60\times 0.8 = 914{,}523\ \text{W or } 914.523\ \text{kW}\ \textbf{Ans.}$$

$$\text{Power factor angle, } \phi = \text{Cos}^{-1}\,0.8 = 36.87^\circ\ \text{(leading)}$$

$$\text{Internal angle } \theta = \text{Tan}^{-1}\frac{X_s}{R_e} = \text{Tan}^{-1}\frac{30}{1} = 88.1^\circ\ \text{(lagging)}$$

$$\text{Impedance per phase, } Z = \sqrt{R_e^2 + X_s^2} = \sqrt{1^2 + 30^2} = 30.2\ \Omega$$

$$\text{Impedance drop per phase, } E_R = IZ = 60\times 30.02 = 1{,}801\ \text{V}$$

$$\text{Induced emf per phase, } E_P = \sqrt{V_P^2 + E_R^2 - 2V_P E_R\cos(\phi+\theta)}$$

$$= \sqrt{6{,}351^2 + 1801^2 - 2\times 6{,}351\times 1{,}801\cos(36.87^\circ + 88.1^\circ)}$$

$$= \sqrt{40{,}355{,}201 + 3{,}243{,}601 + 13{,}111{,}460} = 7.529.3\ \text{V}$$

$$\text{Line value of induced emf, } E_L = \sqrt{3}\times 7{,}529.3 = 13{,}040\ \text{V}\ \textbf{Ans.}$$

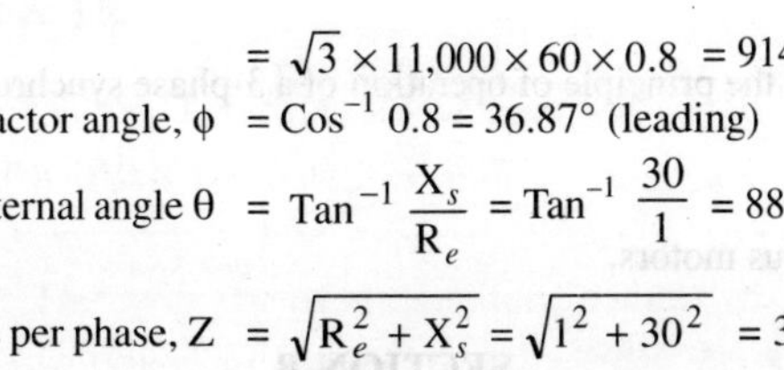

Phasor Diagram For Q. 9.

□□□

INDEX

INDEX

T

V

W

Z

❑❑❑

Win Prizes

ATTENTION READERS

Win Prizes

We request for your frank assessment regarding various aspects of the book as given below :

ELECTRICAL MACHINES–II

By : J.B. GUPTA **(MDU)**

Please Fill up the Sheet in Neat and Clean Writing

(i) What topics of your syllabus, important for your examination point of view, are not covered in this book ?

....

....

....

....

(ii) What are the chapters wherein the treatment of the subject matter is not systematic or organised or updated ?

....

....

....

....

(iii) Have you come across misprints/mistakes/factual inaccuracies in the book ? Please specify the chapters and the page numbers.

....

....

....

....

(iv) What other books on the same subject have you found/heard better than the present book ? Please specify in terms of price and quality.

....

....

....

....

(v) Further suggestions and comments for the improvement of this book.

....

....

....

....

(PS : If need be please attach a separate sheet to write your views)

OTHER DETAILS

(i) Who recommended you this book ? (Please tick (✓) in the box near the option relevant to you.

☐ Teacher ☐ Friends ☐ Bookseller

....

....

....

....

(ii) Name of the recommending teacher, his designation and address:

....

....

....

....

(iii) Name and address of the bookseller you purchased the book from :

....

....

....

....

(iv) Name and address of your institution (Please mention the University or Board as the case may be)

....

....

....

....

(v) Your name and complete address :

....

....

....

....

(vi) Course you are enrolled in :

....

....

....

....

The best assessment will be awarded each month : The award will be in the form of our publications as decided by the Editorial Board.

Please mail the filled up coupon at the vearliest to :

S.K. KATARIA & SONS®

4885/109, Prakash Mahal,

Dr. Subhash Bhargav Lane,

Opposite Delhi Medical Association,

Daryaganj, New Delhi-110002

Phone: +91-11-23243489, 23269324

TeleFax : +91-11-23243489

email : katariabooks@yahoo.com